AF595563

Symmetry Breaking in Cells and Tissues

Symmetry Breaking in Cells and Tissues

Editor

Andrew B. Goryachev

MDPI • Basel • Beijing • Wuhan • Barcelona • Belgrade • Manchester • Tokyo • Cluj • Tianjin

Editor
Andrew B. Goryachev
University of Edinburgh
UK

Editorial Office
MDPI
St. Alban-Anlage 66
4052 Basel, Switzerland

This is a reprint of articles from the Special Issue published online in the open access journal *Cells* (ISSN 2073-4409) (available at: https://www.mdpi.com/journal/cells/special_issues/cells_symmetry-breaking).

For citation purposes, cite each article independently as indicated on the article page online and as indicated below:

LastName, A.A.; LastName, B.B.; LastName, C.C. Article Title. *Journal Name* **Year**, *Volume Number*, Page Range.

ISBN 978-3-0365-0338-7 (Hbk)
ISBN 978-3-0365-0339-4 (PDF)

Contents

About the Editor

Andrew B. Goryachev is a computational cell biologist and biophysicist whose work lies at the interface of experimental biology and theory. He is a Professor at the Centre for Synthetic and Systems Biology, University of Edinburgh. In close collaboration with leading cell and developmental biologists worldwide, his group builds biophysical models of cellular morphogenesis and pattern formation. His main interest is the elucidation of the biophysical mechanisms of symmetry breaking in cells. Andrew is known for his work on the pattern-forming role of small GTPases, cell polarity and the morphogenesis of yeast cells.

Editorial

Symmetry Breaking as an Interdisciplinary Concept Unifying Cell and Developmental Biology

Andrew B. Goryachev

SynthSys, Centre for Synthetic and Systems Biology, Institute for Cell Biology, University of Edinburgh, Edinburg EH9 3BD, UK; Andrew.Goryachev@ed.ac.uk

Received: 22 December 2020; Accepted: 25 December 2020; Published: 7 January 2021

The concept of "symmetry breaking" has become a mainstay of modern biology, yet you will not find a definition of this concept specific to biological systems in Wikipedia. The term first appeared in the early 1960s in theoretical particle physics and rapidly spread throughout the entire domain of physics. When explaining symmetry breaking, physicists like to paint a mental image of a ball precariously perched on the very top of a bump separating two identical wells. The slightest noise is sufficient to push the ball off the top and into either of the wells. After that, only one of them has the ball, and thus the initial unstable symmetry of the system is broken. While visually appealing, the allegory of a ball in a well is not always easy to relate to biology. Furthermore, throughout development, biological organisms often seem to transit from an amorphous asymmetric state, e.g., a clump of dividing cells known as a morula, to the state with a striking apparent symmetry, such as mature hydra polyp. This apparent increase in the symmetry of developing organisms can be misinterpreted as contradicting the applicability of the concept of symmetry breaking in biology.

History can help us resolve this conundrum. The first mention of symmetry breaking in regards to biological and chemical systems dates back, presumably, to the seminal paper by Prigogine and Nicolis [1], who stated that the pattern-forming mechanism proposed by Turing to explain biological morphogenesis [2] is "*symmetry breaking* as it leads from a homogeneous to an inhomogeneous steady state". In the mind of a theoretical physicist, a spatially homogenous chemical system is akin to an unbounded featureless plane, and any geometric translation, mirror reflection, or rotation of such a plane transforms it back into itself, defining it as infinitely symmetric. The appearance of a pattern (such as a hexagonal lattice of Turing spots) breaks this symmetry, leaving few if any geometric degrees of freedom that can transform the now patterned system back into itself. Similarly, an unpolarized cell can be modeled by a perfect sphere, which remains itself after any rotation around any axis that passes through its center. Cellular polarization effectively selects only one such axis around which a cell can be rotated so it is still "transformed into itself". Multicellular development is even harder to interpret in geometric terms, although specific examples of such geometric symmetry-breaking transitions can be identified. Thus, similar to cellular polarization, gastrulation reduces the spherical symmetry of the blastula to the cylindrical symmetry of the gastrula. Instead of looking for specific broken symmetries, it might be more productive to follow in the footsteps of Prigogine and Nicolis and consider all phenomena of biological pattern formation as manifestations of symmetry breaking. This pluralistic approach conceptually unifies efforts to understand biological morphogenesis on both subcellular and multicellular levels and continues to gain popularity [3].

In the interdisciplinary spirit established by the founding fathers of the field, this Special Issue "Symmetry Breaking in Cells and Tissues" presents a collection of 17 reviews, opinions, and original research papers contributed by theoreticians, physicists and mathematicians, as well as experimental biologists, united by the common excitement about biological pattern formation and morphogenesis. In this issue, the contributors discuss diverse manifestations of symmetry breaking in biology and showcase recent developments in experimental and theoretical approaches to biological morphogenesis and pattern formation on multiple scales.

Establishment of cell polarity is, perhaps, one of the best-studied manifestations of symmetry breaking in biology. Unicellular model organisms, yeasts, have been particularly useful for studies of cell polarity due to their ease of genetic manipulation and the spectacular cell surface-localized pattern—the conspicuous micron-scale polarity cluster organized by the activity of Cdc42 and other small GTPases from the Ras and Rho families [4–6]. Moran and Lew discuss the role of differential diffusion of proteins on the plasma membrane in the establishment of cell polarity in budding yeast [7]. Martin and colleagues deploy optogenetics to study the mechanism of Cdc42 cluster formation in fission yeast and suggest the existence of mechanisms inhibiting de novo formation of the polarity cluster on the cell sides [8]. Khalili, Vavylonis, and colleagues continue the topic of fission yeast cell polarity by elaborating a detailed biophysical model that describes not only static but also spatially oscillating patterns of Cdc42 activity [9]. Finally, Daalman et al. bring up an evolutionary aspect of the budding yeast cell polarity network [10].

Par protein systems constitute another fundamental cell polarity paradigm found in diverse cell types of higher eukaryotes [11–14]. While much has been learned about the Cdc42 polarity in fungi and Par polarity in *C. elegans* embryos and epithelial cells, understanding the interplay between these two modules has been notoriously difficult. Seirin–Lee, Gaffney, and Dawes construct and analyze a heuristic model of *C. elegans* embryo polarity to explain the interaction between these two modules [15]. A third fundamental paradigm of cell polarity, the polarization of motile chemotactic cells, is revisited by van Haastert, who proposes a unified model of amoeboid movement applicable to both fast- and slow-moving cells [16]. The topic of cell polarity is rounded off by a comprehensive review of published polarity models by Othmer and colleagues [17].

Recent years have seen a dramatic rise of new topics establishing the mechanisms of biological symmetry breaking, which are distinct from the diffusion-driven instability of the reaction–diffusion systems proposed by Turing. Notable examples of these are mechanical instabilities in active systems consisting of cytoskeletal polymers and molecular motors [18,19] and protein phase separation [20,21]. In this Special Issue, Schwille and colleagues present an in vitro reconstitution of actomyosin cortices, in which they observe symmetry breaking and emergence of directed flows [22]. The theme of actomyosin contractility continues in the contribution from Gerisch and colleagues who discuss unilateral ingression of cleavage furrows in multinucleated cells of *Dictyostelium* amoeba [23]. The theme of protein phase separation and its role in the establishment of cellular memory and stress adaptation is reviewed by Caudron and colleagues [24].

Symmetry breaking on the scale of tissues and organs transcends multiple fields of developmental biology and is an area of research that has both a distinguished past and a rapidly developing present [25–28]. Connecting unicellular and multicellular scales, Yap and colleagues review epithelial cell extrusion, a form of collective behavior of cells within epithelial sheets that plays an important role in normal morphogenesis and development of cancer [29]. A review by Manceau, Bailleul, and Touboul provides an extensive overview of mechanisms and models of multicellular pattern formation [30].

Several contributions focus on theoretical aspects of pattern formation and morphogenesis. Naoz and Gov extend the theme of the pattern-forming role of proteins capable of bending the cell membrane and provide analysis of actin-mediated protrusions on the ventral side of adhered cells [31]. They show that facing hard substratum and adhering to it can stabilize such protrusions or induce their transition into propagating wavelike structures. Formation of protein patterns on the cell membrane in the presence of flows in the cytoplasm or cellular cortex is the focus of contribution by Frey and colleagues, who show that such flows can induce interesting and nontrivial effects modulating protein patterns [32]. An important theoretical question in the field of cellular pattern formation is whether specific mechanisms are required to produce a single structure, such as the one necessary for cellular polarization, or a multitude of similar structures, such as podosomes or microvilli. Goryachev and Leda discuss recent theoretical work exploring this question in the minimal mass-conserved activator–substrate models and conclude that the choice between competition and coexistence of structures is determined by the complex interplay of multiple system parameters, rather than by the type of the molecular mechanism [33]. Developing this theme further, Banerjee and colleagues consider induction of multiple structures in dynamically growing systems [34]. Finally, Beta, Gov, and Yochelis discuss a dynamic pattern, representative of intracellular actin waves, in a minimal activator–inhibitor model with mass conservation [35].

Funding: This research was funded by the Biotechnology and Biological Sciences Research Council of the UK, grant numbers BB/P01190X and BB/P006507.

Conflicts of Interest: The author declares no conflict of interest.

References

1. Prigogine, I.; Nicolis, G. On symmetry-breaking instabilites in dissipative systems. *J. Chem. Phys.* **1967**, *46*, 3542. [CrossRef]
2. Turing, A.M. The chemical basis of morphogenesis. *Philos. Trans. R. Soc. Lond. Ser. B Biol. Sci.* **1952**, *237*, 37–72. [CrossRef]
3. Goryachev, A.B.; Mallo, M. Patterning and Morphogenesis From Cells to Organisms: Progress, Common Principles and New Challenges. *Front. Cell Dev. Biol.* **2020**, *8*, 602483. [CrossRef] [PubMed]
4. Chiou, J.G.; Balasubramanian, M.K.; Lew, D.J. Cell Polarity in Yeast. *Annu. Rev. Cell Dev. Biol.* **2017**, *33*, 77–101. [CrossRef]
5. Goryachev, A.B.; Leda, M. Autoactivation of small GTPases by the GEF-effector positive feedback modules. *F1000Res* **2019**, *8*. [CrossRef]
6. Goryachev, A.B.; Leda, M. Many roads to symmetry breaking: Molecular mechanisms and theoretical models of yeast cell polarity. *Mol. Biol. Cell* **2017**, *28*, 370–380. [CrossRef]
7. Moran, K.D.; Lew, D.J. How Diffusion Impacts Cortical Protein Distribution in Yeasts. *Cells* **2020**, *9*, 1113. [CrossRef]
8. Lamas, I.; Weber, N.; Martin, S.G. Activation of Cdc42 GTPase upon CRY2-Induced Cortical Recruitment Is Antagonized by GAPs in Fission Yeast. *Cells* **2020**, *9*, 2089. [CrossRef]
9. Khalili, B.; Lovelace, H.D.; Rutkowski, D.M.; Holz, D.; Vavylonis, D. Fission Yeast Polarization: Modeling Cdc42 Oscillations, Symmetry Breaking, and Zones of Activation and Inhibition. *Cells* **2020**, *9*, 1769. [CrossRef]
10. Daalman, W.K.; Sweep, E.; Laan, L. The Path towards Predicting Evolution as Illustrated in Yeast Cell Polarity. *Cells* **2020**, *9*, 2534. [CrossRef]
11. Loyer, N.; Januschke, J. Where does asymmetry come from? Illustrating principles of polarity and asymmetry establishment in Drosophila neuroblasts. *Curr. Opin. Cell Biol.* **2020**, *62*, 70–77. [CrossRef] [PubMed]
12. Lang, C.F.; Munro, E. The PAR proteins: From molecular circuits to dynamic self-stabilizing cell polarity. *Development* **2017**, *144*, 3405–3416. [CrossRef] [PubMed]
13. Goehring, N.W. PAR polarity: From complexity to design principles. *Exp. Cell Res.* **2014**, *328*, 258–266. [CrossRef] [PubMed]
14. Hoege, C.; Hyman, A.A. Principles of PAR polarity in Caenorhabditis elegans embryos. *Nat. Rev. Mol. Cell Biol.* **2013**, *14*, 315–322. [CrossRef]
15. Seirin-Lee, S.; Gaffney, E.A.; Dawes, A.T. CDC-42 Interactions with Par Proteins Are Critical for Proper Patterning in Polarization. *Cells* **2020**, *9*, 2036. [CrossRef]
16. Van Haastert, P.J.M. Symmetry Breaking during Cell Movement in the Context of Excitability, Kinetic Fine-Tuning and Memory of Pseudopod Formation. *Cells* **2020**, *9*, 1809. [CrossRef]
17. Cheng, Y.; Felix, B.; Othmer, H.G. The Roles of Signaling in Cytoskeletal Changes, Random Movement, Direction-Sensing and Polarization of Eukaryotic Cells. *Cells* **2020**, *9*, 1437. [CrossRef]
18. Goehring, N.W.; Grill, S.W. Cell polarity: Mechanochemical patterning. *Trends. Cell Biol.* **2013**, *23*, 72–80. [CrossRef]
19. Julicher, F.; Kruse, K.; Prost, J.; Joanny, J.F. Active behavior of the cytoskeleton. *Phys. Rep. Rev. Sec. Phys. Lett.* **2007**, *449*, 3–28. [CrossRef]
20. Hyman, A.A.; Weber, C.A.; Jülicher, F. Liquid-liquid phase separation in biology. *Annu. Rev. Cell Dev. Biol.* **2014**, *30*, 39–58. [CrossRef]

21. Shin, Y.; Brangwynne, C.P. Liquid phase condensation in cell physiology and disease. *Science* **2017**, *357*. [CrossRef] [PubMed]
22. Vogel, S.K.; Wölfer, C.; Ramirez-Diaz, D.A.; Flassig, R.J.; Sundmacher, K.; Schwille, P. Symmetry Breaking and Emergence of Directional Flows in Minimal Actomyosin Cortices. *Cells* **2020**, *9*, 1432. [CrossRef] [PubMed]
23. Bindl, J.; Molnar, E.S.; Ecke, M.; Prassler, J.; Müller-Taubenberger, A.; Gerisch, G. Unilateral Cleavage Furrows in Multinucleate Cells. *Cells* **2020**, *9*, 1493. [CrossRef] [PubMed]
24. Lau, Y.; Oamen, H.P.; Caudron, F. Protein Phase Separation during Stress Adaptation and Cellular Memory. *Cells* **2020**, *9*, 1302. [CrossRef] [PubMed]
25. Čapek, D.; Müller, P. Positional information and tissue scaling during development and regeneration. *Development* **2019**, *146*. [CrossRef] [PubMed]
26. Green, J.B.; Sharpe, J. Positional information and reaction-diffusion: Two big ideas in developmental biology combine. *Development* **2015**, *142*, 1203–1211. [CrossRef]
27. Müller, P.; Schier, A.F. Extracellular movement of signaling molecules. *Dev. Cell* **2011**, *21*, 145–158. [CrossRef]
28. Marcon, L.; Sharpe, J. Turing patterns in development: What about the horse part? *Curr. Opin. Genet. Dev.* **2012**, *22*, 578–584. [CrossRef]
29. Nanavati, B.N.; Yap, A.S.; Teo, J.L. Symmetry Breaking and Epithelial Cell Extrusion. *Cells* **2020**, *9*, 1416. [CrossRef]
30. Bailleul, R.; Manceau, M.; Touboul, J. A "Numerical Evo-Devo" Synthesis for the Identification of Pattern-Forming Factors. *Cells* **2020**, *9*, 1840. [CrossRef]
31. Naoz, M.; Gov, N.S. Cell-Substrate Patterns Driven by Curvature-Sensitive Actin Polymerization: Waves and Podosomes. *Cells* **2020**, *9*, 782. [CrossRef] [PubMed]
32. Wigbers, M.C.; Brauns, F.; Leung, C.Y.; Frey, E. Flow Induced Symmetry Breaking in a Conceptual Polarity Model. *Cells* **2020**, *9*, 1524. [CrossRef] [PubMed]
33. Goryachev, A.B.; Leda, M. Compete or Coexist? Why the Same Mechanisms of Symmetry Breaking Can Yield Distinct Outcomes. *Cells* **2020**, *9*, 2011. [CrossRef] [PubMed]
34. Cornwall Scoones, J.; Banerjee, D.S.; Banerjee, S. Size-Regulated Symmetry Breaking in Reaction-Diffusion Models of Developmental Transitions. *Cells* **2020**, *9*, 1646. [CrossRef] [PubMed]
35. Beta, C.; Gov, N.S.; Yochelis, A. Why a Large-Scale Mode Can Be Essential for Understanding Intracellular Actin Waves. *Cells* **2020**, *9*, 1533. [CrossRef]

Article

Cell-Substrate Patterns Driven by Curvature-Sensitive Actin Polymerization: Waves and Podosomes

Moshe Naoz and Nir S. Gov *

Department of Chemical and Biological Physics, Weizmann Institute of Science, P.O.B. 26, Rehovot 76100, Israel; moshe.naoz@gmail.com

* Correspondence: nir.gov@weizmann.ac.il

Received: 16 February 2020; Accepted: 17 March 2020; Published: 23 March 2020

Abstract: Cells adhered to an external solid substrate are observed to exhibit rich dynamics of actin structures on the basal membrane, which are distinct from those observed on the dorsal (free) membrane. Here we explore the dynamics of curved membrane proteins, or protein complexes, that recruit actin polymerization when the membrane is confined by the solid substrate. Such curved proteins can induce the spontaneous formation of membrane protrusions on the dorsal side of cells. However, on the basal side of the cells, such protrusions can only extend as far as the solid substrate and this constraint can convert such protrusions into propagating wave-like structures. We also demonstrate that adhesion molecules can stabilize localized protrusions that resemble some features of podosomes. This coupling of curvature and actin forces may underlie the differences in the observed actin-membrane dynamics between the basal and dorsal sides of adhered cells.

Keywords: actin waves; curved proteins; dynamic instability; podosomes

---

1. Introduction

The actin cortex of cells is the prominent driver of membrane shape deformations, which exhibit a huge variability, from propagating waves to stable protrusions. It is often observed that the actin-membrane dynamics of adhered cells is very different between the basal and dorsal sides. One of the major differences between the two sides is that on the basal side the membrane is held at close proximity to the solid substrate (in the range of 10–100 nm, depending on the stage of the adhesion [1]), while on the dorsal side the membrane is usually free to deform into the surrounding fluid. In this paper we explore theoretically the actin-membrane dynamics in the presence of the confinement of the substrate, when the actin polymerization is nucleated by curved membrane complexes.

Cells exhibit a variety of propagating waves of actin polymerization on their basal plasma membrane, which are observed under many conditions such as the initial formation of adhesion [2] and during cell motility [3–5]. When these waves propagate on the dorsal side of an adhered cell, or along its perimeter edge, they are accompanied by large membrane deformations. However, when these waves propagate along the basal membrane, at the interface between the cell and the underlying solid substrate, such membrane deformations have not been unambiguously observed. These basal actin waves have been studied intensively [6–9], and many of their features exposed. Mostly these waves have been treated in the framework of reaction-diffusion models [9], where membrane deformations do not play a role.

In previous works [10,11] we have investigated theoretically and experimentally the possible role of curved activators of actin polymerization in the propagation of membrane-actin waves. In these works the positive feedback is in the form of an actin nucleator that has a convex shape (such as the I-BAR protein IRsp53 for example [12,13]), such that it tends to accumulate at the tips of membrane protrusions that are driven by the actin polymerization force. The negative feedback, which is necessary for wave propagation, can be provided by the contractile force of myosin-II [10] or the recruitment of concave-shaped actin nucleators (such as the BAR family proteins, Tuba for example [14]) [11]. More recent work proposed that the negative feedback for propagating basal actin waves arises from the actin network itself [15].

In this paper, we explore the dynamics of the membrane-actin system in the presence of only the convex nucleator, but in the presence of a confining boundary which represents the effect of the solid substrate. When there is no confinement, our model predicts that this system can become unstable and drive the spontaneous initiation of membrane protrusions through a Turing-type instability [16–20], as is indeed observed in experiments [21–24]. We show that in the presence of a confining boundary this system indeed supports protrusions, which are however modified compared to those growing on a free membrane: protrusions may split, and may even convert into propagating rings. These theoretical results may explain some puzzling features observed for actin waves that propagate at the substrate-attached cell surface, such as their tendency to form doublets of concentric actin fronts [9,25,26].

In the last section we demonstrate that by adding adhesion of the membrane to the substrate localized protrusive structures can be stabilized, and these share some features with localized adhesion structures called podosomes.

2. Results

2.1. Expanding Ring of Membrane-Actin Wave

Our model is based on the description of the membrane shape in terms of a single height variable $h(x,y)$, which is appropriate for small membrane deformations (Monge gauge). This is applicable for the present system, where the membrane is adhered to a solid substrate that confines the extent of its normal deflection. on the membrane we consider a density field $n(x,y)$ of "activator" proteins, which are both curved and can recruit the polymerization of actin (such as I-BAR protein IRsp53, for example). These "proteins" may therefore represent bound complexes that contain several proteins that together have this combination of properties. The curved membrane complexes can diffuse on the fluid membrane, as well as form high density aggregations. The spontaneous curvature of membrane proteins spans a large range, from being almost flat [27], to having radii of curvature of the order of 10 nm [12] (we used a value of 100 nm in this work). Please note that in general the membrane-bound protein complex can have a radius of curvature that is different from that of its individual components.

We solve the equations of motion for the two fields (Equations (10) and (11)) numerically using an explicit finite difference scheme with periodic boundary conditions. We first investigate the response of the system to a single small Gaussian perturbation (see Supplementary Movie S1, Figure 1). We choose values for the parameters of the model such that we are in the unstable regime, and protrusions grow spontaneously. However, the numerical values of these parameters are not fitted to any particular experimental measurement, and are simply chosen to demonstrate the qualitative behavior over realistic length and time scales.

As shown in Figure 1a, the perturbation initially grows into a protrusion with a lateral width of the order of the most unstable wavelength λ_c (Equation (12)). During this growth stage, the protrusive force of the actin locally squeezes the layer of long molecules (glycocalix) that buffers the outer surface of the membrane from the substrate (Figure 8). Due to the positive feedback, the density of activators increases at the tip of the growing protrusion (note that throughout the paper the plots of the "activator density" is with respect to the background, uniform density n-n_0).

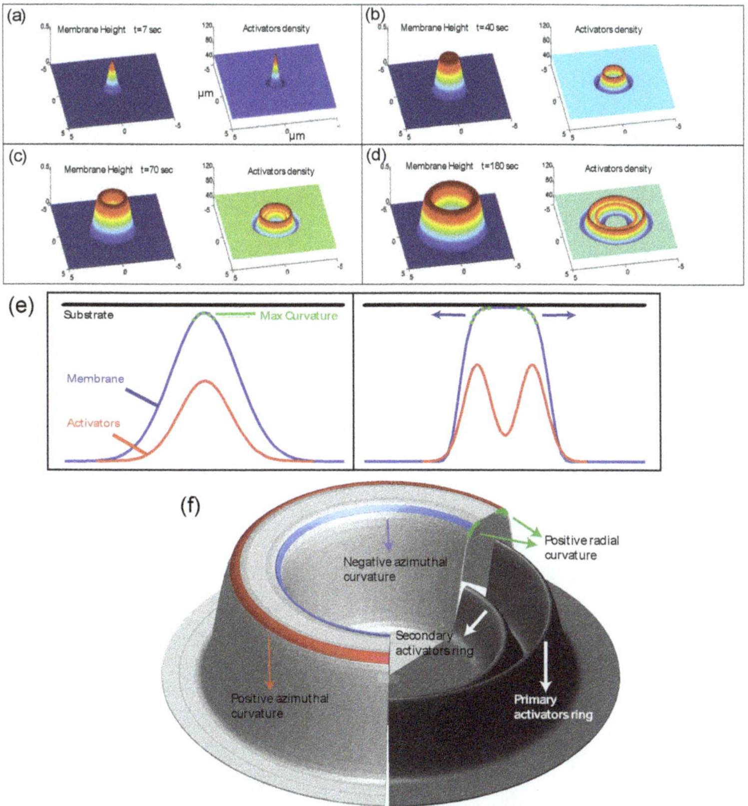

Figure 1. (**a**–**d**) Numerical integration of Equations (10) and (11) over a period of 3 min, over a membrane segment of size 10×10 µm^2. The parameter values used: $D = 0.1$ µm^2/s, $\bar{H} = -10$ µm^{-1}, $n_0 = 50$ µm^{-2}, $n_s = 300$ µm^{-2}, $A = 3.8 \cdot 10^{-5}$ kg µm^5 sec^{-2}, $\kappa = 20\,k_BT$, $\sigma = 8.28 \cdot 10^{-5}$ kg µm^4 sec^{-2}, $\mu = 1.66 \cdot 10^6$ sec µm^{-2} kg^{-1}, $h_{wall} = 0.5$ µm. (**a**) Initial growth of the protrusion, prior to contact with the substrate. Please note that throughout the paper the plots of the "activator density" is with respect to the background, uniform density n-n_0. (**b**) the protrusion after it comes into contact with the substrate and the membrane at the tip becomes flat. Consequently, an activator ring is formed at the edge of the membrane protrusion, where there is large positive curvature. (**c**) the membrane at the disk center has retracted back towards the unperturbed position at $h = 0$, and a secondary inner ring of activators begins to form. (**d**) the membrane ring and two activator rings expand further. (**e**) an illustration of the mechanism that drives the expansion of the protrusion radially outwards. When the membrane reaches the flat substrate its curvature diminishes and the activators are then aggregated at the location of the highest curvature—the shoulders. However, since once the activators aggregate they push the membrane against the substrate which results in the flattening of the shoulders and the formation of new shoulders further away from the protrusion center. (**f**) a diagram of the structure of the expanding ring. Marked in green are the regions high in curvature in the radial direction which is similar in magnitude for both the inner and the outer rings. Marked in red is the outer radius curvature in the azimuthal direction which is positive and marked in blue is the inner radius azimuthal curvature which is negative. Also shown is the concentration of activators which aggregate into two rings at the outer and inner radii of the membrane ring. The concentration of the outer activators ring is higher than the concentration at the inner ring due to the different azimuthal curvatures.

Once the protrusion's height exceeds h_{wall}, the substrate limits further growth and the membrane shape tends to match the contour of the substrate. If the substrate is sufficiently flat the activators which were aggregated at the tip of the protrusion disperse. Please note that we do not consider at this stage any adhesion to the substrate, so that the activators remain mobile on the membrane even when it is in contact with the substrate. The result is a rolling instability where the activators continually aggregate at shoulders of the protrusion (Figure 1b,e), which are the location of the highest convex curvature. The aggregation of activators increases the protrusive force exerted on the shoulders which are therefore results in the membrane deformation moving radially outwards. The rolling instability results in the protrusion developing into an expanding circular structure.

We emphasize that this model includes only normal deformations of the membrane, so the actin force does not directly push the membrane sideways along the substrate. The movement of the protrusion laterally is driven by the flow of the curved activators, and the coupling to the protrusive force of the actin polymerization. Please note that a similar behavior is expected for curved activators that adsorb in a curvature-dependent manner from the cytoplasm [11].

The membrane's initial shape is an expanding cylinder and the activators form a ring at the membrane perimeter of the protrusion (Figure 1b). Once the radius of the membrane cylinder is sufficiently large, the membrane at the center of the cylinder, which is no longer supported by a surplus of actin protrusive force, falls back to the initial height (at $h = 0$). This happens due to the inherent repulsion between the membrane and the substrate, cause by the "cushion" layer of long molecules (glycocalix) that cover the outer surface of the membrane (Figure 8). When the membrane cylinder changes into a ring shape, a secondary inner ring of activators forms at the inner shoulder of the membrane ring (Figure 1c,d,f), where there is high convex curvature.

The amplitude of the activators density at the inner ring is initially considerably smaller then the outer ring amplitude. The reason for the amplitude difference is the difference in the mean curvature between the outer an inner rings. While the radial curvatures (the curvature along the radial coordinate centered at the protrusion center) are very similar, the azimuthal curvature (the curvature along the angular coordinate), which is of the order of $1/R$ (R is the radius of the ring) is positive at the outer radius and negative at the inner radius (Figure 1f). Therefore, at small radii, the difference in the mean curvatures is significant. The higher convex curvature at the outer ring means that the convex activators aggregate there more and the resulting protrusive force exerted on the membrane is stronger. This imbalance results in the continued outwards expansion of the ring. Please note that the inner actin ring does *not* move inwards, since the curved actin activators flow towards increasing mean convex curvature, which decreases if the ring would shift to a smaller radius. Therefore the inner ring is also propagating outwards, at a velocity which is very similar to that of the outer ring, maintaining a roughly constant distance between them.

However, as the ring radius grows larger, the difference between the azimuthal curvatures at the inner and outer rings diminishes and the difference in the amplitudes of the activators rings (and therefore the protrusive force) decreases, which reduces the speed of the ring expansion (Figure 2a,b). In Figure 2c we plot the activators density at the inner and outer rings as function of the local mean curvature, and in Figure 2d we plot the ring velocity as a function of the difference between the density of activators at the inner and outer rings. The plot shows that the velocity is proportional to that difference, i.e., $V_{\text{ring}} \propto \Delta n = n_{\text{outer}} - n_{\text{inner}}$. The results indicate that the ring velocity is indeed proportional to the imbalance in the pushing force of the two actin rings, and explains why it decreases as: $V_{\text{ring}} \sim 1/R$ (Figure 2b).

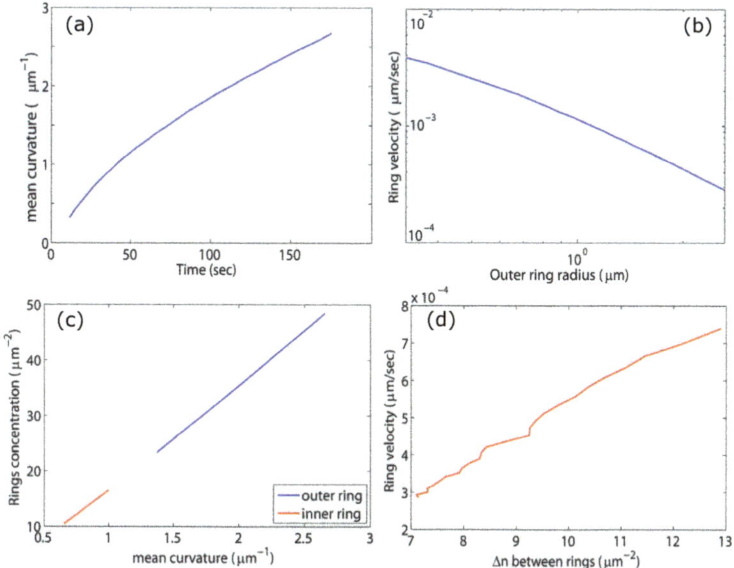

Figure 2. (**a**) the radius of the outer membrane disk (and later ring) as a function of time (all panels correspond to the simulation shown in Figure 1). (**b**) a log plot of the expansion speed of the outer radius vs the radius. We see that the graph is curved at small radii (where the ring is actually a disk) but as the radius grows (and the ring forms) the graph approaches a straight line indicating the power law relation: $V_{\text{ring}} \sim 1/R$. (**c**) the peak values of the differential concentration n-n_0 at the inner (red) and outer (blue) rings as a function of the local membrane curvature. We see the the concentrations are linear in the curvature, as given by Equation (3). (**d**) the ring outwards velocity as a function of the difference in concentrations between the inner and outer activators ring.

We can quantify these observations by the following calculation: If we hold the membrane shape constant, we can solve the steady state concentration profile of the curved activators that corresponds to that shape. By taking $\dot{n} = 0$ in Equation (11) and integrating once we get

$$D\nabla n + \frac{\Lambda\kappa\bar{H}^2}{n_s^2} n\nabla n - \frac{\Lambda\kappa\bar{H}}{n_s} n\nabla^3 h = 0 \tag{1}$$

we then divide by n and integrate again and we are left with

$$D\ln\left(n/n_0\right) + \frac{\Lambda\kappa\bar{H}^2}{n_s^2}(n - n_0) - \frac{\Lambda\kappa\bar{H}}{n_s}\nabla^2 h = 0 \tag{2}$$

For large concentrations we can neglect the first term on the left hand side and get

$$n - n_0 = \frac{\bar{H}}{n_s}\nabla^2 h \tag{3}$$

We therefore find that when the dynamics of the activators is faster than the expansion rate of the membrane deformation, so that the activators' concentration is in a quasi-steady state, we get that the activators amplitude is approximately proportional to the local membrane curvature. In Figure 2c we plot the concentration of the inner and outer rings vs. the mean curvature at these locations. The plot shows the concentrations are a linear function of the mean curvature which confirms that for the parameters used in the calculation, the dynamics of the activators are indeed faster than the membrane dynamics, and the result of Equation (3) is valid.

Since in our model the concentration of actin activators is strongly affected by the local membrane curvature, the dynamics of the ring is sensitive to the topography of the substrate. We illustrate this by simulating the dynamics on a substrate that is roughened with random bumps with an average amplitude of a few tens of nanometers (Figure 3a). Using the same set of equations and the same initial conditions as shown in Figure 1a–d, we now get the result shown in Figure 3b,c. The overall qualitative behavior is similar to the case with a smooth surface, i.e., the formation of an expanding ring-like membrane structure with inner and outer rings of activators. However, the shape of the expanding structure is strongly affected by the substrate roughness and did not retain the circular symmetry it started with. The membrane ring also shows short "finger-like" protrusions extruding radially from the perimeter, which are accompanied by very strong aggregation of activators. The inner activators ring does not extend into these deformations. Due to the surface roughness, and the consequent fluctuations in the membrane curvature, the distribution of the activators inside the membrane ring-like structure becomes very inhomogeneous and fragmented.

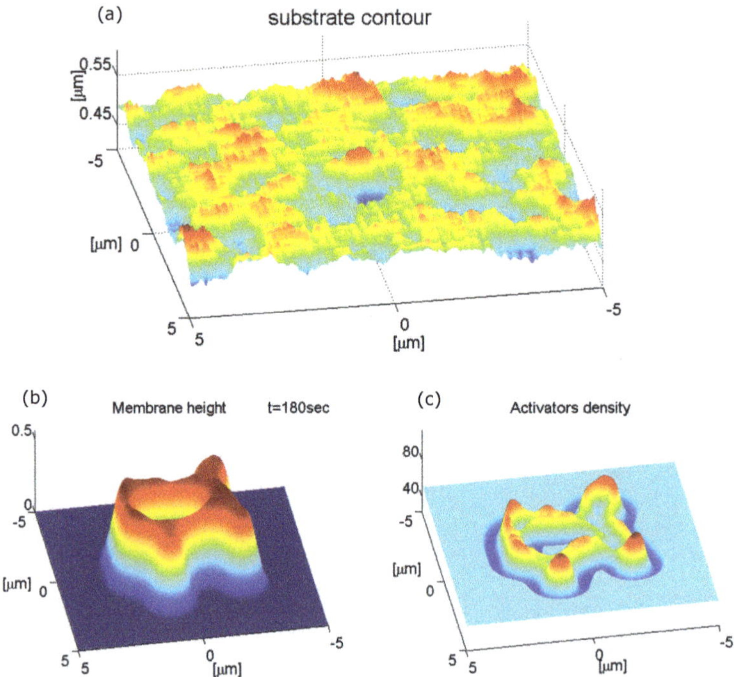

Figure 3. (**a**) Substrate with random roughness, of Gaussian amplitude, with an average amplitude of a few tens of nanometers. (**b**,**c**) Numerical integration over a period of 3 min of a ring expanding over a membrane segment of size $10 \times 10\ \mu m^2$. We used the same parameter values as in Figure 1.

Another illustration of the effects of substrate topography is shown in Figure 4 (Supplementary Movie S2), where a single elongated cylindrical ridge protrudes from the otherwise flat surface (along the x-axis). As can be seen, when the membrane ring first reaches the tip of the ridge it is curved backwards with respect to its expansion direction (top middle panel). As the membrane wraps around the protruding ridge, this membrane part develops negative curvature and the ring of actin activators breaks up at that point (top right panel). Only when the inner activators ring forms, this tip of the bump becomes favorable and concentrates activators that begin to push the membrane ring backwards towards the point of initiation (bottom-right panel). on the two side of the elevated ridge the original membrane ring propagates faster than on the flat substrate. This is due to the higher concentration of actin activators, which is caused by the high positive curvature in the sharp corners that the elevated ridge makes with the flat substrate (bottom middle and right panels).

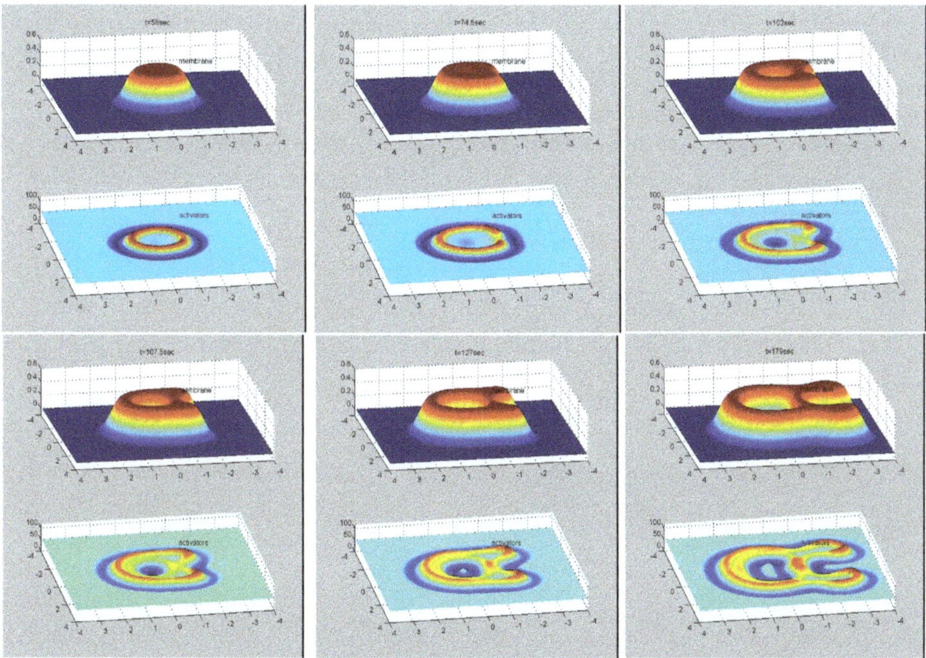

Figure 4. Numerical integration of a ring expanding over a membrane that contains a cylindrical ridge along the x-axis. From top-left to bottom-right, each panel shows the system at increasing time, with the membrane shape and activator density in the top and bottom parts, respectively. We used the same parameter values as in Figure 1.

The simulations in Figures 3 and 4, demonstrate that the distribution of actin activators (and therefore actin filaments) can become highly fragmented due to surface undulations, while the enveloping membrane structure remains continuous and smooth. It is therefore not straightforward to relate the actin signal to the membrane topography when interpreting experimental images of such cell-substrate waves. It also shows that topographic features can cause local direction reversals and break-up of these actin waves.

2.2. *Array of Localized Protrusions*

We now study the conditions that may stabilize localized structures driven by the same curved activator we used so far. In Figure 5 we show that starting with different initial conditions may lead to a quasi-stable array of localized structures. A random noise in the initial membrane height or activators density, instead of a single perturbation, gives rise to multiple protrusions, each with a lateral size comparable to λ_c (Figure 5). When these protrusions reach the substrate they undergo the same process of flattening and expanding that we saw before for a single protrusion. During their expansion, the distance between these protrusions naturally decreases. The interaction between these protrusions is repulsive due to the positive membrane curvature region trapped between neighboring protrusions, similar to those observed in other curved membrane-bound aggregates [28]. Therefore, once they have expanded and reached a distance of order λ_c from each other their expansion is halted, and they stabilize. Over the course of the stabilization, the protrusions expand into all the space that is available by the repulsive interactions between neighbors, leading to the elongation of some of the protrusions.

Furthermore, if two protrusions were initially formed in close proximity, the energy barrier between them is decreased and they can coalesce into a single elongated protrusion (arrows in Figure 5b–d point to such a process).

Please note that in these localized protrusions the actin forms small rings within each protrusion, due to the same process we found in the expanding ring (Figure 1), on a smaller scale. These protrusions, which may seem stable, are highly sensitive to small perturbations, since they are stabilized only by their mutual repulsion due to the local membrane-driven barrier. Over long times we expect noise to cause them to shift and coalesce. Non-linear terms drive coalescence over such barriers, over a long time [28].

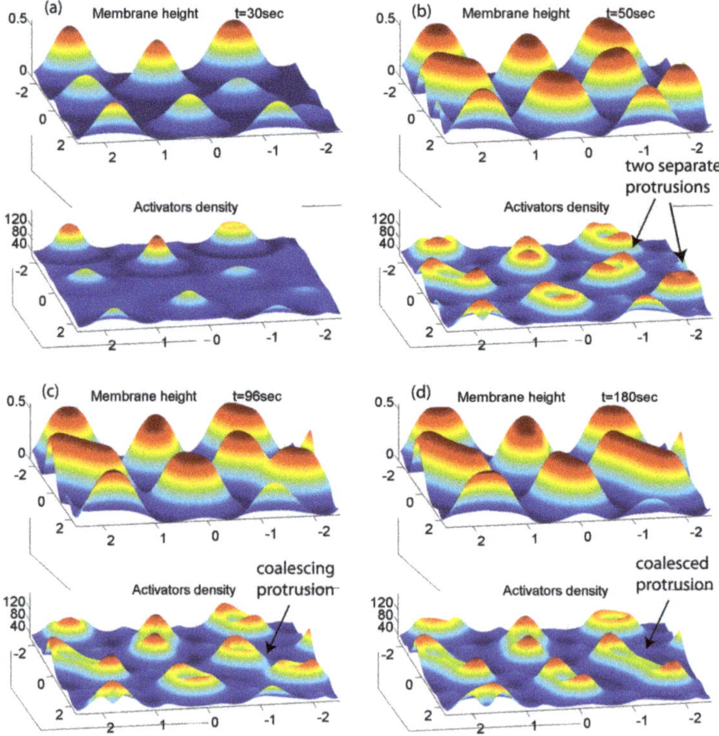

Figure 5. Simulation over a period of 3 min over a membrane segment of size 5×5 μm^2, with Gaussian noise in the initial membrane height with variance of 10nm. The membrane's random initial deformations develop into protrusions of lateral size $\sim \lambda_c$. (**a**) the initial growth period before most of them make contact with the substrate. (**b**,**c**) the stabilization period of the protrusions, which elongate into the available space and may coalesce with very proximal protrusions. (**d**) the final steady state of the system. We used the same parameter values as in Figure 1.

2.3. Adhesion-Stabilized Localized Protrusion: The Podosome

To stabilize an isolated protrusion on the basal side, using the curvature-actin mechanism that we propose, we need to stabilize the localized actin core and prevent the tendency of the protrusion to expand outwards as a ring (Figure 1). An example of a localized adhesion structure on the basal size of many cell types is the podosome [29]. Podosomes are actin-rich protrusive adhesion structures formed on the membrane of several cell types, and have been implicated in the processes of cell migration, tissue invasion and extracellular matrix (ECM) degradation. The mechanisms that give rise to podosome formation, and their large-scale organization in the cell, are still poorly understood and are the subject of ongoing current research [30]. Podosomes are typically formed in monocytic cells such as macrophages, osteoclasts and dendritic cells and similar structures called invadopodia have been observed in carcinoma cells [31–35]. They are relatively dynamic, formed and destroyed in the span of a few minutes and are formed only on the interface between the cell and a substrate. Furthermore, the I-BAR protein IRsp53 was found in podosomes [36,37], suggesting that the basic ingredients of our model may apply for this cellular structure.

The podosome's actin core is surrounded by an adhesion ring, which we did not include so far in the theoretical model, and we therefore suggest that this component may stabilize the core and prevent its ring-like expansion. We propose that the adhesion molecules form a diffusion barrier that greatly inhibits the diffusion of the membrane-bound actin nucleators (Equation (14)), and thereby stabilizes the localized core. In addition, there is evidence that *concave* curved membrane proteins, such as BAR domain proteins, accumulate at the podosome [38], and can further act as a diffusion barrier [39]. we incorporate the adhesion into the model with the same approach that we used to incorporate the actin, namely all the proteins involved in the adhesion process, from plaque proteins that form a scaffold around the actin core to the integrins which connect between the cellular membrane and external ligands, are grouped into a single component which we denote the "adhesion proteins". We consider adhesion proteins to be membrane proteins that have two possible states: a non-adhered, freely diffusing state with concentration g_f and an adhered, immobile state with concentration g_b. The transition from non-adhered to adhered state is only possible when the distance between the membrane and the substrate is small enough, for the membrane-bound integrins to bind to the substrate. Please note that we do not explicitly describe the inter-podosome actin network [34], but rather focus on modeling a single, isolated podosome.

In addition to the geometric constraint, the binding rate of the membrane-bound adhesion proteins depends on the application of tensile forces, as integrins are known to exhibit catch-bond properties [40,41]. In the vicinity of the actin core of the podosome, the tensile force is thought to arise at the outer edge of the core, where the actin filaments flow towards the actin core and apply a pulling and shearing force that facilitates integrin adhesion [42–44] (Figure 6a).

We combine the two properties that affect the adhesion listed above, in a very simplified way, in the following equations for the binding/unbinding rates of the adhesion proteins (used in the first order kinetics Equation (15))

$$k_{on}^{g} = k_{on,0}^{g} f(h) |\nabla n| \tag{4}$$

$$k_{off}^{g} = k_{off,0}^{g} \tag{5}$$

where $f(h)$ has the profile shown in Figure 6b so that binding is only permitted close to the substrate. The last term in Equation (4) describes in the simplest way the fact that the adhesion is dependent on the spatial gradients in the actin force that is applied on the membrane. In addition, we note that the adhesion strength that determines $k_{on,0}^{g}/k_{off,0}^{g}$ is affected by the substrate stiffness and chemical composition.

As before (Figure 1), we investigate the system's behavior due to a small Gaussian perturbation in the membrane height (Figure 7). Numerical integration of Equation (15) shows that the perturbation grows into a protrusion and the curved activators aggregate at the tip of the protrusion. When the protrusion reaches the substrate, the activators disperse from the center towards the shoulders. However, at the same stage, adhesion proteins bind to the substrate around the activators and inhibit their dispersion. If the adhesion proteins aggregate quickly enough they can trap the activators in the protrusion core, despite the negligible curvature that the membrane at the center possess due to the confinement by the flat substrate. However, the activators in the core may still form a small ring due to small changes in membrane curvature (Figure 7). When we choose a shorter binding protein, we get a core of smaller radius, which is uniform, i.e., the activators do not form a ring shape. We therefore demonstrate that the adhesion ring can indeed function as a diffusion barrier that stabilizes the actin core. On long time scales, where the membrane-bound actin nucleators may detach from the membrane, the actin core may decay, and this process could limit the podosome lifetime.

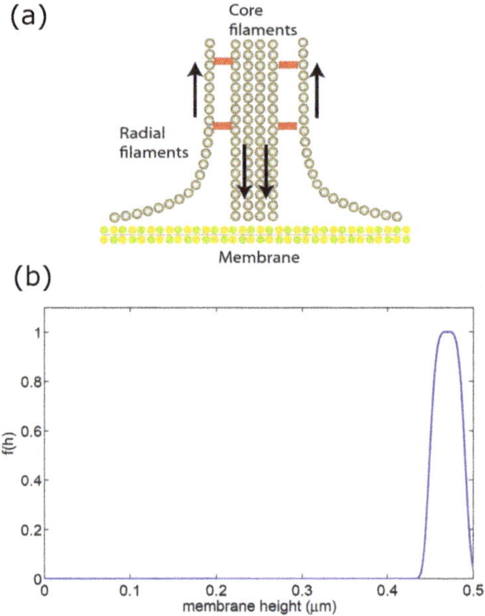

Figure 6. (**a**) Schematic illustration of the actin cables around the core of the podosome, where the treadmilling flow induces shearing and pulling forces near the membrane. These forces are thought to activate integrin-based cell adhesion [30]. (**b**) the function $f(h)$ (Equation (4)) that we used to limit the binding of the adhesion proteins only to within a distance of $2l_0$ from the substrate (here $h_{\mathrm{wall}} = 0.5$ μm).

Overall, the stable podosome-like structure that our model produces exhibits many properties of the podosome. This serves to demonstrate that a model with very few components can give rise to spontaneous formation of membrane protrusions at the basal side of cells, which form an adhesion complex that closely resembles podosomes. It is sometimes observed that the actin core of podosomes may be slightly depleted at its center near the membrane [45], which is a feature that can also appear in our model (Figure 7).

Please note that our model contains a single type of curved actin "activator", while podosomes seem to possess a complex composition of actin filaments [46], as well as a complex and dynamic inter-podosome actin-myosin network [47]. Future modeling of these cellular structures, could involve more of these components, allowing for the simulation of their large-scale dynamics, where podosomes form macro structures comprised of many podosomes [44,48]. For such long-term simulations we will also need to include processes that limit the lifetime of individual podosomes. In most cells, podosomes are organized in a cluster of uniform distribution with a characteristic distance between, undergoing processes of fusion and fission [49]. This organization and dynamics qualitatively resembles the dynamics shown in Section 2.2 (Figure 5).

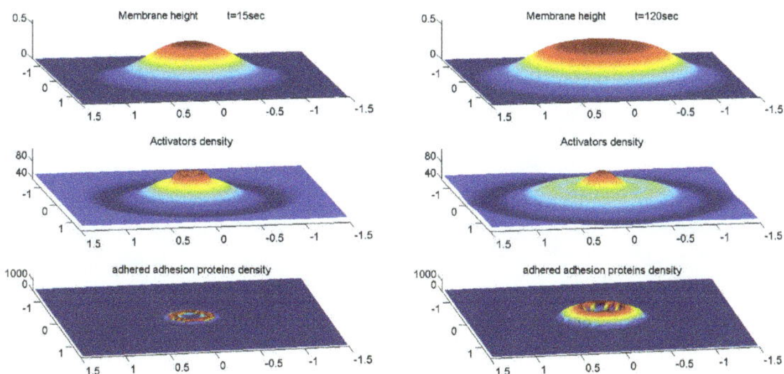

Figure 7. Numerical integration of Equation (15) for the case of constitutive active activators over a period of 2 min over a membrane segment of size 5×5 μm^2. The parameter values used were the same as used previously, with the following changes: $A = 1.9 \cdot 10^{-5}$ kg·μm^5· sec^{-2}, $D_g = 0.1$ μ^2/sec, $k^g_{on,0} = 0.05$ μm^5/sec, $k^g_{off,0} = 1$ μm^2/sec, $l_0 = 0.04$ μm and $g_0 = 50$ μm^{-2} (the initial uniform concentration of the free adhesion proteins). The perturbation develops into a protrusion and when it reaches the wall the activators at the core start expanding into a ring but this expansion is inhibited by the adhesion ring that forms around it. The actin core is then trapped by the adhesion ring and the structure stabilizes.

In several cell types, such as osteoclasts, large collections of podosomes exhibit a transition to an expanding multi-podosome ring structure. The ring is densely populated with podosomes, has a width of a few podosomes and expands at a speed of ~1–2 μm/min [50,51]. The podosomes in the multi-podosome ring are immobile and the ring's outward expansion is achieved by a treadmilling manner: the podosomes decay at the inner part of the multi-podosome ring and form preferentially at its outer edge [49]. These multi-podosome rings move outward and merge with each other until they reach the cell periphery, where they may stabilize as a podosome belt ("sealing zone") [52]. In certain cells the rings seem to originate at the same locations at roughly regular intervals [51].

Within the current model we cannot account for the detailed dynamics of the podosomes within the expanding ring, but we can speculate that its outwards expansion may be related to the mechanism that drives the expansion of the actin ring described in Section 2.1: When a podosome in the ring decays, its constituent proteins diffuse away, and due to the ring-link deformation of the membrane the curved proteins tend to aggregate more strongly at the outer edge of the ring compared to its inner edge (see Figure 1). This mechanism will therefore naturally increase the rate of podosome formation on the outer edge of the ring, compared to its inner edge, and over time cause the outwards expansion of the podosome ring. This expansion will depend on the rate of podosome initiation and decay that enables this effective "podosome treadmill"y.

3. Discussion

The results of our model can give a very natural explanation to several puzzling features observed in experiments. One such feature is that actin waves at the cell-substrate interface are observed to expand as a single ring of actin polymerization, but beyond a certain size an inner ring of actin that follows the outer one appears. The inner ring is often weaker than the outer ring, as our model predicts. This feature was first noted by Vicker [25] in *Dictyostelium discoideum* amoebae, and more recently this feature was studied in great detail [9,26].

The actin fronts observed in these experiments are very often broken and fragmented, with numerous breaks appearing along the ring [25,26]. This feature is a natural consequence in our model of the sensitivity of the actin concentration to the surface topography (Figures 3 and 4), due to the curvature sensitivity of the actin nucleators. Our model further predicts that the actin polymerization will tend to concentrate where the substrate has concave corners, as along the sides of elevated ridges (Figure 4). This prediction is in agreement with observations of crawling *Dicty* on surfaces with patterned ridges [53].

To conclude, we have shown some aspects of the dynamics of the actin-membrane system, when driven by convex actin activators and confined by the substrate. We show that the confinement itself provides a source of negative feedback that can drive propagating fronts. These simulations are simplified and contain several assumptions, which may not apply to all the biological cases. Furthermore, we do not claim that the actin waves along the basal membranes of cells are driven solely by the mechanism that we describe. Clearly complex reaction-diffusion feedbacks play an important role in the propagation of actin waves in cells [4,9,25,26,54–56]. Our work may motivate further studies of models that include both the reaction-diffusion dynamics and the membrane shape, coupled by curved membrane complexes that nucleate actin polymerization [15]. Since reaction-diffusion models lead to rich dynamics, as does the curvature-actin coupling [19], we expect that models with both features could open up new classes of cell membrane dynamics to explore.

Regarding localized structures at the basal membrane, our model predicts that these may be stabilized by the formation of adhesion around the actin core, as observed in podosomes. Furthermore, if the membrane can be maintained in a curved shape at the tip of the protrusion (rather than flatten), we predict that the curved activators will be less strongly dispersed (if at all), and the lifetime of the localized protrusion extended. This prediction is in agreement with observations that podosomes preferentially form along grooves where the membrane naturally has the curvature of the protrusion tip [57,58], and on rough surfaces [59]. When the protrusion is able to penetrate into the substrate, as occurs in invadopodia [60], the curvature at the protrusion tip is also maintained and this can also stabilize it. Future studies could explore the membrane dynamics predicted by this model on curved substrates, during phagocytosis for example.

4. Materials and Methods

4.1. Model Without Membrane-Substrate Adhesion

The model has two variables, $h(\vec{r},t), n(\vec{r},t)$, that describe the local height deformation of the membrane from its uniform state, and the local density of the membrane-bound, and curved, activators of actin polymerization, respectively. We will work in the limit of small membrane deformations, which allow us to treat the elastic energy of the membrane due to tension and bending in the quadratic limit [16,17]. This is an approximation which may be justified due to the presence of the confining boundary that naturally limited the amplitude of the membrane deformations. Non-linear effects arise in our calculations only from the conservation of the activator field n. For simplicity we do not consider the process of binding and unbinding of the actin activators from the membrane, which can be added in the future [11].

We start with the free energy of the membrane and actin activators [16,17]

$$F(h,n) = \int d^2r \left[\frac{1}{2}\kappa \left(\nabla^2 h - \bar{H}\frac{n}{n_s} \right)^2 + \frac{1}{2}\sigma (\nabla h)^2 + k_B T n \ln(n) \right] \quad (6)$$

where σ is the effective membrane tension, κ the bending modulus, $\bar{H}$ the spontaneous curvature of the actin activators and n_s their saturation density. In this treatment the two-dimensional Laplacian $\nabla^2 h$ is the local mean membrane curvature, and we keep the entropic term only at the lowest order, valid for low protein densities $n \ll n_s$.

We model the external barrier (substrate) as a one sided harmonic potential (Figure 8) that affects the membrane if its height coordinate h exceeds the barrier height coordinate h_{wall}. We also subject the membrane to a similar force (though smaller in magnitude) if its height is lower than the initial height (at $h = 0$) to account for the overall average rigidity of the cortical cytoskeleton. The wall and cytoskeleton interactions can be inserted as potentials to the free energy of the system, in the form

$$V_{\text{wall}} = \begin{cases} \frac{1}{2}F_{\text{wall}}\,(h - h_{\text{wall}})^2 & h > h_{\text{wall}} \\ 0 & h \leq h_{\text{wall}} \end{cases} \tag{7}$$

$$V_{\text{cyt}} = \begin{cases} \frac{1}{2}F_{\text{cyt}}h^2 & h < 0 \\ 0 & h \geq 0 \end{cases} \tag{8}$$

where F_{wall},F_{cyt} determine the stiffness of these potentials.

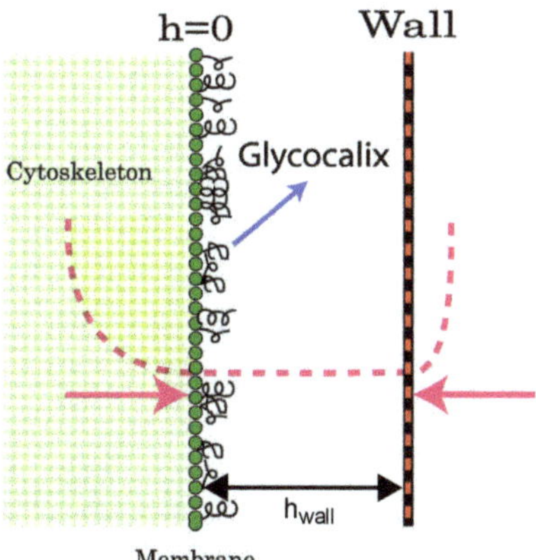

Figure 8. Illustration of the model. The substrate ("wall") acts as a spring with a force $F_{\text{wall}}(h - h_{\text{wall}})$ (right pink arrow) when the membrane height h exceeds the boundary location h_{wall}. The cytoskeleton acts in the same way only in the opposite direction (left pink arrow) and is typically softer, i.e., $F_{\text{cyt}} < F_{\text{wall}}$.

The equation of motion for the membrane height is given by [16,61]

$$\dot{h} = -\int \Gamma(r - r')\frac{\delta F}{\delta h}d^2r \tag{9}$$

where $\Gamma(r - r')$ is the Oseen tensor for the hydrodynamic interactions through the surrounding fluid. We replace this long-range interaction kernel with a local on, as is often used for membranes that are highly confined by the cytoskeleton (and here also by the substrate) [16,17]. We therefore take: $\Gamma(r - r') = \mu\delta(r - r')$, where μ is the drag coefficient of the membrane. We can now use Equations (6) and (9) to write the equation of motion for the membrane height

$$\dot{h} = \mu\left[-\kappa\nabla^4 h + \tfrac{\kappa\bar{H}}{n_s}\nabla^2 n + \sigma\nabla^2 h + A(n - n_0) - F_{\text{wall}}(h - h_{\text{wall}})\theta[h - h_{\text{wall}}] - (1 - \theta[h])F_{\text{cyt}}h\right] \tag{10}$$

where we used the step function $\theta[h]$ to implement the truncated forces of the confining potentials of Equations (7) and (8). The fourth term on the r.h.s. denotes the force due to actin polymerization, which is proportional to the density of actin activators with proportionality factor A. We subtract the average density of the actin cortex in this term to denote the fact that the cell membrane tends to be pushed away from the substrate by a layer of extracellular molecules (called the glycocalix) [1,62]. These molecules act as molecular cushions and maintain weak (non-specific) cell-substrate adhesion, and maintain osmotic pressure on the cell membrane.

The dynamics of the actin activators is given by [16,17]

$$\dot{n} = \Lambda \nabla \left(n \nabla \left(\frac{\delta F}{\delta h} \right) \right) = D \nabla^2 n + \frac{\Lambda \kappa \bar{H}^2}{n_s^2} \nabla \left(n \nabla n \right) - \frac{\Lambda \kappa \bar{H}}{n_s} \nabla \left(n \nabla^3 h \right) \tag{11}$$

where Λ is the mobility of the activators in the membrane, and $D = \Lambda k_B T$ is their diffusion coefficient.

We note that the model presented here considers activators that are permanently bound to the membrane and respond to the curvature by flowing in the membrane. Alternatively, the activators can be considered to adsorb to the membrane from the cytoplasm in a curvature-dependent manner [11].

Linear stability analysis of these equations of motion (Equations (10) and (11)) indicate that the system is unstable to small perturbations in either the membrane shape or activators density, for a range of parameters. For negligible membrane tension the most unstable wavelength is

$$\lambda_c \simeq 2\pi \sqrt{\frac{k_B T \mu}{A \bar{H} n_s}} \tag{12}$$

We chose the parameters to have this wavelength of order 1 μm.

The numerical simulations were done using an explicit finite difference scheme centered in space and forward in time, with periodic boundary conditions. The Cartesian grid used in the simulations was 0.025×0.025 μm in size.

4.2. Model with Membrane-Substrate Adhesion

When including adhesion proteins, the free energy of the system becomes

$$\begin{aligned} F(n,h) = \int dxdy \Bigg[& \frac{1}{2}\kappa \left(\nabla^2 h - \frac{\bar{H}}{n_s} n \right)^2 + \frac{1}{2}\sigma (\nabla h)^2 + g_b \theta (h - h_g) \frac{1}{2} k_d (h - h_{\text{wall}} - l_0)^2 + \\ & k_B T \left(n \ln(n) + g_f \ln(g_f) \right) \Bigg] \end{aligned} \tag{13}$$

where l_0 is the length of the external part of the adhesion protein and h_g is the minimal membrane height required for adhesion to take place. We take this value to be $h_g \simeq h_{\text{wall}} - 2l_0$ to allow for variations in the proteins length or membrane fluctuations.

In addition, we assume that the mobility of the actin activators to decrease in regions that have high concentration of adhered adhesion proteins. The reasoning behind this is that the scaffold of plaque proteins surrounds the actin core and restricts the movement of actin filaments, and forms a "diffusion barrier". If actin activators are attached to actin filaments then they too will be restricted. We therefore take the mobility to decrease with the local density of bound adhesion proteins, as follows

$$\Lambda = \begin{cases} \Lambda_0 - \frac{g_b}{g_{max}}, & \text{for } g_b \leq g_{max} \\ 0, & \text{for } g_b > g_{max} \end{cases} \tag{14}$$

Under these conditions we can write the following equations of motion, including the dynamics of the free and bound adhesion proteins (g_f,g_b respectively)

$$\dot{h} = -\mu\Bigg(\kappa\nabla^4 h + \frac{\kappa\bar{H}}{n_s}\nabla^2 n + \mu\sigma\nabla^2 h + A(n-n_0) - \theta(h-h_g)g_b k_d(h-h_{wall}-l_0) - \tag{15a}$$

$$\theta(h-h_{wall})F_{wall}(h-h_{wall}) - (1-\theta(h))F_{cyt}h\Bigg) \tag{15b}$$

$$\dot{n} = D\nabla^2 n + \frac{\Lambda\kappa\bar{H}^2}{n_s^2}\nabla(n\nabla n) - \frac{\Lambda\kappa\bar{H}}{n_s}\nabla(n\nabla^3 h) \tag{15c}$$

$$\dot{g_b} = \theta(h-h_g)k^g_{on}g_f - k^g_{off}g_b \tag{15d}$$

$$\dot{g_f} = D_g\nabla^2 n_f - \theta(h-h_g)k^g_{on}g_f + k^g_{off}g_b \tag{15e}$$

where D_g is the diffusion coefficient of the free (un-adhered) adhesion proteins in the membrane, and the binding/unbinding rates of the adhesion proteins is given in Equations (4) and (5).

Supplementary Materials: The following are available at http://www.mdpi.com/2073-4409/9/3/782/s1, Movie S1: Movie of numerical integration of Equations (10) and (11) showing a single growing protrusion, that upon collision with the flat substrate is converted into an expanding ring. This movie corresponds to Figure 1 of the paper. Movie S2: Movie of numerical integration of Equations (10) and (11) showing a single growing protrusion, that upon collision with the flat substrate is converted into an expanding ring. The ring then encounters a cylindrical protrusion along the x-axis of the substrate. This movie corresponds to Figure 4 of the paper.

Author Contributions: Conceptualization, M.N. and N.S.G.; numerical simulations, M.N.; writing, M.N. and N.S.G. All authors have read and agreed to the published version of the manuscript.

Funding: This research was funded by Israel Science Foundation grant number 337/05.

Acknowledgments: N.S.G is the incumbent of the Lee and William Abramowitz Professorial Chair of Biophysics. This work is made possible through the historic generosity of the Perlman family. N.S.G. wishes to thank Alessandra Cambi and Carsten Beta for useful discussions.

Conflicts of Interest: The authors declare no conflict of interest. The funders had no role in the design of the study; in the collection, analyses, or interpretation of data; in the writing of the manuscript, or in the decision to publish the results.

References

1. Cohen, M.; Joester, D.; Geiger, B.; Addadi, L. Spatial and temporal sequence of events in cell adhesion: From molecular recognition to focal adhesion assembly. *Chembiochem* **2004**, *5*, 1393–1399. [CrossRef] [PubMed]
2. Gerisch, G.; Bretschneider, T.; Müller-Taubenberger, A.; Simmeth, E.; Ecke, M.; Diez, S.; Anderson, K. Mobile actin clusters and traveling waves in cells recovering from actin depolymerization. *Biophys. J.* **2004**, *87*, 3493–3503. [CrossRef] [PubMed]
3. Bretschneider, T.; Diez, S.; Anderson, K.; Heuser, J.; Clarke, M.; Müller-Taubenberger, A.; Köhler, J.; Gerisch, G. Dynamic actin patterns and Arp2/3 assembly at the substrate-attached surface of motile cells. *Curr. Biol.* **2004**, *14*, 1–10. [CrossRef]
4. Allard, J.; Mogilner, A. Traveling waves in actin dynamics and cell motility. *Curr. Opin. Cell Biol.* **2013**, *25*, 107–115. [CrossRef] [PubMed]
5. Inagaki, N.; Katsuno, H. Actin waves: Origin of cell polarization and migration? *Trends Cell Biol.* **2017**, *27*, 515–526. [CrossRef]
6. Bretschneider, T.; Anderson, K.; Ecke, M.; Müller-Taubenberger, A.; Schroth-Diez, B.; Ishikawa-Ankerhold, H.C.; Gerisch, G. The three-dimensional dynamics of actin waves, a model of cytoskeletal self-organization. *Biophys. J.* **2009**, *96*, 2888–2900. [CrossRef] [PubMed]
7. Schroth-Diez, B.; Gerwig, S.; Ecke, M.; Hegerl, R.; Diez, S.; Gerisch, G. Propagating waves separate two states of actin organization in living cells. *HFSP J.* **2009**, *3*, 412–427. [CrossRef]
8. Gerisch, G. Self-organizing actin waves that simulate phagocytic cup structures. *PMC Biophys.* **2010**, *3*, 7. [CrossRef]

9. Miao, Y.; Bhattacharya, S.; Banerjee, T.; Abubaker-Sharif, B.; Long, Y.; Inoue, T.; Iglesias, P.A.; Devreotes, P.N. Wave patterns organize cellular protrusions and control cortical dynamics. *Mol. Syst. Biol.* **2019**, *15*, e8585. [CrossRef]
10. Shlomovitz, R.; Gov, N.S. Membrane waves driven by actin and myosin. *Phys. Rev. Lett.* **2007**, *98*, 168103. [CrossRef] [PubMed]
11. Peleg, B.; Disanza, A.; Scita, G.; Gov, N. Propagating cell-membrane waves driven by curved activators of actin polymerization. *PLoS ONE* **2011**, *6*, e18635. [CrossRef] [PubMed]
12. Mattila, P.K.; Pykäläinen, A.; Saarikangas, J.; Paavilainen, V.O.; Vihinen, H.; Jokitalo, E.; Lappalainen, P. Missing-in-metastasis and IRSp53 deform PI(4,5)P2-rich membranes by an inverse BAR domain-like mechanism. *J. Cell Biol.* **2007**, *176*, 953–964. [CrossRef] [PubMed]
13. Scita, G.; Confalonieri, S.; Lappalainen, P.; Suetsugu, S. IRSp53: Crossing the road of membrane and actin dynamics in the formation of membrane protrusions. *Trends Cell Biol.* **2008**, *18*, 52–60. [CrossRef]
14. Kovacs, E.M.; Makar, R.S.; Gertler, F.B. Tuba stimulates intracellular N-WASP-dependent actin assembly. *J. Cell Sci.* **2006**, *119*, 2715–2726. [CrossRef]
15. Wu, Z.; Su, M.; Tong, C.; Wu, M.; Liu, J. Membrane shape-mediated wave propagation of cortical protein dynamics. *Nat. Commun.* **2018**, *9*, 136. [CrossRef] [PubMed]
16. Gov, N.S.; Gopinathan, A. Dynamics of membranes driven by actin polymerization. *Biophys. J.* **2006**, *90*, 454–469. [CrossRef]
17. Veksler, A.; Gov, N.S. Phase transitions of the coupled membrane-cytoskeleton modify cellular shape. *Biophys. J.* **2007**, *93*, 3798–3810. [CrossRef]
18. Kabaso, D.; Shlomovitz, R.; Schloen, K.; Stradal, T.; Gov, N.S. Theoretical model for cellular shapes driven by protrusive and adhesive forces. *PLoS Comput. Biol.* **2011**, *7*, e1001127. [CrossRef]
19. Gov, N. Guided by curvature: Shaping cells by coupling curved membrane proteins and cytoskeletal forces. *Philos. Trans. R. Soc. B Biol. Sci.* **2018**, *373*, 20170115. [CrossRef]
20. Fosnaric, M.; Penic, S.; Iglič, A.; Kralj-Iglic, V.; Drab, M.; Gov, N. Theoretical study of vesicle shapes driven by coupling curved proteins and active cytoskeletal forces. *Soft Matter* **2019**, *15*, 5319–5330. [CrossRef]
21. Yang, C.; Hoelzle, M.; Disanza, A.; Scita, G.; Svitkina, T. Coordination of membrane and actin cytoskeleton dynamics during filopodia protrusion. *PLoS ONE* **2009**, *4*, e5678. [CrossRef] [PubMed]
22. Vaggi, F.; Disanza, A.; Milanesi, F.; Di Fiore, P.P.; Menna, E.; Matteoli, M.; Gov, N.S.; Scita, G.; Ciliberto, A. The Eps8/IRSp53/VASP network differentially controls actin capping and bundling in filopodia formation. *PLoS Comput. Biol.* **2011**, *7*, e1002088. [CrossRef] [PubMed]
23. Prévost, C.; Zhao, H.; Manzi, J.; Lemichez, E.; Lappalainen, P.; Callan-Jones, A.; Bassereau, P. IRSp53 senses negative membrane curvature and phase separates along membrane tubules. *Nat. Commun.* **2015**, *6*, 8529. [CrossRef] [PubMed]
24. Begemann, I.; Saha, T.; Lamparter, L.; Rathmann, I.; Grill, D.; Golbach, L.; Rasch, C.; Keller, U.; Trappmann, B.; Matis, M.; et al. Mechanochemical self-organization determines search pattern in migratory cells. *Nat. Phys.* **2019**, *15*, 848–857. [CrossRef]
25. Vicker, M.G. F-actin assembly in *Dictyostelium* cell locomotion and shape oscillations propagates as a self-organized reaction–diffusion wave. *FEBS Lett.* **2002**, *510*, 5–9. [CrossRef]
26. Gerhardt, M.; Ecke, M.; Walz, M.; Stengl, A.; Beta, C.; Gerisch, G. Actin and PIP3 waves in giant cells reveal the inherent length scale of an excited state. *J. Cell Sci.* **2014**, *127*, 4507–4517. [CrossRef] [PubMed]
27. Almeida-Souza, L.; Frank, R.A.; García-Nafría, J.; Colussi, A.; Gunawardana, N.; Johnson, C.M.; Yu, M.; Howard, G.; Andrews, B.; Vallis, Y.; et al. A flat BAR protein promotes actin polymerization at the base of clathrin-coated pits. *Cell* **2018**, *174*, 325–337. [CrossRef]
28. Shlomovitz, R.; Gov, N. Membrane-mediated interactions drive the condensation and coalescence of FtsZ rings. *Phys. Biol.* **2009**, *6*, 046017. [CrossRef] [PubMed]
29. Murphy, D.A.; Courtneidge, S.A. The 'ins' and 'outs' of podosomes and invadopodia: Characteristics, formation and function. *Nat. Rev. Mol. Cell Biol.* **2011**, *12*, 413–426. [CrossRef]
30. Alonso, F.; Spuul, P.; Daubon, T.; Kramer, I.; Génot, E. Variations on the theme of podosomes: A matter of context. *Biochim. Biophys. Acta (BBA) Mol. Cell Res.* **2019**, *1866*, 545–553. [CrossRef]
31. Linder, S.; Aepfelbacher, M. Podosomes: Adhesion hot-spots of invasive cells. *Trends Cell Biol.* **2003**, *13*, 376–385. [CrossRef]

32. Buccione, R.; Orth, J.D.; Mcniven, M.A. Foot and mouth: Podosomes, invadopodia and circular dorsal ruffles. *Nat. Rev. Mol. Cell Biol.* **2004**, *5*, 647–657. [CrossRef] [PubMed]
33. Jurdic, P.; Saltel, F.; Chabadel, A.; Destaing, O. Podosome and sealing zone: Specificity of the osteoclast model. *Eur. J. Cell Biol.* **2006**, *85*, 195–202. [CrossRef] [PubMed]
34. Paterson, E.K.; Courtneidge, S.A. Invadosomes are coming: New insights into function and disease relevance. *FEBS J.* **2018**, *285*, 8–27. [CrossRef] [PubMed]
35. van den Dries, K.; Linder, S.; Maridonneau-Parini, I.; Poincloux, R. Probing the mechanical landscape—New insights into podosome architecture and mechanics. *J. Cell Sci.* **2019**, *132*. [CrossRef]
36. Oikawa, T.; Okamura, H.; Dietrich, F.; Senju, Y.; Takenawa, T.; Suetsugu, S. IRSp53 mediates podosome formation via VASP in NIH-Src cells. *PLoS ONE* **2013**, *8*, e60528. [CrossRef] [PubMed]
37. Meddens, M.B.; van den Dries, K.; Cambi, A. Podosomes revealed by advanced bioimaging: What did we learn? *Eur. J. Cell Biol.* **2014**, *93*, 380–387. [CrossRef]
38. Sánchez-Barrena, M.J.; Vallis, Y.; Clatworthy, M.R.; Doherty, G.J.; Veprintsev, D.B.; Evans, P.R.; McMahon, H.T. Bin2 is a membrane sculpting N-BAR protein that influences leucocyte podosomes, motility and phagocytosis. *PLoS ONE* **2012**, *7*, e52401. [CrossRef]
39. Simunovic, M.; Manneville, J.B.; Renard, H.F.; Evergren, E.; Raghunathan, K.; Bhatia, D.; Kenworthy, A.K.; Voth, G.A.; Prost, J.; McMahon, H.T.; et al. Friction mediates scission of tubular membranes scaffolded by BAR proteins. *Cell* **2017**, *170*, 172–184. [CrossRef]
40. Thomas, W. Catch bonds in adhesion. *Annu. Rev. Biomed. Eng.* **2008**, *10*, 39–57. [CrossRef]
41. Kong, F.; García, A.J.; Mould, A.P.; Humphries, M.J.; Zhu, C. Demonstration of catch bonds between an integrin and its ligand. *J. Cell Biol.* **2009**, *185*, 1275–1284. [CrossRef] [PubMed]
42. Block, M.R.; Badowski, C.; Millon-Fremillon, A.; Bouvard, D.; Bouin, A.P.; Faurobert, E.; Gerber-Scokaert, D.; Planus, E.; Albiges-Rizo, C. Podosome-type adhesions and focal adhesions, so alike yet so different. *Eur. J. Cell Biol.* **2008**, *87*, 491–506. [CrossRef] [PubMed]
43. Luxenburg, C.; Winograd-Katz, S.; Addadi, L.; Geiger, B. Involvement of actin polymerization in podosome dynamics. *J. Cell Sci.* **2012**, *125*, 1666–1672. [CrossRef] [PubMed]
44. Schachtner, H.; Calaminus, S.D.; Thomas, S.G.; Machesky, L.M. Podosomes in adhesion, migration, mechanosensing and matrix remodeling. *Cytoskeleton* **2013**, *70*, 572–589. [CrossRef] [PubMed]
45. Kaverina, I.; Stradal, T.E.; Gimona, M. Podosome formation in cultured A7r5 vascular smooth muscle cells requires Arp2/3-dependent de-novo actin polymerization at discrete microdomains. *J. Cell Sci.* **2003**, *116*, 4915–4924. [CrossRef]
46. Van den Dries, K.; Nahidiazar, L.; Slotman, J.A.; Meddens, M.B.; Pandzic, E.; Joosten, B.; Ansems, M.; Schouwstra, J.; Meijer, A.; Steen, R.; et al. Modular actin nano-architecture enables podosome protrusion and mechanosensing. *Nat. Commun.* **2019**, *10*, 1–16. [CrossRef]
47. Meddens, M.B.; Pandzic, E.; Slotman, J.A.; Guillet, D.; Joosten, B.; Mennens, S.; Paardekooper, L.M.; Houtsmuller, A.B.; Van Den Dries, K.; Wiseman, P.W.; et al. Actomyosin-dependent dynamic spatial patterns of cytoskeletal components drive mesoscale podosome organization. *Nat. Commun.* **2016**, *7*, 13127. [CrossRef]
48. Luxenburg, C.; Addadi, L.; Geiger, B. The molecular dynamics of osteoclast adhesions. *Eur. J. Cell Biol.* **2006**, *85*, 203–211. [CrossRef]
49. Luxenburg, C.; Parsons, J.T.; Addadi, L.; Geiger, B. Involvement of the Src-cortactin pathway in podosome formation and turnover during polarization of cultured osteoclasts. *J. Cell Sci.* **2006**, *119*, 4878–4888. [CrossRef]
50. Destaing, O.; Saltel, F.; Géminard, J.C.; Jurdic, P.; Bard, F. Podosomes display actin turnover and dynamic self-organization in osteoclasts expressing actin-green fluorescent protein. *Mol. Biol. Cell* **2003**, *14*, 407–416. [CrossRef]
51. Collin, O.; Tracqui, P.; Stephanou, A.; Usson, Y.; Clément-Lacroix, J.; Planus, E. Spatiotemporal dynamics of actin-rich adhesion microdomains: Influence of substrate flexibility. *J. Cell Sci.* **2006**, *119*, 1914–1925. [CrossRef] [PubMed]
52. Luxenburg, C.; Geblinger, D.; Klein, E.; Anderson, K.; Hanein, D.; Geiger, B.; Addadi, L. The architecture of the adhesive apparatus of cultured osteoclasts: From podosome formation to sealing zone assembly. *PLoS ONE* **2007**, *2*, e179. [CrossRef]

53. Driscoll, M.K.; Sun, X.; Guven, C.; Fourkas, J.T.; Losert, W. Cellular contact guidance through dynamic sensing of nanotopography. *ACS Nano* **2014**, *8*, 3546–3555. [CrossRef] [PubMed]
54. Khamviwath, V.; Hu, J.; Othmer, H.G. A continuum model of actin waves in *Dictyostelium discoideum*. *PLoS ONE* **2013**, *8*, e64272. [CrossRef]
55. Brzeska, H.; Pridham, K.; Chery, G.; Titus, M.A.; Korn, E.D. The association of myosin IB with actin waves in Dictyostelium requires both the plasma membrane-binding site and actin-binding region in the myosin tail. *PLoS ONE* **2014**, *9*, e94306. [CrossRef] [PubMed]
56. Kruse, K. Cell Crawling Driven by Spontaneous Actin Polymerization Waves. In *Physical Models of Cell Motility*; Aranson, I., Ed.; Springer: Cham, Switzerland, 2016; pp. 69–93.
57. Van den Dries, K.; van Helden, S.F.; Te Riet, J.; Diez-Ahedo, R.; Manzo, C.; Oud, M.M.; van Leeuwen, F.N.; Brock, R.; Garcia-Parajo, M.F.; Cambi, A.; et al. Geometry sensing by dendritic cells dictates spatial organization and PGE 2-induced dissolution of podosomes. *Cell. Mol. Life Sci.* **2012**, *69*, 1889–1901. [CrossRef] [PubMed]
58. Rafiq, N.B.M.; Grenci, G.; Lim, C.K.; Kozlov, M.M.; Jones, G.E.; Viasnoff, V.; Bershadsky, A.D. Forces and constraints controlling podosome assembly and disassembly. *Philos. Trans. R. Soc. B* **2019**, *374*, 20180228. [CrossRef]
59. Shemesh, M.; Addadi, L.; Geiger, B. Surface microtopography modulates sealing zone development in osteoclasts cultured on bone. *J. R. Soc. Interface* **2017**, *14*, 20160958. [CrossRef] [PubMed]
60. Linder, S. The matrix corroded: Podosomes and invadopodia in extracellular matrix degradation. *Trends Cell Biol.* **2007**, *17*, 107–117. [CrossRef]
61. Atilgan, E.; Wirtz, D.; Sun, S.X. Mechanics and dynamics of actin-driven thin membrane protrusions. *Biophys. J.* **2006**, *90*, 65–76. [CrossRef]
62. Loomis, W.F.; Fuller, D.; Gutierrez, E.; Groisman, A.; Rappel, W.J. Innate non-specific cell substratum adhesion. *PLoS ONE* **2012**, *7*, e42033. [CrossRef] [PubMed]

Opinion

How Diffusion Impacts Cortical Protein Distribution in Yeasts

Kyle D. Moran and Daniel J. Lew *

Department of Pharmacology and Cancer Biology, Duke University Medical Center, Durham, NC 27710, USA; kyle.moran@duke.edu

* Correspondence: daniel.lew@duke.edu

Received: 23 March 2020; Accepted: 28 April 2020; Published: 30 April 2020

Abstract: Proteins associated with the yeast plasma membrane often accumulate asymmetrically within the plane of the membrane. Asymmetric accumulation is thought to underlie diverse processes, including polarized growth, stress sensing, and aging. Here, we review our evolving understanding of how cells achieve asymmetric distributions of membrane proteins despite the anticipated dissipative effects of diffusion, and highlight recent findings suggesting that differential diffusion is exploited to create, rather than dissipate, asymmetry. We also highlight open questions about diffusion in yeast plasma membranes that remain unsolved.

Keywords: diffusion; cell polarity; Cdc42

---

1. Introduction

The growth mode of budding yeasts, including the popular model *Saccharomyces cerevisiae*, involves the creation of a new cell (the bud) connected to a mother cell. Bud formation involves sequential changes in the pattern of secretion during the cell cycle, changing from uniform secretion in the mother during G1 phase, to polarized secretion into the bud during S/G2/M phases, to directed secretion towards the mother-bud neck during septation (Figure 1A). This pattern of secretion means that new proteins are delivered to locations that vary depending on when in the cell cycle the proteins are synthesized. For proteins that are linked to the cell wall, where they do not subsequently diffuse, the timing of synthesis determines their spatial distributions, as shown by elegant promoter-swap experiments [1] (Figure 1B). Similar analyses indicated that some "landmark" proteins, which guide the positioning of future bud sites, are also localized by controlling the timing of their expression in the cell cycle [2]. However, it was long assumed that the dissipative effects of diffusion would make this mechanism ineffective for proteins that are not attached to the rigid cell wall.

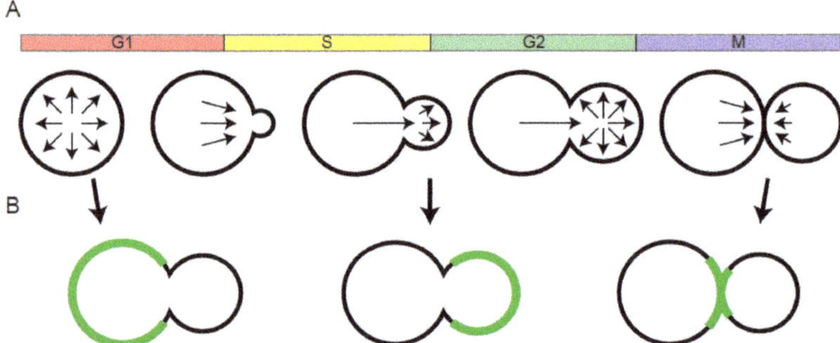

Figure 1. Stereotyped pattern of secretion enables localization dependent on the time of synthesis during the cell cycle. (**A**) Arrows indicate secretion patterns through the cell cycle. (**B**) With no diffusion, cell wall-linked proteins are spatially distributed in a manner that reflects where they were initially secreted. This leads to their accumulation in the mother, the bud, or at sites of cytokinesis, depending on the timing of their synthesis during the cell cycle.

2. Slow Diffusion Restricts Mobility in the Yeast Plasma Membrane

Yeast cells are small, with diameters of 4–6 μm [3]. Over such distances, proteins with diffusion constants typical of most integral membrane proteins in animal and bacterial membranes (~0.1 $\mu m^2/s$) [4–7] would diffuse across the cell in a few minutes. Nevertheless, many yeast membrane proteins accumulate asymmetrically in the mother or bud plasma membrane. Early hypotheses to explain such asymmetries involved a diffusion barrier at the mother-bud neck linked to cytoskeletal septin filaments found at that location [8,9]. However, subsequent work suggested instead that no barrier is needed because diffusion in the yeast plasma membrane is remarkably slow [10].

Fluorescence Recovery After Photobleaching (FRAP) studies showed that single-pass transmembrane proteins in the yeast plasma membrane diffuse with $D < 0.0025$ $\mu m^2/s$, while multipass transmembrane proteins diffuse with $D < 0.0005$ $\mu m^2/s$ [11,12]. Because these studies assumed that all fluorescence recovery was due to diffusion (neglecting other possible recovery pathways), they represent upper bounds on the diffusion constant, so these findings indicate that proteins in the yeast plasma membrane are remarkably immobile. This does not appear to stem from unusual features of the proteins themselves, as similar proteins targeted to a yeast vacuolar membrane or a mammalian plasma membrane are far more mobile (D ~0.1 $\mu m^2/s$) [11]. Consistent with the cited studies, FRAP experiments on a variety of other membrane proteins detected very little recovery on 10 min timescales, in fission yeast as well as budding yeast [13–15]. Thus, it appears that yeast plasma membranes severely restrict diffusion of embedded proteins.

3. Basis for Asymmetric Distribution of Integral Membrane Proteins

Interestingly, several long-lived membrane proteins are enriched in the mother plasma membrane, and this has been linked to the process of aging in yeast [13,16,17]. The proton pump, Pma1, is enriched in the mother and is thought to initiate aging by raising the cytoplasmic pH in the mother cell [16]. Transporters of the multidrug resistance family are also enriched in the mother plasma membrane, and are thought to decay with age, depriving the aging mother cell of detoxification capacity [13].

Given the very restricted lateral mobility of plasma membrane proteins in yeast, one way to obtain mother-enriched protein localization would be to restrict protein synthesis to a specific phase of the cell cycle, as discussed for cell wall proteins (Figure 1B). Indeed, the timing of synthesis combined with slow diffusion appears to account for much of the observed asymmetry for these long-lived proteins [10].

Slow diffusion also has implications for the distribution of short-lived proteins. The yeast pheromone (α-factor) receptor, Ste2, undergoes endocytosis and vacuolar degradation [18,19]. In cells that have not been exposed to pheromone, Ste2 synthesis is constitutive and endocytosis is slow. This causes Ste2 to accumulate in broad zones representing sites of recent delivery that change through the cell cycle [12,20]. In cells exposed to pheromone, Ste2 endocytosis and degradation is rapid, so that only very recently deposited receptor is present on the plasma membrane, accumulating more tightly around the site of polarized secretion [21].

Endocytosis and recycling can also change the distribution of long-lived membrane proteins. In a series of elegant experiments, the polarized distribution of v-SNAREs on the yeast plasma membrane was shown to arise from the combination of polarized delivery, slow diffusion, and rapid endocytosis that recycles v-SNAREs before they diffuse too far from their site of deposition [11] (Figure 2A). Similar mechanisms account for the polarized distribution of the plasma membrane stress sensor Wsc1 [22]. Point mutations that disable the endocytosis of these proteins cause them to accumulate uniformly all over the membrane, while appending a strong endocytosis signal on an otherwise uniformly distributed plasma membrane protein suffices to make its distribution resemble that of v-SNAREs and stress sensors [11].

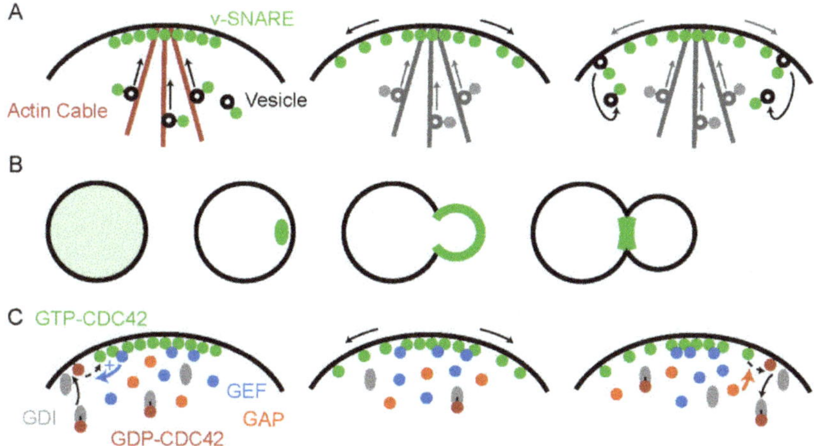

Figure 2. Dynamic localization of integral membrane and peripheral membrane proteins. (**A**) Asymmetric distribution of v-SNAREs arises from vesicle-mediated delivery of v-SNAREs to polarity sites (left), slow diffusion of v-SNAREs on the membrane (middle), and recycling via endocytosis (right). Note that while these processes are highlighted in separate panels for clarity, they all occur continuously. (**B**) Cdc42 localization through the cell cycle. (**C**) Asymmetric distribution of GTP-Cdc42 arises from deposition of GDP-Cdc42 on the membrane by GDI followed by local activation by GEF (left), diffusion of GTP-Cdc42 on the membrane (middle), and inactivation by GAP allowing GDI to pluck GDP-Cdc42 off the membrane, recycling it to the cytoplasm (right). Note that while these processes are highlighted in separate panels for clarity, they all occur continuously. Recruitment of GEF to sites with GTP-Cdc42 enables a positive feedback loop that loads GTP on neighboring Cdc42.

In aggregate, these studies highlight two key aspects of yeast cell biology that enable asymmetric accumulation of integral membrane proteins. First, secretion is directed towards different sites in a stereotypical manner during the cell cycle, allowing targeted delivery of proteins depending on their time of synthesis. Second, slow diffusion in the yeast plasma membrane can allow for asymmetries to persist, either passively for long-lived almost immobile multipass membrane proteins, or via endocytic recycling for more mobile single-pass membrane proteins.

4. Asymmetric Distribution of Peripheral Membrane Proteins: Recycling via the Cytoplasm

Unlike integral membrane proteins, peripheral membrane proteins can detach from the membrane to the cytoplasm, where diffusion is much more rapid (typically 5–15 μm^2/s) [6,7]. For a cell as small as yeast, cytoplasmic diffusion is very effective at dissipating any spatial gradients, as proteins would diffuse throughout the cell in ~1 s. However, selective attachment/detachment from membranes can result in localization to discrete sites. Perhaps the most intensively studied example is the conserved Rho-family GTPase Cdc42, responsible for polarity establishment.

Cdc42 localization changes dramatically through the cell cycle (Figure 2B). Initially dispersed throughout the cell, in late G1 Cdc42 accumulates at a patch on the plasma membrane that defines the site of future bud emergence. Cdc42 is enriched in the bud membrane during bud growth and relocates to the mother-bud neck during cytokinesis [23,24]. Because Cdc42 organizes the downstream polarization of the cytoskeleton, considerable interest has focused on understanding how Cdc42 itself becomes enriched at the polarity site [25].

Cdc42 is prenylated, which promotes its attachment to membranes. An early hypothesis to explain Cdc42 polarization was that, as for the v-SNAREs discussed above, directed secretion combined with endocytosis could mediate asymmetric accumulation of Cdc42 [26] (Figure 2A). However, estimates of the Cdc42 diffusion constant (0.036 μm^2/s: [27]) were far higher than those for v-SNAREs (0.0025 μm^2/s: [11]). With such high mobility, no realistic endocytic pathway could recycle the Cdc42 fast enough to counteract the dissipative effect of diffusion [28,29]. A proposed role for the septin diffusion barrier in maintaining Cdc42 polarity [30] also seems unlikely, as Cdc42 remains well polarized in septin mutants. Instead, current models of Cdc42 localization invoke recycling through detachment of the protein to the cytoplasm.

At the membrane, Cdc42 switches between GTP- and GDP-bound states in a manner assisted by GDP exchange factors (GEFs) and GTPase-activating proteins (GAPs) (Figure 2C). The Cdc42-directed GEF in yeast is primarily cytoplasmic, but associates with other proteins that allow it to be recruited to membrane sites with pre-existing GTP-Cdc42 [31]. This means that locations with some GTP-Cdc42 accumulate GEF, which loads GTP on neighboring Cdc42, creating a positive feedback loop that can drive symmetry breaking polarization. As GTP-Cdc42 diffuses, ubiquitous GAPs promote its conversion to GDP-Cdc42. GDP-Cdc42 can detach from the membrane to the cytoplasm, assisted by a GDP-dissociation inhibitor (GDI) [32] (Figure 2C) or via other poorly characterized pathways [33]. This enables rapid dissipation of any GDP-Cdc42 asymmetry via cytoplasmic diffusion. The preferential detachment of GDP-Cdc42 (compared to GTP-Cdc42) enables recycling of GDP-Cdc42 via the cytoplasm while maintaining a gradient of GTP-Cdc42 at the membrane (Figure 2C) [34–36]. This mechanism resembles the endocytic recycling mechanism of v-SNAREs (Figure 2A), but with rapid cytoplasmic diffusion allowing for much faster recycling that maintains the asymmetry of Cdc42 despite the higher diffusion of this peripheral membrane protein. In addition to cytoplasmic recycling, some studies have suggested that Cdc42 polarization is assisted by differential diffusion of Cdc42 at the plasma membrane.

5. Differential Diffusion of Cdc42 at the Membrane

The first report to consider differential diffusion of Cdc42 suggested that Cdc42 could diffuse more slowly in one part of the plasma membrane than another, perhaps due to membrane domains of different lipid composition [37]. Such location-dependent diffusion would enable diffusion to promote, rather than dissipate, Cdc42 asymmetry at the membrane. The notion that differential diffusion would serve to generate a nonhomogeneous distribution is widely recognized in physics and chemistry. As a simplifying case, let us neglect membrane–cytoplasm exchange and consider a cell that starts with uniformly localized Cdc42 all over the plasma membrane. If Cdc42 diffusion were reduced in a specific zone of the membrane, then molecules entering that zone by rapid lateral diffusion would slow down, and Cdc42 would accumulate in the slow-diffusion zone (Figure 3A). At steady state, Cdc42 concentration would be uniformly high within the zone, and uniformly low outside the zone. At the zone boundary, there would be a balance between escape of the slow-moving Cdc42 and arrival of rapidly-diffusing Cdc42. To maintain that balance, the concentration of Cdc42 within the slow zone must be higher than that outside the zone, by a factor equal to the ratio of diffusion constants.

Evidence for such location-dependent diffusion came from an unexpected observation [37]. GFP-Cdc42 at polarity sites displayed patchy localization, with dark spots in an otherwise bright polarized zone (Figure 3B). The patchiness persisted in mutants lacking the GDI and treated with Latrunculin to disable endocytosis: conditions intended to disable membrane–cytoplasm exchange. Moreover, an iFRAP experiment bleaching all of the GFP-Cdc42 outside of the polarity site led to a gradual loss of fluorescence over 2–4 min while maintaining the patchy pattern at the membrane. How could steep local gradients in GFP-Cdc42 concentration between adjacent patches exist when diffusion should rapidly dissipate such gradients?

The authors proposed that the patches represented zones in which Cdc42 mobility was altered, so that bright patches were slow-mobility zones and dark patches (as well as areas outside the polarity site) were high-mobility zones. In this view, location-dependent diffusion explained the GFP-Cdc42 localization pattern (Figure 3C). Fitting the GFP-Cdc42 localization data yielded diffusion estimates of 0.053 $\mu m^2/s$ (fast zones) and 0.0061 $\mu m^2/s$ (slow zones). The authors speculated that fast and slow diffusion zones might represent patches of membrane with different lipid composition, reminiscent of lipid rafts, although current models for lipid rafts consider much smaller and more transient entities.

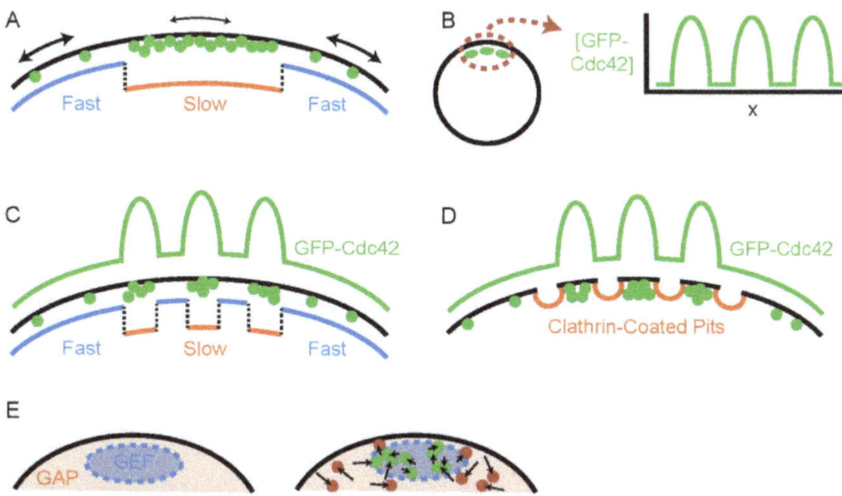

Figure 3. Differential diffusion of Cdc42. (**A**) Location-dependent differential diffusion: zones of fast and slow mobility (blue and orange, respectively) would cause a protein (green) to become concentrated in the slow-diffusion zone. (**B**) GFP-Cdc42 displays patchy localization at polarity sites. (**C**) Patchy localization could be explained by alternating zones of slow and fast mobility. (**D**) Patchy localization of GFP-Cdc42 could also be explained by exclusion of GFP-Cdc42 from endocytic sites (clathrin-coated pits). (**E**) Differential diffusion of GTP-Cdc42 and GDP-Cdc42. Localized GEF activity (blue) and uniform GAP activity (orange) would lead to local activation of Cdc42 in the GEF zone. If GTP-Cdc42 (green) diffuses less than GDP-Cdc42 (red), then Cdc42 accumulates in the GEF zone. Arrows indicate high mobility of GDP-Cdc42 relative to GTP-Cdc42.

An alternative, and in our view more likely, interpretation of the bright and dark patches is based on the observation that fluorescent markers of endocytic sites at the membrane showed a pattern that was anticorrelated with the GFP-Cdc42 signal [37]. We suggest that this could arise if GFP-Cdc42 were excluded from endocytic sites by the clathrin-associated coat proteins, generating the dark patches in the polarity site (Figure 3D). Because Latrunculin treatment blocks internalization of the endocytic coat, endocytic "dark patches" could stably persist in the iFRAP experiment described above. This hypothesis would obviate the need for location-dependent diffusion to explain the existence of stable bright and dark patches. Instead, the gradual dissipation of GFP-Cdc42 could reflect maintenance of asymmetry by GDI-independent recycling via the cytoplasm [33], combating a uniform diffusion of Cdc42.

6. Differential Diffusion of GTP-Cdc42 and GDP-Cdc42

Other studies proposed a distinct scenario in which diffusion of any given Cdc42 molecule is similar at all locations, but different forms of Cdc42 (GTP-Cdc42 vs. GDP-Cdc42) have inherently different mobility [14,38]. Consider a situation in which GDP-Cdc42 diffuses more rapidly in the plane of the membrane than GTP-Cdc42 (Figure 3E). The positive feedback loop mentioned above (Figure 2C) would create a zone where GEF activity is elevated, so that most Cdc42 in that zone would be GTP-bound. Assuming that ubiquitous GAP activity keeps Cdc42 elsewhere mostly GDP-bound, the cell would develop a zone with slow (GTP-)Cdc42 diffusion surrounded by a membrane with rapid (GDP-)Cdc42 diffusion. As with the differential mobility that comes from preferential detachment of GDP-Cdc42 from the membrane to the cytoplasm, this version of differential diffusion enables (rather than dissipating) polarization, and can collaborate with positive feedback via GEF recruitment to concentrate Cdc42 at the polarity site.

Convincing evidence for a stark GDP/GTP-dependent difference in Cdc42 diffusion came from studies in fission yeast, where (unlike in budding yeast) Cdc42 mobility was largely independent of the GDI. GDP-Cdc42 was estimated to diffuse rapidly in the membrane (0.1-0.2 $\mu m^2/s$), while GTP-Cdc42 diffusion was at least 10-fold slower [14]. This difference is sufficient, when combined with local GEF activity, to explain the enrichment of Cdc42 at polarity sites. Although early estimates in budding yeast found no evidence for a difference in diffusion constants for GDP-Cdc42 and GTP-Cdc42 [27], more recent work using single-particle tracking of photo-activatable mEOS-Cdc42 is consistent with an approximately two-fold difference (0.011 $\mu m^2/s$ for GTP-Cdc42 and 0.023 $\mu m^2/s$ for GDP-Cdc42) [38]. These findings suggest that both yeasts exploit differential diffusion of GDP-Cdc42 and GTP-Cdc42 to enhance Cdc42 asymmetric distribution at the membrane.

7. Conclusions and Open Questions

Diffusion is best known as a dissipative process that acts to reduce concentration gradients in cells and membranes. Nevertheless, many proteins in yeast plasma membranes develop markedly asymmetric patterns of accumulation. The studies discussed above suggest that this is enabled by two features. First, protein diffusion in the yeast plasma membrane is unusually slow, which reduces the dissipative effects of diffusion. Second, some proteins, including Cdc42, may diffuse at different rates depending on their conformation. These studies raise several unsolved questions, a few of which are highlighted below.

7.1. Why Is Diffusion So Slow in the Yeast Plasma Membrane?

Mutations that perturb the normal lipid composition of the plasma membrane can modestly increase protein diffusion [11,15]. Thus, an unusual lipid composition may increase membrane viscosity, hence slowing diffusion. However, even quite dramatic changes in lipid composition do not restore diffusion to the levels typical of other membranes, suggesting that non-lipid factors also contribute to immobilization. Because the plasma membrane is in contact with the rigid cell wall, it seems likely that interactions between membrane proteins and the immobile wall may also serve to restrict lateral mobility. Studies in plant cells have shown that even membrane proteins that do not bind to the cell wall can be corralled due to turgor-based apposition of the plasma membrane to the wall [39]. In yeast there are other immobile structures called eisosomes that form stable grooves in the plasma membrane and can impede diffusion of other membrane proteins [40]. Moreover, abundant membrane proteins like the proton pump Pma1 display dynamic mesh-like localization patterns in the membrane [41,42], potentially creating corrals that could slow diffusion of other proteins by confining their movement [43]. Understanding the contributions made by lipids, proteins, and the cell wall would be interesting and informative with regard to the degree to which similar phenomena occur in other systems.

7.2. Why Do So Many Membrane Proteins Accumulate Asymmetrically?

With the exception of the bud-site selection landmarks [2] and perhaps the stress sensors [22], we do not understand why a cell would benefit by accumulating many membrane proteins asymmetrically. Indeed, in several cases it seems counterproductive to do so. Asymmetric accumulation of proton pumps and transporters has been linked to deterioration of mother cells during aging [13,16], and it is unclear what benefit might outweigh that cost. Asymmetric accumulation of pheromone receptors has the potential to mislead mating cells about the location of a mating partner, requiring corrective mechanisms to provide accurate detection [12]. It would seem quite simple to avoid these problems, and there are instances where endocytic recycling has been tuned so as to counteract the asymmetry derived from timed synthesis in the cell cycle [10]. Why, then, are apparently unproductive asymmetries maintained?

7.3. What Is the Basis for the Differential Diffusion of GDP-Cdc42 and GTP-Cdc42?

The structures of GDP- and GTP-bound Cdc42 differ in well-appreciated ways [44], but it is not clear why that conformational change would significantly alter the protein's mobility in the membrane. Perhaps the simplest hypothesis would be that there is a much lower-mobility integral membrane protein that interacts selectively with GTP-Cdc42, slowing its diffusion. Single-particle tracking experiments in plant and animal as well as yeast cells identified a subset of "immobile" molecules, and high variability in the trajectories of individual molecules, consistent with interactions that transiently immobilize the protein. Other potential reasons for such transient immobilization invoke roles for lipids and nanoclusters of Cdc42 [38,45–47]. Overall, it seems we still have much to learn about how various proteins diffuse in the yeast plasma membrane.

Author Contributions: The authors contributed equally to this work. All authors have read and agreed to the published version of the manuscript.

Funding: This work was funded by NIH/NIGMS grant R35GM122488 to D.J.L.

Acknowledgments: We thank Tim Elston, Sam Ramirez, Nick Buchler, Masayuki Onishi, Steve Haase, and Amy Gladfelter, as well as members of the Lew lab for comments on the manuscript.

Conflicts of Interest: The authors declare no conflicts of interest.

References

1. Smits, G.J.; Schenkman, L.R.; Brul, S.; Pringle, J.R.; Klis, F.M. Role of Cell Cycle-Regulated Expression in the Localized Incorporation of Cell Wall Proteins in Yeast. *Mol. Biol. Cell* **2006**, *17*, 3267–3280. [CrossRef] [PubMed]
2. Schenkman, L.R.; Caruso, C.; Pagé, N.; Pringle, J.R. The Role of Cell Cycle–Regulated Expression in the Localization of Spatial Landmark Proteins in Yeast. *J. Cell Biol.* **2002**, *156*, 829–841. [CrossRef] [PubMed]
3. Klis, F.M.; de Koster, C.G.; Brul, S. Cell Wall-Related Bionumbers and Bioestimates of Saccharomyces cerevisiae and Candida albicans. *Eukaryot. Cell* **2014**, *13*, 2–9. [CrossRef] [PubMed]
4. Harms, G.S.; Cognet, L.; Lommerse, P.H.; Blab, G.A.; Kahr, H.; Gamsjaeger, R.; Spaink, H.P.; Soldatov, N.M.; Romanin, C.; Schmidt, T. Single-Molecule Imaging of L-Type Ca(2+) Channels in Live Cells. *Biophys. J.* **2001**, *81*, 2639–2646. [CrossRef]
5. Crane, J.M.; Verkman, A.S. Long-Range Nonanomalous Diffusion of Quantum Dot-Labeled Aquaporin-1 Water Channels in the Cell Plasma Membrane. *Biophys. J.* **2008**, *94*, 702–713. [CrossRef]
6. Kumar, M.; Mommer, M.S.; Sourjik, V. Mobility of Cytoplasmic, Membrane, and DNA-Binding Proteins in Escherichia coli. *Biophys. J.* **2010**, *98*, 552–559. [CrossRef]
7. Mika, J.T.; Poolman, B. Macromolecule Diffusion and Confinement in Prokaryotic Cells. *Curr. Opin. Biotechnol.* **2011**, *22*, 117–126. [CrossRef]

8. Barral, Y.; Mermall, V.; Mooseker, M.S.; Snyder, M. Compartmentalization of the Cell Cortex by Septins Is Required for Maintenance of Cell Polarity in Yeast. *Mol. Cell* **2000**, *5*, 841–851. [CrossRef]
9. Takizawa, P.A.; DeRisi, J.L.; Wilhelm, J.E.; Vale, R.D. Plasma Membrane Compartmentalization in Yeast by Messenger RNA Transport and a Septin Diffusion Barrier. *Science* **2000**, *290*, 341–344. [CrossRef]
10. Sugiyama, S.; Tanaka, M. Distinct Segregation Patterns of Yeast Cell-Peripheral Proteins Uncovered by a Method for Protein Segregatome Analysis. *Proc. Natl. Acad. Sci. USA* **2019**, *116*, 8909–8918. [CrossRef]
11. Valdez-Taubas, J.; Pelham, H.R. Slow Diffusion of Proteins in the Yeast Plasma Membrane Allows Polarity to Be Maintained by Endocytic Cycling. *Curr. Biol.* **2003**, *13*, 1636–1640. [CrossRef] [PubMed]
12. Henderson, N.T.; Pablo, M.; Ghose, D.; Clark-Cotton, M.R.; Zyla, T.R.; Nolen, J.; Elston, T.C.; Lew, D.J. Ratiometric GPCR Signaling Enables Directional Sensing in Yeast. *PLoS Biol.* **2019**, *17*, e3000484. [CrossRef] [PubMed]
13. Eldakak, A.; Rancati, G.; Rubinstein, B.; Paul, P.; Conaway, V.; Li, R. Asymmetrically Inherited Multidrug Resistance Transporters Are Recessive Determinants in Cellular Replicative Ageing. *Nat. Cell Biol.* **2010**, *12*, 799–805. [CrossRef] [PubMed]
14. Bendezú, F.O.; Vincenzetti, V.; Vavylonis, D.; Wyss, R.; Vogel, H.; Martin, S.G. Spontaneous Cdc42 Polarization Independent of GDI-Mediated Extraction and Actin-Based Trafficking. *PLoS Biol.* **2015**, *13*, e1002097. [CrossRef]
15. Singh, P.; Ramachandran, S.K.; Zhu, J.; Kim, B.C.; Biswas, D.; Ha, T.; Iglesias, P.A.; Li, R. Sphingolipids Facilitate Age Asymmetry of Membrane Proteins in Dividing Yeast Cells. *Mol. Biol. Cell* **2017**, *28*, 2712–2722. [CrossRef]
16. Henderson, K.A.; Hughes, A.L.; Gottschling, D.E. Mother-Daughter Asymmetry of pH Underlies Aging and Rejuvenation in Yeast. *eLife* **2014**, *3*, e03504. [CrossRef]
17. Thayer, N.H.; Leverich, C.K.; Fitzgibbon, M.P.; Nelson, Z.W.; Henderson, K.A.; Gafken, P.R.; Hsu, J.J.; Gottschling, D.E. Identification of Long-Lived Proteins Retained in Cells Undergoing Repeated Asymmetric Divisions. *Proc. Natl. Acad. Sci. USA* **2014**, *111*, 14019–14026. [CrossRef]
18. Hicke, L.; Riezman, H. Ubiquitination of a Yeast Plasma Membrane Receptor Signals Its Ligand-Stimulated Endocytosis. *Cell* **1996**, *84*, 277–287. [CrossRef]
19. Hicke, L.; Zanolari, B.; Riezman, H. Cytoplasmic Tail Phosphorylation of the Alpha-Factor Receptor Is Required for Its Ubiquitination and Internalization. *J. Cell Biol.* **1998**, *141*, 349–358. [CrossRef]
20. Emmerstorfer-Augustin, A.; Augustin, C.M.; Shams, S.; Thorner, J. Tracking Yeast Pheromone Receptor Ste2 Endocytosis Using Fluorogen-Activating Protein Tagging. *Mol. Biol. Cell* **2018**, *29*, 2720–2736. [CrossRef]
21. Ayscough, K.R.; Drubin, D.G. A Role for the Yeast Actin Cytoskeleton in Pheromone Receptor Clustering and Signalling. *Curr. Biol.* **1998**, *8*, 927–930. [CrossRef]
22. Piao, H.L.; Machado, I.M.; Payne, G.S. NPFXD-Mediated Endocytosis Is Required for Polarity and Function of a Yeast Cell Wall Stress Sensor. *Mol. Biol. Cell* **2007**, *18*, 57–65. [CrossRef] [PubMed]
23. Ziman, M.; Preuss, D.; Mulholland, J.; O'Brien, J.M.; Botstein, D.; Johnson, D.I. Subcellular Localization of Cdc42p, a Saccharomyces cerevisiae GTP-Binding Protein Involved in the Control of Cell Polarity. *Mol. Biol. Cell* **1993**, *4*, 1307–1316. [CrossRef] [PubMed]
24. Richman, T.J.; Sawyer, M.M.; Johnson, D.I. Saccharomyces cerevisiae Cdc42p Localizes to Cellular Membranes and Clusters at Sites of Polarized Growth. *Eukaryot. Cell* **2002**, *1*, 458–468. [CrossRef] [PubMed]
25. Chiou, J.G.; Balasubramanian, M.K.; Lew, D.J. Cell Polarity in Yeast. *Annu. Rev. Cell Dev. Biol.* **2017**, *33*, 77–101. [CrossRef]
26. Wedlich-Soldner, R.; Altschuler, S.; Wu, L.; Li, R. Spontaneous Cell Polarization Through Actomyosin-Based Delivery of the Cdc42 GTPase. *Science* **2003**, *299*, 1231–1235. [CrossRef]
27. Marco, E.; Wedlich-Söldner, R.; Li, R.; Altschuler, S.J.; Wu, L.F. Endocytosis Optimizes the Dynamic Localization of Membrane Proteins that Regulate Cortical Polarity. *Cell* **2007**, *129*, 411–422. [CrossRef]
28. Layton, A.T.; Savage, N.S.; Howell, A.S.; Carroll, S.Y.; Drubin, D.G.; Lew, D.J. Modeling Vesicle Traffic Reveals Unexpected Consequences for Cdc42p-Mediated Polarity Establishment. *Curr. Biol.* **2011**, *21*, 184–194. [CrossRef]
29. Savage, N.S.; Layton, A.T.; Lew, D.J. Mechanistic Mathematical Model of Polarity in Yeast. *Mol. Biol. Cell* **2012**, *23*, 1998–2013. [CrossRef]

30. Orlando, K.; Sun, X.; Zhang, J.; Lu, T.; Yokomizo, L.; Wang, P.; Guo, W. Exo-Endocytic Trafficking and the Septin-Based Diffusion Barrier Are Required for the Maintenance of Cdc42p Polarization During Budding Yeast Asymmetric Growth. *Mol. Biol. Cell* **2011**, *22*, 624–633. [CrossRef]
31. Kozubowski, L.; Saito, K.; Johnson, J.M.; Howell, A.S.; Zyla, T.R.; Lew, D.J. Symmetry-Breaking Polarization Driven by a Cdc42p GEF-PAK Complex. *Curr. Biol.* **2008**, *18*, 1719–1726. [CrossRef] [PubMed]
32. Garcia-Mata, R.; Boulter, E.; Burridge, K. The 'Invisible Hand': Regulation of RHO GTPases by RHOGDIs. *Nat. Rev. Mol. Cell Biol.* **2011**, *12*, 493–504. [CrossRef] [PubMed]
33. Woods, B.; Lai, H.; Wu, C.F.; Zyla, T.R.; Savage, N.S.; Lew, D.J. Parallel Actin-Independent Recycling Pathways Polarize Cdc42 in Budding Yeast. *Curr. Biol.* **2016**, *26*, 2114–2126. [CrossRef] [PubMed]
34. Goryachev, A.B.; Pokhilko, A.V. Dynamics of Cdc42 Network Embodies a Turing-Type Mechanism of Yeast Cell Polarity. *FEBS Lett.* **2008**, *582*, 1437–1443. [CrossRef]
35. Johnson, J.M.; Jin, M.; Lew, D.J. Symmetry Breaking and the Establishment of Cell Polarity in Budding Yeast. *Curr. Opin. Genet. Dev.* **2011**, *21*, 740–746. [CrossRef]
36. Woods, B.; Lew, D.J. Polarity Establishment by Cdc42: Key Roles for Positive Feedback and Differential Mobility. *Small GTPases* **2019**, *10*, 130–137. [CrossRef]
37. Slaughter, B.D.; Unruh, J.R.; Das, A.; Smith, S.E.; Rubinstein, B.; Li, R. Non-Uniform Membrane Diffusion Enables Steady-State Cell Polarization Via Vesicular Trafficking. *Nat. Commun.* **2013**, *4*, 1380. [CrossRef]
38. Sartorel, E.; Ünlü, C.; Jose, M.; Massoni-Laporte, A.; Meca, J.; Sibarita, J.B.; McCusker, D. Phosphatidylserine and GTPase Activation Control Cdc42 Nanoclustering to Counter Dissipative Diffusion. *Mol. Biol. Cell* **2018**, *29*, 1299–1310. [CrossRef]
39. Martinière, A.; Lavagi, I.; Nageswaran, G.; Rolfe, D.J.; Maneta-Peyret, L.; Luu, D.T.; Botchway, S.W.; Webb, S.E.; Mongrand, S.; Maurel, C.; et al. Cell Wall Constrains Lateral Diffusion of Plant Plasma-Membrane Proteins. *Proc. Natl. Acad. Sci. USA* **2012**, *109*, 12805–12810. [CrossRef]
40. Moseley, J.B. Eisosomes. *Curr. Biol.* **2018**, *28*, R376–R378. [CrossRef]
41. Malínská, K.; Malinsky, J.; Opekarova, M.; Tanner, W. Visualization of Protein Compartmentation within the Plasma Membrane of Living Yeast Cells. *Mol. Biol. Cell* **2003**, *14*, 4427–4436. [CrossRef] [PubMed]
42. Spira, F.; Mueller, N.S.; Beck, G.; von Olshausen, P.; Beig, J.; Wedlich-Söldner, R. Patchwork Organization of the Yeast Plasma Membrane into Numerous Coexisting Domains. *Nat. Cell Biol.* **2012**, *14*, 640–648. [CrossRef] [PubMed]
43. Kusumi, A.; Suzuki, K.G.; Kasai, R.S.; Ritchie, K.; Fujiwara, T.K. Hierarchical Mesoscale Domain Organization of the Plasma Membrane. *Trends Biochem. Sci.* **2011**, *36*, 604–615. [CrossRef] [PubMed]
44. Feltham, J.L.; Dötsch, V.; Raza, S.; Manor, D.; Cerione, R.A.; Sutcliffe, M.J.; Wagner, G.; Oswald, R.E. Definition of the Switch Surface in the Solution Structure of Cdc42Hs. *Biochemistry* **1997**, *36*, 8755–8766. [CrossRef] [PubMed]
45. Das, S.; Yin, T.; Yang, Q.; Zhang, J.; Wu, Y.I.; Yu, J. Single-Molecule Tracking of Small GTPase Rac1 Uncovers Spatial Regulation of Membrane Translocation and Mechanism for Polarized Signaling. *Proc. Natl. Acad. Sci. USA* **2015**, *112*, E267–E276. [CrossRef] [PubMed]
46. Remorino, A.; De Beco, S.; Cayrac, F.; Di Federico, F.; Cornilleau, G.; Gautreau, A.; Parrini, M.C.; Masson, J.B.; Dahan, M.; Coppey, M. Gradients of Rac1 Nanoclusters Support Spatial Patterns of Rac1 Signaling. *Cell Rep.* **2017**, *21*, 1922–1935. [CrossRef] [PubMed]
47. Platre, M.P.; Bayle, V.; Armengot, L.; Bareille, J.; Marquès-Bueno, M.D.M.; Creff, A.; Maneta-Peyret, L.; Fiche, J.B.; Nollmann, M.; Miège, C.; et al. Developmental Control of Plant Rho GTPase Nano-Organization by the Lipid Phosphatidylserine. *Science* **2019**, *364*, 57–62. [CrossRef]

Review

Protein Phase Separation during Stress Adaptation and Cellular Memory

Yasmin Lau, Henry Patrick Oamen and Fabrice Caudron *

School of Biological and Chemical Sciences, Queen Mary University of London, Mile End Road, London E1 4NS, UK; y.e.y.lau@qmul.ac.uk (Y.L.); h.oamen@qmul.ac.uk (H.P.O.)
* Correspondence: f.caudron@qmul.ac.uk

Received: 21 April 2020; Accepted: 21 May 2020; Published: 23 May 2020

Abstract: Cells need to organise and regulate their biochemical processes both in space and time in order to adapt to their surrounding environment. Spatial organisation of cellular components is facilitated by a complex network of membrane bound organelles. Both the membrane composition and the intra-organellar content of these organelles can be specifically and temporally controlled by imposing gates, much like bouncers controlling entry into night-clubs. In addition, a new level of compartmentalisation has recently emerged as a fundamental principle of cellular organisation, the formation of membrane-less organelles. Many of these structures are dynamic, rapidly condensing or dissolving and are therefore ideally suited to be involved in emergency cellular adaptation to stresses. Remarkably, the same proteins have also the propensity to adopt self-perpetuating assemblies which properties fit the needs to encode cellular memory. Here, we review some of the principles of phase separation and the function of membrane-less organelles focusing particularly on their roles during stress response and cellular memory.

Keywords: stress; cellular memory; phase separation; prions

---

1. Introduction

Phase separation is the demixing of a homogeneous mixture in solution to two separated phases, one of which can take the form of an assembled and detectable structure. These assembled structures have been given various nomenclatures including biomolecular super-assemblies, condensates, quinary structures or membraneless organelles, which suggests that a wide array of structures could be included in this classification [1]. One such example of a phase-separated structure is P-granules and is involved in specifying germ cells in *Caenorhabditis elegans* [2,3]. Since phase separation was found to play an important role into the biology of *C. elegans*, a plethora of structures forming through phase separation have been identified and examples of these include nucleolus for ribosome assembly, paraspeckles and nuclear speckles which regulate gene expressions, P-bodies for RNA storage and processing, Cajal bodies for regulation of small nuclear and small nucleolar RNA genes, stress granules for storage of stress-halted proteins and RNA [4–11]. In addition to phase separation being fundamental in cell physiology, its role and perturbation in diseases has emerged as a novel focus to understand the mechanisms of some pathologies such as amyotrophic lateral sclerosis (ALS). Some of these assemblies can be very dynamic; they condensate and dissolve very quickly. For example, the poly(A)-binding protein (Pab1) of *Saccharomyces cerevisiae*, has been demonstrated to phase separate in vivo and in vitro to form hydrogels in response to physiological heat stress [12]. In this context, phase separation of Pab1 is suspected to function in regulating translation of heat stress-related mRNAs [13]. Like Pab1, the yeast Sup35, a translation termination factor, is also capable of forming liquid/gel-like condensates in vivo during energy depletion-induced pH stress which completely dissolves once conditions return to normal [13]. The dynamic nature of these structures is very interesting for rapid response to quickly changing environments. Interestingly, some of these structures can in addition lock into a stable prion state which is an appealing property for a protein complex to establish and maintain cellular memory. Prions are self-templating, altered, heritable forms of normal cellular proteins which are often associated with neurodegenerative diseases and more recently adaptation to environmental changes [14,15]. Phase separating proteins and prions are both enriched in low complexity regions biased for specific amino acids. The difference, however, is that prions form stable structures while structures arising from phase separation are more dynamic [15].

In this review, we discuss some of the principles of protein phase separation and how cells use phase separation to adapt to stress conditions. We also examine how phase separation can support cellular memory in different organisms.

2. Phase Separation Drives the Formation of Membraneless Organelles

Phase separation may involve a single protein (simple coacervation) or it could require two or more proteins or RNA forming a complex (complex coacervation) [16]. A feature often found in the sequence of proteins that undergo phase separation is the presence of intrinsically disordered regions (IDRs) [17]. They are characterised by low complexity regions enriched in repetitive amino acid sequences or motifs [18]. These motifs are multivalent and allow the formation of a network of crosslinking caused by specific motifs within the same protein (Figure 1) [19]. Multivalency refers to specific intermolecular interactions resulting from the presence of multiple domains or motifs within a protein [16]. Some studies have described proteins with multivalency as quinary structures, which are protein structures that are shaped and selected by evolution rather than randomly formed, misfolded aggregates [20–22]. Because such repeats in IDRs are enriched with amino acid side chains possessing biased properties (uncharged, charged or aromatic), specific secondary structure driving forces such as hydrogen bonding, dipole–dipole, pi–pi interactions and pi–cation interactions lead to a variety of contacts within the same protein structure [23,24].

IDRs enriched in positively charged amino acids (arginine/glutamine) have the tendency to bind negatively charged RNA and phase separate in the process [16]. For example, Fused in sarcoma (FUS), an RNA-binding protein involved in a range of essential processes including transcription, splicing, mRNA transport and translation, undergoes reversible phase separation in vitro between a dispersed, liquid droplet or hydrogel state under low salt concentration. Mutations in FUS induce pathological protein assemblies which are associated with diseases such as familial amyotrophic lateral sclerosis (fALS) and frontotemporal lobar degeneration (FTLD). This shows why it is important to investigate the structure and dynamics of IDR-containing proteins [24,25]. FUS contains an RGG domain enriched in arginine/glycine which displays a high content of beta-strand structure [26]. The presence of RNA in solution with the FUS protein accelerates its self-assembly, because the RNA-binding domain provides a scaffold for self-assembly of the protein [27]. Together, IDRs, RGG domains of FUS and RNA form essential components of stress granules and concomitantly, multivalency influences the capability of a protein to phase separate [27].

Many proteins and in many organisms across kingdoms contain IDRs. For example, the budding yeast *S. cerevisiae* proteome contains over 20% of proteins with IDRs, and in silico screening of IDR proteins in 31 eukaryotic genomes showed that 52–67% have long (at least 40 consecutive residues) IDRs [28–32]. However, some proteins are more prone to phase separate than others. Currently, it is unclear what really determines the phase separation tendency of an IDR-containing protein solely from their sequences. An answer may come from analysis of the steady-state phase separation of Whi3 in a multinucleate organism such as *Ashbya gossypii.* Whi3 is an mRNA binding protein involved in regulating nuclear division and polarity by interacting with various mRNAs [33]. These interactions affect the structural organisation of Whi3, which consequently allows it to regulate these processes [33]. Whi3 forms droplets in vitro and in vivo. In vivo, phase separation of Whi3 is largely promoted by two mRNAs *CLN3* and *BNI1* and interestingly, the biophysical properties of the droplets are specific to the bound mRNA [33]. Local concentrations of Whi3, *CLN3* mRNA and *BNI1* mRNA could therefore serve as a way to regulate various droplets associated with different cellular processes at distinct locations within the same cytoplasm. This could be a mechanism by which constitutive phase separation and a switch between two states is regulated. Phosphoregulation may be another reason that could be responsible for specifying the function of various Whi3 condensates [34]. Thus, it is important to note that the assembly of membraneless organelles is the collective effect of multivalency and other factors such as RNA which can aid or expedite their assembly. This may underline that multivalency is an essential criterion for phase transition and that IDRs are only the tip of the iceberg of protein sequences promoting phase separation [35,36].

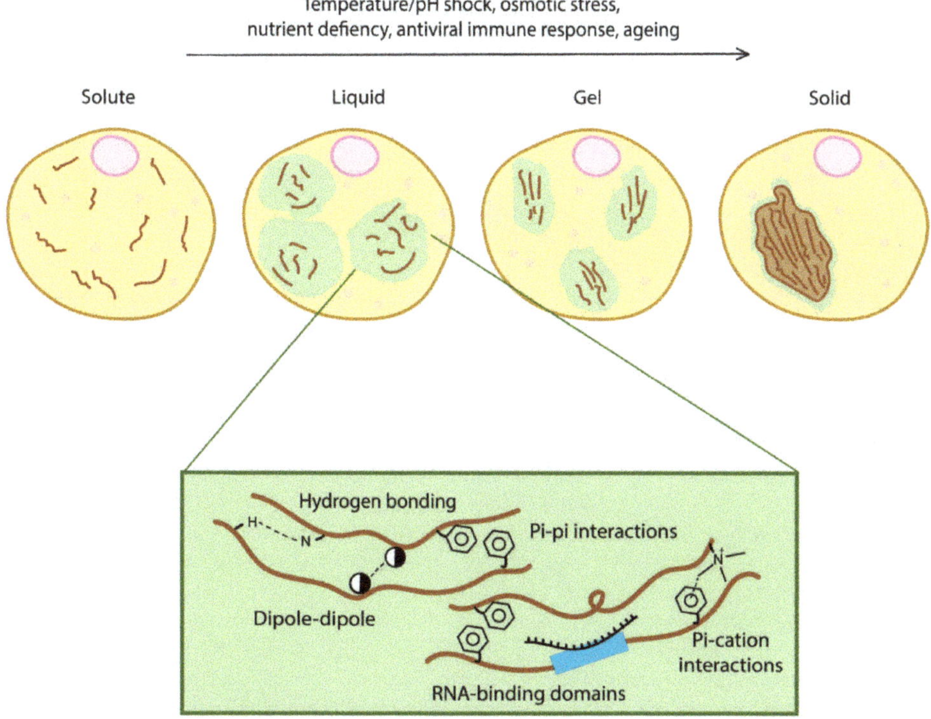

Figure 1. Different levels of phase separation. Multivalent interactions between intrinsically disordered regions (such as hydrogen bonding, pi–pi interactions, pi–cation interactions and dipole–dipole interactions) are necessary for phase separation.

As a result of different combinations of intermolecular interactions, multivalency can give rise to a range of phase separated structures. Here, two case studies in which disparate structures arise from two different IDR-containing proteins can be considered: the regulation of P granules assembly by the RNA helicase LAF-1 in *C. elegans* and a protein involved in ALS, TDP-43. P granules are *C. elegans* germline granules consisting of RNA and protein condensates within the cell and are key in determining the properties of germ cells [37,38]. Phase separation of P granules is induced by the *N*-terminal low-complexity RGG domain of LAF-1 [3]. The fact that the intermolecular interactions governed by the LAF-1 RGG domain are influenced by salt concentration underscores that such interactions are weak enough to be destabilised by an increase in electric charge [3]. This is important as these weak interactions in liquid droplets formed by LAF-1 render them reversible, unlike TDP-43 aggregates.

TDP-43 is an aggregation-prone protein which largely influences the onset of sporadic ALS and frontotemporal dementia, owing to the presence of ALS-linked mutations in its C-terminus [39].

Similar to LAF-1, TDP-43 contains an arginine/glycine rich domain. However, unlike LAF-1, TDP-43 has the propensity to form solid deposits. The presence of stress granules has been shown to enhance TDP-43 aggregate formation due to the sequestration of proteostasis factors such as HDAC6. Indeed, HDAC6 inhibition correlated with an increase in TDP-43 size [40]. The enhancement of TDP-43 aggregation by stress granule association has also previously been proposed to be a result of increased ubiquitylation and reduced splicing activity [41]. Remarkably, the point mutations found in ALS patients accelerate the formation of TDP-43 solid inclusions in vitro [39]. Interestingly, different morphologies were also observed depending on the amino acid mutated, such as defined filamentous or amorphous structures [39]. The fact that IDRs can induce different types of assemblies could be explained by the conditions; it is possible that in physiological conditions, these proteins favour one form over another depending on, for example, the presence of RNA or a protein concentration threshold. Thus, each type of structure may have specific assembly requirements. It is not surprising that similar IDRs lead to a range of structures, as this is characteristic of multivalency and weak interactions in quinary structures, as well as the pleiotropy in some protein assemblies [19,22,42]. However, the biochemistry of such structures is likely to be far more complicated inside cells than what is currently experimentally observed, and there may be many more factors influencing them.

3. Phase Separation as an Adaptation Mechanism in Cells

Specific environmental cues including temperature, pH, osmotic and nutrient changes trigger phase separation of proteins and RNA. Here, we discuss a series of adaptation mechanisms that rely on protein phase separation.

3.1. Polyadenylate Binding Protein (Pab1) Droplet Formation in Temperature Sensing

The temperature threshold at which most cellular processes are able to be carried out is limited. Beyond these thresholds, physiology begins to deteriorate as a consequence of protein denaturation and disruption of cell membrane integrity. Therefore, living organisms need to execute an ensemble of adaptation mechanisms to try to survive to temperature stress. Stress granules (SGs) are formed in response to unfavourable changes in environmental conditions including temperature. While numerous stress-granule-associated proteins have been identified, Pab1 is unique in its exceptional temperature-sensitivity. Indeed, Pab1 displays a thermal-responsive self-assembly rate 160-fold higher than typical biological reactions [12]. Like many SG-associated proteins, Pab1 has been hypothesized to upregulate mRNA translation by forming higher-order protein assemblies (Figure 2) [12]. The structural requirements for Pab1 to phase separate as an adaptive response to heat shock has been previously probed in vitro, where the significance of multivalency in the context of Pab1 demixing has been highlighted. It was found that while the P domain of Pab1 (the proline-rich IDR) can modulate the extent of phase separation, it is not necessary for phase separation to occur. The deletion of the proline-rich domain of Pab1 showed reduced phase separation [43]. However, the additional knockouts of 3 of its RRMs (RNA-recognition motifs) exhibited the greatest increase in the temperature boundary required for its phase separation [43]. This implicates that in the absence of each of its six domains, Pab1 was still able to phase separate at different temperatures, but absence of the RRMs showed the most impairment to de-mix. Therefore, the IDR of Pab1 is not required for phase separation and the RRMs contain major molecular determinants for demixing. This suggests that Pab1 demixing is more reliant on multivalent interactions of the RRM domain with the IDR serving as an additional tuner, rather than the IDR being the key player.

While being an unparalleled temperature-sensing system, the P-domain of Pab1 is highly evolutionarily conserved across fungal species [1,12]. Remarkably, the frequency of usage of aliphatic residues within the domain is reflected by a fitness advantage. Amongst numerous fungal species, proline-rich regions are almost identical, while alternating patterns between aliphatic residues such as methionine, valine and isoleucine is observed [12]. Interestingly, a negative relationship was found between the frequency of aliphatic amino acids and their hydrophobicity in the interchangeable sections of the P-domain across these species [12], while this trend was different albeit consistent across the rest of the protein sequence, the yeast proteome and disordered regions. The composition of these flexible regions of the P-domain has been naturally selected according to hydrophobicity; Pab1 phase separation is adaptive and has been fine-tuned on an evolutionary time scale. From this, we can conclude that Pab1 acts as a temperature sensor by using its RRMs, and to a lesser extent its P-domain, to tailor mRNA translation levels according to temperature changes within and around the cell through phase separation.

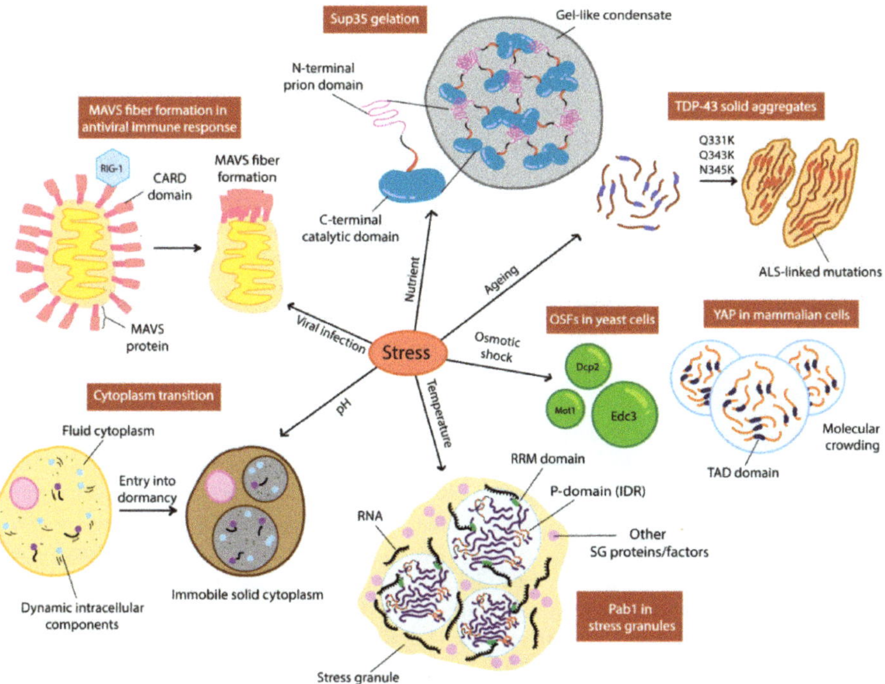

Figure 2. Different stress signals trigger phase separation of intracellular components in yeast (OSFs, Pab1, cytoplasm, Sup35) and mammalian cells (MAVS, TDP-43, YAP). Schemes adapted from [12,14,44].

3.2. *pH Sensing with Poly-Uridylate Binding Protein (Pub1) and the Cytoplasm*

Temperature change in the cell is often coupled with fluctuation in pH [43]. Stress response pathways overlap in both conditions, such as the formation of stress granules. Along with Pab1, the protein Pub1 is also a canonical component of stress granules [45] and as the name suggests, it is also an RNA-binding protein [46]. In situations of glucose starvation, cellular ATP levels drop and the proton-exporting activity by ATPases is reduced, which in turn induces a decrease in cytosolic pH [47]. In these conditions, Pub1 forms condensates and acts as a pH sensor [48].

Is there any common ground in proteins that sense stress? Both Pab1 and Pub1 are SG-related and RNA-binding proteins and the efficiency of phase separation in both proteins is perturbed by the presence of RNA [12,48]. Moreover, in both cases, the RRM domains play a significant role in influencing phase separation. In another study, imaging of cells expressing full-length Pub1, Pub1-RRM (Pub1 with only RRMs) or Pub1-LC (Pub1 with only the low-complexity domain) showed that only the former two variants formed condensates after energy depletion-induced pH change [48]. Considering the data on the P domain in Pab1, such IDRs/LC domains seem to serve a secondary role in regulating phase separation via the RRMs, which act as the primary modulators [48]. While further studies are needed to pinpoint the canonical phase separating features of SG-proteins, these studies on Pab1 and Pub1 tell us that their IDRs act to regulate solubility of the protein which provides specificity of phase separation morphology. Such morphologies are likely to be evolutionarily refined to detect different thresholds of stress.

In more extreme cases of pH stress, a widespread phase transition can also occur in the cell as a result of entry into dormancy [49]. This shifts the cell into a state of cell cycle arrest and reduced metabolic activity involving the packing of proteins into higher order structures, caused by a transition of the cytoplasm from a liquid to a glass-like state upon manipulation of cytosolic pH in yeast and bacteria [49–51] (Figure 2). This is a more stringent mechanism of phase separation as it is triggered in response to more extreme environmental stresses. This cytoplasmic freezing has been observed in budding yeast upon reduction of cytosolic pH, which was achieved by depleting 95% of cellular ATP, inducing dormancy. Unlike conventional techniques used such as fluorescence microscopy to discern condensate formation, this study placed its focus on the mechanical stability and mobility of macromolecules in the cell using micro-rheological techniques [49]. To measure the fluidity of the cytoplasm, a bead-like foreign tagged particle was introduced into the cells and its movement within the cell was tracked. It was found that lowering cytosolic pH alone in the presence of glucose (no starvation) is sufficient to display a reduced particle mobility phenotype, thus reduced cytoplasmic fluidity [51]. This reinforces the idea that proteins and macromolecules respond to simple physicochemical cues both on an individual and global scale. Interestingly, upon screening of the isoelectric points of the entire yeast proteome, the majority were found to overlap with the pH value corresponding to conditions of starvation [51]. Because the solubility of proteins decreases abruptly when their surrounding pH converges with their isoelectric point [52,53], it is highly likely that such proteins phase-separate to form higher-order assemblies in instances of energy depletion. As a result, the cell enters a state of dormancy when this occurs in a widespread manner under precipitous conditions, which could explain the overall change from fluid to solid-like state of the cytoplasm [51]. This is biologically relevant in terms of phase separation, as cytoplasmic freezing appears to be a mechanism by which proteins, as well as RNA, are reorganised into such a state that likely also modulates translational activity to reduce metabolism.

Molecular crowding was found to also play a role in the cytoplasmic freezing of yeast cells in conditions of glucose starvation, ultimately leading to a decrease of mechanical fluidity of cellular components [49]. Bacterial cytoplasmic freezing has been recorded as a mechanism to organise cellular components in a size-dependent manner to regulate metabolic activity [50]. This demonstrates that glass-like states may be induced by a variety of stimuli and it begs the question of whether this event is confined to unicellular organisms, or in other words, do multicellular organisms regulate metabolic activity through cytoplasmic freezing? While it has been reported that cells of metazoans such as the marine brine shrimp also enter dormancy in a pH-dependent manner [54], it is unclear whether they do so with the same mechanism. However, it is likely that single cells utilize stimuli-induced changes in global material properties of the cytoplasm more commonly than multicellular organisms to sense and adapt to simple physicochemical changes in an individual manner. This is probably owing to the fact that multicellular organisms have alternative routes of sensing stress, such as the sensing of temperature stress in *C. elegans* through thermosensory neurons to regulate metabolic activity and initiate an organismal response [55].

3.3. Osmotic Shock Foci (OSF) Formation in Osmotic Stress

A ubiquitous environmental challenge that cells experience in all organisms is a change in extracellular ion/salt concentration, which often leads to hyperosmotic shock to cells [56]. Consequently, this causes cell shrinkage and water to move out of the cell [57]. Similar to a change in pH, this process also disrupts the electrostatic charge of the intracellular area, which can concomitantly influence the multivalent interactions (such as protonation of side chains) within proteins and affect solubility, inducing phase separation.

To investigate proteins that phase separate in response to osmotic shock in yeast, the formation of P-bodies and sequestering of chaperones has been recently characterised; these proposed liquid-liquid phase separated droplets were termed as OSFs (osmotic shock foci) [57]. Highly dynamic and reversible OSFs that were found to be the chaperones Ssa1, Hsp104 and Hsp42 formed in vivo upon induction of hyperosmotic shock, which disappeared swiftly upon stress removal [56]. The speed and reversibility of this appearance supports the notion that such chaperone OSFs are liquid droplets rather than stable aggregates. Interestingly, along with the reversible P-body proteins Dcp2 and Edc3, the overexpression of other proteins with amyloidogenic domains which have been reported to form stable aggregates (Sup35, Mot3 and Pan1) (Figure 2) [56], were all able to assemble and disassemble swiftly following shock and relaxation respectively. This exhibits a drastically different phenotype to amyloid structures which are reportedly fairly stable given that they usually require solubilizing agents such as guanidine hydrochloride and sodium sulfate to be reversed [58,59]. Moreover, this indicates that the formation of these alternative, unstable liquid-like droplets appears to be favoured over stable amyloid aggregates in specific hyperosmotic conditions.

The main driving factor of phase separation as a result of hyperosmotic stress could be explained by molecular crowding [60]. In another recent study, the protein YAP (yes-associated protein) in mammalian cells was also shown to form liquid droplets seconds after hyperosmotic shock (Figure 2), and to localise to areas of the cell with high concentrations of accessible chromatin domains as a result of cell shrinkage and molecular crowding [60]. YAP is a transcriptional factor that normally binds to enhancers of specific target genes [61]. Phase separation of YAP relies on its IDR (the transcription activation domain), as mutants lacking this domain failed to form droplets under crowding conditions [60]. The fact that liquid droplet formation can be induced directly by crowding implicates that such proteins act as sensors of mechanical stress produced by crowding, thus phase separating and localising specifically to adjust and regulate gene expression. Phase separation could function as a rapid mechanism to respond to mechanical stress, and YAP serves as a starting point to expanding the library of mechanical-sensing phase separating proteins that we already know, such as TAZ and spindroin-like proteins in spider silk [62–64].

3.4. *Sup35 Gelation in Nutrient Deficiency*

One of the most well-characterized prion-like proteins in yeast is Sup35, a translation termination factor which, upon inactivation via a conformational switch to its prion form [*PSI+*], leads to STOP codon read-through [13,52]. Recent studies have revealed the reversible, non-prion structure of Sup35, as a result of energy depletion-induced gelation (Figure 2) [13]. Sup35 can also be considered as a stress granule protein due to its ability to sequester with Pab1 after phase separation and to regulate translation termination [52]. It is unclear whether this alternative form of Sup35 directly results in its inactivation and translation read-through. A study on Sup35 has demonstrated that, unlike prion formation, gelation only requires its GTPase catalytic domain which is responsible for translation termination in vivo [13]. Although this is the case, the IDR of Sup35 (NM domain) was shown to be important for governing the droplet properties of Sup35 gels [13]. The fact that the catalytic domain alone was able to form droplets could further suggest that gelation of Sup35 can be achieved solely with the multivalent interactions within this domain rather than requiring the IDR of the NM domain. However, the NM domain is important for rescuing the catalytic domain from potential stress-induced damage by enhancing droplet reversibility. It is important to note that Sup35 prions are also reversible, although this often requires curing with chaperones, while gel structures appear to be reversed simply through a conditional feedback process [13,65].

Sup35 is structurally versatile in that it can phase separate into gel-like structures, but it is also a prion which can form amyloid-like structures. Sup35 is present across many fungal species, but [*PSI+*] is predominantly found in *S. cerevisiae* [66,67]. Interestingly, phase separation of Sup35 is conserved among distant yeast relatives, while this is not the case for prion formation [13]. In contrast to *S. cerevisiae*, Sup35 is unable to propagate [*PSI+*] in *Schizosaccharomyces pombe* in a Hsp104-dependent manner, while both have been shown to be able to form stress-induced Sup35 condensates [13]. However, whether endogenous [*PSI+*] can be propagated in *S. pombe* has not yet been tested and it is possible that prion formation is mediated by other chaperone machineries. This suggests that the temporary phase transition as a stress response mechanism may evolutionarily precede prion propagation and induction. The reason why *S. cerevisiae* possesses numerous prion-forming proteins while only one was found in *S. pombe* so far [68], but forms other condensates, is still unclear. This is especially interesting as *S. cerevisiae* and *S. pombe* both share a wide range of chaperone systems that are responsible for prion propagation [68,69]. The difference in their prion-forming tendencies may be related to asymmetric cell division of *S. cerevisiae*, or the possession of other cytoplasmic determinants in budding yeast that are open to be discovered in the future.

Taking this data into consideration, we can hypothesize that Sup35 phase separation serves a distinctly different role to prion formation due to its induction under specific stress conditions such as nutrient deficiency. Upon recovery to healthy conditions, such condensates return to their soluble state in the cell [13]. Contrarily, Sup35 prion formation has been proposed to occur as a means to maintain a basal level of mRNA translation regulation and to pass on stable aggregates to future generations [52]. This is interesting because Sup35 gelation appears to be a frequently occurring, stress-sensing process that will happen in all cells, while Sup35 prion formation is unlikely to happen. Conversion to [*PSI+*] has been proposed to be a bet-hedging device in terms of translation fidelity, such that prion formation is positively selected for stressful conditions in exchange for a reduced overall fitness [70,71]. Nonetheless, both mechanisms are orchestrated for cell survival in fluctuating environments.

3.5. Mitochondrial Antiviral Signalling Protein (MAVS) Fiber Formation in Antiviral Immune Responses

The implementation of phase separating proteins has also been reported in binary decision-making and signal transduction in innate immune response. In humans, the mitochondrial antiviral signalling protein (MAVS) is activated by the viral RNA-detecting protein RIG-1 [44] and polymerizes into amyloid fibers [72] (Figure 2). MAVS contains a CARD (caspase activation recruitment) domain at its *N*-terminus, which belongs to the death domain superfamily. The CARD domain underlies filament formation [73–75] and shares features with prion fibers [73]. Formation of filaments is required for the activation of the downstream effectors NF-kB and IRF3 in the RIG-1 immune response pathway [75,76]. Therefore, MAVS amyloid fibers and prion-like behaviour is a physiological mechanism.

Interestingly, the CARD domain found in MAVS is able to functionally replace the NM domain of Sup35 in yeast [44] and remarkably, the NM domain can also function in place of the MAVS CARD domain in an NM-MAVS construct in mammalian cells. NM-MAVS filament formation was promoted by transfecting cells with NM fibers inducing downstream signalling, which was measured by activation of interferon beta 1a [44]. This highlights the remarkable interchangeability of Sup35 NM and MAVS CARD domains, which seem to mostly require filament assembly and self-templating activity. This, contrary to the different faces of Sup35 we have discussed before, may suggest that prion behaviour can be selected for during evolution and is not simply a by-product of phase separation. The relationship between multivalent interactions providing a biophysical basis for phase separation and self-templating activities displayed by prions is a fascinating topic to be addressed in the future.

3.6. Std1 and Carbon Source Sensing

S. cerevisiae cells are craving glucose as a primary carbon source for the production of energy. The presence of glucose in the culture medium represses the expression of genes involved in alternative carbon source utilisation, a phenomenon termed glucose repression [77]. Cells must therefore fine tune their ability to sense and respond to low glucose levels if they are to survive. Snf1, a heterotrimeric protein kinase of the Snf1/AMP-activated protein kinase family, primarily ensures survival under limiting glucose conditions by activating genes involved in utilisation of sucrose, galactose or ethanol. Snf1 is also involved in cellular developmental processes such as meiosis, sporulation and ageing as well as response to other stress conditions such as oxidative stress, heat stress and regulation of various metabolic enzymes. Activation of Snf1, under glucose limitation is mediated by Std1 [78–81]. Under glucose replete condition, Std1 undergoes phase separation and localises to cytoplasmic foci. When glucose is limiting, foci formed by Std1 dissolve, releasing Std1 which can translocate to the nucleus and activate Snf1. Through this mechanism, cells are able to respond appropriately to limiting glucose levels and probably buffer noise in the concentration or readout of the concentration of sugar [80]. Phase separation is therefore a very interesting mechanism for both sensing and adapting to stresses, as demonstrated by the few examples we have presented here. Interestingly, Std1 when over-expressed in *S. cerevisiae* promotes the formation of [*GAR*+] prion, generating an adaptive phenotype where the cells are not reliant on glucose as primary energy source [82]. Again, just as in the case of Sup35, this demonstrates how the same protein can promote the formation of structurally and physically distinct structures depending on different environmental cues.

4. Phase Separation and Cellular Memory

In addition to adaptation, there have been examples of protein-based memory which may rely on phase separation. Whi3, in *S. cerevisiae*, is involved in encoding memory of deceptive courtship and cytoplasmic polyadenylation element-binding proteins (CPEB) are involved in long term potentiation during courtship in *Drosophila*.

There are two mating types in haploid yeast cells, *MAT*a and *MAT*α, which release pheromone as a and α-factor respectively. Mating factors are sensed via the cell surface receptors Ste2 (*MAT*a) and Ste3 (*MAT*α) [83–85]. Cells exposed to α-factor respond by arresting in the G1 phase of the cell cycle and developing cytoplasmic projections (called a shmoo) in order to grow towards the source of pheromone, which is supposedly a mating partner. However, when there is no mating partner, cells escape pheromone arrest and proceed to division [86]. This type of failed mating attempts can arise when cells cannot reach their mating partner in time, which could happen for example if several cells try to fuse with the same one [83]. After escape from pheromone arrest, cells maintain a memory of these deceptive mating encounters in the form of Whi3 condensates into super-assemblies. Therefore, like Sup35, Whi3 appears to be able to form or join very different condensates because it was found in stress granules, super-assemblies and age-induced aggregates [87,88]. Whi3 is an mRNA binding protein which regulates cell size and cell cycle progression during the G1 phase of the cell cycle [89,90]. These super-assemblies asymmetrically partition into mother cells when the cells divide after escaping pheromone arrest (Figure 3). Indeed, mother cells do not shmoo any longer in the presence of pheromone, but daughter cells emanating from these refractory mothers do not inherit this memory and shmoo as soon as they are born [83,87,91].

Whi3 from *S. cerevisiae* possesses low complexity regions rich in Q/N residues, which have been shown to mediate super-assemblies formation and memory [83,90]. Since Whi3 super-assemblies are asymmetrically inherited, we suspect that such asymmetric segregation could be a powerful mechanism to establish cell fate and maintain asymmetries. One observation however is that these self-templating proteins have the tendency to adopt detrimental conformations during ageing. For example, some variants of Sup35 accumulate specifically at age-induced protein deposits [87]. Conformational flexibility offered by the PrD comes at a cost when cellular physiology changes during ageing. For example, pH is not regulated or controlled as tightly as in young cells and this may have as a consequence that conformationally flexible proteins are suddenly locked into a more stable and solid state. Whi3 also forms foci during ageing that result in a pheromone insensitivity of old yeast cells [92]. Interestingly, cells harbouring a mutant form of Whi3 (*whi3-ΔpQ*) which do not adopt the super-assembly conformation have been shown to live slightly longer than wild type cells [87,93]. It is not clear whether the asymmetric segregation of Whi3 super-assemblies in mother cells is a mechanism to induce ageing and limit lifespan. Furthermore, it is not clear whether the variants of Sup35 which accumulate at age-induced deposits have a function or not. These questions are still open for investigation.

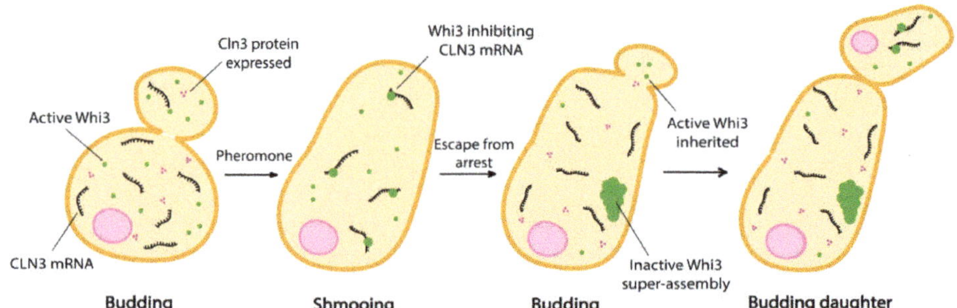

Figure 3. The pheromone refractory state in budding yeast is propagated by Whi3 super-assembly formation. Cln3 protein expression is constitutive when *CLN3* mRNA is free from inhibition. Inactive Whi3 is confined in the mother cell, while daughter cells containing active Whi3 continue to shmoo.

The involvement of prion-like proteins in formation and maintenance of memory is not a specificity of budding yeast. Cytoplasmic polyadenylation element-binding proteins (CPEBs) are mRNA binding proteins that exists in different cells types in a range of organisms such as sea slugs (*Aplysia californica*, AsCPEB), flies (*D. melanogaster*, Orb2) and mice (CPEB3) where they regulate protein synthesis. Isoforms of CPEBs found in neurons particularly differ from those found in other cell types by the fact that their *N*-terminal is rich in glutamine which is a typical characteristic of yeast prion domains [94,95]. CPEBs exist in two different conformational states, a monomeric form and a prion-like self-sustaining amyloidogenic aggregated form that behaves like a prion [95]. Aggregated states of CPEBs have been linked to persistence and regulation of memory [96,97]. For example, in *Drosophila*, an already mated female fly refuses to mate again. After prolonged rejection, a male fly learns to suppress its desire to mate with already mated females, but not with virgin ones, for days because it develops a long-term memory that it had been rejected before [98]. Flies which have significantly reduced aggregation of Orb2 also have impaired retention of long-term memory of previous rejection. Orb2 has two isoforms, a shorter, very rare and poorly expressed Orb2A and a longer, abundant and constitutively expressed Orb2B. Orb2A is required for aggregation of Orb2B, a process necessary for the formation of long-term memory. Both isoforms possess prion-like sequences in their *N*-terminal domain [98]. In vitro studies have shown that the aggregated form of Orb2 is an activator of translation while the monomeric form acts as a repressor of target genes associated with long term memory [99]. The monomer binds to the CG13928 protein, which is responsible for repression, while the oligomer binds to CG4612, which contributes to activation (Figure 4) [99]. This observation is not only limited to Orb2. Interestingly, this phenomenon has also been reported for AsCPEB where the active state is the aggregated prion state when it actively binds mRNA in vitro [95]. In neurons, AsCPEB exclusively form prion-like punctate structures where they also regulate long term synaptic facilitation. It was suggested that these structures may have formed from protein–protein interactions [97]. Protein–protein interactions between one or more proteins also promotes liquid-liquid phase separation [16]. To investigate the exact role of the prion form of AsCPEB in vivo, Kausik et al. [96] proceeded to overexpress AsCPEB in neurites and observed formation of puncta outside synaptic areas which had no effect on normal synaptic functions but retained ability to bind mRNAs. They therefore suggested that these prion forms may actually play other active roles in this conformation. Conformational switches in classical prions normally induce loss of function phenotype [73]. It appears therefore that apart from being significant in phase separation, conformational switches may actually uncover activity in certain prion-like proteins.

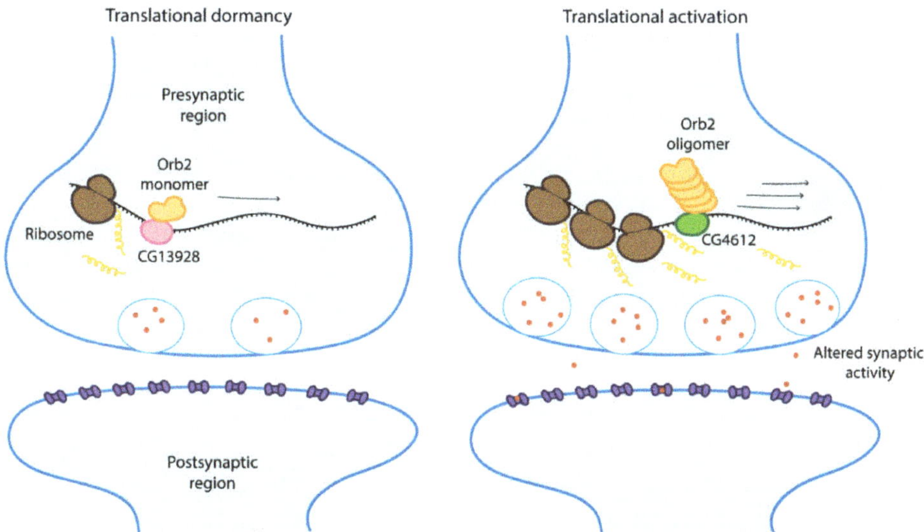

Figure 4. The repression and activation of mRNA translation by Orb2 monomer and oligomer respectively, resulting in altered synaptic activity and memory stabilisation. Scheme adapted from [99].

This observation opens up a new paradigm of looking at how Whi3 regulates memory in yeast. Our current understanding is that conformational change to the super-assembly form transforms Whi3 from an active inhibitor of *CLN3* mRNA translation to an inactive protein. However, it could well be that super-assembled Whi3 is activating or enhancing translation of *CLN3* mRNA. Altogether, this could suggest that conformational switches of prion-like proteins may have a conserved functional role.

Another parallel that can be drawn between Whi3 and AsCPEB is that both proteins undergo conformational changes specifically in the presence of pheromone and serotonin respectively. This suggests that different memory encoding substances may be activated differently in a manner that is not characteristic of prion proteins. We propose that, added to phase separation, prion-like characteristics such as self-templating conformations of these proteins may be necessary for memory. Confinement of this activity to the mother cells or activated synapse suggests that there is some form of control. The mechanism behind this, however, is yet to be uncovered.

5. Conclusions

What precise role could phase separation play in ageing? Is there possibly an interplay between ageing and memory acquisition? Ageing is a gradual decrease in the physiology of cells with time or number of divisions. During ageing, cells have difficulties to deal with accumulated aberrant protein aggregates which results partly from weakened mitochondrial activity. Ageing also affects proper regulation of gene expression which in turn can result in the distortion of protein concentrations and stoichiometry. Therefore cells have to cope with the increasing need to repair or degrade damaged proteins that accumulate as they undergo complex metabolic and physiological activities [100,101]. The overall consequence of this is a deviation from physiological states and ultimately death. Changes in RNA concentration has been shown to affect phase separation of RNA-binding proteins and subsequently the properties of biomolecular condensates [33]. While IDR-bearing proteins such as FUS and TDP-43 have been shown to fuse into normal stress granules, mutants of these proteins contribute to irreversible phase transitions of stress granules into pathological stress granules which are hallmarks of age-related diseases such as Alzheimer's disease [101,102]. The clearance of older stress granules as well as pathological stress granules involves the autophagic machinery [103,104]. Defective autophagic machinery has been linked with increase in prion formation and neurodegenerative diseases such as ALS [105–107]. It is however not clear how pathological stress granules evade cellular clearance machineries. We have noted previously that Whi3 and Sup35 aggregate into a range of different structures including age-induced deposits. It seems that IDRs have evolved to allow for structural flexibility which enable proteins to phase separate into different assemblies where their exact function may not yet be very clear.

In summary, we have highlighted how certain prion-like proteins, driven by multivalency, undergo phase separation and adopt different physical states under various stress conditions, including those implicated in memory establishment. We suspect that phase separation could be a mechanism by which single cell organisms such as yeast and bacteria, as well as a number of multicellular higher organisms, cope with stress conditions on a cellular scale. This process has also been refined by natural selection across these organisms on the primary sequence level for each IDR-bearing protein in specific stress scenarios, giving a spectrum of structures and components involved. While we have described several well-studied stress-response systems, we predict that findings on phase separation in memory acquisition and possibly ageing such as for Whi3 and CPEBs are only at the beginning of their emergence.

Funding: This work was supported by a Biotechnology and Biological Sciences Research Council Project grant (BB/S001204/1) to H.P.O. and F.C., and QMUL to Y.L., H.P.O. and F.C.

Conflicts of Interest: The authors declare no conflict of interest.

References

1. Yoo, H.; Triandafillou, C.; Drummond, D.A. Cellular sensing by phase separation: Using the process, not just the products. *J. Biol. Chem.* **2019**, *294*, 7151–7159. [CrossRef]
2. Elbaum-Garfinkle, S. Matter over mind: Liquid phase separation and neurodegeneration. *J. Biol. Chem.* **2019**, *294*, 7160–7168. [CrossRef]
3. Elbaum-Garfinkle, S.; Kim, Y.; Szczepaniak, K.; Chen, C.C.H.; Eckmann, C.R.; Myong, S.; Brangwynne, C.P. The disordered P granule protein LAF-1 drives phase separation into droplets with tunable viscosity and dynamics. *Proc. Natl. Acad. Sci. USA* **2015**, *112*, 7189–7194. [CrossRef]
4. Mitrea, D.M.; Kriwacki, R.W. Phase separation in biology; functional organization of a higher order. *Cell Commun. Signal.* **2016**, *14*, 1–20. [CrossRef] [PubMed]
5. Uversky, V.N. Intrinsically disordered proteins in overcrowded milieu: Membrane-less organelles, phase separation, and intrinsic disorder. *Curr. Opin. Struct. Biol.* **2017**, *44*, 18–30. [CrossRef] [PubMed]
6. Frottin, F.; Schueder, F.; Tiwary, S.; Gupta, R.; Körner, R.; Schlichthaerle, T.; Cox, J.; Jungmann, R.; Hartl, F.U.; Hipp, M.S. The nucleolus functions as a phase-separated protein quality control compartment. *Science* **2019**, *365*, 342–347. [CrossRef] [PubMed]
7. Feric, M.; Vaidya, N.; Harmon, T.S.; Mitrea, D.M.; Zhu, L.; Richardson, T.M.; Kriwacki, R.W.; Pappu, R.V.; Brangwynne, C.P. Coexisting Liquid Phases Underlie Nucleolar Subcompartments. *Cell* **2016**, *165*, 1686–1697. [CrossRef]
8. Brangwynne, C.P.; Eckmann, C.R.; Courson, D.S.; Rybarska, A.; Hoege, C.; Gharakhani, J.; Jülicher, F.; Hyman, A.A. Germline P Granules Are Liquid Droplets That Localize by Controlled Dissolution/Condensation. *Science* **2009**, *324*, 1729–1732. [CrossRef]
9. Guillén-Boixet, J.; Kopach, A.; Holehouse, A.S.; Wittmann, S.; Jahnel, M.; Schlüßler, R.; Kim, K.; Trussina, I.R.E.A.; Wang, J.; Mateju, D.; et al. RNA-Induced Conformational Switching and Clustering of G3BP Drive Stress Granule Assembly by Condensation. *Cell* **2020**, *181*, 346.e17–361.e17. [CrossRef]
10. Fox, A.H.; Nakagawa, S.; Hirose, T.; Bond, C.S. Paraspeckles: Where Long Noncoding RNA Meets Phase Separation. *Trends Biochem. Sci.* **2018**, *43*, 124–135. [CrossRef]
11. Luo, Y.; Na, Z.; Slavoff, S.A. P-Bodies: Composition, Properties, and Functions. *Biochemistry* **2018**, *57*, 2424–2431. [CrossRef] [PubMed]
12. Riback, J.A.; Katanski, C.D.; Kear-Scott, J.L.; Pilipenko, E.V.; Rojek, A.E.; Sosnick, T.R.; Drummond, D.A. Stress-Triggered Phase Separation Is an Adaptive, Evolutionarily Tuned Response. *Cell* **2017**, *168*, 1028.e19–1040.e19. [CrossRef] [PubMed]

13. Franzmann, T.M.; Jahnel, M.; Pozniakovsky, A.; Mahamid, J.; Holehouse, A.S.; Nüske, E.; Richter, D.; Baumeister, W.; Grill, S.W.; Pappu, R.V.; et al. Phase separation of a yeast prion protein promotes cellular fitness. *Science* **2018**, *359*, eaao5654. [CrossRef] [PubMed]
14. Alberti, S.; Halfmann, R.; King, O.; Kapila, A.; Lindquist, S. A Systematic Survey Identifies Prions and Illuminates Sequence Features of Prionogenic Proteins. *Cell* **2009**, *137*, 146–158. [CrossRef] [PubMed]
15. Chakravarty, A.K.; Jarosz, D.F. More than Just a Phase: Prions at the Crossroads of Epigenetic Inheritance and Evolutionary Change. *J. Mol. Biol.* **2018**, *430*, 4607–4618. [CrossRef] [PubMed]
16. Alberti, S. Phase separation in biology. *Curr. Biol.* **2017**, *27*, 1097–1101. [CrossRef]
17. Oldfield, C.J.; Dunker, A.K. Intrinsically Disordered Proteins and Intrinsically Disordered Protein Regions. *Annu. Rev. Biochem.* **2014**, *83*, 553–584. [CrossRef]
18. Wootton, J.C.; Federhen, S. Analysis of compositionally biased regions in sequence databases. *Methods Enzymol.* **1996**, *266*, 554–571. [CrossRef]
19. Boeynaems, S.; Alberti, S.; Fawzi, N.L.; Mittag, T.; Polymenidou, M.; Rousseau, F.; Schymkowitz, J.; Shorter, J.; Wolozin, B.; Van Den Bosch, L.; et al. Protein Phase Separation: A New Phase in Cell Biology. *Trends Cell Biol.* **2018**, *28*, 420–435. [CrossRef]
20. Chien, P.; Gierasch, L.M. Challenges and dreams: Physics of weak interactions essential to life. *Mol. Biol. Cell* **2014**, *25*, 3474–3477. [CrossRef]
21. Wallace, E.W.J.; Kear-Scott, J.L.; Pilipenko, E.V.; Schwartz, M.H.; Laskowski, P.R.; Rojek, A.E.; Katanski, C.D.; Riback, J.A.; Dion, M.F.; Franks, A.M.; et al. Reversible, Specific, Active Aggregates of Endogenous Proteins Assemble upon Heat Stress. *Cell* **2015**, *162*, 1286–1298. [CrossRef] [PubMed]
22. Kroschwald, S.; Alberti, S. Gel or Die: Phase Separation as a Survival Strategy. *Cell* **2017**, *168*, 947–948. [CrossRef] [PubMed]
23. Vernon, R.M.C.; Chong, P.A.; Tsang, B.; Kim, T.H.; Bah, A.; Farber, P.; Lin, H.; Forman-Kay, J.D. Pi-Pi contacts are an overlooked protein feature relevant to phase separation. *Elife* **2018**, *7*, 1–48. [CrossRef] [PubMed]
24. Qamar, S.; Wang, G.Z.; Randle, S.J.; Ruggeri, F.S.; Varela, J.A.; Lin, J.Q.; Phillips, E.C.; Miyashita, A.; Williams, D.; Ströhl, F.; et al. FUS Phase Separation Is Modulated by a Molecular Chaperone and Methylation of Arginine Cation-π Interactions. *Cell* **2018**, *173*, 720–734. [CrossRef] [PubMed]
25. Murakami, T.; Qamar, S.; Lin, J.Q.; Schierle, G.S.K.; Rees, E.; Miyashita, A.; Costa, A.R.; Dodd, R.B.; Chan, F.T.S.; Michel, C.H.; et al. ALS/FTD Mutation-Induced Phase Transition of FUS Liquid Droplets and Reversible Hydrogels into Irreversible Hydrogels Impairs RNP Granule Function. *Neuron* **2015**, *88*, 678–690. [CrossRef]
26. Schwartz, J.C.; Wang, X.; Podell, E.R.; Cech, T.R. RNA Seeds Higher-Order Assembly of FUS Protein. *Cell Rep.* **2013**, *5*, 918–925. [CrossRef]
27. Fan, A.C.; Leung, A.K.L. RNA granules and diseases: A case study of stress granules in ALS and FTLD. *Adv. Exp. Med. Biol.* **2016**, *907*, 263–296. [CrossRef]
28. Coletta, A.; Pinney, J.W.; Solís, D.Y.W.; Marsh, J.; Pettifer, S.R.; Attwood, T.K. Low-complexity regions within protein sequences have position-dependent roles. *BMC Syst. Biol.* **2010**, *4*, 43. [CrossRef]
29. Wootton, J.C. Sequences with "unusual" amino acid compositions. *Curr. Opin. Struct. Biol.* **1994**, *4*, 413–421. [CrossRef]
30. Uversky, V.N. Intrinsically disordered proteins and their "Mysterious" (meta)physics. *Front. Phys.* **2019**. [CrossRef]
31. Oldfield, C.J.; Cheng, Y.; Cortese, M.S.; Brown, C.J.; Uversky, V.N.; Bunker, A.K. Comparing and combining predictors of mostly disordered proteins. *Biochemistry* **2005**, *44*, 1989–2000. [CrossRef] [PubMed]
32. Dunker, A.K.; Obradovic, Z.; Romero, P.; Garner, E.C.; Brown, C.J. Intrinsic protein disorder in complete genomes. *Genome Inform. Ser. Workshop Genome Inform.* **2000**, *11*, 161–171. [CrossRef] [PubMed]
33. Zhang, H.; Elbaum-Garfinkle, S.; Langdon, E.M.; Taylor, N.; Occhipinti, P.; Bridges, A.A.; Brangwynne, C.P.; Gladfelter, A.S. RNA Controls PolyQ Protein Phase Transitions. *Mol. Cell* **2015**, *60*, 220–230. [CrossRef] [PubMed]
34. Gerbich, T.M.; McLaughlin, G.A.; Cassidy, K.; Gerber, S.; Adalsteinsson, D.; Gladfelter, A.S. Phosphoregulation provides specificity to biomolecular condensates in the cell cycle and cell polarity. *J. Cell Biol.* **2020**, *219*, e201910021. [CrossRef]

35. Li, P.; Banjade, S.; Cheng, H.C.; Kim, S.; Chen, B.; Guo, L.; Llaguno, M.; Hollingsworth, J.V.; King, D.S.; Banani, S.F.; et al. Phase transitions in the assembly of multivalent signalling proteins. *Nature* **2012**, *483*, 336–340. [CrossRef]
36. Banjade, S.; Rosen, M.K. Phase transitions of multivalent proteins can promote clustering of membrane receptors. *Elife* **2014**, *3*, e04123. [CrossRef]
37. Wang, J.T.; Seydoux, G. P granules. *Curr. Biol.* **2014**, *24*, R637–R638. [CrossRef]
38. Marcello, M.R.; Singson, A. Germline determination: Don't mind the P granules. *Curr. Biol.* **2011**, *21*, R155–R157. [CrossRef]
39. Johnson, B.S.; Snead, D.; Lee, J.J.; McCaffery, J.M.; Shorter, J.; Gitler, A.D. TDP-43 is intrinsically aggregation-prone, and amyotrophic lateral sclerosis-linked mutations accelerate aggregation and increase toxicity. *J. Biol. Chem.* **2009**, *284*, 20329–20339. [CrossRef]
40. Chen, Y.; Cohen, T.J. Aggregation of the nucleic acid-binding protein TDP-43 occurs via distinct routes that are coordinated with stress granule formation. *J. Biol. Chem.* **2019**, *294*, 3696–3706. [CrossRef]
41. Hans, F.; Glasebach, H.; Kahle, P.J. Multiple distinct pathways lead to hyperubiquitylated insoluble TDP-43 protein independent of its translocation into stress granules. *J. Biol. Chem.* **2020**, *294*, 3696–3706. [CrossRef] [PubMed]
42. Edelstein, S.J. Patterns in the quinary structures of proteins. Plasticity and inequivalence of individual molecules in helical arrays of sickle cell hemoglobin and tubulin. *Biophys. J.* **1980**, *32*, 347–360. [CrossRef]
43. Weitzel, G.; Pilatus, U.; Rensing, L. Similar dose response of heat shock protein synthesis and intracellular pH change in yeast. *Exp. Cell Res.* **1985**, *59*, 252–256. [CrossRef]
44. Hou, F.; Sun, L.; Zheng, H.; Skaug, B.; Jiang, Q.X.; Chen, Z.J. MAVS forms functional prion-like aggregates to activate and propagate antiviral innate immune response. *Cell* **2011**, *146*, 448–461. [CrossRef]
45. Buchan, J.R.; Muhlrad, D.; Parker, R. P bodies promote stress granule assembly in Saccharomyces cerevisiae. *J. Cell Biol.* **2008**, *183*, 441–455. [CrossRef]
46. Matunis, M.J.; Matunis, E.L.; Dreyfuss, G. PUB1: A major yeast poly(A)+ RNA-binding protein. *Mol. Cell. Biol.* **1993**, *13*, 6114–6123. [CrossRef]
47. Orij, R.; Brul, S.; Smits, G.J. Intracellular pH is a tightly controlled signal in yeast. *Biochim. Biophys. Acta* **2011**, *1810*, 933–944. [CrossRef]
48. Kroschwald, S.; Munder, M.C.; Maharana, S.; Franzmann, T.M.; Richter, D.; Ruer, M.; Hyman, A.A.; Alberti, S. Different Material States of Pub1 Condensates Define Distinct Modes of Stress Adaptation and Recovery. *Cell Rep.* **2018**, *23*, 3327–3339. [CrossRef]
49. Lennon, J.T.; Jones, S.E. Microbial seed banks: The ecological and evolutionary implications of dormancy. *Nat. Rev. Microbiol.* **2011**, *9*, 119–130. [CrossRef]
50. Parry, B.R.; Surovtsev, I.V.; Cabeen, M.T.; O'Hern, C.S.; Dufresne, E.R.; Jacobs-Wagner, C. The bacterial cytoplasm has glass-like properties and is fluidized by metabolic activity. *Cell* **2014**, *156*, 183–194. [CrossRef]
51. Munder, M.C.; Midtvedt, D.; Franzmann, T.; Nüske, E.; Otto, O.; Herbig, M.; Ulbricht, E.; Müller, P.; Taubenberger, A.; Maharana, S.; et al. A pH-driven transition of the cytoplasm from a fluid- to a solid-like state promotes entry into dormancy. *Elife* **2016**, *5*, e09347. [CrossRef]
52. Lyke, D.R.; Dorweiler, J.E.; Manogaran, A.L. The three faces of Sup35. *Yeast* **2019**, *36*, 465–472. [CrossRef] [PubMed]
53. Shaw, K.L.; Grimsley, G.R.; Yakovlev, G.I.; Makarov, A.A.; Pace, C.N. The effect of net charge on the solubility, activity, and stability of ribonuclease Sa. *Protein Sci.* **2001**, *10*, 1206–1215. [CrossRef] [PubMed]
54. Busa, W.B.; Crowe, J.H. Intracellular pH Regulates Transitions Between Dormancy and Development of Brine Shrimp (*Artemia salina*) Embryos. *Science* **1983**, *221*, 366–368. [CrossRef] [PubMed]
55. Prahlad, V.; Cornelius, T.; Morimoto, R.I. Regulation of the cellular heat shock response in Caenorhabditis elegans by thermosensory neurons. *Science* **2008**, *320*, 811–814. [CrossRef] [PubMed]
56. Stears, R.L.; Gullans, S.R. Transcriptional response to hyperosmotic stress. In *Cell and Molecular Response to Stress*; Storey, K.B., Storey, J.M., Eds.; Elsevier Science B.V.: Amsterdam, The Netherlands, 2000; pp. 129–139.
57. Alexandrov, A.I.; Grosfeld, E.V.; Dergalev, A.A.; Kushnirov, V.V.; Chuprov-Netochin, R.N.; Tyurin-Kuzmin, P.A.; Kireev, I.I.; Ter-Avanesyan, M.D.; Leonov, S.V.; Agaphonov, M.O. Analysis of novel hyperosmotic shock response suggests 'beads in liquid' cytosol structure. *Biol. Open* **2019**, *8*, bio044529. [CrossRef] [PubMed]

58. Coalier, K.A.; Paranjape, G.S.; Karki, S.; Nichols, M.R. Stability of early-stage amyloid-β(1-42) aggregation species. *Biochim. Biophys. Acta* **2013**, *1834*, 65–70. [CrossRef] [PubMed]
59. Ramirez-Alvarado, M.; Merkel, J.S.; Regan, L. A systematic exploration of the influence of the protein stability on amyloid fibril formation in vitro. *Proc. Natl. Acad. Sci. USA* **2000**, *97*, 8979–8984. [CrossRef]
60. Cai, D.; Feliciano, D.; Dong, P.; Flores, E.; Gruebele, M.; Porat-Shliom, N.; Sukenik, S.; Liu, Z.; Lippincott-Schwartz, J. Phase separation of YAP reorganizes genome topology for long-term YAP target gene expression. *Nat. Cell Biol.* **2019**, *21*, 1578–1589. [CrossRef]
61. Li, Z.; Zhao, B.; Wang, P.; Chen, F.; Dong, Z.; Yang, H.; Guan, K.L.; Xu, Y. Structural insights into the YAP and TEAD complex. *Genes Dev.* **2010**, *24*, 235–240. [CrossRef]
62. Lu, Y.; Wu, T.; Gutman, O.; Lu, H.; Zhou, Q.; Henis, Y.I.; Luo, K. Phase separation of TAZ compartmentalizes the transcription machinery to promote gene expression. *Nat. Cell Biol.* **2020**, *22*, 453–464. [CrossRef]
63. Franklin, J.M.; Guan, K.-L. YAP/TAZ phase separation for transcription. *Nat. Cell Biol.* **2020**, *22*, 357–358. [CrossRef] [PubMed]
64. Lemetti, L.; Hirvonen, S.P.; Fedorov, D.; Batys, P.; Sammalkorpi, M.; Tenhu, H.; Linder, M.B.; Aranko, A.S. Molecular crowding facilitates assembly of spidroin-like proteins through phase separation. *Eur. Polym. J.* **2019**, *112*, 539–546. [CrossRef]
65. Park, Y.-N.; Morales, D.; Rubinson, E.H.; Masison, D.; Eisenberg, E.; Greene, L.E. Differences in the Curing of [PSI+] Prion by Various Methods of Hsp104 Inactivation. *PLoS ONE* **2012**, *7*, e37692. [CrossRef] [PubMed]
66. Edskes, H.K.; Khamar, H.J.; Winchester, C.L.; Greenler, A.J.; Zhou, A.; McGlinchey, R.P.; Gorkovskiy, A.; Wickner, R.B. Sporadic distribution of prion-forming ability of sup35p from yeasts and fungi. *Genetics* **2014**, *198*, 605–616. [CrossRef] [PubMed]
67. True, H.L.; Lindquist, S.L. A yeast prion provides a mechanism for genetic variation and phenotypic diversity. *Nature* **2000**, *407*, 477–483. [CrossRef] [PubMed]
68. Sideri, T.; Yashiroda, Y.; Ellis, D.A.; Rodríguez-López, M.; Yoshida, M.; Tuite, M.F.; Bähler, J. The copper transport-associated protein Ctr4 can form prion-like epigenetic determinants in Schizosaccharomyces pombe. *Microb. Cell* **2017**, *4*, 16–28. [CrossRef]
69. Reidy, M.; Sharma, R.; Masison, D.C. Schizosaccharomyces pombe disaggregation machinery chaperones support saccharomyces cerevisiae growth and prion propagation. *Eukaryot. Cell* **2013**, *12*, 739–745. [CrossRef]
70. Newby, G.A.; Lindquist, S. Blessings in disguise: Biological benefits of prion-like mechanisms. *Trends Cell Biol.* **2013**, *23*, 251–259. [CrossRef]
71. Halfmann, R.; Alberti, S.; Lindquist, S. Prions, protein homeostasis, and phenotypic diversity. *Trends Cell Biol.* **2010**, *20*, 124–133. [CrossRef]
72. Cai, X.; Chen, J.; Xu, H.; Liu, S.; Jiang, Q.X.; Halfmann, R.; Chen, Z.J. Prion-like polymerization underlies signal transduction in antiviral immune defense and inflammasome activation. *Cell* **2014**, *156*, 1207–1222. [CrossRef] [PubMed]
73. Cai, X.; Chen, Z.J. Prion-like polymerization as a signaling mechanism. *Trends Immunol.* **2014**, *35*, 622–630. [CrossRef] [PubMed]
74. Sanders, D.W.; Kaufman, S.K.; Holmes, B.B.; Diamond, M.I. Prions and Protein Assemblies that Convey Biological Information in Health and Disease. *Neuron* **2016**, *89*, 433–448. [CrossRef] [PubMed]
75. Wu, B.; Hur, S. How RIG-I like receptors activate MAVS. *Curr. Opin. Virol.* **2015**, *12*, 91–98. [CrossRef]
76. Wang, Z.; Ji, J.; Peng, D.; Ma, F.; Cheng, G.; Qin, F.X.-F. Complex Regulation Pattern of IRF3 Activation Revealed by a Novel Dimerization Reporter System. *J. Immunol.* **2016**, *24*, R637–R638. [CrossRef]
77. Kayikci, Ö.; Nielsen, J. Glucose repression in Saccharomyces cerevisiae. *FEMS Yeast Res.* **2015**, *15*, fov068. [CrossRef]
78. Kuchin, S.; Vyas, V.K.; Kanter, E.; Hong, S.P.; Carlson, M. Std1p (Msn3p) positively regulates the Snfl kinase in Saccharomyces cerevisiae. *Genetics* **2003**, *163*, 507–514.
79. Ghillebert, R.; Swinnen, E.; Wen, J.; Vandesteene, L.; Ramon, M.; Norga, K.; Rolland, F.; Winderickx, J. The AMPK/SNF1/SnRK1 fuel gauge and energy regulator: Structure, function and regulation. *FEBS J.* **2011**, *278*, 3978–3990. [CrossRef]
80. Simpson-Lavy, K.; Xu, T.; Johnston, M.; Kupiec, M. The Std1 Activator of the Snf1/AMPK Kinase Controls Glucose Response in Yeast by a Regulated Protein Aggregation. *Mol. Cell* **2017**, *68*, 1120.e3–1133.e3. [CrossRef]
81. Hedbacker, K.; Carlson, M. SNF1/AMPK pathways in yeast. *Front. Biosci.* **2008**, *13*, 2408–2420. [CrossRef]

82. Brown, J.C.S.; Lindquist, S. A heritable switch in carbon source utilization driven by an unusual yeast prion. *Genes Dev.* **2009**, *23*, 2320–2332. [CrossRef]
83. Caudron, F.; Barral, Y. A Super-Assembly of Whi3 encodes memory of deceptive encounters by single cells during yeast courtship. *Cell* **2013**, *155*, 1244–1257. [CrossRef] [PubMed]
84. Hagen, D.C.; McCaffrey, G.; Sprague, G.F. Evidence the yeast STE3 gene encodes a receptor for the peptide pheromone a factor: Gene sequence and implications for the structure of the presumed receptor. *Proc. Natl. Acad. Sci. USA* **1986**, *83*, 1418–1422. [CrossRef] [PubMed]
85. Jenness, D.D.; Burkholder, A.C.; Hartwell, L.H. Binding of α-factor pheromone to yeast a cells: Chemical and genetic evidence for an α-factor receptor. *Cell* **1983**, *35*, 521–529. [CrossRef]
86. Moore, S.A. Yeast cells recover from mating pheromone α factor-induced division arrest by desensitization in the absence of α factor destruction. *J. Biol. Chem.* **1984**, *259*, 1004–1010.
87. Caudron, F. Protein aggregation triggers a declining libido in elder yeasts that still have a lust for life. *Microb. Cell* **2017**, *4*, 200–202. [CrossRef] [PubMed]
88. Holmes, K.J.; Klass, D.M.; Guiney, E.L.; Cyert, M.S. Whi3, an S. cerevisiae RNA-binding protein, is a component of stress granules that regulates levels of its target mRNAS. *PLoS ONE* **2013**, *8*, e84060. [CrossRef] [PubMed]
89. Nash, R.; Tokiwa, G.; Anand, S.; Erickson, K.; Futcher, A.B. The WHI1+ gene of Saccharomyces cerevisiae tethers cell division to cell size and is a cyclin homolog. *EMBO J.* **1988**, *7*, 4335–4346. [CrossRef]
90. Garí, E.; Volpe, T.; Wang, H.; Gallego, C.; Futcher, B.; Aldea, M. Whi3 binds the mRNA of the G1 cyclin CLN3 to modulate cell fate in budding yeast. *Genes Dev.* **2001**, *15*, 2803–2808. [CrossRef]
91. Caudron, F.; Barral, Y. Mnemons: Encoding memory by protein super-assembly. *Microb. Cell* **2014**, *1*, 100–102. [CrossRef]
92. Schlissel, G.; Krzyzanowski, M.K.; Caudron, F.; Barral, Y.; Rine, J. Aggregation of the Whi3 protein, not loss of heterochromatin, causes sterility in old yeast cells. *Science* **2017**, *355*, 1184–1187. [CrossRef] [PubMed]
93. Clay, L.; Caudron, F.; Denoth-Lippuner, A.; Boettcher, B.; Frei, S.B.; Snapp, E.L.; Barral, Y. A sphingolipid-dependent diffusion barrier confines ER stress to the yeast mother cell. *Elife* **2014**, *3*, e01883. [CrossRef]
94. Heinrich, S.U.; Lindquist, S. Protein-only mechanism induces self-perpetuating changes in the activity of neuronal Aplysia cytoplasmic polyadenylation element binding protein (CPEB). *Proc. Natl. Acad. Sci. USA* **2011**, *108*, 2999–3004. [CrossRef]
95. Si, K.; Lindquist, S.; Kandel, E.R. A Neuronal Isoform of the Aplysia CPEB Has Prion-Like Properties. *Cell* **2003**, *115*, 879–891. [CrossRef]
96. Si, K.; Choi, Y.B.; White-Grindley, E.; Majumdar, A.; Kandel, E.R. Aplysia CPEB Can Form Prion-like Multimers in Sensory Neurons that Contribute to Long-Term Facilitation. *Cell* **2010**, *140*, 421–435. [CrossRef]
97. Majumdar, A.; Cesario, W.C.; White-Grindley, E.; Jiang, H.; Ren, F.; Khan, M.R.; Li, L.; Choi, E.M.L.; Kannan, K.; Guo, F.; et al. Critical role of amyloid-like oligomers of Drosophila Orb2 in the persistence of memory. *Cell* **2012**, *148*, 515–529. [CrossRef]
98. Li, L.; Sanchez, C.P.; Slaughter, B.D.; Zhao, Y.; Khan, M.R.; Unruh, J.R.; Rubinstein, B.; Si, K. A Putative Biochemical Engram of Long-Term Memory. *Curr. Biol.* **2016**, *26*, 3143–3156. [CrossRef] [PubMed]
99. Khan, M.R.; Li, L.; Pérez-Sánchez, C.; Saraf, A.; Florens, L.; Slaughter, B.D.; Unruh, J.R.; Si, K. Amyloidogenic Oligomerization Transforms Drosophila Orb2 from a Translation Repressor to an Activator. *Cell* **2015**, *163*, 1468–1483. [CrossRef] [PubMed]
100. López-Otín, C.; Blasco, M.A.; Partridge, L.; Serrano, M.; Kroemer, G. The hallmarks of aging. *Cell* **2013**, *153*, 1194–1217. [CrossRef] [PubMed]
101. Alberti, S.; Hyman, A.A. Are aberrant phase transitions a driver of cellular aging? *BioEssays* **2016**, *38*, 959–968. [CrossRef] [PubMed]
102. Niaki, A.G.; Sarkar, J.; Cai, X.; Rhine, K.; Vidaurre, V.; Guy, B.; Hurst, M.; Lee, J.C.; Koh, H.R.; Guo, L.; et al. Loss of Dynamic RNA Interaction and Aberrant Phase Separation Induced by Two Distinct Types of ALS/FTD-Linked FUS Mutations. *Mol. Cell* **2020**, *77*, 82.e4–94.e4. [CrossRef] [PubMed]
103. Buchan, J.R.; Kolaitis, R.-M.; Taylor, J.P.; Parker, R. Eukaryotic Stress Granules Are Cleared by Autophagy and Cdc48/VCP Function. *Cell* **2013**, *153*, 1461–1474. [CrossRef] [PubMed]
104. Marrone, L.; Poser, I.; Casci, I.; Japtok, J.; Reinhardt, P.; Janosch, A.; Andree, C.; Lee, H.O.; Moebius, C.; Koerner, E.; et al. Isogenic FUS-eGFP iPSC Reporter Lines Enable Quantification of FUS Stress Granule

Pathology that Is Rescued by Drugs Inducing Autophagy. *Stem Cell Rep.* **2018**, *10*, 375–389. [CrossRef] [PubMed]

105. Speldewinde, S.H.; Doronina, V.A.; Grant, C.M. Autophagy protects against de novo formation of the [PSI+] prion in yeast. *Mol. Biol. Cell* **2015**, *26*, 4541–4551. [CrossRef] [PubMed]
106. Li, Y.R.; King, O.D.; Shorter, J.; Gitler, A.D. Stress granules as crucibles of ALS pathogenesis. *J. Cell Biol.* **2013**, *201*, 361–372. [CrossRef]
107. Barbosa, M.C.; Grosso, R.A.; Fader, C.M. Hallmarks of aging: An autophagic perspective. *Front. Endocrinol. (Lausanne)* **2019**, *9*, 790. [CrossRef]

Review

Symmetry Breaking and Epithelial Cell Extrusion

Bageshri Naimish Nanavati, Alpha S. Yap * and Jessica L. Teo

Division of Cell and Developmental Biology, Institute for Molecular Bioscience, The University of Queensland, St. Lucia, Brisbane, QLD 4072, Australia; b.nanavati@uq.edu.au (B.N.N.); j.teo@imb.uq.edu.au (J.L.T.)

* Correspondence: a.yap@uq.edu.au

Received: 14 May 2020; Accepted: 4 June 2020; Published: 7 June 2020

Abstract: Cell extrusion is a striking morphological event found in epithelia and endothelia. It is distinguished by two symmetry-breaking events: a loss of planar symmetry, as cells are extruded in either apical or basal directions; and loss of mechanochemical homogeneity within monolayers, as cells that are fated to be extruded become biochemically and mechanically distinct from their neighbors. Cell extrusion is elicited by many diverse events, from apoptosis to the expression of transforming oncogenes. Does the morphological outcome of extrusion reflect cellular processes that are common to these diverse biological phenomena? To address this question, in this review we compare the progress that has been made in understanding how extrusion is elicited by epithelial apoptosis and cell transformation.

Keywords: apoptotic extrusion; oncogenic extrusion; contractility; actomyosin

1. Introduction

Cell extrusion is a distinctive morphological phenomenon where cells are physically expelled from tissues. This expulsion process, also described as delamination, is strikingly evident where the affected cells appear to pop out of their tissue of origin (Figure 1A,B). Characteristically, extrusion occurs in epithelia and endothelia tissues [1] which consist of polarized cells linked together by cell–cell junctions. For simplicity, in this article we will principally refer to epithelia, where much of the work has been done to date.

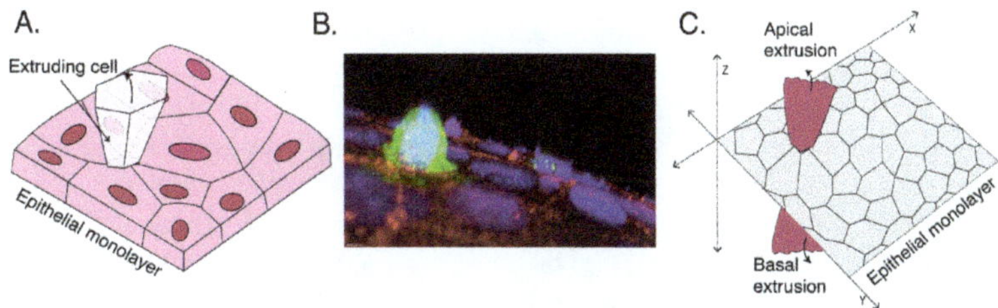

Figure 1. Symmetry breaking in cell extrusion. (**A**) Epithelia face diverse challenges to their integrity and homeostasis, including cell death, transformation, and overcrowding. One striking homeostatic response is for cells to be physically expelled from the monolayer in either an apical or basal direction. This process is called cell extrusion. (**B**) Immunofluorescence image of a cell expressing H-Ras^{V12} being apically extruded from a Caco-2 colon epithelial monolayer. (Green: GFP-H-Ras^{V12}; red: N-WASP marking cell–cell contacts; purple: DAPI; courtesy of Dr. Selwin Wu.) (**C**) During the process of cell extrusion, the overall geometric symmetry of tissue is broken as the cell is expelled out in Z dimension either apically or basally. Moreover, the affected cell is biochemically and often mechanically different from its neighboring healthy cells, breaking the planar symmetry or homogeneity of the monolayer.

Extrusion involves two forms of symmetry breaking (Figure 1C). First, extrusion breaks the planar symmetry of the host tissue by expelling cells in a direction that is orthogonal to the plane of its monolayer. This can be understood as a form of geometric symmetry breaking (Figure 1C). Cell expulsion has been described to occur in either apical or basal directions, which would generally lead to the expelled cells being directed into the external environment or towards entering the body, respectively. For simplicity, we will focus on apical extrusion in this article. Secondly, it is increasingly evident that important biochemical and biomechanical differences distinguish the cells destined for expulsion from their surrounding neighbors. Thus, at a first generalization we can consider extrusion to disrupt the biochemical and mechanical homogeneity of the tissue at the interface between the extruded cell and its neighbors (Figure 1C). Ultimately, any adequate description of the extrusion process must explain how the geometric event of expulsion arises from the biochemical and mechanical differences between the expelled cell and its neighbors.

A striking feature of the extrusion phenomenon is the diverse range of biological processes in which it has been implicated. For example, it is elicited when epithelial cells undergo apoptosis (apoptotic extrusion [2]) or when they express oncogenes (oncogenic extrusion [3]). Extrusion also occurs when epithelia become overcrowded [1,4] or when they are infected with intracellular organisms, such as salmonella [5,6]. In some of these cases, extrusion can be understood as a way of preserving tissue homeostasis, but in other circumstances extrusion is involved in cell differentiation [7]. This suggests that extrusion may be a final common response to very different biological processes. What is not clear, though, is whether the common morphological features of extrusion reflect biochemical and mechanobiological mechanisms that are shared between these different processes, or whether different forms of extrusion may be better understood as distinct phenomena. In this review, we endeavor to consider this question by comparing what is currently known about the two best-studied forms of extrusion: apoptotic and oncogenic extrusion.

2. Defining Extrusion

To begin, it is useful to consider how we define cell extrusion. The term is often used to refer to the morphological process of expulsion or delamination. This is understandable, given the dramatic morphology of this process. However, diverse mechanisms may be responsible for similar morphological events. For example, the basal egress of transformed cells from an epithelium has often been ascribed to epithelial–mesenchymal transitions and local degradation of the basement membrane [8] as well as to basal extrusion. On purely morphological grounds alone, it is difficult to distinguish what may be mechanistically distinct phenomena.

In addition, to focus on expulsion alone runs the risk of overlooking a key feature of the extrusion process, namely the necessary involvement of the epithelial cells which surround the cell that will be extruded ("neighbor cells" for simplicity). This is exemplified by the role of E-cadherin, the classical cadherin that is a major component of adherens junctions (AJ) in epithelia. Both apoptotic [9] and oncogenic extrusion [3] are inhibited when E-cadherin is depleted in the epithelium surrounding the cells that are to be extruded. Although it is not yet clear why E-cadherin is required in the neighbors, this role for cell–cell interactions between neighbor cells implies that extrusion is a cell nonautonomous phenomenon. The concept of cell nonautonomy in the surrounding epithelium is reinforced by evidence that cell shape [2], the cytoskeleton [10,11], and cell signaling pathways [12] are altered in the neighbor cells—often reflecting the biochemical symmetry breaking described earlier—and these changes may be confined to the local environment of the extruded cells (only one or a few cell diameters away).

We therefore suggest that it is useful to define extrusion as encompassing both (a) the phenomenon of cell expulsion; and (b) the active contribution of the surrounding monolayer, both to expel the extruded cell and, as we shall see, to preserve the epithelial barrier itself. Taking this perspective, we will frame our review of apoptotic and oncogenic extrusion around the following questions: (1) What are the relevant biochemical and/or mechanical changes that occur in the cells to be extruded? What changes in these cells are responsible for triggering the extrusion process? (2) How do the neighbor cells participate in the extrusion response? And what are the underlying cell biological and mechanical changes that characterize the neighbor response? 3) How are changes in the extruded cells coordinated with changes in their neighbors? Is there cell-to-cell communication between these two cell populations? As we shall see, the answers to these questions are provisional and may not be the same for apoptotic and oncogenic extrusion.

3. Apoptotic Cell Extrusion

Apoptosis is a form of programmed cell death and a major cause of tissue turnover during development and in postdevelopmental life [13]. Indeed, apoptosis is estimated to be responsible for the daily turnover of ~150 billion cells in the healthy human body [14]. However, this burden of cell turnover presents a homeostatic challenge for the body: to prevent the release of cellular contents that can provoke inflammatory and autoimmune reactions. In the first instance, this is achieved by apoptotic cells becoming fragmented into membrane-bound apoptotic bodies [15,16]. The immunological inertness of the apoptotic process can be maintained so long as those membranes remain intact. However, membrane integrity can eventually be lost through secondary necrosis [17]. Therefore, additional mechanisms are required to eliminate apoptotic corpses and fragments before their membrane integrity becomes compromised. The best-understood of these secondary mechanisms is the phagocytosis of apoptotic bodies (efferocytosis) by macrophages/monocytes and nonprofessional phagocytes (including epithelial cells) [18,19].

However, epithelia and endothelia can also eliminate apoptotic cells by extrusion. This has been observed in model systems [20], but also occurs in pure cell culture [2,11,20] indicating that it is an epithelium-intrinsic process that does not require other cells of the innate immune system. Indeed, apical extrusion has a potential design advantage: since the apical surface of epithelial cells typically faces to the external environment, apical extrusion allows apoptotic cells to be directly eliminated from the body. Of note, extrusion appears to be engaged as a relatively early response to apoptosis: it begins—and is sometimes completed [21]—before markers of late apoptosis (such as the ectofacial presentation of phosphatidylserine) become evident [2,11]. Thus, apical extrusion may constitute a first-line response to apoptosis in epithelia, one that can complement, or even prevent, the need for efferocytosis. It should be noted, though, that there are species-specific differences: whereas apoptotic extrusion typically occurs in an apical direction in vertebrates, in Drosophila, apoptotic cells are commonly extruded basally [22–27].

The extrusion process can also address another challenge that apoptosis presents for epithelia. The potential for apoptotic bodies to break down and provoke inflammation is one that is probably shared by all tissues. But epithelia have the additional responsibility of providing major physiological barriers in the body. In particular, epithelia provide the barrier separating the body from its external environment, controlling the flux of ions and fluid, absorbing nutrients, and preventing the entry of infective and noxious agents [28]. Functioning epithelial barriers require both intact cells and also specialized cell–cell junctions, both of which are potentially compromised by apoptosis. Thus, it is striking that epithelial barrier function is preserved even when apoptosis is enhanced by agents such as etoposide [2,9]. Here, the neighbor cell response appears to critically preserve the barriers. During apoptotic extrusion, neighbor cells elongate and extend lamellipodial extensions underneath the apoptotic cell [29,30]. It is thought that these responses allow tight and adhesive junctions to be preserved between the neighbor cells, thereby maintaining the barrier whilst the apoptotic cell is eliminated. Indeed, both AJ and desmosomes are preserved during the process of apoptotic extrusion [9,31]. This reinforces the notion that active neighbor cell responses are a critical part of the apoptotic extrusion phenomenon.

How does the apoptotic cell induce extrusion? Apoptotic cells undergo both chemical and mechanical changes that may trigger the extrusion process. Apoptotic cells produce a wide range of chemical signals that have often been implicated in aspects of efferocytosis [32]. Of these signals, the soluble lipid sphingosine-1-phosphate (S1P) has also been implicated in apoptotic extrusion. Immunostaining revealed that S1P was initially evident in the apoptotic cells and later appeared to accumulate in their neighbors [20]. Furthermore, extrusion was blocked by inhibiting sphingosine-1-kinase, the enzyme responsible for synthesis of S1P [20]. This suggested that S1P can be released from apoptotic cells to trigger extrusion.

Apoptotic cells also change their mechanical properties. Cells become hypercontractile when they undergo apoptosis [15,33]. In many, but not all [21], cases, this is thought to be due to the activation of procontractile kinases by apoptotic caspases. For example, Rho Kinase (ROCK) is autoinhibited by an intramolecular interaction between its N-terminus and its C-terminus [34]. However, the N-terminus of ROCK1 is cleaved by caspase-3, yielding a fragment that can constitutively activate downstream contractile signaling [15,33], to promote Myosin II and LIM kinase activation. Apoptotic epithelial cells show features of hypercontractility, such as blebbing and an increase in cortical actomyosin [9,11,15,21,33]. Conversely, the sporadic expression of truncated, constitutively active ROCK2 mutants in epithelia can induce extrusion, even though the transgene-expressing cells do not undergo apoptosis [9,11]. Together, these observations imply that enhanced contractility within apoptotic cells can provide a mechanical stimulus for extrusion.

How do neighbor cells expel apoptotic cells? To address this question, we focus on the mechanical processes that allow neighbor cells to expel apoptotic cells and their underlying cell biological mechanisms. Two mechanisms are currently known that can allow neighbor cells to apply compressive forces onto, and therefore expel, dying, apoptotic cells.

First, the contractile cortex can be reinforced within neighbor cells, especially at their interface with the apoptotic cell. Both F-actin and Myosin II have been observed to accumulate at this site during the extrusion process. This generates a cup-like actomyosin network with increased contractile tension [11] within the neighbor cells specifically at its junction with the apoptotic cell. Contraction in this enhanced actomyosin cortex could effectively apply compressive forces to expel the apoptotic cell. It should be noted that in this model, apical extrusion would imply that neighbor cell contraction is principally applied at their basal regions, creating a net compression that directs the apoptotic cell apically. Indeed, this has been observed in many cases [9–11].

In the second mechanism, compression can be generated by the lamellipodial protrusions that neighbor cells make as they crawl underneath the apoptotic cells [35]. It has been suggested that cortical contractility may operate for single cells, whereas lamellipodia may be involved where larger cells, or small groups of cells, are being extruded [29]. These mechanisms are not necessarily exclusive, as cryptic lamellipodia which extend underneath migrating epithelial cells can also form AJ [36,37].

Both of these processes imply that the cytoskeleton is altered within neighbor cells in response to signals from the apoptotic cell. This would entail intracellular signaling pathways in the neighbor cell that ultimately regulate the cytoskeleton, especially the actomyosin apparatus. To date, the RhoA GTPase, a canonical activator of the actomyosin cytoskeleton, is the signal that has been most extensively studied in this process. Levels of the active GTP-loaded form of RhoA have been reported to increase in neighbor cells and selective inhibition of RhoA by microinjection of C3-transferase into the neighbor cells blocked apoptotic extrusion [2,9]. RhoA is activated by guanine nucleotide exchange factors (GEFs) and, consistent with this, p115 RhoGEF also localized to the basal regions of neighbor cells during apoptotic extrusion [10]. Many other cytoskeletal effectors and signals are involved in both cell contractility [11] and lamellipodial activity, so it is likely that other important pathways remain to be identified.

Importantly, there are pathways for these chemical and mechanical apoptotic signals to elicit the cytoskeletal responses of their neighbors. Extrusion would then arise from intercellular communication between these two cell populations (Figure 2A).

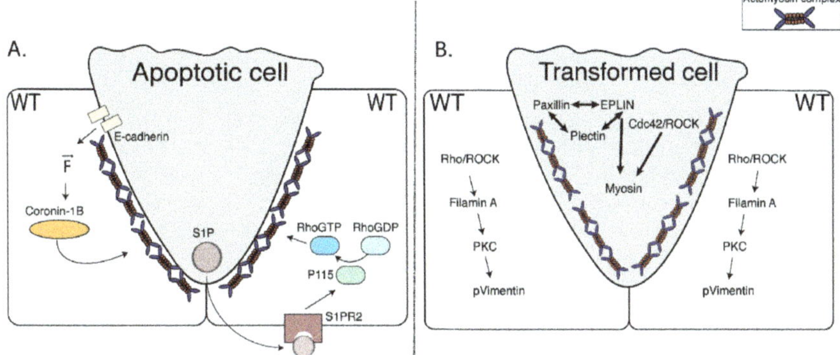

Figure 2. Types of extrusion. (**A**) Apoptotic cell extrusion. There are two known pathways for apoptotic cells to elicit extrusion responses in their nonapoptotic neighbors. [Left side] When a cell undergoes apoptosis, it becomes hypercontractile. This contractile force is sensed by its healthy neighbors through E-cadherin which leads, in turn, to assembly of an actomyosin complex through coronin-1B. [Right side] The apoptotic cell releases S1P. S1P binds to S1P$_2$ receptors present on the healthy neighbors of the dying cell; this ultimately activates RhoA through a signal transduction pathway that involves p115 RhoGEF. The active form of RhoA mediates assembly of actomyosin complex in the neighbors. (**B**) Oncogenic cell extrusion. Depending on the type of oncogene expressed, a transformed cell elicits distinct signaling pathways which potentially converge to generate biomechanical changes. Activation of the Cdc42/ROCK signaling pathway and upregulation of cytoskeletal-interacting proteins (e.g., EPLIN, paxillin, and plectin) enhance contractility in the transformed cell. Simultaneously, in the neighboring wild-type cells, activation of the RhoA/ROCK signaling pathway upregulates other cytoskeletal proteins (e.g., Filamin A, vimentin). Mechanisms triggered in both the transformed and wild-type cells display mutual coregulation.

The capacity for cell-to-cell communication is best understood for S1P, where extrusion could be blocked by antagonizing the S1P receptor specifically in the neighbor cells [20]. Of the four classes of S1P receptors present in vertebrates (S1P$_1$ to S1P$_4$), the S1P$_2$ receptor appears to be specifically involved, as extrusion was blocked with antagonist drugs and RNAi specific for this molecule [20]. However, extrusion was not inhibited by depleting S1P$_2$ in the apoptotic cell itself [20], which made it unlikely that S1P was working in an autocrine fashion. Instead, extrusion was blocked when S1P$_2$ was depleted in the surrounding epithelium [20]. This implied that the S1P$_2$ receptor was only required in the neighbor cells, yielding a paracrine model where S1P secreted by the apoptotic cell binds to S1P$_2$ on its neighbors to elicit the extrusion response (Figure 2A). As S1P$_2$ is a G-protein coupled receptor that can activate p115 RhoGEF via its associated Gα12/13 subclass of G-proteins [10], it could anchor a paracrine pathway to ultimately activate actomyosin in the neighbor cells via RhoA.

Adherens junctions also provide the potential for mechanical signals from apoptotic cells to be transmitted to their neighbors. Indeed, a number of tension-sensitive mechanisms have recently been identified to associate with E-cadherin [38–42] which could respond to the enhanced contractility of apoptotic cells. They would also explain why apoptotic extrusion requires E-cadherin to be present in the neighbor cells. Interestingly, one of these, which utilizes Myosin VI as a mechanosensor [42], also activates RhoA via p114 RhoGEF, a close relative of p115 RhoGEF. Although tantalizing, though, a role for mechanotransduction has yet to be tested in apoptotic extrusion.

4. Oncogenic Cell Extrusion

Cell extrusion also occurs when oncogenes are sporadically expressed in epithelia [43–46]. In vertebrates, oncogenic extrusion commonly occurs in the apical direction, although basal extrusion has also been observed [3,47–49] (Figure 2B). Oncogenic extrusion can be elicited in cultured epithelial monolayers [3,45,50–53], indicating that it is an epithelium-intrinsic process like apoptotic extrusion. Extrusion typically occurs only when oncogenes are expressed in single cells or small groups of cells that are surrounded by nonexpressing epithelium. Indeed, extrusion did not occur when transforming Ras mutants were expressed ubiquitously in MDCK monolayers [3].

Extrusion can be elicited by a diverse range of oncogenes. In addition to Ras, these include Src [54], Cdc42 [55], Yap1 [56], and ERBB2 [46]. In addition to a range of cultured epithelia, oncogenic extrusion occurs in Drosophila [3], zebrafish [57], and mouse models [58]. Thus, it has been suggested that extrusion represents a fundamental response of epithelia to early transformation. The functional consequence of extrusion is less clear. One possibility is that apical extrusion represents a mechanism to eliminate newly transformed cells from the body (and it has accordingly been described as an epithelial defense against cancer, EDAC) [59–61]. Akin to apoptotic extrusion, apical extrusion would be considered to expel newly transformed cells into the external environment. However, apical extrusion may also allow transformed cells to proliferate, by removing them from inhibitory influences in the host epithelium [46], something that may be especially relevant where the apical compartment is relatively enclosed. It should also be noted that experimental studies to elicit extrusion have typically expressed strong oncogenes. It will be interesting to test whether extrusion is elicited or altered by the loss of tumor suppressor genes or as driver mutations accumulate during the natural history of cancer.

What changes in the oncogene-expressing cells? The oncogenes tested to date include small GTPases (Ras) and tyrosine protein kinases (e.g., Src, ERBB2) that are capable of engaging a diverse range of downstream signaling pathways. Therefore, it is necessary to identify the pathway(s) and their effector targets that are responsible for eliciting extrusion.

Indeed, a variety of signaling pathways are activated in cells that mosaically express oncogenes such as Ras or Src, but their relative contributions can differ. For example, Cdc42 activity increases in cells that express constitutively active H-Ras^{V12} and extrusion is inhibited by coexpression of a dominantly interfering Cdc42 mutant [3]. Furthermore, expression of a constitutively active Cdc42 mutant could itself induce extrusion [55]. Thus, Cdc42 appears to be a key mediator of extrusion downstream of Ras. However, ERK/MAPK activity also appeared to be necessary for Ras to induce extrusion, although expression of constitutively active Raf, which mediates signaling from Ras to ERK, did not induce extrusion by itself [3]. Therefore, although signals like Cdc42 may play critical roles in oncogene expressing, extrusion is likely to reflect their interaction with other signaling pathways.

A number of potential effector targets have also been identified in oncogene-expressing cells, whose effects can provide insight into the mechanism of cell expulsion. For example, levels of EPLIN, plectin, and paxillin were upregulated in cells mosaically expressing H-Ras^{V12} and these appeared to have secondary impacts on membranes and the cytoskeleton [50–52,57,59]. EPLIN appeared to upregulate phosphomyosin levels while paxillin and plectin were implicated in increasing tubulin and acetylated tubulin levels in H-Ras^{V12} cells, perhaps by inhibiting the α-tubulin deacetylase, HDAC [51,62]. Interestingly, atomic force microscopy measurements suggested that H-Ras^{V12} cells become stiffer and more viscous [3], but whether these cytoskeletal changes are responsible for that mechanical change remains to be tested. It might also be anticipated that apical extrusion would require the de-adhesion of cells from their underlying extracellular matrix. Thus, it is interesting that matrix metalloproteinase expression is increased in cells that mosaically express constitutively active Cdc42 or ERBB2 [46,55]. Interestingly, MMP inhibition reduced the extrusion of ERBB2-expressing cells and, indeed, mosaic overexpression of MT1-MMP could itself support extrusion. Although the picture remains incomplete, these findings suggest that multiple cytoskeletal and adhesive changes occur in oncogene-expressing cells which collaborate to support extrusion.

How do the neighbor cells change? In contrast, it is much less clear how neighbor cells are altered during oncogenic extrusion. Whereas an enhanced actomyosin cortex is commonly seen in neighbor cells during apoptotic extrusion, it has seldom been observed during oncogenic extrusion [3]. Even though RhoA/ROCK signaling was required in neighbor cells, myosin activity was dispensable [59]. Thus, oncogenic extrusion does not appear to require a consistent contractile response in the neighbor cells, as is the case for apoptosis. Instead, other cytoskeletal changes have been reported to occur in the neighbors of oncogene-expressing cells. These include upregulation of filamin, an F-actin cross-linker which promotes cell surface tension [63], whose accumulation in neighbor cells depends on RhoA/ROCK signaling [59]. As well, intermediate filament components are upregulated in neighbor cells [59]. Keratin 5/8 was enhanced around H-Ras^{V12} cells and vimentin accumulated in the neighbors of v-Src cells. Intermediate filaments are capable of strain stiffening, to support the mechanical integrity of cells and couple to desmosomes to promote cell–cell adhesion and tissue integrity [64,65]. However, how this may contribute to the extrusion of oncogene-expressing cells is not yet clear, and its elucidation is likely to require a clearer understanding of the mechanical events that drive oncogenic extrusion (see below).

Is there cell–cell communication in oncogenic extrusion? Are these neighbor cell changes occurring in response to signals from the oncogene cell? The answer to this question is much less clear than that for apoptosis. The S1P receptor, S1P$_2$, has been reported to be required in neighbor cells for H-Ras^{V12}-induced extrusion [47,48,61], but whether this is responding to an S1P signal from the H-Ras^{V12} cells is unclear. Indeed, extrusion was not affected by inhibiting cellular production of S1P by sphingosine-1-kinase [61]. As well, S1P levels were reduced by autophagy when K-Ras^{V12} cells underwent basal extrusion [48].

Further, the nature of cell-to-cell signaling appears to be much more complex for oncogenic extrusion than is the case for apoptotic extrusion. For example, active GTP-Cdc42 levels were increased in cells that mosaically expressed H-Ras^{V12} (i.e., those whose neighbors did not express Ras), but not when Ras was ubiquitously expressed in the monolayers [3]. Thus, oncogene-free neighbors were necessary for Cdc42 to be activated in the H-Ras^{V12} cells. Similarly, ephrin-A in neighbor cells has been reported to activate its cognate receptor, EphA2, in H-Ras^{V12} cells, to support F-actin and Myosin II levels in the oncogene-expressing cells [66,67]. Together, these observations imply that there is signaling cross-talk between both oncogene cells and their oncogene-free neighbors that contributes to the nonautonomous nature of this extrusion phenomenon.

5. Thoughts for the Future

So, where does this leave us in our efforts to identify commonalities between apoptotic and oncogenic extrusion? Overall, we appear to have a reasonable working model to help guide investigation of apoptotic extrusion (Figure 2A). Here, signals from the apoptotic cell elicit cytoskeletal responses in its neighbor cells that ultimately generate compressive forces that expel the apoptotic cell from the epithelium. These apoptotic signals may be chemical and/or mechanical and the cytoskeletal responses in the neighbors can be contractile and/or lamellipodial. It will therefore be important to test if these options are alternatives that are used in different contexts or may function as complementary pathways. Indeed, considering the second possibility, it is interesting to consider how a diffusible apoptotic signal, such as S1P, can elicit a localized contractile response in the immediate neighbors of the apoptotic cell [2,11,20]. One possibility is that mechanotransduction cooperates with S1P to refine the site where RhoA is activated. Nonetheless, even with these open questions, the current picture provides a framework to understand why apoptotic extrusion is a cell nonautonomous phenomenon. It also begins to explain how the cell biological changes in apoptotic cells and neighbors account for the biomechanical process of extrusion.

In contrast, a unifying model for oncogenic extrusion is less evident on present evidence (Figure 2B). In particular, despite the wealth of biochemical changes that have been documented to occur in both the oncogene-expressing cell and its neighbors, the mechanics of oncogenic extrusion have yet to be characterized in sufficient depth to explain how these biochemical changes lead to the morphogenetic event of extrusion. Tissue mechanics do affect the efficacy of oncogenic extrusion. For example, at the interface between epithelial cells, mechanical tension is greater at the apical zonula adherens (ZAs) than in the lateral cell–cell contact surfaces found basal to the ZA [53]. However, increasing lateral tension at the cell–cell junctions compromised extrusion. On a larger length scale, increasing tension within epithelial monolayers also antagonizes extrusion [68].

However, to understand how these hypertensile changes exert their impact, we will need to understand the mechanobiology of oncogenic extrusion. It seems likely that expulsion of oncogene-expressing cells requires compression from its surrounding epithelium, but this has to be confirmed experimentally, and then we need to elucidate the responsible cellular mechanism. In particular, it will be important to understand whether and how the mechanical changes of extrusion arise from the biochemical cross-talk that occurs between oncogene cells and their neighbors. This knowledge will help us compare these two forms of extrusion and better define their commonalities and differences. It will also provide a basis to characterize the diverse other forms of extrusion that have been documented.

Finally, it will be important to understand the degree to which epithelial integrity is preserved in different kinds of extrusion and how this is accomplished. As noted above, apoptotic extrusion preserves the epithelial barrier. How this may be coordinated with the process of apoptotic corpse expulsion is less clear, but it is interesting to consider whether the contractile forces of apoptotic extrusion coordinate the assembly of junctions between neighbors. Both contractility within apoptotic cells [2] and cortical contraction in the neighbors [11] could help to bring the surfaces of neighbor cells together, zippering them up as the apoptotic cell is expelled. In contrast, whether the epithelial barrier is preserved during oncogenic extrusion has not been thoroughly analyzed. Junctional proteins, such as E-cadherin and β-catenin, have been reported to redistribute to the interface between oncogene-expressing cells and their neighbors [54,55,57], a process that may involve cell contractility [54]. But it will be important to test directly whether or not the epithelial barrier remains intact, as has been demonstrated for apoptotic extrusion [2], and, if so, how interactions between neighbors are preserved to achieve this. Alternatively, if the epithelial barrier is not preserved during oncogenic extrusion, this might be another clue to suggest that it is biologically distinct from apoptotic extrusion.

Author Contributions: Original draft preparation, B.N.N., J.L.T.; review and editing, A.S.Y.; supervision, A.S.Y. All authors have read and agreed to the published version of the manuscript.

Funding: National Health and Medical Research Council of Australia: 1140090.

Acknowledgments: The authors were supported by grants and fellowships from the National Health and Medical Research Council of Australia (grant 1140090 and fellowship APP1139592) to A.S.Y. J.L.T. was the recipient of an Equity Trustees PhD Scholarship in Medical Research. B.N.N. was the recipient of UQ Research Training Scholarship.

Conflicts of Interest: The authors declare no conflict of interest.

References

1. Eisenhoffer, G.T.; Loftus, P.D.; Yoshigi, M.; Otsuna, H.; Chien, C.-B.; Morcos, P.A.; Rosenblatt, J. Crowding induces live cell extrusion to maintain homeostatic cell numbers in epithelia. *Nature* **2012**, *484*, 546–549. [CrossRef] [PubMed]
2. Rosenblatt, J.; Raff, M.C.; Cramer, L.P. An epithelial cell destined for apoptosis signals its neighbors to extrude it by an actin- and myosin-dependent mechanism. *Curr. Biol.* **2001**, *11*, 1847–1857. [CrossRef]
3. Hogan, C.; Dupre-Crochet, S.; Norman, M.; Kajita, M.; Zimmermann, C.; Pelling, A.E.; Piddini, E.; Baena-Lopez, L.A.; Vincent, J.P.; Itoh, Y.; et al. Characterization of the interface between normal and transformed epithelial cells. *Nat. Cell Biol.* **2009**, *11*, 460–467. [CrossRef] [PubMed]
4. Marinari, E.; Mehonic, A.; Curran, S.; Gale, J.; Duke, T.; Baum, B. Live-cell delamination counterbalances epithelial growth to limit tissue overcrowding. *Nature* **2012**, *484*, 542–545. [CrossRef] [PubMed]
5. Sellin, M.E.; Müller, A.A.; Felmy, B.; Dolowschiak, T.; Diard, M.; Tardivel, A.; Maslowski, K.M.; Hardt, W.-D. Epithelium-Intrinsic NAIP/NLRC4 Inflammasome Drives Infected Enterocyte Expulsion to Restrict Salmonella Replication in the Intestinal Mucosa. *Cell Host Microbe* **2014**, *16*, 237–248. [CrossRef]

6. Saadat, I.; Higashi, H.; Obuse, C.; Umeda, M.; Murata-Kamiya, N.; Saito, Y.; Lu, H.; Ohnishi, N.; Azuma, T.; Suzuki, A.; et al. Helicobacter pylori CagA targets PAR1/MARK kinase to disrupt epithelial cell polarity. *Nature* **2007**, *447*, 330–333. [CrossRef]
7. Miroshnikova, Y.A.; Le, H.Q.; Schneider, D.; Thalheim, T.; Rubsam, M.; Bremicker, N.; Polleux, J.; Kamprad, N.; Tarantola, M.; Wang, I.; et al. Adhesion forces and cortical tension couple cell proliferation and differentiation to drive epidermal stratification. *Nat. Cell Biol.* **2018**, *20*, 69–80. [CrossRef]
8. Yilmaz, M.; Christofori, G. EMT, the cytoskeleton, and cancer cell invasion. *Cancer Metastasis Rev.* **2009**, *28*, 15–33. [CrossRef]
9. Lubkov, V.; Bar-Sagi, D. E-cadherin-mediated cell coupling is required for apoptotic cell extrusion. *Curr. Biol.* **2014**, *24*, 868–874. [CrossRef]
10. Slattum, G.; McGee, K.M.; Rosenblatt, J. P115 RhoGEF and microtubules decide the direction apoptotic cells extrude from an epithelium. *J. Cell Biol.* **2009**, *186*, 693–702. [CrossRef]
11. Michael, M.; Meiring, J.C.; Acharya, B.R.; Matthews, D.R.; Verma, S.; Han, S.P.; Hill, M.M.; Parton, R.G.; Gomez, G.A.; Yap, A.S. Coronin 1B Reorganizes the Architecture of F-Actin Networks for Contractility at Steady-State and Apoptotic Adherens Junctions. *Dev. Cell* **2016**, *37*, 58–71. [CrossRef] [PubMed]
12. Takeuchi, Y.; Narumi, R.; Akiyama, R.; Vitiello, E.; Shirai, T.; Tanimura, N.; Kuromiya, K.; Ishikawa, S.; Kajita, M.; Tada, M.; et al. Calcium Wave Promotes Cell Extrusion. *Curr. Biol.* **2020**, *30*, 670–681.e676. [CrossRef] [PubMed]
13. Kerr, J.F.; Wyllie, A.H.; Currie, A.R. Apoptosis: a basic biological phenomenon with wide-ranging implications in tissue kinetics. *Br. J. Cancer* **1972**, *26*, 239–257. [CrossRef]
14. Bianconi, E.; Piovesan, A.; Facchin, F.; Beraudi, A.; Casadei, R.; Frabetti, F.; Vitale, L.; Pelleri, M.C.; Tassani, S.; Piva, F.; et al. An estimation of the number of cells in the human body. *Ann. Hum. Biol.* **2013**, *40*, 463–471. [CrossRef] [PubMed]
15. Coleman, M.L.; Sahai, E.A.; Yeo, M.; Bosch, M.; Dewar, A.; Olson, M.F. Membrane blebbing during apoptosis results from caspase-mediated activation of ROCK I. *Nat. Cell Biol.* **2001**, *3*, 339–345. [CrossRef] [PubMed]
16. Green, D.R. Apoptotic Pathways: The Roads to Ruin. *Cell* **1998**, *94*, 695–698. [CrossRef]
17. Orning, P.; Lien, E.; Fitzgerald, K.A. Gasdermins and their role in immunity and inflammation. *J. Exp. Med.* **2019**, *216*, 2453–2465. [CrossRef]
18. Silva, M.; Correia-Neves, M. Neutrophils and Macrophages: the Main Partners of Phagocyte Cell Systems. *Front. Immunol.* **2012**, *3*. [CrossRef]
19. Monks, J.; Rosner, D.; Jon Geske, F.; Lehman, L.; Hanson, L.; Neville, M.C.; Fadok, V.A. Epithelial cells as phagocytes: apoptotic epithelial cells are engulfed by mammary alveolar epithelial cells and repress inflammatory mediator release. *Cell Death Differ.* **2005**, *12*, 107–114. [CrossRef]
20. Gu, Y.; Forostyan, T.; Sabbadini, R.; Rosenblatt, J. Epithelial cell extrusion requires the sphingosine-1-phosphate receptor 2 pathway. *J. Cell Biol.* **2011**, *193*, 667–676. [CrossRef]
21. Kuipers, D.; Mehonic, A.; Kajita, M.; Peter, L.; Fujita, Y.; Duke, T.; Charras, G.; Gale, J.E. Epithelial repair is a two-stage process driven first by dying cells and then by their neighbours. *J. Cell Sci.* **2014**, *127*, 1229–1241. [CrossRef] [PubMed]
22. Kiehart, D.P.; Galbraith, C.G.; Edwards, K.A.; Rickoll, W.L.; Montague, R.A. Multiple Forces Contribute to Cell Sheet Morphogenesis for Dorsal Closure in Drosophila. *J. Cell Biol.* **2000**, *149*, 471–490. [CrossRef] [PubMed]
23. Toyama, Y.; Peralta, X.G.; Wells, A.R.; Kiehart, D.P.; Edwards, G.S. Apoptotic Force and Tissue Dynamics during Drosophila Embryogenesis. *Science* **2008**, *321*, 1683. [CrossRef] [PubMed]
24. Meghana, C.; Ramdas, N.; Hameed, F.M.; Rao, M.; Shivashankar, G.V.; Narasimha, M. Integrin adhesion drives the emergent polarization of active cytoskeletal stresses to pattern cell delamination. *Proc. Natl. Acad. Sci. USA* **2011**, *108*, 9107. [CrossRef]
25. Muliyil, S.; Krishnakumar, P.; Narasimha, M. Spatial, temporal and molecular hierarchies in the link between death, delamination and dorsal closure. *Development* **2011**, *138*, 3043. [CrossRef]
26. Sokolow, A.; Toyama, Y.; Kiehart, D.P.; Edwards, G.S. Cell Ingression and Apical Shape Oscillations during Dorsal Closure in Drosophila. *Biophys. J.* **2012**, *102*, 969–979. [CrossRef]
27. Monier, B.; Gettings, M.; Gay, G.; Mangeat, T.; Schott, S.; Guarner, A.; Suzanne, M. Apico-basal forces exerted by apoptotic cells drive epithelium folding. *Nature* **2015**, *518*, 245–248. [CrossRef]

28. Duszyc, K.; Gomez, G.A.; Schroder, K.; Sweet, M.J.; Yap, A.S. In life there is death: How epithelial tissue barriers are preserved despite the challenge of apoptosis. *Tissue Barriers* **2017**, *5*, e1345353. [CrossRef]
29. Kocgozlu, L.; Saw, T.B.; Le, A.P.; Yow, I.; Shagirov, M.; Wong, E.; Mège, R.-M.; Lim, C.T.; Toyama, Y.; Ladoux, B. Epithelial Cell Packing Induces Distinct Modes of Cell Extrusions. *Curr. Biol.* **2016**, *26*, 2942–2950. [CrossRef]
30. Fadul, J.; Rosenblatt, J. The forces and fates of extruding cells. *Curr. Opin. Cell Biol.* **2018**, *54*, 66–71. [CrossRef]
31. Thomas, M.; Ladoux, B.; Toyama, Y. Desmosomal Junctions Govern Tissue Integrity and Actomyosin Contractility in Apoptotic Cell Extrusion. *Curr. Biol.* **2020**, *30*, 682–690.e685. [CrossRef] [PubMed]
32. Seong, S.-Y.; Matzinger, P. Hydrophobicity: an ancient damage-associated molecular pattern that initiates innate immune responses. *Nat. Rev. Immunol.* **2004**, *4*, 469–478. [CrossRef] [PubMed]
33. Sebbagh, M.; Renvoizé, C.; Hamelin, J.; Riché, N.; Bertoglio, J.; Bréard, J. Caspase-3-mediated cleavage of ROCK I induces MLC phosphorylation and apoptotic membrane blebbing. *Nat. Cell Biol.* **2001**, *3*, 346–352. [CrossRef] [PubMed]
34. Riento, K.; Guasch, R.M.; Garg, R.; Jin, B.; Ridley, A.J. RhoE Binds to ROCK I and Inhibits Downstream Signaling. *Mol. Cell. Biol.* **2003**, *23*, 4219. [CrossRef] [PubMed]
35. Tamada, M.; Perez, T.D.; Nelson, W.J.; Sheetz, M.P. Two distinct modes of myosin assembly and dynamics during epithelial wound closure. *J. Cell Biol.* **2007**, *176*, 27–33. [CrossRef] [PubMed]
36. Ozawa, M.; Hiver, S.; Yamamoto, T.; Shibata, T.; Upadhyayula, S.; Mimori-Kiyosue, Y.; Takeichi, M. Adherens junction serves to generate cryptic lamellipodia required for collective migration of epithelial cells. *bioRxiv* **2020**. [CrossRef]
37. Jain, S.; Cachoux, V.M.L.; Narayana, G.H.N.S.; de Beco, S.; D'Alessandro, J.; Cellerin, V.; Chen, T.; Heuzé, M.L.; Marcq, P.; Mège, R.-M.; et al. The role of single-cell mechanical behaviour and polarity in driving collective cell migration. *Nat. Phys.* **2020**. [CrossRef]
38. Yonemura, S.; Wada, Y.; Watanabe, T.; Nagafuchi, A.; Shibata, M. α-Catenin as a tension transducer that induces adherens junction development. *Nat. Cell Biol.* **2010**, *12*, 533–542. [CrossRef]
39. le Duc, Q.; Shi, Q.; Blonk, I.; Sonnenberg, A.; Wang, N.; Leckband, D.; de Rooij, J. Vinculin potentiates E-cadherin mechanosensing and is recruited to actin-anchored sites within adherens junctions in a myosin II–dependent manner. *J. Cell Biol.* **2010**, *189*, 1107–1115. [CrossRef]
40. Buckley, C.D.; Tan, J.; Anderson, K.L.; Hanein, D.; Volkmann, N.; Weis, W.I.; Nelson, W.J.; Dunn, A.R. The minimal cadherin-catenin complex binds to actin filaments under force. *Science* **2014**, *346*, 1254211. [CrossRef]
41. Yao, M.; Qiu, W.; Liu, R.; Efremov, A.K.; Cong, P.; Seddiki, R.; Payre, M.; Lim, C.T.; Ladoux, B.; Mège, R.-M.; et al. Force-dependent conformational switch of α-catenin controls vinculin binding. *Nat. Commun.* **2014**, *5*, 4525. [CrossRef] [PubMed]
42. Acharya, B.R.; Nestor-Bergmann, A.; Liang, X.; Gupta, S.; Duszyc, K.; Gauquelin, E.; Gomez, G.A.; Budnar, S.; Marcq, P.; Jensen, O.E.; et al. A Mechanosensitive RhoA Pathway that Protects Epithelia against Acute Tensile Stress. *Dev. Cell* **2018**, *47*, 439–452.e6. [CrossRef] [PubMed]
43. Hogan, C.; Kajita, M.; Lawrenson, K.; Fujita, Y. Interactions between normal and transformed epithelial cells: Their contributions to tumourigenesis. *Int. J. Biochem. Cell Biol.* **2011**, *43*, 496–503. [CrossRef] [PubMed]
44. Gu, Y.; Shea, J.; Slattum, G.; Firpo, M.A.; Alexander, M.; Mulvihill, S.J.; Golubovskaya, V.M.; Rosenblatt, J. Defective apical extrusion signaling contributes to aggressive tumor hallmarks. *eLife* **2015**, *4*, e04069. [CrossRef] [PubMed]
45. Kon, S.; Ishibashi, K.; Katoh, H.; Kitamoto, S.; Shirai, T.; Tanaka, S.; Kajita, M.; Ishikawa, S.; Yamauchi, H.; Yako, Y.; et al. Cell competition with normal epithelial cells promotes apical extrusion of transformed cells through metabolic changes. *Nat. Cell Biol.* **2017**, *19*, 530–541. [CrossRef]
46. Leung, C.T.; Brugge, J.S. Outgrowth of single oncogene-expressing cells from suppressive epithelial environments. *Nature* **2012**, *482*, 410–413. [CrossRef]
47. Hendley, A.M.; Wang, Y.J.; Polireddy, K.; Alsina, J.; Ahmed, I.; Lafaro, K.J.; Zhang, H.; Roy, N.; Savidge, S.G.; Cao, Y.; et al. p120 Catenin Suppresses Basal Epithelial Cell Extrusion in Invasive Pancreatic Neoplasia. *Cancer Res.* **2016**, *76*, 3351. [CrossRef]

48. Slattum, G.; Gu, Y.; Sabbadini, R.; Rosenblatt, J. Autophagy in Oncogenic K-Ras Promotes Basal Extrusion of Epithelial Cells by Degrading S1P. *Curr. Biol.* **2014**, *24*, 19–28. [CrossRef]
49. Villeneuve, C.; Lagoutte, E.; de Plater, L.; Mathieu, S.; Manneville, J.-B.; Maître, J.-L.; Chavrier, P.; Rossé, C. aPKCi triggers basal extrusion of luminal mammary epithelial cells by tuning contractility and vinculin localization at cell junctions. *Proc. Natl. Acad. Sci. USA* **2019**, *116*, 24108. [CrossRef]
50. Kadeer, A.; Maruyama, T.; Kajita, M.; Morita, T.; Sasaki, A.; Ohoka, A.; Ishikawa, S.; Ikegawa, M.; Shimada, T.; Fujita, Y. Plectin is a novel regulator for apical extrusion of RasV12-transformed cells. *Sci Rep.* **2017**, *7*, 44328. [CrossRef]
51. Kasai, N.; Kadeer, A.; Kajita, M.; Saitoh, S.; Ishikawa, S.; Maruyama, T.; Fujita, Y. The paxillin-plectin-EPLIN complex promotes apical elimination of RasV12-transformed cells by modulating HDAC6-regulated tubulin acetylation. *Sci Rep.* **2018**, *8*, 2097. [CrossRef] [PubMed]
52. Ohoka, A.; Kajita, M.; Ikenouchi, J.; Yako, Y.; Kitamoto, S.; Kon, S.; Ikegawa, M.; Shimada, T.; Ishikawa, S.; Fujita, Y. EPLIN is a crucial regulator for extrusion of RasV12-transformed cells. *J. Cell Sci.* **2015**, *128*, 781. [CrossRef] [PubMed]
53. Wu, S.K.; Gomez, G.A.; Michael, M.; Verma, S.; Cox, H.L.; Lefevre, J.G.; Parton, R.G.; Hamilton, N.A.; Neufeld, Z.; Yap, A.S. Cortical F-actin stabilization generates apical–lateral patterns of junctional contractility that integrate cells into epithelia. *Nat. Cell Biol.* **2014**, *16*, 167–178. [CrossRef] [PubMed]
54. Kajita, M.; Hogan, C.; Harris, A.R.; Dupre-Crochet, S.; Itasaki, N.; Kawakami, K.; Charras, G.; Tada, M.; Fujita, Y. Interaction with surrounding normal epithelial cells influences signalling pathways and behaviour of Src-transformed cells. *J. Cell Sci.* **2010**, *123*, 171. [CrossRef] [PubMed]
55. Grieve, A.G.; Rabouille, C. Extracellular cleavage of E-cadherin promotes epithelial cell extrusion. *J. Cell Sci* **2014**, *127*, 3331–3346. [CrossRef] [PubMed]
56. Chiba, T.; Ishihara, E.; Miyamura, N.; Narumi, R.; Kajita, M.; Fujita, Y.; Suzuki, A.; Ogawa, Y.; Nishina, H. MDCK cells expressing constitutively active Yes-associated protein (YAP) undergo apical extrusion depending on neighboring cell status. *Sci Rep.* **2016**, *6*, 28383. [CrossRef]
57. Saitoh, S.; Maruyama, T.; Yako, Y.; Kajita, M.; Fujioka, Y.; Ohba, Y.; Kasai, N.; Sugama, N.; Kon, S.; Ishikawa, S.; et al. Rab5-regulated endocytosis plays a crucial role in apical extrusion of transformed cells. *Proc. Natl. Acad. Sci. USA* **2017**, *114*, E2327. [CrossRef]
58. Sasaki, A.; Nagatake, T.; Egami, R.; Gu, G.; Takigawa, I.; Ikeda, W.; Nakatani, T.; Kunisawa, J.; Fujita, Y. Obesity Suppresses Cell-Competition-Mediated Apical Elimination of RasV12-Transformed Cells from Epithelial Tissues. *Cell Rep.* **2018**, *23*, 974–982. [CrossRef]
59. Kajita, M.; Sugimura, K.; Ohoka, A.; Burden, J.; Suganuma, H.; Ikegawa, M.; Shimada, T.; Kitamura, T.; Shindoh, M.; Ishikawa, S.; et al. Filamin acts as a key regulator in epithelial defence against transformed cells. *Nat. Commun.* **2014**, *5*, 4428. [CrossRef]
60. Yako, Y.; Hayashi, T.; Takeuchi, Y.; Ishibashi, K.; Kasai, N.; Sato, N.; Kuromiya, K.; Ishikawa, S.; Fujita, Y. ADAM-like Decysin-1 (ADAMDEC1) is a positive regulator of Epithelial Defense Against Cancer (EDAC) that promotes apical extrusion of RasV12-transformed cells. *Sci. Rep.* **2018**, *8*, 9639. [CrossRef]
61. Yamamoto, S.; Yako, Y.; Fujioka, Y.; Kajita, M.; Kameyama, T.; Kon, S.; Ishikawa, S.; Ohba, Y.; Ohno, Y.; Kihara, A.; et al. A role of the sphingosine-1-phosphate (S1P)–S1P receptor 2 pathway in epithelial defense against cancer (EDAC). *Mol. Biol. Cell* **2015**, *27*, 491–499. [CrossRef] [PubMed]
62. Deakin, N.O.; Turner, C.E. Paxillin inhibits HDAC6 to regulate microtubule acetylation, Golgi structure, and polarized migration. *J. Cell Biol.* **2014**, *206*, 395–413. [CrossRef] [PubMed]
63. Stossel, T.P.; Condeelis, J.; Cooley, L.; Hartwig, J.H.; Noegel, A.; Schleicher, M.; Shapiro, S.S. Filamins as integrators of cell mechanics and signalling. *Nat. Rev. Mol. Cell Biol.* **2001**, *2*, 138–145. [CrossRef] [PubMed]
64. Janmey, P.A.; Euteneuer, U.; Traub, P.; Schliwa, M. Viscoelastic properties of vimentin compared with other filamentous biopolymer networks. *J. Cell Biol.* **1991**, *113*, 155–160. [CrossRef]
65. Hatzfeld, M.; Keil, R.; Magin, T.M. Desmosomes and Intermediate Filaments: Their Consequences for Tissue Mechanics. *Cold Spring Harb Perspect. Biol* **2017**, *9*, a029157. [CrossRef]
66. Hill, W.; Hogan, C. Normal epithelial cells trigger EphA2-dependent RasV12 cell repulsion at the single cell level. *Small Gtpases* **2019**, *10*, 305–310. [CrossRef]

67. Porazinski, S.; de Navascués, J.; Yako, Y.; Hill, W.; Jones, M.R.; Maddison, R.; Fujita, Y.; Hogan, C. EphA2 Drives the Segregation of Ras-Transformed Epithelial Cells from Normal Neighbors. *Curr. Biol.* **2016**, *26*, 3220–3229. [CrossRef]
68. Teo, J.L.; Gomez, G.A.; Noordstra, I.; Verma, S.; Tomatis, V.; Acharya, B.R.; Balasubramaniam, L.; Katsuno-Kambe, H.; Templin, R.; McMahon, K.-A.; et al. Caveolae set levels of epithelial monolayer tension to eliminate tumor cells. *bioRxiv* **2019**, 632802. [CrossRef]

Article

Symmetry Breaking and Emergence of Directional Flows in Minimal Actomyosin Cortices

Sven K. Vogel [1], Christian Wölfer [2], Diego A. Ramirez-Diaz [1,3], Robert J. Flassig [2,4], Kai Sundmacher [2,5] and Petra Schwille [1,*]

1 Max Planck Institute of Biochemistry, Am Klopferspitz 18, D-82152 Martinsried, Germany; svogel@biochem.mpg.de (S.K.V.); daramirez@g.harvard.edu (D.A.R.-D.)
2 Max Planck Institute for Dynamics of Complex Technical Systems, Sandtorstr. 1, D-39106 Magdeburg, Germany; woelfer@mpi-magdeburg.mpg.de (C.W.); flassig@th-brandenburg.de (R.J.F.); sundmacher@mpi-magdeburg.mpg.de (K.S.)
3 Graduate School of Quantitative Biosciences, Ludwig-Maximilians-Universität, Feodor-Lynen-Str. 25, D-81377 Munich, Germany
4 Department of Engineering, Brandenburg University of Applied Sciences, Magdeburger Str. 50, D-14770 Brandenburg, Germany
5 Institute for Process Engineering, Otto von Guericke University Magdeburg, Universitätsplatz 2, D-39106 Magdeburg, Germany
* Correspondence: schwille@biochem.mpg.de

Received: 31 March 2020; Accepted: 31 May 2020; Published: 9 June 2020

Abstract: Cortical actomyosin flows, among other mechanisms, scale up spontaneous symmetry breaking and thus play pivotal roles in cell differentiation, division, and motility. According to many model systems, myosin motor-induced local contractions of initially isotropic actomyosin cortices are nucleation points for generating cortical flows. However, the positive feedback mechanisms by which spontaneous contractions can be amplified towards large-scale directed flows remain mostly speculative. To investigate such a process on spherical surfaces, we reconstituted and confined initially isotropic minimal actomyosin cortices to the interfaces of emulsion droplets. The presence of ATP leads to myosin-induced local contractions that self-organize and amplify into directed large-scale actomyosin flows. By combining our experiments with theory, we found that the feedback mechanism leading to a coordinated directional motion of actomyosin clusters can be described as asymmetric cluster vibrations, caused by intrinsic non-isotropic ATP consumption with spatial confinement. We identified fingerprints of vibrational states as the basis of directed motions by tracking individual actomyosin clusters. These vibrations may represent a generic key driver of directed actomyosin flows under spatial confinement in vitro and in living systems.

Keywords: bottom-up synthetic biology; motor proteins; pattern formation; self-organization

1. Introduction

In animal cells, cortical actomyosin motions including actomyosin flows have been proposed to drive cell locomotion, cytokinesis, left-right symmetry breaking during embryonic development of multicellular organisms, and cellular or tissue chirality [1–7]. Despite the omnipresent functions and implications of cortical actomyosin motions and flows, the exact molecular origins and fundamental determinants of these phenomena, in particular how local symmetry breaking within apparently isotropic actin cortices scales up into these flows, are far from being understood. In many eukaryotic model systems, myosin motors acting on actin filaments are proposed to play a key role in driving actomyosin dynamics in cytokinetic rings and cortical actomyosin flows [1,5,8–10]. Distinct manipulation of myosin and actin independent of other cellular processes is challenging as they both are functionally highly integrated cellular proteins. To elucidate whether directional flows may result spontaneously from active processes without being guided by the structural anisotropy of the cellular architecture, it is mandatory to explore these phenomena in well-controlled reconstituted systems [11,12]. Recent studies used an approach of encapsulating frog egg cell extracts in droplets, so manipulation of the protein players may be easier when compared with living systems [13,14]. Nevertheless, due to the compositional complexity of cell extracts is comparable to living model systems, it may be difficult to pinpoint the minimal set of necessary proteins to produce a certain phenomenon. High density regimes of nematic (filament alignment in parallel lines) cytoskeletal in vitro systems showed pattern formation and dynamics in a collective manner based on motor driven sliding mechanisms between long cytoskeletal filaments [15,16]. In contrast, actomyosin cortices and rings in living systems consist of non-polar and disordered actin filament networks and are coupled to the cell membrane [17,18]. By a combination of bottom-up in vitro experiments and theory, we identified a generic mechanism of how large-scale directional flow-like actomyosin motions spanning several hundred micrometers are generated by myosin motors in a disordered, membrane coupled, and isotropic cytoskeletal system in a confined spherical environment.

2. Results

To investigate the emergence of coordinated large-scale flows from spontaneous contractions in quasi-isotropic cortices, we made use of recently developed minimal actin cortices (MACs) [19,20] and confined them in water-in-oil droplets using microfluidic emulsification on Polydimethylsiloxan (PDMS) chips. The chip design is shown in Figure 1A. (Note that we also performed emulsification simply by manually mixing the components, in order to rule out any suspicion of flows caused by the pneumatic microfluidic setup.) 1,2-di-(9Z-octadecenoyl)-*sn*-glycero-3-phosphocholine (DOPC) and biotinylated lipids were dissolved in mineral oil, and for droplet formation we used a pneumatic microfluidic system where the flow rate of each channel is individually controlled (Movie S1). The aqueous solution contained dissolved actin monomers, neutravidin, and myosin in a salt buffer system. To test the proper formation of a lipid monolayer that includes biotinylated lipids (DSPE-PEG-2000-Biotin), we started with the encapsulation of fluorescently labeled neutravidin (Oregon Green 488 Neutravidin) (Figure 1B–D). By analyzing the fluorescence intensity distribution throughout the droplets, we showed that neutravidin only binds to the lipid monolayer when the lipid oil mixture also contains biotinylated lipids (Figure 1C,D). This indicates that neutravidin does not bind non-specifically to the water/oil interface. Co-encapsulation of myosin motors with the actin monomer and anchor system in the presence of ATP resulted in the formation of a MAC. Inside the droplets, actin monomers started to polymerize and bind to the lipid monolayer that contained biotinylated lipids, thus forming an initially isotropic, i.e., spatially symmetric actomyosin cortex (Figure 1B,E, Movie S2). Concurrently, ATP-induced contraction of the myosin motors resulted in the symmetry breakage of the isotropic actin carpet and in the formation of actomyosin clusters (Figure 2A–C, Movie S3). Strikingly, within minutes, a large-scale directional movement emerged from these spherically confined actomyosin clusters, which we refer to as cortical actomyosin motion (CAM) (Figures 1E and 2B; Movies S3 and S5). Please note that in the absence of ATP or in the presence of a motor inhibitor no CAMs occurred (data not shown). The CAMs could last more than one hour and cover distances of several hundred micrometers before eventually ceasing, reaching a state with only minimal movements (Movies S3 and S5). Interestingly, these flow-like motions somewhat resemble cortical actomyosin flow-like motions in vivo, e.g., in the *C. elegans* embryo. We characterized the velocity of the actomyosin cluster movements by using particle image velocimetry [21] (Figure 2C; Movie S4). In order to test parameters that change the velocity of CAMs, we co-encapsulated a crowding agent (methylcellulose) at various concentrations and found a respective increase of the velocities in the presence of methylcellulose (Figure 2D, compare Movies S3 and S5). By increasing the effective concentration of actin and myosin at the lipid monolayer membrane interface, the overall cluster velocity was increased more than five times.

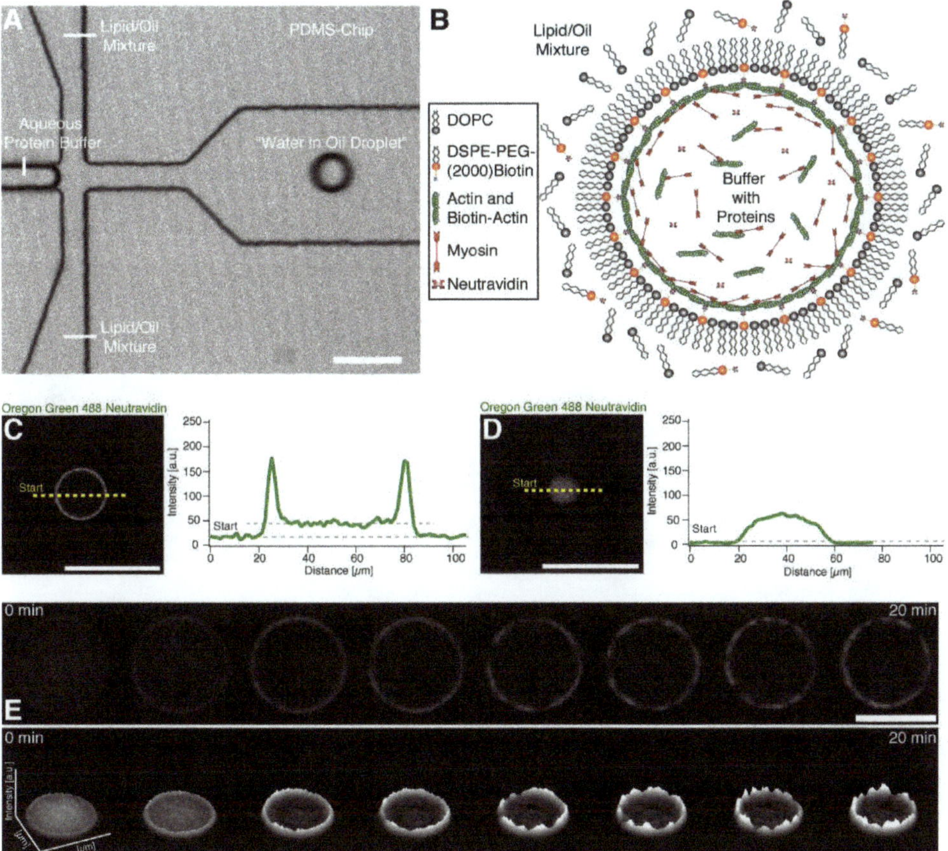

Figure 1. Encapsulation and actin cortex formation inside water in oil droplets. (**A**) Confocal image of a PDMS chip where the encapsulation of the buffer-protein system (see (**B**)) and formation of the water in oil droplet is shown (Movie S1). (B) Illustration depicting the formation of an actomyosin cortex. Neutravidin with its four binding sites may bind several actin filaments and biotinylated lipids. (**C**) Confocal image of the equatorial plane of a droplet with a lipid monolayer containing biotinylated lipids showing that encapsulated Oregon green labeled neutravidin binds to the lipid monolayer. Line profile of the fluorescence signal of the Oregon green labeled neutravidin shows two peaks which indicate binding of the neutravidin to the lipid monolayer interface of the droplet (right). (**D**) In contrast, Oregon green labeled neutravidin does not bind to the lipid monolayer and is distributed throughout the lumen of the droplet in the absence of biotinylated lipids. The fluorescence line profile shows no peaks in the absence of biotinylated lipids indicating the absence of unspecific binding to the lipid monolayer (right). (**E**) Confocal time-lapse images of encapsulated Alexa-488 labeled actin and myosin motors in the presence of ATP at the droplet equator plane. The formation of an actin cortex and actomyosin clusters is shown (upper row) (Movie S2). The respective fluorescent intensity profile indicates the formation of actomyosin clusters and shows their dynamics (lower row). Scale bars, 100 µm.

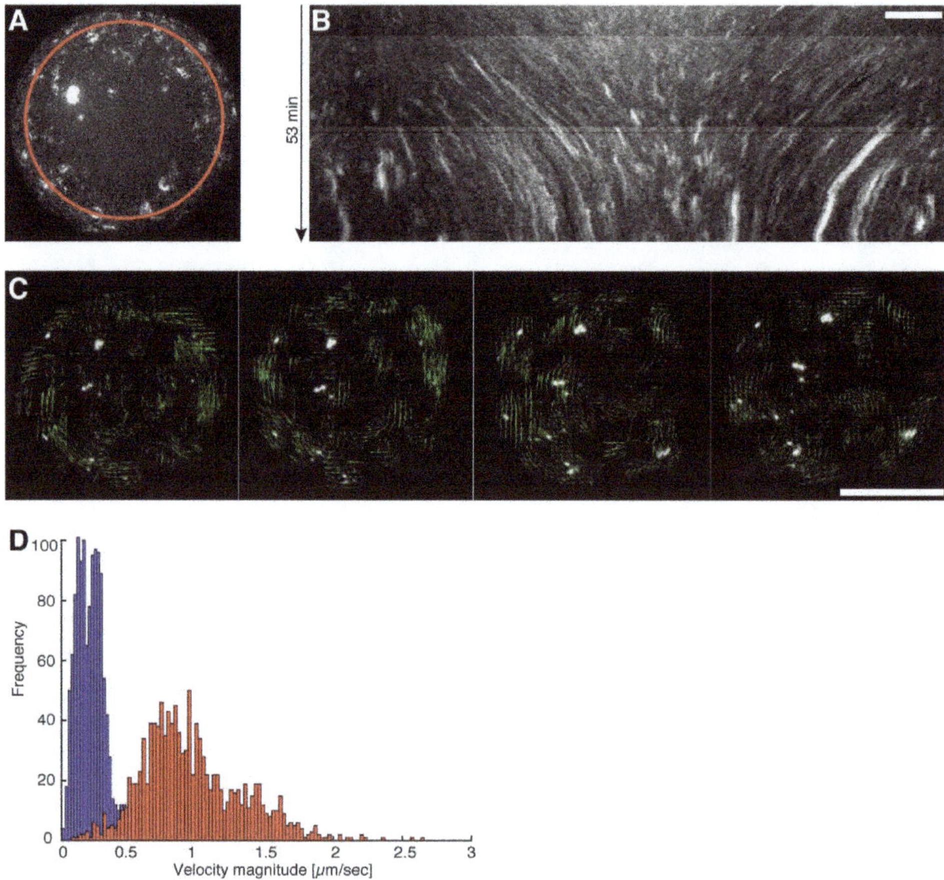

Figure 2. Directed movement of actomyosin clusters upon ATP dependent actomyosin contractions. (**A**) Maximum intensity projection from a half droplet confocal z-stack where Alexa-488 labeled actin clusters are visible (Movie S3). The red circle indicates the path of the generated kymograph. (**B**) A kymograph of the maximum intensity projected half sphere is shown where directed movements of individual clusters represented by distinct lines are visible. Scale bar, 10 μm. (**C**) Confocal time-lapse image sequence used for particle image velocimetry (PIV) by (*21*). Vectors (green arrows) indicate the flow direction of the directed movement of the actomyosin clusters (Movie S4). Scale bar, 50 μm. (**D**) Velocity profiles measured by PIV of droplets with (red) and without a crowding agent (blue) are shown (Movies S3 and S5).

Theoretical Model of the Cortical Actomyosin Motions

We then aimed to elucidate the mechanism behind myosin driven CAMs and complemented our experiments with a theoretical model based on biophysical first principles, i.e., without assuming an explicit mechanism *a priori*. We modeled the MAC droplet as a continuous, isotropic fluid. The interaction between actin and myosin was modeled by a simplified myosin cross-bridge cycle [22–24] where the force generating conformational change (r_1) of the myosin head (M) occurs immediately after the binding of filamentous actin (A) (Figure 3A). The actomyosin complex dissociates after binding and ATP hydrolysis reloads the myosin head into the active state (r_2). Typically, considered intermediate species only have a very short lifetime [25] and are therefore neglected. As the MACs were formed by polymerization in the absence of actin-regulating proteins [26–28], a rudimentary polymerization cycle [25] was added to the model. F-actin depolymerizes (r_3) into monomeric G-actin with bound ADP (G_D), followed by a spontaneous nucleotide exchange (r_4). The resulting ATP bound G-actin (G_T) binds stronger to F-actin (r_5) than G_D-actin (r_6), and therefore G_T-actin polymerization is dominant [29]. Both reaction systems are necessary to reasonably describe the local ATP consumption by polymerization and contraction. F-actin, cross-linked by myosin II molecules, forms a mesh-like cortex, causing an active viscoelastic material behavior with a rheological property combination of the Maxwell and Kelvin-Voigt models [30] considered in the momentum equation [31] (Supplemental Equation (S14)). To model force generation according to the myosin cross-bridge sub-model, the active stress was developed from a stress term of our previous study [32] considering the observed medium ATP-dependency of myosin pulls in MACs [20]. The distributed reaction network is described by a system of partial differential equations (PDEs) assuming diffusion (G_D, G_T, ATP) or combined convection and diffusion flux (A, A_M, M) (Supplemental Equations (S1)–(S6)). We reduced our analysis to a one-dimensional ring topology, modeling a section of the spherical droplet to improve and simplify the interpretation of the simulation results. To adopt experimental conditions of spontaneous symmetry breaking, the PDE system was simulated with an initially slightly inhomogeneous myosin distribution and revealed merging of small clusters into a gradually bigger main cluster, similar to experimental observations (Figure 3B,C) for the used parameter space, which is largely based on literature values (Table S2). The final non-symmetric cluster starts to move directionally in the one-dimensional ring system like a propagating wave (Figure 3B,C). Hereinafter, two feedback characteristics are described which are presumably essential for this movement. First, a non-symmetric contraction profile is preserved by an unbalanced consumption of ATP at the leading and trailing edge of the propagating cluster. Because of the locomotion and the ongoing network depolymerization, the cluster traces non-convective G_D-actin like a comet tail (Figure S1). In the G_D-actin tail, ATP is additionally consumed for the nucleotide exchange of monomeric actin (r_4). Thus, less ATP is available for the force-generating myosin cross-bridge cycle at the trailing edge of the wave and hence, less contractile stress is generated compared to the leading edge (Figure 3D). This local asymmetry is maintained despite consideration of the high diffusivity of ATP in an aqueous medium (Table S2). A reduction of the F-actin depolymerization leads to the loss of the G_D-actin tail the termination of the cluster locomotion (Figure S7). Additionally, locally reduced ATP distributions induce a change of locomotion direction as soon as the cluster reaches a reduced region. Here, more ATP is available at the trailing edge than at the leading edge and the asymmetry is reversed (Figure S8). Second, examining the fine structure of the simulated moving cluster reveals a vibration. It is driven by periodic depletion of the local ATP concentration, leading to oscillating active and passive mechanical stresses inside the cluster (Figure S3A). The cluster vibration passes through the following repetitive phases (Figure 3F):

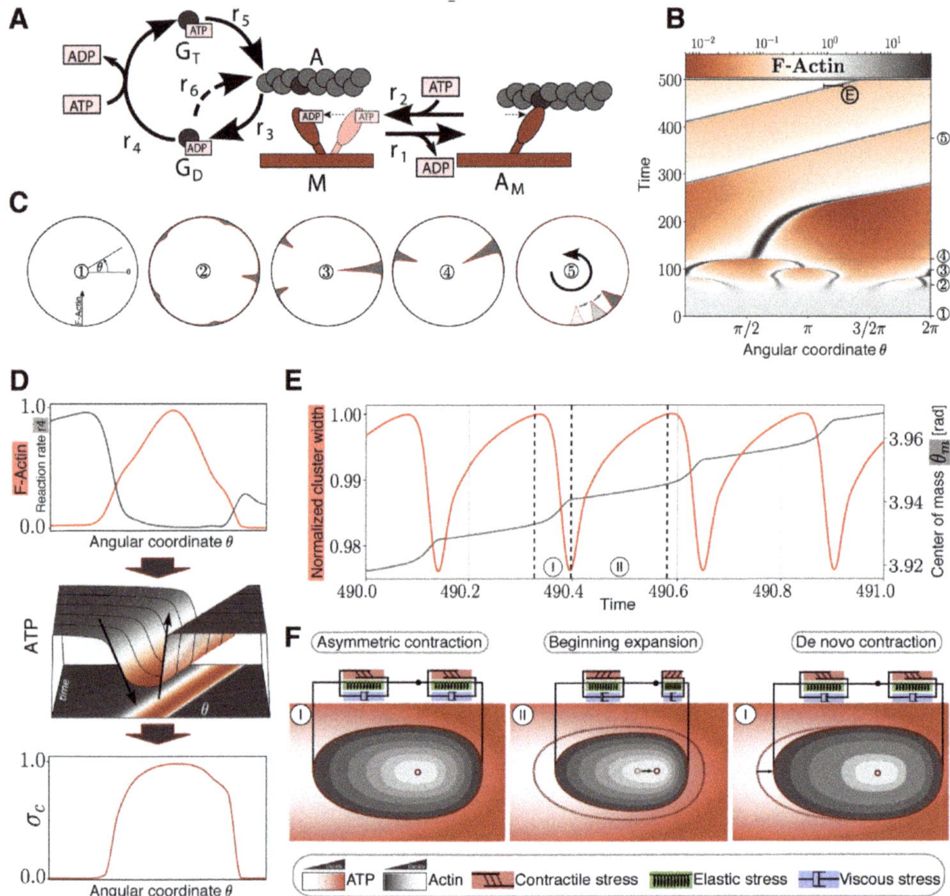

Figure 3. Modeling and simulation of actomyosin cluster motions reveal a propagation mechanism through active matter vibrations. (**A**) Kinetic reaction network with F-actin polymerization cycle and simplified myosin cross-bridge model. (**B**) Temporal development of a one-dimensional spatial F-actin distribution with color-coded local concentration. (**C**) Circular representation of distributed F-actin concentration according to the model topology for selected time points (marked in (B) with circles). (**D**) Non-symmetric contraction of a propagating cluster. Upper graph: Normalized distribution of reaction rate 4 (gray) compared to normalized F-actin cluster location (red). Middle graph: Non-symmetric ATP distribution around the cluster. Lower graph: Normalized asymmetric contractile stress pattern with color code representing ATP concentration according to the projection. (**E**) Vibration of an F-actin cluster. Evolution of normalized cluster width (red) and displacement of center of mass θ_m (gray) of F-actin in radiant. Cluster boundaries are defined as the points where the F-actin concentration exceeds the mean concentration. (**F**) Sketch of the cortical actin cluster migration mechanism with qualitative color-coded Actin and ATP gradients shows repetitive asymmetric contractions (I) and an expansion phase (II) resulting in a shift of the center of mass according to (D) indicated by a black arrow (middle panel) and displacement of the cluster indicated by an arrow (right panel).

I. Contraction Phase: In the beginning of the contraction phase, a high ATP concentration inside the cluster results in an increase of the contractile stress accompanied by an increase of elastic stress. With progressing cluster compaction, the actomyosin species gets locally accumulated, causing stronger contractile stress and increased local ATP consumption.

II. Expansion Phase: Increasing local ATP depletion and network strain leads to a local dominance of elastic stress resulting in an expansion of the cluster. Because of the reduced network densities inside of the cluster and therefore reduced energy consumption, ATP flows in diffusively and increases the local ATP level, followed by a de novo contraction phase.

In addition, the asymmetric contraction pattern causes an accelerated displacement of the center of mass during the contraction phase (Figure 3E,F). Owing to viscoelastic material properties and asymmetric creep, the center of mass is not pushed back to the initial position during the expansion phase, a result of the asymmetric contraction pattern. To investigate the wave propagation mechanism, we assumed a constant diffusive ATP supply. Since the active stress is ATP-driven, the wave propagation ceases, owing to the global ATP depletion (Figure S2). Hence, cluster vibrations due to local ATP depletion and asymmetric cluster contractions are the main drivers of single cluster motions. Parameter studies show that these motions occur in a large parameter space, encompassing several orders of magnitude, and are therefore robust for the assumed parameters (Figure S9). In contrast to previous theoretical studies [33,34], where the required asymmetry of a self-propagating cluster is caused by polar actin bundles, our suggested model is also able to explain cluster migrations of disordered actin filaments that can be expected in an isotropic cortex.

To find experimental evidence for the theoretically predicted vibrations of the actomyosin clusters, we automatically tracked their directional movement (Figure 4A). In addition, by analyzing the directed movement of the individual actomyosin clusters, we noticed that rotation of clusters around their centers of mass correlated with a change in direction displacement (Figure 4A,B; Movies S6 and S7). This can be explained with torque produced by an imbalance of forces perpendicular to the translational trajectory. On the other hand, the model predicts vibrations with a certain oscillation period (Figure 3E). Evidence for such vibrations in our experimental conditions can be found independently of the sampling rate. For example, a sinusoidal signal with a specific frequency can be measured at under-sampling conditions (Figure S5). The reconstructed signals still conserve the periodic behavior reflected by the Fourier analysis, yet not maintaining the original frequency (Figure S5). To find evidence for vibrations inside the clusters as predicted by our theory, we analyzed image sequences to measure fluctuations in the distance between the geometrical and the intensity-weighted center of mass of several clusters. The geometrical center of mass was computed by regular segmentation and binarization defining the geometry of the cluster. In contrast, the intensity-weighted center of mass was calculated over the same geometry but having the camera-intensity as the statistical weight of every pixel. Since clusters have various geometries, we defined a characteristic length for the clusters as the square root of the area. This quantity allowed us to measure changes between the geometrical and intensity centers of mass with respect to their characteristic length. As a result, we were able to combine fluctuations of myosin clusters with different sizes and experimental conditions (Figure S6). Fourier analysis of these combined fluctuations strikingly indicates the existence of vibrational states (Figure 4D, red, blue and yellow curves) which are clearly distinguishable from static conditions (Figure 4D, grey curve) or acquisition and camera artifacts by measuring the background between moving clusters (data not shown). We interpret these vibrational states as fingerprints of the theoretically predicted vibrations that provide a mechanism for the directed and rotational large-scale movements of the actomyosin clusters in silico and in vitro. Under crowding conditions where CAMs show the highest velocities, we found higher amplitudes for the same observed frequencies in comparison to the other tested conditions (Figure 4D). This finding implies that larger amplitudes correlate with higher velocities as expected for vibrations-based movements. Since the crowder increases the effective concentration of actin and myosin at the lipid monolayer membrane interface, we conclude that higher actomyosin concentrations may lead to higher cluster velocities and therefore to larger amplitudes of the vibrations.

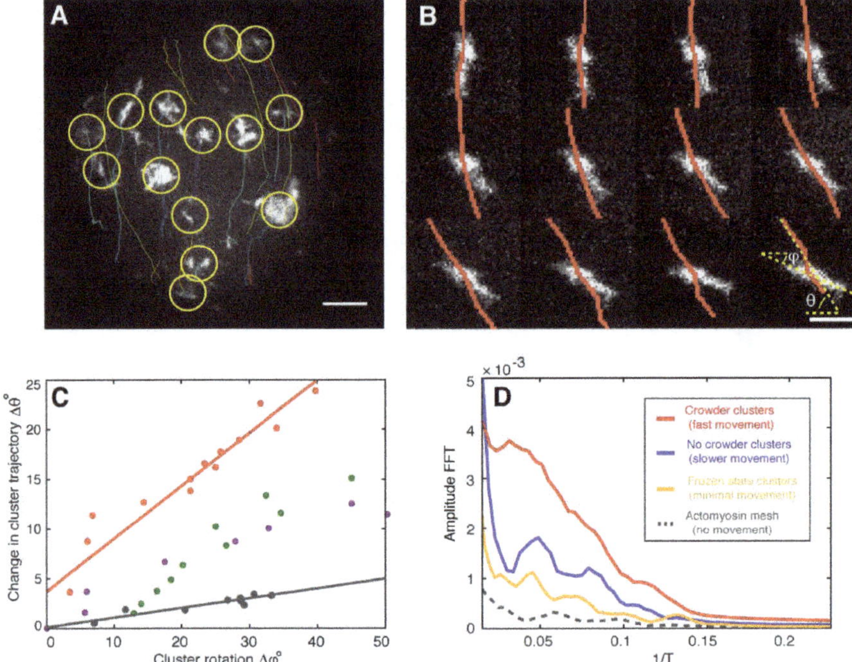

Figure 4. Fingerprints of actomyosin cluster vibrations during their directed movements. (**A**) A maximum intensity projection of confocal images of a half sphere during cortical actomyosin cluster movements is shown. Tracked actin clusters are marked by circles (yellow) and their trajectories shown (various colors, Movie S6). (**B**) Changes in the angle trajectory (red lines, θ) agree with cluster rotations along its center of mass (ϕ) (left, Movie S7). Scale bar, 4 µm. (**C**) Analysis of four independent clusters (various colors) suggest that rotation influences cluster steering and then a change in trajectory (Δθ). The red data points correspond to the montage shown in (B). (**D**) The fast Fourier transformation (FFT) indicates the presence of vibrational states for moving actomyosin clusters (red, blue, and yellow) different from static systems and acquisition artifacts (grey). The higher amplitude for the crowder condition with the highest cluster velocities of all measured systems implies that larger amplitudes correlate with higher velocities. Scale bar, 10 µm.

3. Discussion

Based on our suggested mechanism for the directional motion of an individual cluster, we consider the possibility of an alignment of motion of a group of clusters in one direction to result from potential cluster interactions via the cortical actin network, thereby explaining the observed flow-like group behavior of the actomyosin clusters. Alternatively, a macroscopically homogeneous quasi-symmetric actin carpet is initially formed, as shown in our experiments (see Figure 1E, Movies S2 and S3). Since the resulting clusters are initially connected via the isotropic actin mesh, the random asymmetry of the largest cluster would determine the direction of movement of the entire cluster group, in the absence of coordinating processes. Each cluster of the group is thus impressed with a direction of movement or an asymmetric gradient profile originating from the respective monomer tail. In conclusion, our theory suggests that the direct translational and rotational movements of actomyosin clusters originate from an imbalance of oscillatory contractile stresses within the individual actomyosin clusters. Fourier analysis of our experimental data indicates the existence of vibrational states that drive the directional movements of individual actomyosin clusters and the formation of flow-like CAMs in the spherical confinement of the active matter droplets. Hence, for future work it will be of great interest to investigate whether these vibrational states can be also identified in cellular model systems that exhibit myosin driven actomyosin motions, e.g., in cytokinetic actomyosin rings or in cortical actomyosin flows.

4. Methods

Additional Information and details of the model as well as all methods and experimental procedures and movie legends can be found in the Supporting Information Appendix.

Supplementary Materials: The following are available online at http://www.mdpi.com/2073-4409/9/6/1432/s1, Supporting Information (word document), Movie S1: Encapsulation and water in oil droplet formation, Movie S2: Actin cortex formation and myosin contraction, Movie S3: Directional actomyosin flows, Movie S4: Particle Image Velocimetry (PIV) of moving actomyosin clusters, Movie S5: Enhanced directional actomyosin flows through crowding, Movie S6: Tracking of individual actomyosin clusters for determining fingerprints of cluster vibrations, Movie S7: Rotation of actomyosin clusters.

Author Contributions: Conceptualization experiments: S.K.V. and P.S.; Conceptualization theory: C.W. and R.J.F. Methodology experiments: S.K.V., D.A.R.-D.; Methodology theory: C.W. and R.J.F.; Software experiments: S.K.V and D.A.R.-D.; Software theory: C.W. and R.F.F.; Validation: S.K.V., C.W., D.A.R.-D. and R.J.F.; Formal analysis experiments: S.K.V., D.A.R.-D.; Formal analysis theory: C.W. and R.J.F.; Investigation experiments: S.K.V.; Investigation theory: C.W.; Resources, S.K.V.; Writing—original draft preparation: S.K.V., C.W., R.J.F., D.A.R.-D.; writing—review and editing, P.S. and K.S.; Visualization: S.K.V. and C.W.; Funding acquisition: P.S. and K.S. All authors have read and agreed to the published version of the manuscript.

Funding: This research was funded by MaxSynBio, a joint research network sponsored by the German Federal Ministry of Education and Research (BMBF) and the MaxPlanck Society (MPG).

Acknowledgments: We thank Chase Broedersz for discussions and critical reading of the manuscript. We are grateful for the financial support by the MaxSynBio consortium jointly funded by the Federal Ministry of Education and Research of Germany and the Max Planck Society, the Daimler und Benz foundation (Project Grant PSBioc 8216), the Gottfried Wilhelm Leibniz-Program of the DFG (SCHW 716/8-1) and the support of the Graduate School of Quantitative Biosciences Munich.

Conflicts of Interest: The authors declare no conflict of interest.

References

1. Bray, D.; White, J.G. Cortical flow in animal cells. *Science* **1988**, *239*, 883–888. [CrossRef]
2. Sedzinski, J.; Biro, M.; Oswald, A.; Tinevez, J.Y.; Salbreux, G.; Paluch, E. Polar actomyosin contractility destabilizes the position of the cytokinetic furrow. *Nature* **2011**, *476*, 462–466. [CrossRef]
3. Khaliullin, R.N.; Green, R.A.; Shi, L.Z.; Gomez-Cavazos, J.S.; Berns, M.W.; Desai, A.; Oegema, K. A positive-feedback-based mechanism for constriction rate acceleration during cytokinesis in Caenorhabditis elegans. *Elife* **2018**, *7*, e36073. [CrossRef] [PubMed]
4. Naganathan, S.R.; Middelkoop, T.C.; Furthauer, S.; Grill, S.W. Actomyosin-driven left-right asymmetry: From molecular torques to chiral self organization. *Curr. Opin. Cell Biol.* **2016**, *38*, 24–30. [CrossRef] [PubMed]
5. Wollrab, V.; Thiagarajan, R.; Wald, A.; Kruse, K.; Riveline, D. Still and rotating myosin clusters determine cytokinetic ring constriction. *Nat. Commun.* **2016**, *7*, 11860. [CrossRef] [PubMed]
6. Callan-Jones, A.C.; Voituriez, R. Actin flows in cell migration: From locomotion and polarity to trajectories. *Curr. Opin. Cell Biol.* **2016**, *38*, 12–17. [CrossRef]
7. Klughammer, N.; Bischof, J.; Schnellbacher, N.D.; Callegari, A.; Lenart, P.; Schwarz, U.S. Cytoplasmic flows in starfish oocytes are fully determined by cortical contractions. *PLoS Comput. Biol.* **2018**, *14*, e1006588. [CrossRef] [PubMed]
8. Munro, E.; Nance, J.; Priess, J.R. Cortical flows powered by asymmetrical contraction transport PAR proteins to establish and maintain anterior-posterior polarity in the early C. elegans embryo. *Dev. Cell* **2004**, *7*, 413–424. [CrossRef]
9. Naganathan, S.R.; Furthauer, S.; Nishikawa, M.; Julicher, F.; Grill, S.W. Active torque generation by the actomyosin cell cortex drives left-right symmetry breaking. *Elife* **2014**, *3*, e04165. [CrossRef]
10. Nishikawa, M.; Naganathan, S.R.; Julicher, F.; Grill, S.W. Controlling contractile instabilities in the actomyosin cortex. *Elife* **2017**, *6*, e19595. [CrossRef]
11. Fletcher, D. Which biological systems should be engineered? *Nature* **2018**, *563*, 177–179. [CrossRef]

12. Schwille, P. Bottom-up synthetic biology: Engineering in a tinkerer's world. *Science* **2011**, *333*, 1252–1254. [CrossRef]
13. Suzuki, K.; Miyazaki, M.; Takagi, J.; Itabashi, T.; Ishiwata, S. Spatial confinement of active microtubule networks induces large-scale rotational cytoplasmic flow. *Proc. Natl. Acad. Sci. USA* **2017**, *114*, 2922–2927. [CrossRef] [PubMed]
14. Tan, T.H.; Malik-Garbi, M.; Abu-Shah, E.; Li, J.; Sharma, A.; MacKintosh, F.C.; Keren, K.; Schmidt, C.F.; Fakhri, N. Self-organized stress patterns drive state transitions in actin cortices. *Sci. Adv.* **2018**, *4*, eaar2847. [CrossRef] [PubMed]
15. Schaller, V.; Weber, C.; Semmrich, C.; Frey, E.; Bausch, A.R. Polar patterns of driven filaments. *Nature* **2010**, *467*, 73–77. [CrossRef] [PubMed]
16. Keber, F.C.; Loiseau, E.; Sanchez, T.; DeCamp, S.J.; Giomi, L.; Bowick, M.J.; Marchetti, M.C.; Dogic, Z.; Bausch, A.R. Topology and dynamics of active nematic vesicles. *Science* **2014**, *345*, 1135–1139. [CrossRef]
17. Svitkina, T.M.; Borisy, G.G. Correlative light and electron microscopy of the cytoskeleton of cultured cells. *Methods Enzymol.* **1998**, *298*, 570–592.
18. Stachowiak, M.R.; Laplante, C.; Chin, H.F.; Guirao, B.; Karatekin, E.; Pollard, T.D.; O'Shaughnessy, B. Mechanism of cytokinetic contractile ring constriction in fission yeast. *Dev. Cell* **2014**, *29*, 547–561. [CrossRef]
19. Vogel, S.K.; Heinemann, F.; Chwastek, G.; Schwille, P. The design of MACs (minimal actin cortices). *Cytoskeleton* **2013**, *70*, 706–717. [CrossRef]
20. Vogel, S.K.; Petrasek, Z.; Heinemann, F.; Schwille, P. Myosin Motors Fragment and Compact Membrane-Bound Actin Filaments. *eLife* **2013**, *2013*, e00116. [CrossRef]
21. Thielicke, W. PIVlab—Towards User-friendly, Affordable and Accurate Digital Particle Image Velocimetry in MATLAB. *J. Open Res. Softw.* **2014**, *2*, e30. [CrossRef]
22. Huxley, H.E. The mechanism of muscular contraction. *Science* **1969**, *164*, 1356–1365. [CrossRef] [PubMed]
23. Lymn, R.W.; Taylor, E.W. Mechanism of adenosine triphosphate hydrolysis by actomyosin. *Biochemistry* **1971**, *10*, 4617–4624. [CrossRef] [PubMed]
24. Spudich, J.A. The myosin swinging cross-bridge model. *Nat. Rev. Mol. Cell Biol.* **2001**, *2*, 387–392. [CrossRef] [PubMed]
25. Howard, J. (Ed.) *Mechanics of Motor Proteins and the Cytoskeleton*; Sinauer Associates Inc.: Sunderland, MA, USA, 2001.
26. Pollard, T.D.; Borisy, G.G. Cellular motility driven by assembly and disassembly of actin filaments. *Cell* **2003**, *112*, 453–465. [CrossRef]
27. Pollard, T.D.; Cooper, J.A. Actin, a central player in cell shape and movement. *Science* **2009**, *326*, 1208–1212. [CrossRef]
28. Blanchoin, L.; Boujemaa-Paterski, R.; Sykes, C.; Plastino, J. Actin dynamics, architecture, and mechanics in cell motility. *Physiol. Rev.* **2014**, *94*, 235–263. [CrossRef]
29. Pollard, T.D.; Goldberg, I.; Schwarz, W.H. Nucleotide exchange, structure, and mechanical properties of filaments assembled from ATP-actin and ADP-actin. *J. Biol. Chem.* **1992**, *267*, 20339–20345.
30. Mogilner, A. Mathematics of cell motility: Have we got its number? *J. Math. Biol.* **2009**, *58*, 105–134. [CrossRef]
31. Lewis, O.L.; Guy, R.D.; Allard, J.F. Actin-myosin spatial patterns from a simplified isotropic viscoelastic model. *Biophys. J.* **2014**, *107*, 863–870. [CrossRef]
32. Wolfer, C.; Vogel, S.K.; Mangold, M. A curvilinear Model Approach: Actin Cortex Clustering Due to ATP-induced Myosin Pulls. *IFAC Pap.* **2016**, *49*, 103–108. [CrossRef]
33. Kruse, K.; Camalet, S.; Julicher, F. Self-propagating patterns in active filament bundles. *Phys. Rev. Lett.* **2001**, *87*, 138101. [CrossRef] [PubMed]
34. Kreten, F.H.; Hoffmann, C.; Riveline, D.; Kruse, K. Active bundles of polar and bipolar filaments. *Phys. Rev. E* **2018**, *98*, 012413. [CrossRef] [PubMed]

Review

The Roles of Signaling in Cytoskeletal Changes, Random Movement, Direction-Sensing and Polarization of Eukaryotic Cells

Yougan Cheng [1], Bryan Felix [2] and Hans G. Othmer [2,*]

[1] Bristol Myers Squibb, Route 206 & Province Line Road, Princeton, NJ 08543, USA; yougan.cheng@gmail.com
[2] School of Mathematics, University of Minnesota, Minneapolis, MN 55445, USA; felix077@umn.edu
* Correspondence: othmer@umn.edu

Received: 21 April 2020; Accepted: 29 May 2020; Published: 10 June 2020

Abstract: Movement of cells and tissues is essential at various stages during the lifetime of an organism, including morphogenesis in early development, in the immune response to pathogens, and during wound-healing and tissue regeneration. Individual cells are able to move in a variety of microenvironments (MEs) (A glossary of the acronyms used herein is given at the end) by suitably adapting both their shape and how they transmit force to the ME, but how cells translate environmental signals into the forces that shape them and enable them to move is poorly understood. While many of the networks involved in signal detection, transduction and movement have been characterized, how intracellular signals control re-building of the cyctoskeleton to enable movement is not understood. In this review we discuss recent advances in our understanding of signal transduction networks related to direction-sensing and movement, and some of the problems that remain to be solved.

Keywords: cell motility; signal transduction; actin dynamics; intracellular waves; polarization; direction sensing; symmetry-breaking; biphasic responses; reaction-diffusion; membrane and cortical tension

---

1. Introduction

Active movement of cells, either individually or collectively, is essential in early development, in the immune response, and in a variety of other processes [1]. Evolution has led to a number of different modes of active movement of single-cell organisms, ranging from crawling to swimming. Certain bacteria, such as *E. coli*, use flagella to swim, while paramecia use cilia, but each can only use one mode of movement. However, some motile eukaryotic cells are more flexible and can adopt the mode used to the environment in which they find themselves, swimming in some MEs and crawling in others [2]. In addition, whether cells move individually or collectively can depend on the density of the extracellular matrix (ECM) in their ME [3]. This plasticity or adaptability has significant implications for understanding cell motility for it implies that the focus must be on the behavior of the integrated system, not simply on its component parts.

Cells use complex signaling pathways to control the local structure of actin networks, which comprise both branched and linear filaments, and the contraction of myosin-II (myo-II) motors, which are embedded both in the cortical cytoskeleton—the cross-linked actin network that is linked to the membrane—and in the remaining intracellular network of actin filaments, microtubules and other structures. Together these networks, which we call the cytoskeleton (CSK), produce the forces that drive cell shape changes and movement, whether they are random in spacetime or spatially directed in response to signals in the environment. In order to move, the forces must be transmitted to the environment, and the ease or difficulty of movement in a given context produces feedback that is used by cells to control their movement. Understanding how signaling networks and the mechanical responses are integrated to produce shape changes and movement, and how external signals—either in the form of imposed spatially or temporally varying signals or those generated by movement—affect the responses, remains a major challenge in biology. The complexity of these processes is such that mathematical models are essential for synthesizing what is known to unify observations, and for making predictions that can guide further experimental work.

In this review, we focus on the processes that control single-cell motility, which we categorize as (i) signal transduction, (ii) actin network dynamics and intracellular waves, (iii) direction-sensing and polarization, and (iv) integration of signaling and structural changes. Each of these is a major topic in itself, and while individual processes have been reviewed elsewhere [4–7], our objective is to give a broader overview of their interdependence. We begin with a brief sketch of this that hints at how they are integrated.

In the absence of directional signals in their ME, many cell types, including neutrophils and *Dictyostelium discoideum* (Dicty), explore their environment more-or-less randomly [8,9], and therefore the intracellular signaling networks that control the shape changes must be tuned to produce signals that generate this movement. Thus, a first objective is to understand how the pathways that control actin network dynamics can produce random extensions of the membrane, whether in the form of filopodia, pseudopodia or lamellipodia. To this end, it is necessary to determine whether the known pathways can at least generate random actin waves that might trigger such protrusions, ignoring whether the membrane deformations needed for a protrusion emerge from these actin structures. Some have suggested that an integrated model for direction-sensing, adaptation, and signal-independent actin waves is comprised of two components—a signal-transduction excitable network (STEN) coupled to a CSK oscillatory network (CON) [10]. In Section 3 we review the signal-transduction networks in Dicty and neutrophils and discuss the dynamics of the Ras-PI3K-PTEN pathway. In Section 4 we discuss a number of models for actin waves that have been developed and show that a recent, detailed model of frustrated phagocytosis can replicate the experimentally observed waves in this system.

In the presence of a chemotactic, durotactic or other directionally biased signal in the environment the cells must orient or re-orient themselves appropriately, and this involves both direction-sensing and polarization. This is a two-step process, the former defined as determining the most favorable direction of movement, whether up the gradient of an attractant or down that of a repellent. This is a classical problem and it is well understood what a cell must do, and in Section 5.2 we describe a model for direction-sensing in Dicty that is based on extensive experimental data. The second step of the process is polarization—often referred to as symmetry-breaking [11]—in which the cell establishes an internal directional bias in the cytoskeletal structure. Simply put, this amounts to establishing a front and a back of a motile cell. However, polarization is not restricted to migrating cells—epithelial cells and budding yeast cells can become polarized without moving, the former to distinguish the 'top' from the 'bottom' and the latter to establish the budding site. The dynamics of the integrated signaling networks and their role in generating polarization in an external signal is discussed in Section 6.

2. The Primary Modes of Cell Movement

Since different types of cells use vastly different modes of movement that involve different modes of control of the CSK, we begin with a brief description of the various modes. An extended review of cell motility is given elsewhere [2].

The two major modes of eukaryotic cell movement are called mesenchymal and amoeboid [12,13]. Mesenchymal movement is used by fibroblasts and various tumor cells, and usually involves strong adhesion to the substrate and extension of relatively flat lamellipodia at the leading edge (Figure 1). The construction of lamellipodia involves nucleation of filaments at the membrane that then treadmill as in solution. The densely branched structure of the network arises via Arp2/3-controlled nucleation of branches on existing filaments [14]. Transmission of force to the environment involves integrin-mediated focal adhesions that are connected to the CSK via stress fibers, and this mode often involves proteolysis of the ECM to create a pathway for the cell [15].

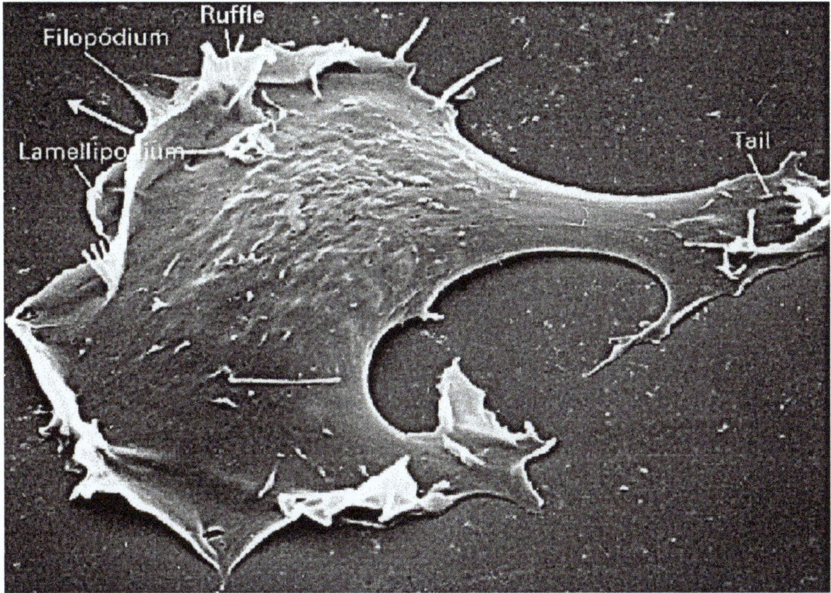

Figure 1. A fibroblast cell on a surface.

The amoeboid mode of movement is based on a less-structured CSK and typically involves less adhesion to the substrate. In the amoeboid mode cells adopt a more rounded cell shape and often have a highly contractile 'tail' called the uropod [16]. There are several distinct types of amoeboid motion that have been identified. In the first type cells generate a rearward flow of actin in the cortex, which leads to a reactive tension gradient in the membrane that propels the cell forward. This is called the tension- or friction-driven mode [17]. In the second type cells use actin-rich protrusions called pseudopodia at the leading edge, or by blebbing, in which cycles of extension of the front and retraction of the rear as shown in Figure 2b are used. In a third mode Dicty cells on a substrate move by deforming their shape in a wave-like manner [18], similar to a swimming mode described below.

In an environment less favorable to mesenchymal movement, eg., due to changes in the stiffness or adhesiveness of the substrate [19,20], cells compensate by undergoing a 'mesenchymal-to-amoeboid' transition (MAT) [21,22]. For example, leukocytes can use the mesenchymal mode when moving in a tissue, but can also migrate in vivo without using integrins, instead using a 'flowing and squeezing' mechanism [16]. In a cyclic AMP (cAMP) gradient on a rigid substrate, Dicty moves either by extending pseudopodia or by blebbing, and determines which mode to use by monitoring the stiffness of the surroundings. Pseudopodia are used in a compliant medium and blebbing is used in stiffer media [23]. While pseudopods and blebs involve very different actin dynamics, with the former based on a highly branched dendritic network, whereas the latter involves high contractility of the cortex that produces a high intracellular pressure and detachment of the membrane from the cortex at the leading edge, they can coexist at the leading edge [23]. Furthermore, blebs may transform into pseudopods by continued actin polymerization at the cortex, while pseudopods can spawn blebs at their edges [23,24]. Thus there can be a delicate balance between them. Finally, some cells move only by blebbing. Certain types of carcinoma cells are immobile on 2D substrates, but polarize spontaneously, form blebs, and move efficiently in a confined environment [25].

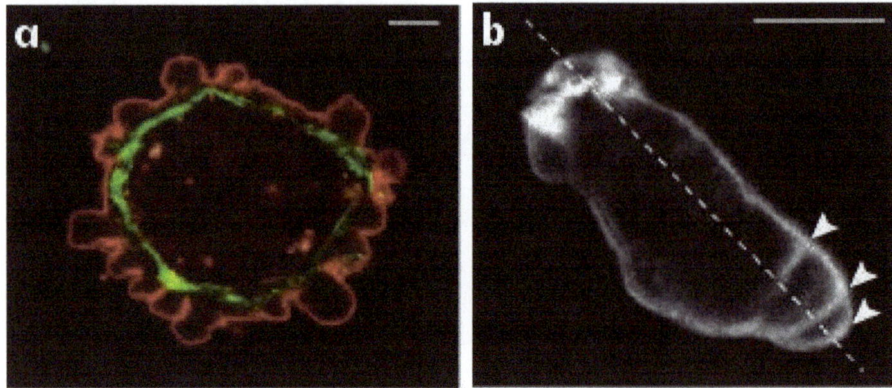

Figure 2. (**a**) Blebbing on a melanoma cell: myosin (green) localizes under the blebbing membrane (red) (**b**) The actin cortex of a Dicty cell migrating to the lower right. Arrowheads indicate the successive blebs and arcs of the actin cortex. The scale bar is 5 μm is each panel. From [26].

While it is less frequently used as a mode of movement, Dicty and neutrophils can swim in a fluid [27], and presumably use this mode to move through fluid-filled voids in their environment. Figure 3a shows a schematic that illustrates how Dicty moves by propagating protrusions down its length, and Figure 3b shows a time sequence of shape changes that Dicty executes as it swims toward the site of an attractant. This has been modeled and analyzed [28], and it was shown how characteristics of the protrusions, such as their height, affect the swimmer's speed and efficiency. In addition, it is also known that Dicty cells can swim for several cell lengths without shape changes [29], and it has been shown that they can do so by creating an axial tension gradient in the membrane [17].

Swimming and crawling are two very different strategies for movement, and raise the problem of understanding how mechano-chemical sensing of the environment and transduction of the information to the intracellular networks is used to control the structure of the CSK [30], which is clearly different in a fibroblast from that in a swimming cell. Protrusions and other shape changes require forces that must be correctly orchestrated in space and time to produce net motion—those on cells in Figure 2a are not, while those in Figure 2b are—and to understand this orchestration one must couple the intracellular dynamics with the state of the surrounding fluid or ECM. Tension in and curvature of the membrane and cortex have emerged as important determinants in the orchestration, whether in the context of undirected cell movement, or in movement in response to environmental cues [17].

Finally, since cells can be motile in the absence of extracellular signals, the autonomous dynamics of the actin network governing un-stimulated movement must be understood separately from the stimulated response. The fact that different modes can coexist in cells such as Dicty suggests that the balance between factors or pathways that determine the modes may be delicate.

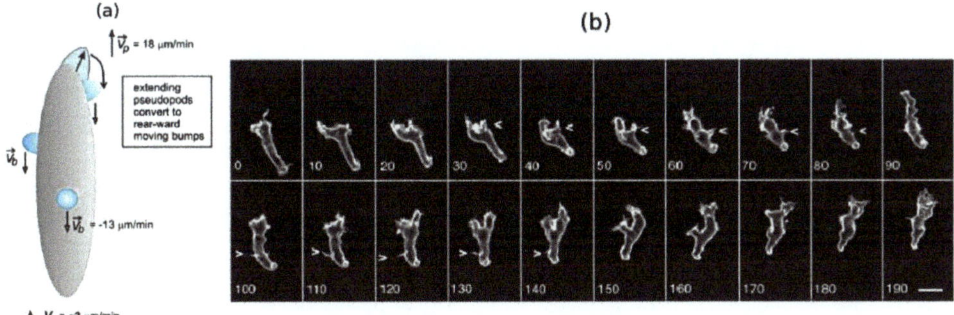

Figure 3. How Dicty amoebae swim by protrusions and shape changes. (**a**) a schematics of a swimming cell with 3 protrusions. From [31]; (**b**) the shape of a Dicty cell as it swims. Scale bar 10 μm. From [27].

3. The Signal-Transduction Network in Eukaryotic Cells

Prokaryotes such as *E. coli* are too small to measure the spatial gradient of signals across their body length, and thus developed a 'run-and-tumble' strategy in which they execute a random walk with persistence when searching for a favorable environment or trying to leave an unfavorable one. To implement this strategy *E. coli* has developed a sophisticated signal-transduction network that controls the rotational bias of flagella that propel the cell [32,33]. Motile eukaryotic cells such as neutrophils, fibroblasts and Dicty have also developed search strategies that involve execution of a persistent random walk [8,9], but since they are large enough to measure signal differences over their body length, the mechanism for implementing the search strategy is quite different. For instance, cAMP is a chemoattractant for Dicty, but in the absence of an external signal, cells spontaneously form and extend pseudopods [34–36], which involves localized re-building of the actin network. These new pseudopodia can either be retracted or can attach to the substrate, and in the latter case the cell adopts a polarized shape and moves in the new direction with a persistence time of about 9 min [37]. Of course the question is which intracellular signaling pathways control the location of a new pseudopod and the remodeling of the cortex and CSK to form a pseudopod, and how is this system biased by an external signal. Here the current state of knowledge for eukaryotes is far behind that in *E. coli*. Due to its genetic and biochemical tractability, Dicty is a widely used model system for studying these questions, and is to date the best understood eukaryotic system [38,39].

3.1. The Signal-Transduction Networks in Dictyostelium and Neutrophils

The small GTPases in the Ras superfamily, of which there are 150 human members and orthologs in yeast, Dicty and other species [40], are essential components of the pathways controlling the CSK in eukaryotic cells. These are grouped into five families—Ras, Rho, Rab, Ran and Arf—of which the first two are of primary interest here. The GTPases act as molecular Boolean switches in signaling pathways, with the on-off state determined by whether they are GTP-bound ('on') or GDP-bound ('off'). The binding state is controlled by GEFs (GTP exchange factors) or GAPs (GTPase-activating factors (Figure 4). The state of the switch can be controlled by controlling the GEFs and GAPs, which in turn can be controlled by other factors, and thus there are at least a two levels of control involved. Active GTPases act on downstream effectors to control network structure and dynamics by controlling two classes of actin nucleators, WASP and SCAR/WAVE proteins in one and Diaphanous-related formins in the other. The first controls production of branched dendritic networks, and the other long, frequently bundled, linear filaments. By controlling their localization with membranes in the presence of different signals, the spatial location of different network types can be controlled in a cell.

Three Rho GTPases—Rho, Rac and Cdc42—all activated by Ras, control three pathways in neutrophils that control actin network contraction, extension of filopodia [41], and lamellipodia [15,42], resp. Cdc42 and Rac control dendritic network formation by activation of scaffold proteins of the WASP family, which when activated facilitate actin polymerization by regulating Arp2/3 [43]. When activated, RhoA facilitates formation of actin bundles and stress fibers by activating the contraction of myo-II, which is done by deactivating MLCPase, an inhibitor of myosin contraction [44]. A sketch of these pathways in Dicty, which lacks both Rho and Cdc42, but uses Akt/PKB instead, is shown in Figure 5. Figure 5a shows the five main pathways in Dicty that are involved in transducing an extracellular cAMP signal to changes in the actin network.

The first step is binding of cAMP to one of the G-protein-coupled cAR receptors, which activates the G-proteins. G-proteins consist of an α subunit that contains a GTP/GDP binding domain as well as intrinsic GTPase activity, and a complex of a β and a γ subunit. The α and $\beta\gamma$ subunits dissociate after activation, and each can activate downstream signaling pathways as shown in Figure 5a. A major one is via $G_{\beta\gamma}$, RasG,D and PI(4,5)P2/PI(3,4,5)P3 (PIP2/PIP3), to Rac1, adenylate cyclase and cAMP, another is via Ras C and the TOR pathway, also to Rac1, and other pathways are driven by PLA2, by guanylate cyclase (GC), and by Ca^{+2}. While many components are shown there, the diagram only contains representatives of the principal actors and pathways. For example, there are a number of G_{α}s, and five different Ras proteins, three of which, RasG, Ras D and RasC are shown and are principals in the chemotaxis pathways. RasG is a primary regulator via localization of phosphatidylinositol-3 kinase (PI3K), which converts PIP2 into PIP3, while RasC regulates activity of the target of rapamycin complex 2 (TORC2), a parallel pathway that regulates chemotaxis. Figure 5b shows an expanded version of the PIP2-PIP3 component, which is central to the waves described later [45,46]. A mechanistic description of the PLC and CRAC pathways is given elsewhere [46,47].

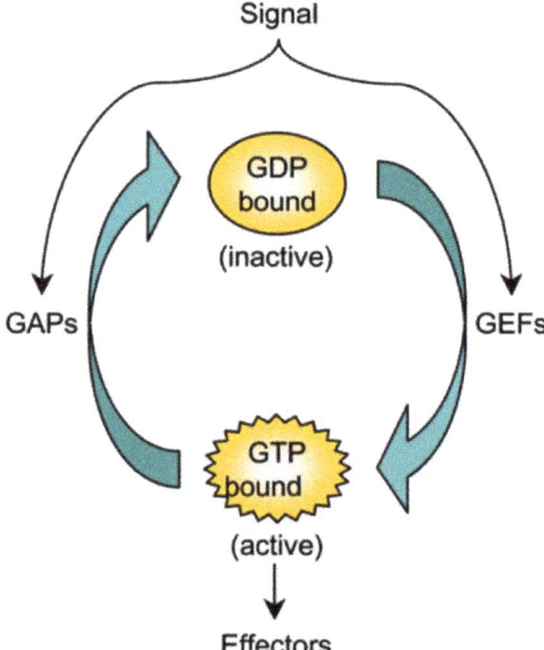

Figure 4. The molecular switch for a RHo GTPase—Rho is 'on' or active when GTP-bound, and 'off' or inactive when GDP-bound. From Charest et al. [43].

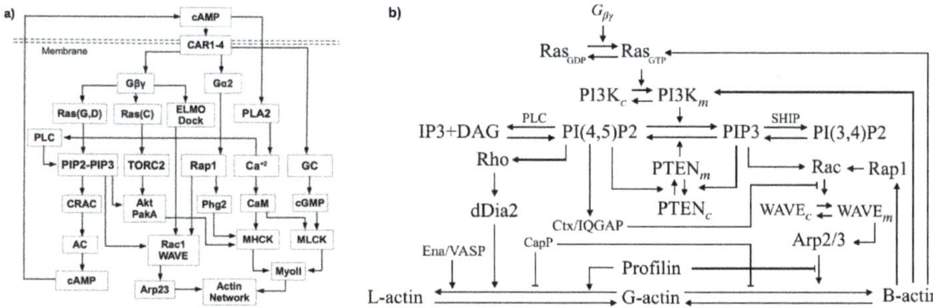

Figure 5. (**a**) The principal pathways in cAMP signal transduction in Dicty. CAR1-4: the cAMP receptors, G_{α_2} and $G_{\beta\gamma}$: components of the G-protein used for the transduction of the cAMP signal, Ras, Rac1: small GTPases, PIP2 and PIP3; membrane components that are inter-converted via phosphorylation and de-phosphorylation, IP_3 and DAG: products of PIP2 degradation, GC: guanylate cyclase—the enzyme that produces cyclic GMP (cGMP), AC: adenylate cyclase—the enzyme that produces cAMP, Myo-II: a motor protein involved in contraction of the actin network. Arrows indicate an effect, but not whether it is activating or inhibiting, and feedback steps are not shown. (**b**) Details of the PIP2-PIP3 subnetwork.

Assembly of the motor protein myo-II is controlled in part by PAKa via its effect on MHCK [48], and contraction is stimulated via the cGMP pathway by deactivation of an inhibitor of myo-II contraction [44]. The balance between the Rac1, Rap1 and GC pathways, in conjunction with other factors such as membrane tension, determine whether dendritic network formation (B-actin) or bundling of long filaments (L-actin) and myo-II-controlled contraction dominates, and as will be seen later the competition between them can lead to complex patterns of traveling waves.

Not all steps are shown in Figure 5, and other feedback interactions will be discussed later. Mutual inhibition between these pathways may ensure that the mesenchymal and amoeboid modes are mutually exclusive in some cells, but it is not absolute, since Dicty can use a mixed-mode strategy that involves either pseudopodia or blebbing [24]. High-level models of some of the interactions shown in Figure 5 are reviewed in [49–51].

3.2. The Dynamics of the Ras-PI3K-PTEN Pathway

In the absence of cAMP stimuli Dicty cells plated on glass extend pseudopods in random directions, but under spatially uniform cAMP stimuli aggregation-competent cells first respond by suppressing existing pseudopods and rounding up (the 'cringe response'), which occurs within about 20 s and lasts about 30 s [52]. This first phase of the response is characterized by uniform and transient membrane localization of PH_{CRAC}-GFP , a marker for PIP3, along the cell periphery within 10 s [53,54]. This fast phase of PIP3 accumulation is less affected by PI3K inhibition, which suggests that another pathway may be involved. Given that ElmoE interacts directly with $G_{\beta\gamma}$ [55], one could speculate that ElmoE might be essential in the first phase rise that occurs on a faster time-scale. Under uniform stimuli there is a second phase characterized by localized patches of labeled CRAC (Figure 6), extension of pseudopods in various directions, and an increase in the motility [56–58]. The second-phase rise is probably due to other signaling pathways, possibly involving F-actin (actin filaments of either type), that react on a slower time-scale. A localized application of cAMP elicits the cringe response followed by a localized extension of a pseudopod near the point of application of the stimulus [59].

A model described in Section 5.2 shows how the cell can use Ras activation to determine the direction in which the signal is largest, but how it organizes the motile machinery to polarize and move in that direction is still a major question from both the experimental and theoretical viewpoint. A subsequent step downstream of Ras is the generation of pleckstrin homology (PH) binding sites by the phosphorylation of the membrane lipid PIP2 by phosphoinositide 3-kinases (PI3Ks) to produce PIP3, which in turn is dephosphorylated to produce PtdIns(3,4)P2 (Figure 5b). In Dicty, PIP3 is produced by a class IA type kinase (PI3K1 and PI3K2) and a class IB type, kinase designated PI3K$_\gamma$ [60,61]. The former are activated via cytosolic tyrosine kinases, and thus may contribute to basal activity, whereas the latter consists of a catalytic unit and binds to F-actin via the N-terminal region. The latter fact may explain why the fast phase of the response to a uniform stimulus is PI3K insensitive. Both PIP3 and PI(3,4)P2 provide binding sites for various cytosolic proteins containing PH domains (PHPs) and recruitment is rapid: localization of PHPs at the membrane peaks 5–6 s after global stimulation with cAMP [60,62]. Both PIP3 and PI(3,4)P2 are tightly regulated by the phosphatases PTEN and SHIP, and within 10–15 s following uniform cAMP increases the PHPs return to the cytoplasm [60,63]. This rapid increase of PIP3 at the membrane couples the extracellular signal to actin polymerization via Rac1-WAVE-Arp2/3 (Figure 5a), which creates a feedback loop that leads to increased PI3K binding and increased PIP3 production. The level of activated G-proteins in continuously stimulated cells reaches a stimulus-dependent level, while membrane-associated CRAC first increases, but then returns to its basal level. Therefore, adaptation of the PIP3 and cAMP responses, as well as directional sensing, is downstream of $G_{\beta\gamma}$and upstream of PIP3 and CRAC [64,65].

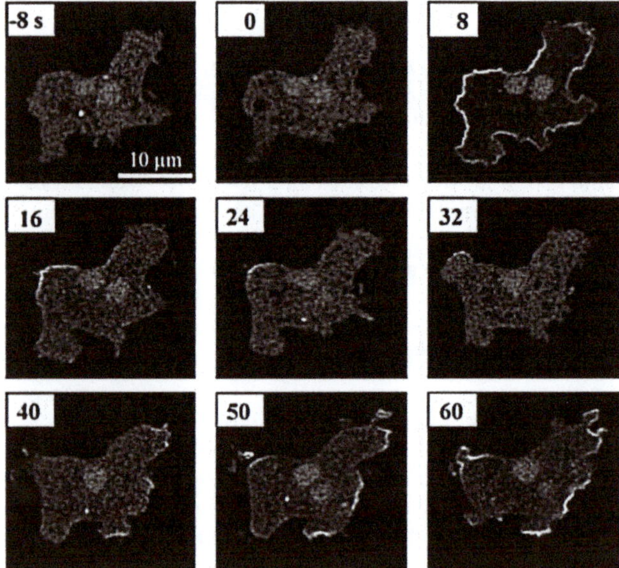

Figure 6. GFP in the cytosol before cAMP stimulation, a transient translocation of GFP to the entire boundary of the cell at 8 s after addition of cAMP, and patches of GFP at the boundary after 40 s. From Postma et al. [53].

The ratio of PIP2 to PIP3 has a significant role in the blebbing vs. pseudopod dichotomy mentioned earlier. This ratio dictates if either detachment of the membrane from the cortex or, B-actin formation at the membrane occurs. A reduction of PIP2 increases blebbing, possibly via its effect on membrane-cortex adhesion [66], whereas the absence of PIP2 conversion in *pi3k*$^-$ leads to greatly reduced production of blebs compared with wild-type cells [23,24]. One can see in Figure 5 that another balance, that between the Ras-independent and Ras-dependent pathways may be an essential factor in resolving the blebbing-pseudopod competition.

4. Intracellular Actin Waves in the Absence of Directional Signals

As remarked earlier, cells can execute a persistent random walk in a signal-free environment, and it has been found that the intracellular components of the network exhibit a variety of spatio-temporal wave patterns under such conditions. The first to observe actin waves in Dicty were Vicker et al. [67–69], and Vicker [70] was the first to suggest that these waves were generated by an excitable reaction-diffusion system involving actin dynamics. Since then such waves have been observed in Dicty, neutrophils and other cell types [8,9,71–77]. Inagaki et al. [78] provide a broad overview of waves and their role in various aspects of cell dynamics.

The actin waves in Dicty and macrophages arise during re-building of the actin network following treatment with latrunculin A (LatA), which sequesters G-actin and leads to disintegration of the network and rounding of the cells. After removal of the drug the cells return to their pre-stimulus state, but in the interim there are distinct domains of the membrane that is in contact with the surface in which different actin structures exist (Figure 7). In one PIP3, Ras and Arp2/3 are high and the network is dendritic, whereas in the other PIP3 is low, PIP2 and cortexillin are high, and F-actin is is linear and bundled. The existence of two distinct domains separated by a propagating actin wave suggests that the underlying dynamics are bistable, with one state in which PIP3 is high and PIP2 is low, and the other in which the roles are reversed. The waves that exist between domains of high and low PIP3 are usually closed and of varying shape, and actin recovery after bleaching shows that they propagate by treadmilling [79]. Myosin-IB, a membrane-cortex linker protein [80,81], is found at the front of a wave, and a dense dendritic network is found in the high PIP3 domain. Other components that are found where PIP3 is low include coronin, which inhibits filament nucleation and indirectly regulates cofilin activity via dephosphorylation [82], and cortexillin, which organizes actin filaments into anti-parallel bundles.

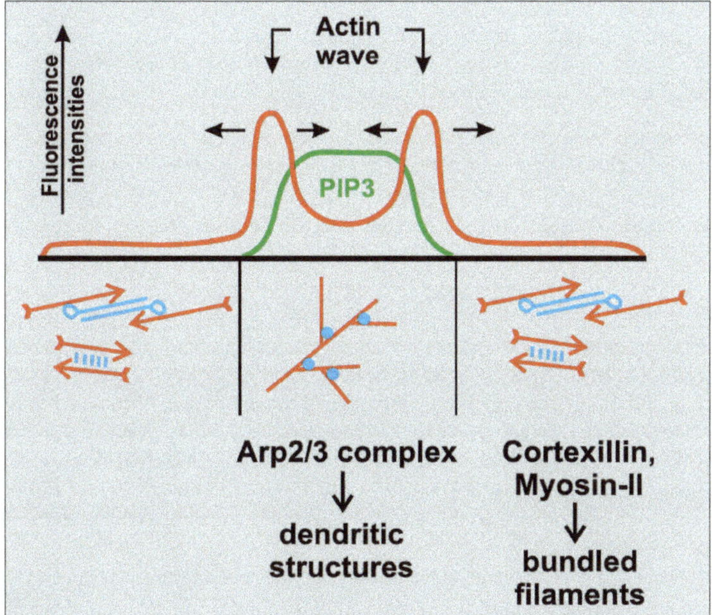

Figure 7. A cross-section of the waves in LatA-treated Dicty cells. From Gerisch [79].

Because it is the balance between the Rac1, Rap1 and GC pathways that determines whether formation of dendritic networks or formation of linear actin dominates, experimentalists believed that the complex patterns of traveling actin waves in the cortex that are observed in the absence of directional signals may be the result of competition between them. Figure 7 shows that Arp2/3 is prevalent in the inner region while bundled filaments are dominant in the outer region, but the dichotomy may not be so clean. Recent work shows that formins, which nucleate and elongate actin filaments, are distributed throughout the inner and outer regions, but the type of formin varies , and the waves disappear when cells are treated with a formin inhibitor [83]. Figure 8A shows that formin A is high outside the wave, reduced in the inner region, but high in the wave front and back. Figure 8B shows the coexistence of formin B (green) and Arp2/3 (red) along the wave front. Thus the actin wave is undoubtedly a mix of long filaments and branched network. This changes the picture of how wave formation is controlled significantly, since it indicates that the formin-controlled pathway is essential.

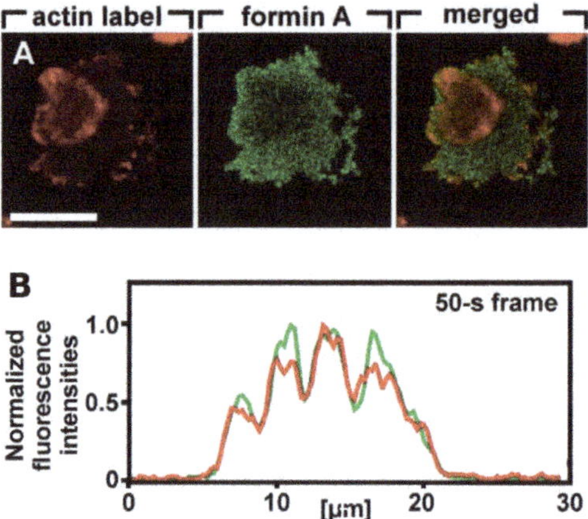

Figure 8. (**A**) Individual and overlay images of actin and formin A in a cell recovering from treatment with 5 µM latrunculin, showing actin and formins on the substrate-attached cell surface. Scale bar 10 µm. (**B**) The spatial distribution of formin (green) and Arp2/3 (red) along the wavefront. From Ecke et al. [83].

Oscillations in the local network dynamics, as well as waves in the corresponding reaction-diffusion systems, often originate from a certain balance between positive feedback and slow inhibition in the network. The simplest mathematical model that can produce the different wave types leads to the following equation, which stems from reaction and diffusion in a two-phase medium with rapid interphase transport.

$$\frac{\partial u}{\partial t} + (f_0 + f_1 u)\frac{\partial u}{\partial x} = \frac{\partial^2 u}{\partial x^2} + g(u) \tag{1}$$

The second term on the left side represents an active or convective transport process in the fluid phase and $g(u)$ is qualitatively a cubic nonlinearity with zeroes $u_1 \leq u_2 \leq u_3$. By adjusting the parameters f_0 and f_1 one can obtain a propagating transition wave from $u_1 \rightarrow u_3$ or $u_3 \rightarrow u_1$, periodic and damped oscillatory waves, or transition waves [84]. Furthermore, if a second mechanism controls one of these parameters, one can make the waves stall, as in [85] or reverse the waves at the boundary. The transition waves are stable on the line, while the others are not, but this is not relevant to the scale of a cell. However, this is only a cartoon description—the underlying mechanism is much more complicated.

While the observed patterns of wave initiation and propagation suggest that the waves are governed by an excitable system, it has been difficult to identify a minimal set of components of the network shown in Figure 5 responsible for them. For example, it is known that the Ras activation step in isolation is not excitable—it responds proportionately to any stimulus and adapts, and thus if Ras is part of an excitable STEN there must be downstream feedback on Ras. However, Ras and PI3K can still be activated in $g_{\beta\gamma}$-null cells, thus eliminating the effect of extracellular cAMP [34]. These authors and others [86] suggest that there is a feedback loop from F-actin to Ras, as shown in Figure 5b, but the feedback may arise from components further up the pathway. Another question stems from the fact that it is often difficult to obtain the long-term evolution of the waves in order to determine whether they are transition waves between two distinct steady states, or pulse waves that begin and end at the same steady state. A model of frustrated phagocytosis described later supports both types in different parameter regimes [87].

4.1. Models of Intracellular Waves in Dicty

It was reported that Ras waves exist in un-stimulated Dicty cells in the absence of both PI3K and F-actin feedback, and it was concluded that Ras waves drive PIP3 waves [88]. The authors suggest that spontaneous Ras wave formation is possible without any downstream feedback and that this drives PI3K waves, and they develop a model in which there is feedback between the RasGTP and RasGAP to obtain waves. Computations using a model described in Section 5.2 show that without such feedback Ras alone cannot generate waves.

An important feature of the waves is that typically, but not always, the domains described above are well separated, in that PIP3 is low where PIP2 is high and vice versa. There are several models proposed to explain the PIP3/PTEN dynamics [73,89,90], in which the reduction in PTEN/PIP2 plays the main inhibitory role such that the scheme follows the "substrate-depletion"-type mechanism in line with the fact that PIP3 and PTEN appear to be anti-phase. The authors find that F-actin is not required to generate PIP3/PTEN waves and propose a model based on the mechanism shown in Figure 9. The computational results shown in Figure 10 show the out-of-phase relationship found in some, but not all experiments. However, there are several concerns about the model. Concerning the effect of effect of PIP3 on PTEN, it has been pointed out that neither an increase in PIP3 nor a decrease in PIP3 levels influenced the membrane-binding of PTEN [91]. In addition, the sampling techniques may lead to erroneous conclusions. Gerisch et al. [92] show that when PIP3 and PTEN are sampled over the entire attached surface the results appear to be consistent with the anti-phase conclusion, but when sampled in a very small region the results are different. Figure 10B shows that there are regions in which PIP3 decreases sharply even as PTEN continues to decrease, which is contrary to the model predictions in Figure 10A. The switch from rise to fall of PIP3 is therefore unlikely to be caused by a depletion of PTEN, and the authors suggest that other factors are involved in the PIP3-PTEN dynamics. The model proposes that PIP3 might negatively regulate PTEN recruitment and positively regulate PI3K recruitment [73,90], but this has not been experimentally confirmed. It has been observed that PIP3 regulation of PI3K recruitment is F-actin dependent, and that there is no PI3K recruitment in LatA-treated cells [93,94].

One of those additional factors may be the actin network, for it is well established that the membrane-binding and activation of PI3-kinases depends on F-actin [93]. However, the role of actin or actin waves in the generation of the PIP3/PTEN patterns is still controversial. It is reported that PIP3/PTEN patterns disappear at a higher dose of LatA treatment (10 μM) [75,90], while PIP3 waves can still be observed in mild LatA treatments (0.5–2.0 μM). Nishikawa et al. [75] also reported that PIP3 waves reappear with addition of 1 nM cAMP under 10 μM LatA treatment. On the other hand, Arai et al. [73] report that PIP3/PTEN patterns are formed in the presence of 5 μM latrunculin A, a concentration that the authors considered to be sufficient for the complete inhibition of actin polymerization.

Another unknown in establishing a network concerns other factors affecting PIP2. In addition to the PTEN-regulated supply from PIP3, PIP2 can be supplied by PTEN-independent pathways [87,95–97], and as shown in Figure 5b, PIP2 can be degraded through calcium-dependent PLC activity and via PI3K- and PLC-independent pathways [4,95]. Moreover, PTEN interacts with lipids through several binding and catalytic domains [98–102], and it has been proposed that positively charged residues in the PIP2-binding and C2 domains can recruit PTEN to the plasma membrane through associations with negatively charged membrane lipid head groups [102–106], which suggests that there is a positive feedback loop in the PIP2-PTEN interaction.

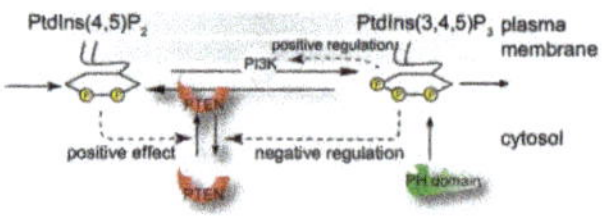

PtdIns(4,5)P$_2$
PtdIns(3,4,5)P$_3$ plasma membrane
positive regulation
PI3K
PTEN
positive effect
negative regulation
cytosol
PTEN
PH domain

$$\frac{\partial[\mathrm{PIP3}]}{\partial t} = -R_{\mathrm{PTEN}} + R_{\mathrm{PI3K}} - \lambda_{\mathrm{PIP3}}[\mathrm{PIP3}] + D\nabla^2[\mathrm{PIP3}],$$

$$\frac{\partial[\mathrm{PIP2}]}{\partial t} = R_{\mathrm{PTEN}} - R_{\mathrm{PI3K}} + k - \lambda_{\mathrm{PIP2}}[\mathrm{PIP2}] + D\nabla^2[\mathrm{PIP2}],$$

$$\frac{\partial[\mathrm{PTEN}]}{\partial t} = V_{ass}[\mathrm{PTEN}]_{cyt}\frac{[\mathrm{PIP2}]}{K_{\mathrm{PIP2}} + [\mathrm{PIP2}]} \frac{K_\alpha + \alpha[\mathrm{PIP3}]}{K_\alpha + [\mathrm{PIP3}]} - \lambda_{\mathrm{PTEN}}[\mathrm{PTEN}],$$

$$R_{\mathrm{PTEN}} = V_{\mathrm{PTEN}}[\mathrm{PTEN}]\frac{[\mathrm{PIP3}]}{K_{\mathrm{PTEN}} + [\mathrm{PIP3}]},$$

$$R_{\mathrm{PI3K}} = V_{\mathrm{PI3K}}\frac{\beta K_\beta + [\mathrm{PIP3}]}{K_\beta + [\mathrm{PIP3}]}\frac{[\mathrm{PIP2}]}{K_{\mathrm{PI3K}} + [\mathrm{PIP2}]}.$$

Figure 9. The model of the PIP2-PIP3 reaction and the governing equations. From Arai et al. [73].

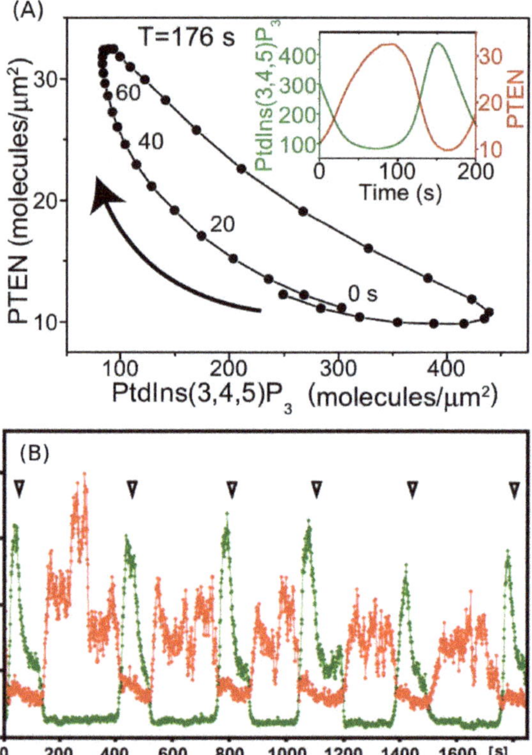

Figure 10. (**A**) The predicted PIP3-PTEN phase plane. From Arai et al. [73]. (**B**) A plot of PTEN (red) and PIP3 (green) vs. time taken in a small spot on the membrane. From Gerisch et al. [92].

A more detailed model of the observed waves that incorporates more molecular details of the PI3K pathway and the actin network dynamics has been proposed by Khamviwath et al. [107] (Figure 11A). Stimulation with a localized pulse of activated receptors, leads to a a single pulse, whose magnitude grows in time while the pulse spreads in both the x- and z-directions, the latter representing the height of the actin network (Figure 11B). A threshold stimulus is needed for initiation of a wave because the uniform rest state is stable, and this suggests that the model is excitable. The amplitude in the center decays later, and the pulse splits into two symmetric pulses, which is consistent with experimental observations [71].

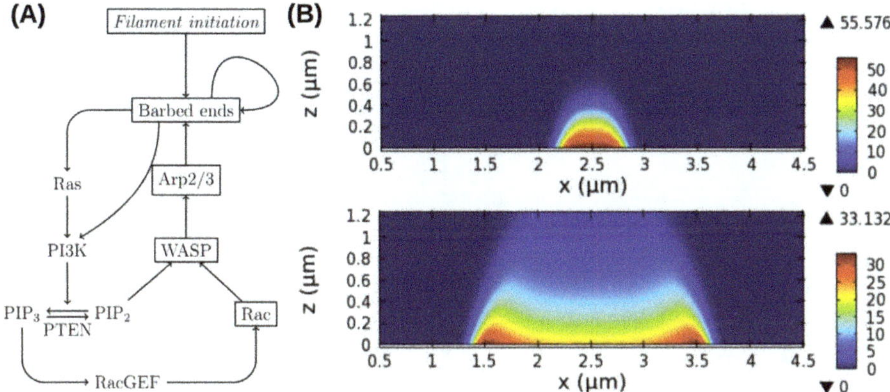

Figure 11. (**A**) A schematic of the network structure and molecular interactions in the model. (**B**) Two snapshots in time of an actin wave initiated at x = 2.5, showing the network density (color) as a function of space (*x*-axis) and network height (z-axis). From Khamviwath et al. [107].

Furthermore, Rac, which is a proxy for PIP3 in the model, shows no peaks, which is also consistent with the experimental observations [71]. Another prediction of the model is that the inclusion of PTEN leads to reversal of the waves, which agrees with the observations that the waves often propagate to the cell edge and then reverse direction. Other wave models are reviewed elsewhere [51], but much remains to be done to understand the internal structure of waves. For example, it has been found in macrophages described later that PIP(3,4)2 is enriched in the wave center, rather than PIP3, as in Dicty [76].

While the foregoing models involve PTEN and other components of the signaling network, it has been shown that SCAR/WAVE, Arp 2/3 and actin-binding proteins can generate rapid, localized oscillatory SCAR/WAVE-actin foci in Dicty cells lacking G_β and PTEN [10], and more recently it was shown that $G_{\beta\gamma}$ has important effects on the dynamics. Knockout of $G_{\beta\gamma}$ completely blocks chemotaxis and CSK dynamics [4,34], but recently a G_β sequestration technique to study the effect of G_β on the spatial interaction of the foci was developed [108]. It was found that sequestration of G_β induces large-scale oscillations of LimE-GFP, a reporter for F-actin, due to long-range coupling of actin foci, and that very few G_β-null cells display LimE-GFP oscillations. The global coupling of the local oscillators interferes with the sensing of extracellular signals and the changes in local actin dynamics needed to produce protrusions, but how this is effected remains to be explained.

4.2. Intracellular Waves in Frustrated Phagocytosis

Another system in which waves are observed involves macrophages that are in contact with a surface, undergoing a process called 'frustrated phagocytosis. Phagocytosis is a process in which phagocytes such as lymphocytes or macrophages engulf and destroy foreign bodies called pathogens in a tissue, and it is initiated when a cell of the immune system detects antibodies carried by a pathogen via receptors in the membrane. Signaling mechanisms that control the changes of the cellular cytoskeleton needed for engulfment of the pathogen lead to large mechanical deformations of the cell, and thus a mathematical model of the entire process would be extremely complicated. Recent experiments have used an experimental technique similar that used in LatA-treated Dicty cells in which the membrane does not deform, but rather, signaling triggers actin waves that propagate along the boundary of the cell [76].

This eliminates the large-scale deformations and facilitates modeling of the wave dynamics. A model of the actin dynamics observed in frustrated phagocytosis that can replicate the experimental observations has been developed [87], and the key components that control the actin waves have been identified. Figure 12 shows the relative positions of different components in the wave found experimentally [76], and these should be compared with those shown for actin waves in Dicty in Figure 7.

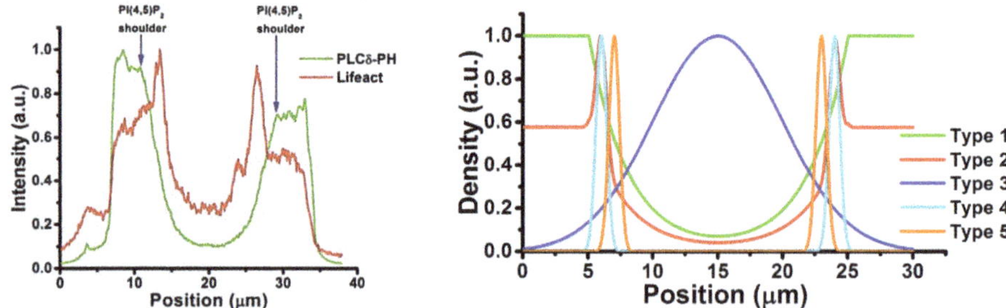

Figure 12. Left: A cross-section of an actin wave, showing actin (red) and PIP2 (green). The drop in PIP2 occurs at the boundary of the cell. Right: The spatial profiles of components in the wave, all normalized to their maximum value in the wave. Components were classified into 5 types, based on their pattern of in the waves: Type 1—Cortical actin, PI(4,5)P2, PI5K; Type 2—Total F-actin; Type 3—PI(3,4)P2, DAG; Type 4—Branched actin, N-WASP Type 5–SHIP2. Please note that PIP2 is constant in front of the wave. Taken from Masters et al. [76].

The signaling network is controlled by the FcγR receptor, and it is known that receptor activation following binding of the antibody immunoglobulin leads to a sequence of spatial and temporal changes in phosphoinositides, Rho-family GTPases and actin nucleation-promoting factors [109]. The spatio-temporal dynamics of these molecules control processes such as remodeling of the cytoskeleton, membrane fusion and the production of reactive oxygen intermediates that are necessary for particle internalization. However, it is not clear how the molecular scale activation of FcγR's leads to the observed micron scale patterns of activation and inactivation of network components reflected in the propagating actin waves, and the network shown in Figure 13 was developed to address this issue [87]. Only membrane-localized components are shown, and all are placed in their approximate order of activation, with red faster than blue. The internal structure of the wave is captured in the model, and two snapshots of the time-evolution following a stimulus are shown in Figure 14.

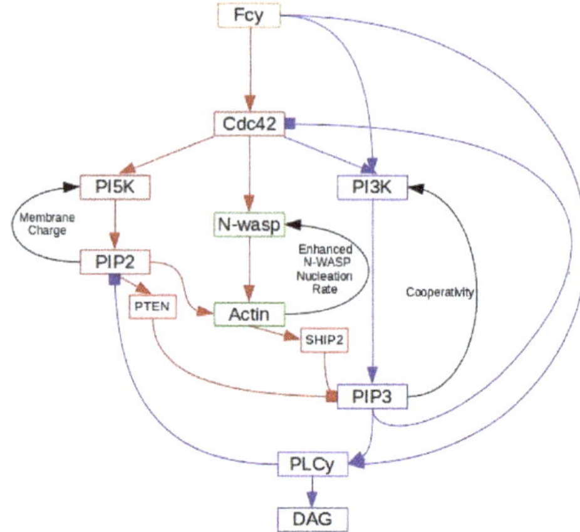

Figure 13. The signal-transduction steps following FcγR activation. From Ponce de Leon & Othmer [87].

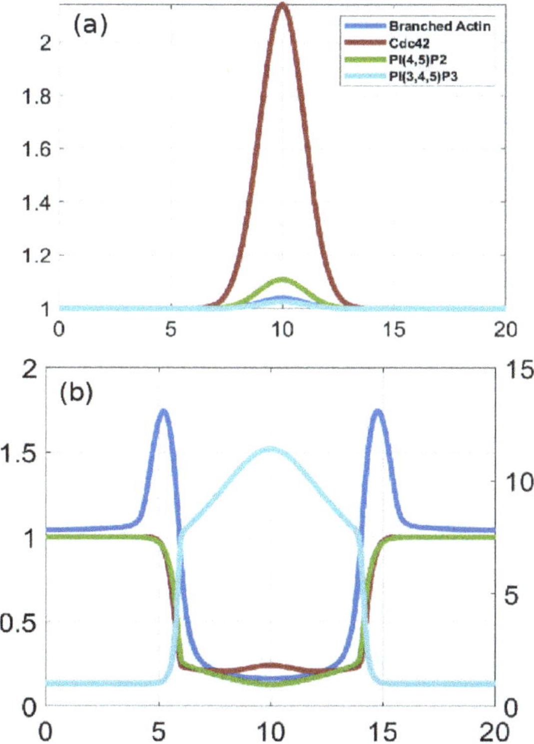

Figure 14. Branched actin wave initialization and propagation. Shown are the profiles of branched actin, PI(4,5)P2, Cdc42, and PIP3 at $t = 5$ s (**a**), and 180 s (**b**) after a perturbation. The concentrations are normalized by the original steady state. The right axis on (**b**) is for PI(3,4)P2, which is shown in light cyan.

There are a number of interesting predictions that emerge from the model. In particular, one is that diffusion coefficients of membrane-bound species must be larger behind the wavefront than in front to replicate the internal structure of the waves.

5. Models for Polarization and Direction Sensing

Cells in vivo are never either spatially homogeneous or geometrically symmetric at the molecular level, and thus the commonly used term *polarized* must be defined more loosely. Usually cell polarity refers to a spatial distribution of a protein, a lipid, the CSK, or another component that exhibits an identifiable spatial gradient. The ability to polarize at the cellular level is an essential property for cell division, cell-cell interactions such as mating in yeast cells, and the ability to move in a favorable direction or against an unfavorable one, and most cell types exhibit some form of polarity, which enables them to carry out these specialized functions. Polarity is very dynamic and can be very short-lived, as in the signaling patches that may generate actin waves and lead to small local membrane deformations, or longer-lived, such as the morphological polarity visible in cells such as neurons, in migrating fibroblasts, or in epithelial cells, which have a well-defined apical-basal polarity [110]. Given that polarity may be evanescent or persistent, a major question is how these spatio-temporal events are generated, and what distinguishes those that are evanescent from those that are imprinted for a longer time. For example, the actin patches described earlier are generally short-lived, and what determines the threshold that distinguishes a short-lived patch from a new pseudopod is not known. In the context of the problems described herein, the imprinting is rarely permanent, but does involve some architectural changes in the CSK on the time scale of interest. The molecular underpinnings of polarity in yeast cells are reasonably well understood, and mathematical models have been developed to investigate the role of different pathways in the polarization process [111–116]. However, the underlying networks are more complex in motile cells such as Dicty and neutrophils, and progress has been slower.

While polarity at the molecular or CSK level can arise in the absence of external signals, direction-sensing involves the detection of spatial differences in an external signal over the cell surface, thereby determining what the most favorable direction is according to some criterion. This typically leads to the localization of some components in the membrane that can initiate structural polarization at the level of the CSK, or if the best direction of the signal changes, lead to either re-polarization of the cell or re-orientation of it. Typically, there is a threshold level of the signal needed to initiate a response, and this may vary depending on the history of exposure to the signal, since some cell types adapt to constant signal levels. In addition, many cell types can detect low levels of a signal, but then amplify them internally to initiate the appropriate intracellular response.

5.1. *Mathematical Models for Polarization*

5.1.1. Localization of 'Hotspots' for Wave Initiation and Polarization

We have seen that a variety of membrane waves exist in Dicty and other cell types, but how they are initiated is not well understood. Mathematically speaking, it is known that different perturbations of an excitable system can lead to waves, but how the perturbation that triggers the event in question arises spontaneously in the membrane environment is not known. Here we describe two different mechanisms that may be involved—a spontaneous coagulation mechanism that creates a spatially distinct region or 'hotspot', and a positive feedback mechanism that has a similar effect. In either case the objective is to create a localized nanostructure in the membrane that initiates the appropriate activity. Important questions concern how large the nanostructure must be in spatial extent, how long it must persist in time to create observable events such as propagating waves, and how long its effects persist. In the case of PIP2-PIP3 waves the effect may be relatively short-lived, since once the wave has passed the system may relax to the unperturbed state, whereas in other cases it may persist over a much longer time scale.

Modeling of symmetry-breaking, which usually means establishment of polarity in a cell, addresses either the question of 'long-term' or imprinted polarization, as in the budding yeast or a cell migrating along the gradient of a signal, or relatively short-term polarization, either spontaneous or in response to a fluctuating extracellular signal. In persistent polarization the initial response to the event may stimulate reinforcing events, such as modification of the CSK that prolong the asymmetry or polarity for a longer period. This will be discussed in the context of direction-sensing in response to an external cu—here we discuss establishment of localized nanostructures.

The general mechanism of cluster formation is illustrated in Figure 15, which shows how diffusion on the membrane can lead to clusters of proteins. The model arose in the context of cell polarization from the observation that surface-bound Cdc42 forms nanoclusters in the membrane in budding yeast [117]. The clusters diffuse more slowly than single molecules and are larger at the cell poles and thus they tend to localize there [118]. Clearly interactions with the membrane and other proteins must lead to a reduction in the free energy of the system, else there would be no clustering, and it is found that the cluster size depends on both the scaffold protein Bem1 and the lipid environment, in particular phosphatidylserine levels [118].

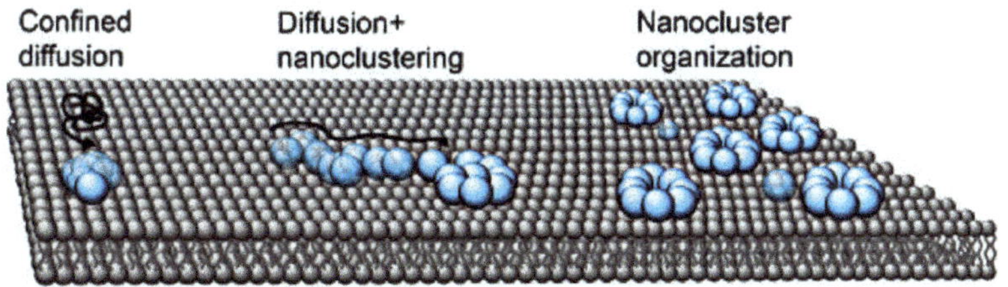

Figure 15. Formation of naonclusters on a lipid bilayer. From Sartorel et al. [117].

At sufficiently high densities the process can be described by the Smoluchowski equation

$$\frac{\partial f_n}{\partial t}(x,t) = D(n)\Delta f_n(x,t) + Q_1^n(f)(x,t) - Q_2^n(f)(x,t). \tag{2}$$

Here f_n is the membrane density of clusters of size n, $D(n)$ is the diffusion coefficient of such clusters, and Δ is the Laplacian on the surface. The term

$$Q_1^n(f)(x,t) = \frac{1}{2}\sum_{m=1}^{n-1} \alpha(m, n-m) f_m(x,t) f_{n-m}(x,t), \tag{3}$$

represents creation of clusters, where α is the coefficient of creation of a cluster of size n from clusters of size m and $n-m$. Similarly, the loss term is given by

$$Q_2^n(f)(x,t) = f_n(x,t)\sum_{m=1}^{\infty} \beta(m,n) f_m(x,t). \tag{4}$$

Of course treating the aggregation process using continuum densities may not be valid in general, in which case one must revert to a stochastic simulation of Equation (2). This was done for general reaction-diffusion equations in Hu et al. [119] and in the context of aggregation on membranes in Turner et al. [120] and Richardson et al. [121].

$$\frac{\partial u}{\partial t}(x,t) = \frac{1}{2}D\Delta u + k_{on}(1-h) + k_{fb}(1-h)u - k_{off}u, \tag{5}$$

Another mechanism closely related to models described in the next section involves reinforcement of binding from the cytosol by previously bound ligands on the membrane. The simplest model of this involves a single species that is either bound to the membrane or is freely diffusing in the cytosol [122]. The molecular species shuttles between the cytosol and membrane in a simple on-off step, but the rate of binding to the membrane can also be increased by membrane-bound species, as shown in Figure 16 (top). The governing equation for such a reinforced-binding mechanism in which diffusion on the membrane is allowed is where h is the membrane-bound fraction of the total number of molecules—both in the cytosol and on the membrane [122].

Again the question of whether a continuum description is appropriate arises, and the authors tested a stochastic simulation of Equation (5) and found that the results depend strongly on the total number of particles, as shown in Figure 16 (bottom). When there are many particles the entire membrane is covered and the distribution of signaling molecules on the membrane converges to a homogeneous steady state. However, when there are few particles—1000 under the conditions used—the model with a positive feedback alone is sufficient to create and maintain a single localization site of membrane-bound molecules. This model has also been applied by Houk et al. [123] to explain how membrane tension maintains cell polarity by confining signals to the leading edge during neutrophil migration.

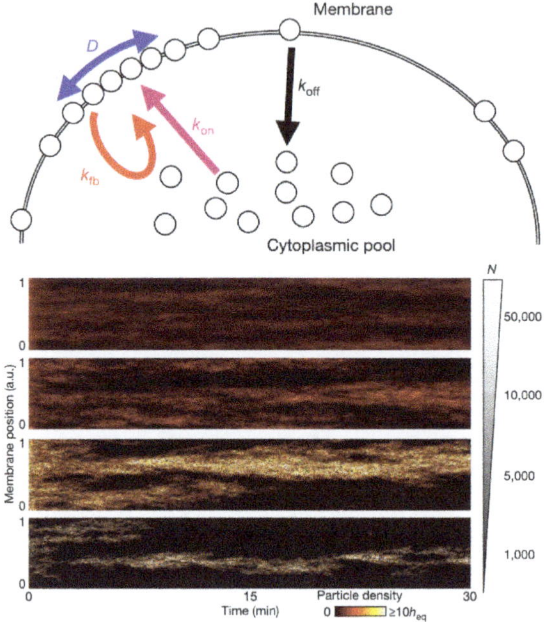

Figure 16. The reinforced-binding model. Modified from Altschuler et al. [122].

A different model for localization depicted in Figure 17 was proposed by Marco et al. [124]. In this model localization stems from the balance of three processes: diffusion along the membrane, transport to the membrane along actin or microtubules, and recycling to the cytoplasm via endocytic uptake and membrane recycling. In this model local polarization of the CSK is assumed to entail active transport, but the objective is to show how proteins can be localized on the membrane.

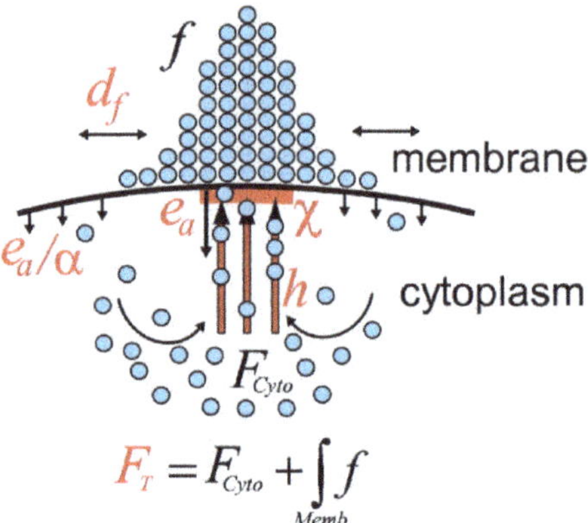

Figure 17. Schematic diagram of the model. From Marco et al. [124].

Let f denote the density distribution of the protein on the membrane, and let F_{cyto} be the homogeneous cytoplasmic concentration of the protein. Then the evolution of f is governed by

$$\frac{\partial f}{\partial t} = d_f \Delta f - \left(e_a \chi + \frac{e_a}{\alpha}(1-\chi)\right) f + hF_{cyto}\frac{\chi}{\int \chi}, \tag{6}$$

where d_f is the diffusion constant on the membrane and χ is the directed transport window function defining the region of the plasma membrane to which cytoskeletal tracks are attached. Furthermore, h is the directed transport rate along cytoskeletal tracks, e_a is the endocytosis rate within the directed transport region, α is the ratio of the endocytosis rates within and away from the directed transport region, and F_{cyto} is the cytoplasmic pool of f which is homogeneous due to the fast dispersion in the cytosol. Their experimental results corroborate the assumption that co-localization of endocytosis with actin patches in the polarized region is important for cell polarity. Of particular significance is the theoretical result that endocytosis rates can regulate dynamically balanced systems to optimize the asymmetric localization of membrane-protein distributions. Endocytosis will reappear later in the context of regulating the spatial distribution of WASP in Dicty.

Which of the three mechanisms might be involved in wave formation as described earlier? The reinforced-binding mechanism and the endocytosis model are directed more to creation of permanent 'poles' in a cell, rather than an ephemeral 'hot spot', since there is no explicit mechanism for deconstruction of the localization by turning off the positive feedback on binding, while localization in the cluster-formation model is more likely to be disrupted due to membrane fluctuations.

5.1.2. Reaction-Diffusion Models for Gradient Establishment

In the clustering and reinforced-binding mechanisms polarity arises independently of the CSK, but when motility is involved bidirectional interaction of signaling and the CSK is essential, as seen in the model of phagocytosis. When there is a link the most complete model description involves both the signaling networks and the mechanical effects of membrane deformations and pseudopod growth, and there is as yet no mechanistic model of these interactions. At the other extreme, there are many mathematical models that have been formulated that contain no link between signaling and the CSK, and the objective in these is to establish a gradient of a signaling molecule involved in polarization. Meinhardt [125] suggested an activator-inhibitor model that incorporates a third species that functions as a local inhibitor. Small external differences are amplified via a Turing instability in the activator-inhibitor system, and the slower in-activator suppresses the primary activation. It was shown that transient maxima of the internal signals arise at random locations in the absence of external signals, and for suitable parameters the model can generate stable cell polarization. This model is an interesting high-level description of the process, but has no direct relation with the underlying biochemistry in any system.

A class of more recent models for gradient establishment and direction-sensing are also based on an activator-inhibitor mechanism. These so-called LEGI—local excitation and global inhibition—models are used to explain direction sensing and adaptation in a constant chemoattractant field [126]. The models incorporate a fast-responding but slowly diffusing activator and a slow-acting, rapidly diffusing inhibitor, similar to what is used in a Turing mechanism, to set up an internal gradient of activity that tracks the extracellular gradient. The usefulness of such models is limited because of the oversimplification of the signal-transduction network, and the need for a wide disparity in the diffusion coefficients of the inhibitor and activator to establish an intracellular gradient.

Other models have been built around two-component systems of reaction-diffusion equations that involve binding of a cytosolic species to the membrane, and in these the difference in diffusion rates arises from the fact that one component diffuses in the cytosol and the other on the membrane. Such models are typically described by a system of the form,

$$\begin{aligned} \frac{\partial u}{\partial t} &= D_u \Delta u + f(u,v) \\ \frac{\partial v}{\partial t} &= D_v \Delta v + g(u,v). \end{aligned} \tag{7}$$

where u is the cytosolic concentration and v is the areal density on the membrane. It is assumed that u is constant in the direction normal to the surface, which is appropriate for a membrane under a thin layer of fluid, but not in general, and thus $f(u,v)$ represents the binding step. Furthermore it is usually assumed that binding is the only reaction, and that $f(u,v) = -g(u,v)$, i.e., v and u are membrane-bound and cytosolic forms of the same species that are converted point-wise in space. However, this is not strictly correct, since the volumetric change in u is not equivalent to the areal change in v [127].

In general, $f(u,v)$ incorporates both a nonlinear feedback component that reflects reinforcement of binding, i.e., a form of autocatalytic binding, as in the model described earlier, as well as saturation. Otsuji et al. [128] have used several different forms of the nonlinearity, *viz.*,

$$f(u,v) = a - 1\left\{\frac{u+v}{[a_2 S(u+v)+1]^2} - v\right\} \text{ or } f(u,v) = -a_1(u+v)[(\alpha u+v)(u+v)-a_2], \tag{8}$$

where $D_u = \alpha D_v$, S is intensity of stimulation, and a_1, a_2 are model parameters whose biological meanings are obscure. In a model due to Mori et al. [129] f is chosen as

$$f(u,v) = \left(k_0 + \gamma\frac{u^n}{u_0^n+u^n}\right)v - \delta u, \tag{9}$$

where k_0 is the rate of activation of u from v and δ is the rate of inactivation of u to v. Here a Hill function in u alone, as distinct from the first form in Equation (8), was used to describe self-activation or binding of u. The authors show that a wave initiated at one boundary can stall or be 'pinned' under suitable conditions, thus leading to stable polarization.

While these models provide some insight into cell polarization, they are in general too simplistic to make significant predictions concerning polarization in a given system. The most significant limitation of all models of the type at Equation (8) is that the stimulus is restricted to one point on the boundary. As a result, they cannot be used for understanding how a cell in reacts to a graded signal, since there is no extracellular signal except at the point of stimulation. A second problem concerns how the structure of the nonlinearities used might arise from a mechanistic description. Figure 18 shows the steps in a mechanistic description of yeast polarization, which would lead to a complex system of equations for components on the membrane that would be difficult to describe with two variables. Other models of the type in Equation (8) are reviewed in [130].

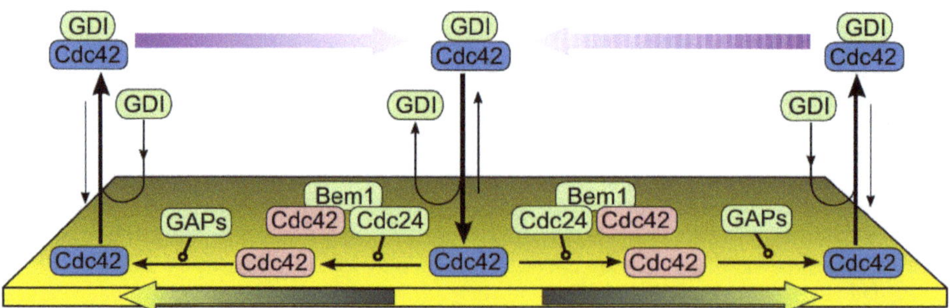

Figure 18. The components of a model for self-organized clustering of activated Cdc42 in yeast. Blue (pink) represents inactive (active) Cdc42 in the cluster. Inactive Cdc42 is activated at the membrane by a complex of active Cdc43, Bem1 and Cdc24. The guanine dissociation inhibitor (GDI) shuttles inactive Cdc42 to and from the membrane. From Goryachev and Pokhilko [111].

5.2. A Model for Direction-Sensing in Dictyostelium in cAMP Gradients

For a Dicty cell to align with the local gradient in a noisy chemotactic field, it must measure the local cAMP concentration at its surface and determine the direction in which to move. A precise choice is not necessary—a mathematical model predicts that cells can aggregate as long as they choose their direction within a cone of $\pm 135^\circ$ of the correct direction, but they aggregate more slowly [131]. A computational model of the $G_{\beta\gamma}$-AC-cAMP part of the network in Figure 5 shows that a for a sufficient length of time a cell experiences a significant difference in the front-to-back ratio of cAMP when a neighboring cell signals [132]. It follows from this that other components in the signal-transduction pathway will sustain similar front-to-back differences in a gradient, and experiments have shown that this holds for PIP3, PI3K, and PTEN.

A more recent model for the first downstream steps in the signal-transduction pathway in Dicty incorporates more of the underlying biochemistry and can replicate a number of experimental observations. These include amplification at the level of RasG (hereafter simply Ras), the observed biphasic response to graded stimuli, the existence of a refractory period for repeated stimuli, and 'memory' of the up-gradient direction in a wave [133]. In LatA-treated cells [134,135] the feedback effect from the actin cytoskeleton on Ras is eliminated, and the model is based on these experiments. Figure 5 shows that activated Ras activates PI3K and other downstream steps to actin polymerization, but the model was restricted to the Ras dynamics in response to cAMP because there is no known direct feedback to Ras from downstream steps between Ras and the actin cytoskeleton.

The model involves three main processes: signal detection via CAR1, transduction based on activation of $G_{\alpha_2\beta\gamma}$, and activation of Ras (Figure 19). The key components in the model are $G_{\alpha_2\beta\gamma}$, Ric8, (a GEF that activates G_{α_2} [136]), Ras, and RasGEF and RasGAP. All components except G_{α_2} cycle between the membrane and the cytosol. RasGEF and RasGAP are activated at the membrane by free $G_{\beta\gamma}$, and the translocation of RasGEF from the cytosol is enhanced by the activated form of G_{α_2}.

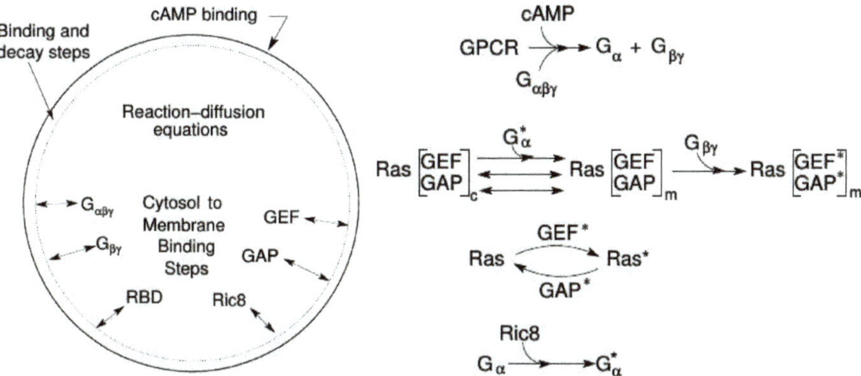

Figure 19. A schematic of the major processes in the model (**left**), and the primary steps in the network (**right**). From Cheng and Othmer [133].

It has been observed [134] that adaptation to constant cAMP stimuli occurs at the level of Ras, whose activity is controlled by a balance between RasGEF* and RasGAP*—none of the upstream components adapt. At low stimuli adaptation is near perfect, but at higher stimuli adaptation is imperfect. The model is able to capture the dose-dependent Ras activation and various patterns such rectification and refractoriness under uniform stimuli. It can be shown that $G^*_{\alpha 2}$ contributes to the observed imperfect adaptation in a uniform stimulus due to the asymmetrical translocation of RasGEF. Earlier we noted that the cell-level response to a uniform stimulus is a 'cringe', which appears within about 20 s and lasts about 30 s—comparable to the time-scale for adaptation of the Ras response.

Another experimental observation under uniform stimuli is that cells exhibit a refractory period after stimulation [10]. A short delay following a stimulus leads to a small response, and the response increases if the delay is increased. This is often taken as an indicator of excitability, but there is no indication that there is a threshold stimulus in the experiments or the model—the maximum Ras* response increases monotonically over four orders of magnitude of the stimulus. This can be explained by considering the ratio RasGEF*/RasGAP*. For short delays the slower inactivation of RasGAP* reduces the amount of Ras that can be activated. Under uniform stimuli Ric8 plays a minor role and *Ric8-null* cells respond essentially as WT cells. However, it plays a major role under graded stimuli.

Under graded stimuli the response in LatA-treated cells is biphasic: on a short time-scale (10 s) Ras is activated over the entire membrane, the activation decays within 20 s, and this is followed by a persistent polarization of Ras activation that is high at the high point of the gradient. The model reveals that the fast time-scale of $G_{\beta\gamma}$ mediated RasGEF and RasGAP activation induces the first transient Ras activation on the entire membrane, while the slow time-scale of overall equilibration—which includes redistribution due to diffusion, membrane localization and positive feedback between Ric8 and G_{α}—induces the delayed secondary response that produces the symmetry breaking. Figure 20a,b show that the biphasic response—initially uniform around the cell, followed by symmetry-breaking later—in a graded stimulus is captured.

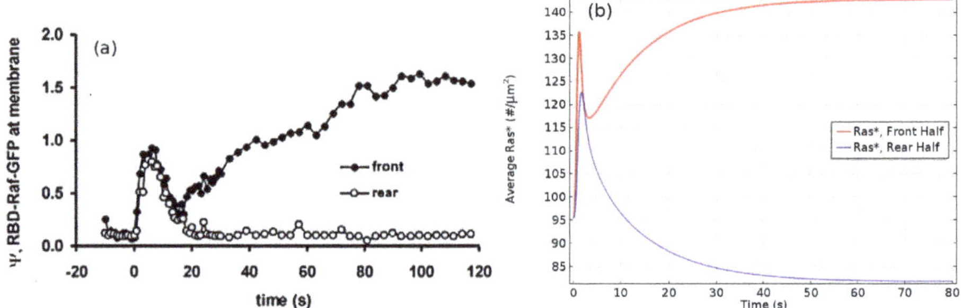

Figure 20. The levels of activated Ras* at the front and back in a static cAMP gradient as a function of time measured experimentally (**a**) From Kortholt et al. [135] and the model prediction (**b**) [133] From Cheng and Othmer [133].

Important insights into the role of diffusion emerge from the model. While all cytosolic components diffuse at the same rate, the model predicts the observed symmetry-breaking, and analysis shows that diffusion of both RasGEF, the activator, and RasGAP, the inhibitor, is necessary. Simulations also show that there is no symmetry-breaking in g_{α}-null cells and there is no direction-sensing in Ric8-null cells when exposed to a shallow gradient or a steep gradient with high mean concentration. Finally, slow diffusion of components on the membrane enhances, but is not necessary, for symmetry-breaking. As shown in Figure 21 left, only unrealistically high diffusion rates on the membrane removes the biphasic response. In particular, symmetry-breaking does not require a disparity between the diffusion coefficients of the activator (RasGEF) and the inhibitor (RasGAP), as is required in LEGI models.

Well-polarized cells are able to detect and respond to chemoattractant gradients with a 2% concentration difference between the anterior and posterior of the cell [63], and in Figure 22 we show the results of the model predictions for the response to the Ras gradient in unpolarized cells. One sees that the amplification is significant for a 2% and 20% difference, but less for a very steep gradient. Amplification of a cAMP gradient stems from two outputs of the network. First, the $G_{\alpha 2}^{*}$ concentration on the membrane is highest where the cAMP concentration is highest, and this produces higher localization and activation of Ric8, which reactivates $G_{\alpha 2}$ and further promotes RasGEF localization there. Secondly, faster $G_{\alpha_2\beta\gamma}$ re-association at the rear because Ric8 is lower there, which leads to lower $G^*_{\alpha 2}$ and creates gradients of $G_{\alpha_2\beta\gamma}$ and $G_{\beta\gamma}$, the former high at the rear and low at the front, and conversely for the latter, as is observed experimentally [137]. Furthermore, Ric8 contributes to the amplification of Ras activity by regulating $G_{\alpha 2}$ dynamics: the reactivation of $G_{\alpha 2}$ by Ric8 induces further asymmetry in $G_{\alpha_2\beta\gamma}$ dissociation, which in turn amplifies the Ras activity. Thus $G_{\alpha_2\beta\gamma}$ cycling modulated by Ric8 drives multiple phases of Ras activation and leads to direction-sensing and signal amplification in cAMP gradients. The biphasic response can be understood as follows. Initially the cAMP stimulus produces a nearly uniform response due to rapid diffusion of $G_{\beta\gamma}$ in the cytosol, but on a slower time-scale symmetry-breaking is driven by an 'indirect' positive feedback between Ric8 and G_{α_2} (activated G_{α_2} promotes Ric8 binding at the membrane, and activated Ric8 promotes reactivation of G_{α_2}). Increasing diffusion of membrane components reduces the spatial asymmetry this produces.

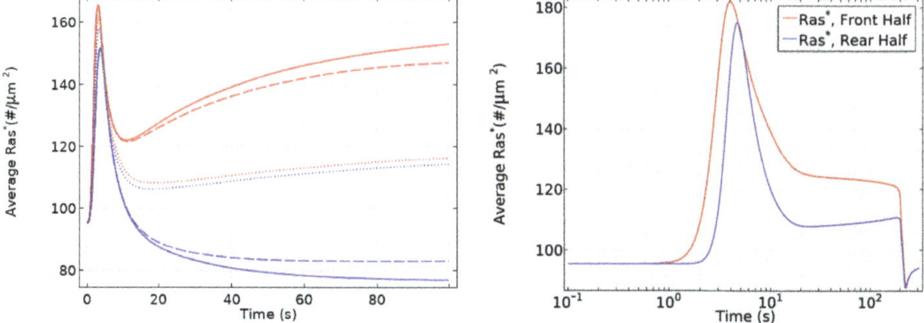

Figure 21. (**Left**) The effect of diffusion on the biphasic response. Solid, dashed and dotted lines are for D = 0, 0.1 and 10 μm²/s resp.—cytosolic diffusion of all components: D = 30 μm²/s. (**Right**) The average Ras* in the front and rear halves in response to a passing triangle wave. From Cheng and Othmer [133].

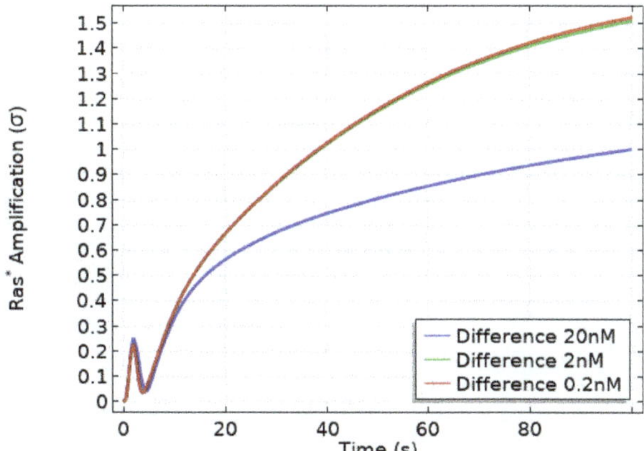

Figure 22. Amplification, as defined in [133], of the Ras signal in an unpolarized cell in a linear gradient with mean 10 nM and front-to-back differences as labeled.

In an imposed triangular wave of height 1 μm and wavelength 1 mm [138], Ras* at the front is always larger than at the rear throughout passage of the wave (Figure 21 right), which reflects a form of 'memory' of the point at which the cell first received the signal. This shows that symmetry-breaking at the level of Ras encodes sufficient 'memory' to maintain directional orientation during a passing wave and thus provides a solution to the 'back-of-the-wave' problem, in that cells do not turn to follow the cAMP gradient after the wave has passed, despite the fact that the spatial gradient reverses as the wave passes over the cell [10,139]. It should be emphasized that the model was built on the rounded LatA-treated cells that have no intrinsic polarity, which suggests that polarity is not necessary for the persistence of direction-sensing at the natural wave speed, even at the level of Ras activity.

6. The Integration of Signaling, Polarization and Structural Changes in the CSK

In general, establishment of polarity in an un-stimulated, unpolarized eukaryotic cell at rest that is exposed to a graded, time-independent external signal involves three major steps.

1. Detection of the chemical and mechanical signals in the ME with membrane receptors, adhesive sites, and other detection mechanisms.
2. Transduction of the extracellular signals into spatially biased intracellular signals that reflect the external signals and activate one or more downstream signaling pathways.
3. Translation of the output of these signaling pathways into the changes in the CSK needed to begin directed motion.

Similar steps occur in an already-polarized cell, but in that case the last step also involves the decision to change direction if necessary, or to simply continue motion.

While these steps may appear to involve simple feed-forward processes, there are numerous feedback loops between the signaling pathways (Figure 5) and significant overlap in their downstream effects, and thus the balances between them determine the response when all are functional. The complexity of the CSK [140] and the fact that the same cell type can use very different modes of motion in different MEs makes it difficult to translate what is known about steps 1 and 2 into a set of 'rules' for carrying out step 3. Moreover, we have thus far focused on chemical signals to the exclusion of mechanical signals, but Dicty, neutrophils and other cell types continuously monitor their ME and adapt their mode of movement to it. For example, in a fluid Dicty swims, while in other environments it moves either by extending pseudopods and contracting the rear, by blebbing, or by a combination of these. The evolutionary advantage of this flexibility is clear, but it also means that determining the rules for implementing step 3 remains a major challenge. However, we can identify components of what is involved in implementing step 3 under chemotactic gradients, which is done in the context of Dicty next. Moreover, the component parts are fairly universal [4], and there is evidence that mechanical stimuli act through the same pathways as chemical signals in Dicty [141].

6.1. *How Graded Chemical Signals Lead to Polarization*

Since cAMP receptors remain uniformly distributed on the membrane following stimulation [137], polarization first occurs at the level of Ras, G_{α_2}, and $G_{\beta\gamma}$ (Figure 5), followed by adaptation in Ras activation. Activated Ras activates PI3K, which leads to a local increase in PIP3 production and a local increase in PI3K, the latter dependent on actin polymerization [86]. Thus, without any interaction with other pathways, there is a front-to-rear (Hereafter we refer to the region on the membrane that receives the highest stimulus as the 'front', and the antipodal part the 'rear'). decrease in activated Ras, PI3K and PIP3. Since all points on the cell receive the cAMP signal, the signal transduction network is active over the entire cell and the gradients that arise are the global composite of local changes and diffusive and other types of transport.

PIP3 has a PH domain that serves as a docking site for cytoplasmic proteins such as the the GTPase Rac1 and the kinase Akt. The increase in PIP3 leads to rapid binding and activation of Rac1 via a GEF, and rapid localization and activation of Akt, which is essential for CSK polarization and chemotaxis—mutants lacking Akt cannot polarize the CSK properly in a chemotactic gradient and the cells move slowly [142]. Experimentally it is found that rapid withdrawal of the gradient leads to the return of PTEN and PH_{CRAC}-GFP (labeled CRAC) to their pre-stimulus distribution, but reapplication of a uniform cAMP stimulation produces a clear PH_{CRAC}-GFP translocation to the rear, but not to the front [143]. This indicates that a stronger 'inhibition' of polarization is maintained at the front of a polarized cell. It was shown that this inhibition is not caused by PTEN, $G\alpha_1$ or $G\alpha_9$, but the observations remain to be explained.

The SCAR/WAVE regulatory complex (WRC) is a five-protein complex that binds both activated Rac1 (Rac1GTP) and Arp2/3, and thus provides a link between the two that leads to formation of branched actin [144]. Another member of the WAVE family, WASP, also binds Rac1GTP, and can activate Arp2/3 and produce pseudopods in the absence of WAVE, but plays other roles when WAVE is expressed [145]. The protein complex DGap1/cortexillin also binds Rac1GTP, but apparently only acts to sequester it [146]. Cortexillin is known to bind to PIP2, which increases down-regulation of Rac1 at the rear.

Because PTEN docks to PIP2, the reduction of PIP2 due to conversion into PIP3, coupled with possible inhibition of PTEN localization by PIP3 [147], reduces the membrane-attached PTEN, which produces a reverse gradient in bound PTEN and further increases PIP3 at the leading edge. In addition, an increase of PTEN at the rear decreases PIP3 there, further amplifying the front-to-rear PIP3 gradient. Thus, one of the second steps in polarization is establishment of the front-to-rear gradients in PIP3, AKT, and the SCAR/WAVE regulatory complex (WRC) and the reverse gradient in PTEN.

Myo-II has several effects in the cortex. One is to stabilize it by associating with anti-parallel linear actin filaments to produce actomyosin, and the other is contraction of the filaments needed both in movement by blebbing and via pseudopods. The motor activity of myo-II, independent from its cross-linking function, is up-regulated by myosin light chain kinases (MLCK). In a pathway parallel to the $G_{\beta\gamma}$ pathways, G_{α_2} activates Rap1 (Figure 5a) and a downstream effector, the kinase Phg2. This localizes and activates the heavy-chain kinase MHCK, which leads to myo-II disassembly [148] and in turn reduces the cortical density and facilitates branched actin polymerization and pseudopod extension. PakA inhibits the cGMP-promoted MLCK activation of motor activity and hence reduces contractility [43], and together this leads to a front-to-back gradient of free myo-II, which can lead to an increase of its L-actin-binding at the rear. Thus, the spatio-temporal balance of the effectors of the cGMP, Ras and Rap1 pathways controls actin polymerization and actomyosin assembly, as well as their spatial localization [4].

It is known that myo-II is localized at the rear of migrating Dicty cells [149], but whether PTEN controls its localization is not known. It has been shown that PTEN localization at the sides and the rear of cells occurs prior to myo-II localization there [150], and it was suggested that PTEN may be involved in a positive feedback loop in which contraction enhances accumulation of PTEN and myo-II [150]. Since PI(4,5)P2 promotes membrane-binding of PTEN, the gradient of PIP2 increases its posterior localization [151], but PTEN is not the sole controller of myo-II localization—it still localizes in *pten*$^-$ cells. This may involve the cGMP pathway in Dicty [152,153], and in other cells myo-II preferentially binds to actin filaments in tension, and a reduction in the tension leads to release of myo-II [154].

In the presence of diffusion of components on the membrane and in the cytosol, the composite effect of the processes described would be to produce smooth variation of the components on the membrane and those in the cytosol. If we define 'frontness' by a propensity to produce predominately branched actin and pseudopods, whereas 'rearness' is characterized by a preponderance of linear actin and actomyosin, how does a cell polarize into a well-defined front and rear? Do these characteristics vary smoothly in proportion to the gradients described above, or are there additional steps that sharpen the distributions? Experimental images of tagged components suggest the latter, but this can be misleading because there is always a threshold in detection of labeled components. Assuming that the separation is quite sharp, how can the gradients be amplified locally? For instance, if there is cooperative binding similar to that in models described earlier, will the frontness and rearness be more clearly separated? Since activated Rac1GTP and WAVE are key components in branched actin production, can the WRC and Rac1GTP be localized more sharply at the front?

WAVE binds to both Rac1GTP and PIP3, and a possible step in this direction is shown in Figure 23, where the WRC-Rac1GTP-PIP3 units form clusters that produce branched actin more rapidly than the sum of the individual units. If complex formation between WRC-Rac1GTP and PIP3 evolves according to

$$\frac{dC}{dt} = F(\text{WRC-Rac1GTP, PIP3}) \cdot \text{WRC} - kC \tag{10}$$

where C is the WRC-Rac1GTP-PIP3 complex, then localized C will result for an appropriate F provided diffusion in the membrane is not too rapid. For instance, if F is increasing in both arguments and reflects cooperativity in Rac1GTP and PIP3, either separately or jointly, then formation of the complex will be restricted to regions in which both Rac1GTP and PIP3 are large. This is not a mechanistic description, but rather a qualitative argument of what a more detailed mechanism could produce. Moreover, this is not the complete story in Dicty, for in the absence of WASP, WRC-RacGTP accumulates at the rear of the cell [155]. In addition to activating Arp2/3, WASP is also thought to remove Rac1GTP from the membrane, thus depleting active Rac1GTP at the rear. DGap1/cortexillin complexes may have a similar role. The combination of these steps can lead to a relatively sharp variation between the region in which formation of branched actin dominates and that in which linear actin and actomyosin prevail.

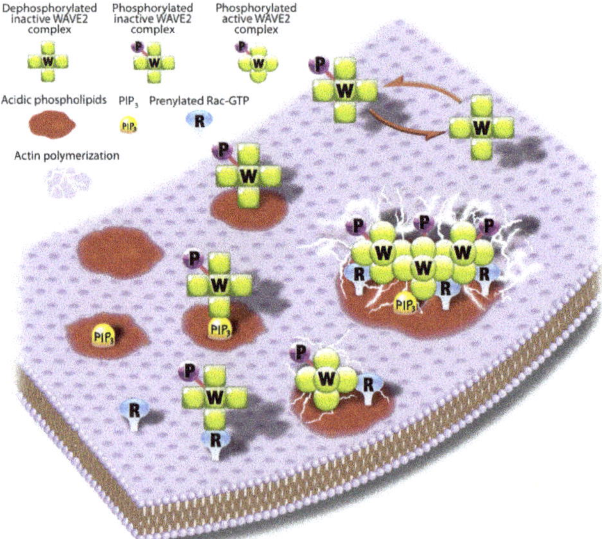

Figure 23. A possible step in the localization of WRC. From Lebensohn and Mitchison [144] with permission.

An alternate approach to generating the separation between frontness and rearness has been suggested in the context of cancer cells [156]. In that approach a sharp demarcation is achieved with a network in which RhoA and Rac1 are linked by a double-negative feedback loop. This leads to a spatial distribution of RhoA, Rac, and the inhibitor PAK, and the boundary between frontness and rearness occurs at particular values of PAK at which there is a spatial discontinuity in the RHoA and Rac distributions. Such discontinuities would be difficult to sustain in the presence of diffusion, but the effects of diffusion are not considered by the authors.

Other factors may also play a role in polarization. For example, cofilin promotes breakup of actin filaments, and suppression of its expression results in re-localization of Arp2/3 to one pole and protrusions from only that pole [157]. Myo-IB, the membrane-cortex linker protein [81], preferentially binds to PIP2, and thus is released when PIP2 is converted to PIP3. Another potential factor is profilin, which increases formin-mediated elongation rates in a concentration-dependent manner [158]. At the same time, profilin-bound monomers inhibit the polymerase activity of WH domains of SCAR/WAVE and WASP by competing for G-actin monomers [159]. Although recent studies demonstrate that WAVE contains a proline-rich domain, which is capable of delivering free actin monomers to barbed ends in vitro [160], its activity could be slow compared to that of filament elongators such as formins.

6.2. The Role of Membrane and Cortical Tension in Polarization

A question that arises in the context of the preceding models in which diffusion is the primary transport mechanism is whether diffusion is fast enough to change the polarity of a cell in response to changes in the signal. In Dicty directional changes of a shallow gradient induce polarized cells to turn, whereas large changes lead to large-scale disassembly of motile components and creation of a new 'leading edge' directed toward the stimulus [161]. In Figure 24 one sees that fibroblasts require 40 min to re-orient 90°, whereas a Dicty cell can re-polarize in 40 s. Computational experiments based on the model in Section 5.2 show that Ras activation can be reversed in 50–60 s in response to large-amplitude reversals of the cAMP gradient, but diffusion alone may not suffice, since reversal becomes much slower when exposed to a weaker reversed gradient. Moreover, it has not been demonstrated that the necessary rearrangements of factors controlling the CSK can redistribute rapidly enough via diffusion. In fact, it has been shown that a diffusion-based polarization mechanism cannot provide long-range inhibition of secondary pseudopods in neutrophils, and it was suggested that membrane tension may be involved [123]. Since the membrane is generally modeled as elastic or viscoelastic, changes in tension propagate much more rapidly than diffusion-propagated signals, and may be involved in suppression of pseudopods toward the rear in both for Dicty and neutrophils.

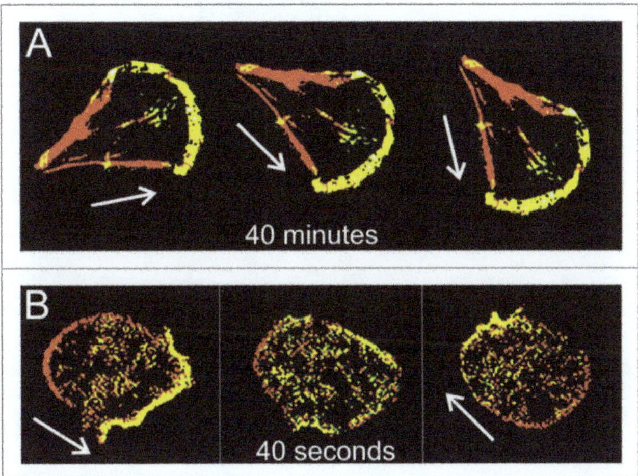

Figure 24. Re-orientation of fibroblasts (**A**) and Dicty (**B**). From Faix et al. [162].

Membrane tension plays a role in other contexts as well. Numerous proteins that contain a BAR domain can associate with curved membranes because they are sensitive to curvature [163]. Elevated membrane tension reduces the local curvature and can reduce the binding of such proteins [164]. This might regulate the membrane-binding of GEFs and GAPs that regulate the GTPase switches, which in turn provides feedback between curvature and actin dynamics. Cells such as fibroblasts sense the rigidity of the ECM via stress transmitted through integrin-mediated focal adhesions, which can lead to conformational changes in proteins within the complex. For example, in the case of BCAR1 proteins, force applied to the adhesion complex leads to exposure of phosphorylation sites for SRC-family kinases that can recruit signaling proteins and up-regulate the activity of Rac1 and Rap1.

In another example of mechanical effects, Dicty cells in a fluid flow establish a protruding front directed against the flow and a retracting rear, as indicated by labels for polymerized actin and myo-II markers at the front and rear, resp [165]. At a shear stress of ∼2.1 Pa the cell becomes polarized with an actin-enriched front upstream, and when the flow is reversed quickly, cells reverse their polarity in several phases. First, actin disassembles at the previous front between 0∼60 s after flow reversal. Then polymerization of a new front upstream begins at 30 s and stabilizes by ∼90 s. In the interim the amount of actin in the cortex decreases, which means that polarity reversal entails a significant re-building of the entire cortex. How shear stress is transduced into control of actin polymerization is not known, but as remarked earlier, it is thought that the signaling pathways are the same for both chemical and mechanical signals. Interestingly, the authors noted that similar patterns of front and rear inter-conversion were observed in cells re-orienting in strong gradients of cAMP.

Recent work has also shown that some cell types use strong cortical flows to propel themselves, and the intracellular actin flows that are generated polarize the cell and could move other signaling molecules axially. Ruprecht et al. [166] show that a stable non-polarized blebbing cell can be converted into a permanently polarized shape by increasing the contractility in cells. They also report cortical flow rates of 10's of μms/min (Figure 25), which would induce an anterior-to-posterior cytoplasmic flow near the cortex, and thus a posterior-to-anterior flow in the center, as shown in Figure 25. The authors suggest that there is a high growth rate of the cortex at the front of a cell and a high disassembly rate at the rear, which would require a very different set of controls for the actin network.

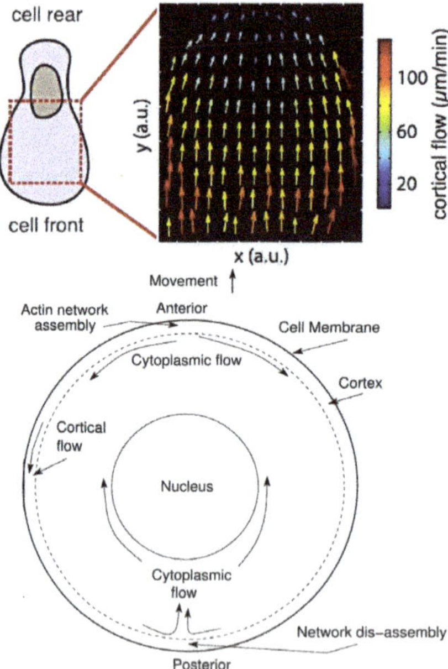

Figure 25. The measured cortical flow (**top**) (From Ruprecht et al. [166]), and the postulated intracellular flows (**bottom**).

A second stable-bleb type is more cylindrical and has a large uropod [167]. This also involves high myo-II activity and strong retrograde actin flow, and arises when slow mesenchymal cells undergo a MAT under low adhesion and confinement between plates. Evidence for involvement of the cortex in both cases is the fact that blebbistatin, an inhibitor of myo-II contractility, and LatA both inhibit polarization. It is thought that a gradient of cortical density and myo-II generates both the cortical flow and an axial pressure gradient in both morphologies, but what initiates the flow remains undetermined [168]. Computations reported in Wu et al. [17] show that tension gradients in the cortex can generate large-scale flows sufficient to carry monomers anteriorly, and thus can provide another mechanism for polarization by segregating components via the flow.

7. Epilogue and Open Problems

Our objective in this review was to describe some of the wide range of problems that arise in trying to understand cell motility. In the previous sections we discussed recent advances in understanding the dynamics of intracellular biochemical networks and how they are involved in actin waves, direction-sensing and polarization, but the problem of understanding how chemical and mechanical signals are used to control movement is both broader and deeper. Broader in the sense that other pathways not touched upon are involved, and deeper in the sense that our knowledge of the details of transduction of mechanical signals is shallow in many respects. As a result, there are many open questions that remain to be solved. Several that are closely related to topics discussed earlier are as follows.

- What is the minimal set of components of the network shown in Figure 5 that can control the random initiation of intracellular waves in un-stimulated cells? Experimental work described earlier suggests that a minimal set in Dicty may be SCAR/WAVE, Arp 2/3 and actin-binding proteins, but there are presently no models that can replicate the experimental results. A related question is what controls the initiation sites for pseudopodia. Is it randomness in the wave generation, or are there randomly located sites of decreased membrane tension that facilitate membrane deformation, or both?
- A question raised earlier concerns how cells establish a sharp demarcation between 'frontness' and 'backness' in the presence of an extracellular signal. This involves the spatial distribution of numerous species, and a minimal set of components to produce the demarcation is not yet known. A related question is how the strength of the signal determines whether the cell turns in response to a change in direction of the signal, or whether it completely rebuilds the CSK.
- There are as yet no models that integrate mechanical and chemical pathways to predict actin flows and structural changes in the CSK—even within a fixed cell shape. In the previous sections we simply described how some of the separate components may be involved in polarization, but their integration remains to be addressed.
- A larger question is how these pathways control the mode of migration used by a cell. Cells moving on flat surfaces often use lamellipodia, but movement in confined spaces can prevent the extension of lateral membrane protrusions, which may account in part for the use of blebs in confined spaces. The coexistence of blebs and pseudopods in Dicty suggests that the balance can be subtle, but there are experimental conditions under which one or the other dominates. Since cells often move in a spatially variable environment, the feedback from the ME can affect the mode of movement dynamically, and far more work is needed to understand how the cell-ME interaction controls the mode of movement. Significant progress has been made on simpler systems such as keratocytes moving on a flat surface [169], and recent techniques that can capture more dynamic shape changes in 3D via interface tracking shows promise [170], but much remains to be done. In the context of swimmers such as shown in Figure 3, a model in which protrusions propagate along the body length can replicate swimming speeds under various conditions [28], but how extension of protrusions is controlled by local fluid properties and other factors is not yet known.

Concerning pathways not described, the protein-calcium pathway in Figure 5 has attracted much less attention than other pathways in the context of chemotaxis, but it may play a significant role there. Early work suggested that calcium is not essential for chemotaxis [171], but other work shows that it plays an important role. Lusche et al. [172] show that extracellular calcium acts as a chemoattractant in parallel with cAMP, while other research shows that calcium is vital in cell migration of various cell types [173–178]. In macrophages and glial cells calcium influx plays a major role in maintaining the structure of the leading edge during migration [175,179]. In Dicty cAMP both stimulates and inhibits PLC activity via G_{α_2} and G_{α_1} protein subunits, resp. [180,181]. Kortholt et al. [182] reported that plc-null cells are resistant to the PI3K inhibitor LY294002 and produce little PIP3 after cAMP stimulation, while PLC over-expression increases PIP3, which affects chemotaxis similar to loss of PTEN. The dynamics of intracellular calcium range from individual stochastic events to global phenomena like waves and oscillations following stimulation [183–185], and given the excitability of the IP3-Calcium module, PIP2-PLC-Calcium and PIP3-PI3K-PTEN triangles can potentially inherit the excitability. Thus, an open problem is to investigate whether integration of the PIP2-PLC-Calcium triangle with the PIP2-PIP3-PI3K triangle could shed more light on the self-organization mechanisms.

Another aspect that deserves more attention concerns the role of stochastic fluctuations at various steps of the signaling pathways and network dynamics. Estimates made earlier of the signal noise in cAMP receptors in Dicty shows that noise may be important at low signal levels [47], but stochastic simulations of the exterior reaction-diffusion system are needed to make this more precise. Separately, given the more detailed models of intracellular signaling that are now available, an analysis of how cells cope with noise in the signals is feasible. For example, it was noted earlier that Dicty cells need not be very precise in their detection of the chemotactic gradient to aggregate, but imprecision carries the cost of less efficient aggregation [132]. Related to the question of how the random initiation of intracellular waves is controlled is the question of stochastic effects on the location of actin puncti at potential sites of protrusion. A stochastic model using a simplified signaling network shows how random actin spots can shrink and die or develop into full-fledged propagating waves [186], but further work on this is needed.

In summary, it is safe to say that a deep understanding of how the nanomachines that we call cells move is still is the future.

Author Contributions: All authors contributed equally to the writing and editing of this article. All authors have read and agreed to the published version of the manuscript.

Funding: Supported in part by NSF Grant DMS # 178743, NSF award CON-75851, project 00074041, and by NIH Grants # GM29123 and 54-CA-210190. Any opinions, findings, and conclusions or recommendations expressed in this material are those of the authors and do not necessarily reflect the views of the National Science Foundation or of their current employers, nor do they necessarily represent the official views of the National Institutes of Health.

Conflicts of Interest: The authors declare no conflict of interest.

Abbreviations

The following abbreviations are used in this manuscript:

F-actin	Branched and linear actin or either
CON	CSK oscillatory network
Dicty	Dictyostelium discoideum
$G_{\beta\gamma}$	G beta-gamma complex
$G_{\alpha 2}$	G-protein alpha-subunit
GEF	GTP exchange factor
GAP	GTPase-activating factor
DGAP1	IQGAP-related protein
PHP	PH domain proteins
Akt	PI3 kinase and protein kinase B
WRC	SCAR/WAVE regulatory complex
SHIP	SH2-containing inositol 5-phosphatase
WAVE	WASP-family verprolin-homologous protein
WASP	Wiskott–Aldrich syndrome protein
Arp2/3	actin-related protein 2 and 3 complex
AC	adenylate cyclase
B-actin	branched actin
Ca^2	calcium
CaM	calmodulin
CapP	capping proteins
Ctx	cortexilin
CAR	cyclic AMP receptor
cGMP	cyclic GMP
cAMP	cyclic AMP
CSK	cytoskeleton
CRAC	cytosolic regulator of adenylyl cyclase
DAG	diacylglycerol
ELMO	eukaryotic engulfment and cell motility proteins
ECM	extracellular matrix
G-actin	free actin monomer
GFP	green-fluorescent-protein
GDI	guanine dissociation inhibitor
GDP	guanosine diphosphate
GTP	guanosine triphosphate
GC	guanylate cyclase
IP3	inositol 1,4,5-trisphosphate
LatA	latrunculin A
L-actin	linear actin
LEGI	local excitation and global inhibition
MAT	mesenchymal-to-amoeboid transition
ME	microenvironment
MHCK	myosin heavy-chain kinase
MLCK	myosin light chain kinase
myo-IB	myosin-IB
myo-II	non-muscle myosin-II
PakA	p21-activated kinase A
PTEN	phosphatase and tensin homologue
PIP3	phosphatidylinositol 3,4,5-trisphosphate
PI3K	phosphatidylinositol-3 kinase
PIP2	phosphatidylinositol-4,5-diphosphate
PI5K	phosphatidylinositol-5 kinase
PLA2	phospholipase A_2
PLC	phospholipase C
PH	pleckstrin homology
STEN	signal-transduction excitable network
TORC2	target of rapamycin complex 2
WT	wild-type

References

1. Nürnberg, A.; Kitzing, T.; Grosse, R. Nucleating actin for invasion. *Nat. Rev. Cancer* **2011**, *11*, 177–187. [CrossRef] [PubMed]
2. Othmer, H.G. Eukaryotic cell dynamics from crawlers to swimmers. *Wiley Interdiscip. Rev. Comput. Mol. Sci.* **2019**, *9*, e1376. [CrossRef]
3. Haeger, A.; Wolf, K.; Zegers, M.M.; Friedl, P. Collective cell migration: Guidance principles and hierarchies. *Trends Cell Biol.* **2015**, *25*, 556–566. [CrossRef] [PubMed]
4. Artemenko, Y.; Lampert, T.J.; Devreotes, P.N. Moving towards a paradigm: Common mechanisms of chemotactic signaling in Dictyostelium and mammalian leukocytes. *Cell. Mol. Life Sci.* **2014**, *71*, 3711–3747. [CrossRef] [PubMed]
5. Charras, G.; Sahai, E. Physical influences of the extracellular environment on cell migration. *Nat. Rev. Mol. Cell Biol.* **2014**, *15*, 813–824. [CrossRef] [PubMed]
6. Hind, L.E.; Vincent, W.J.; Huttenlocher, A. Leading from the back: The role of the uropod in neutrophil polarization and migration. *Dev. Cell* **2016**, *38*, 161–169. [CrossRef] [PubMed]
7. Petri, B.; Sanz, M.J. Neutrophil chemotaxis. *Cell Tissue Res.* **2018**, *371*, 425–436. [CrossRef]
8. Weiner, O.D.; Marganski, W.A.; Wu, L.F.; Altschuler, S.J.; Kirschner, M.W. An actin-based wave generator organizes cell motility. *PLoS Biol.* **2007**, *5*, 2053–2063. [CrossRef]
9. Asano, Y.; Nagasaki, A.; Uyeda, T.Q.P. Correlated waves of actin filaments and PIP3 in Dictyostelium cells. *Cell Motil. Cytoskelet.* **2008**, *65*, 923–934. [CrossRef]
10. Huang, C.H.; Tang, M.; Shi, C.; Iglesias, P.A.; Devreotes, P.N. An excitable signal integrator couples to an idling cytoskeletal oscillator to drive cell migration. *Nat. Cell Biol.* **2013**, *15*, 1307–1316. [CrossRef]
11. Goryachev, A.B.; Leda, M. Many roads to symmetry breaking: Molecular mechanisms and theoretical models of yeast cell polarity. *Mol. Biol. Cell* **2017**, *28*, 370–380. [CrossRef] [PubMed]
12. Friedl, P.; Wolf, K. Plasticity of cell migration: A multiscale tuning model. *J. Cell Biol.* **2010**, *188*, 11–19. [CrossRef]
13. Binamé, F.; Pawlak, G.; Roux, P.; Hibner, U. What makes cells move: Requirements and obstacles for spontaneous cell motility. *Mol. BioSyst.* **2010**, *6*, 648–661. [CrossRef] [PubMed]
14. Pollard, T.D.; Blanchoin, L.; Mullins, R.D. Molecular mechanisms controlling actin filament dynamics in nonmuscle cells. *Ann. Rev. Biophys. Biomol. Struct.* **2000**, *29*, 545–576. [CrossRef] [PubMed]
15. Sanz-Moreno, V.; Gadea, G.; Ahn, J.; Paterson, H.; Marra, P.; Pinner, S.; Sahai, E.; Marshall, C.J. Rac activation and inactivation control plasticity of tumor cell movement. *Cell* **2008**, *135*, 510–523. [CrossRef]
16. Lämmermann, T.; Bader, B.L.; Monkley, S.J.; Worbs, T.; Wedlich-Söldner, R.; Hirsch, K.; Keller, M.; Förster, R.; Critchley, D.R.; Fässler, R.; et al. Rapid leukocyte migration by integrin-independent flowing and squeezing. *Nature* **2008**, *453*, 51–55. [CrossRef]
17. Wu, H.; de Leon, M.A.P.; Othmer, H.G. Getting in shape and swimming: The role of cortical forces and membrane heterogeneity in eukaryotic cells. *J. Math. Biol.* **2018**, 1–32. [CrossRef]
18. Driscoll, M.K.; McCann, C.; Kopace, R.; Homan, T.; Fourkas, J.T.; Parent, C.; Losert, W. Cell shape dynamics: From waves to migration. *PLoS Comp. Biol.* **2012**, *8*. [CrossRef]
19. Friedl, P.; Alexander, S. Cancer Invasion and the Microenvironment: Plasticity and Reciprocity. *Cell* **2011**, *147*, 992–1009. [CrossRef]
20. van Zijl, F.; Krupitza, G.; Mikulits, W. Initial steps of metastasis: Cell invasion and endothelial transmigration. *Mutat. Res. Mutat. Res.* **2011**, *728*, 23–34. [CrossRef]
21. Wolf, K.; Mazo, I.; Leung, H.; Engelke, K.; Andrian, U.H.V.; Deryugina, E.I.; Strongin, A.Y.; Bröcker, E.B.; Friedl, P. Compensation mechanism in tumor cell migration: Mesenchymal-amoeboid transition after blocking of pericellular proteolysis. *J. Cell Biol.* **2003**, *160*, 267–277. [CrossRef] [PubMed]
22. Friedl, P.; Wolf, K. Tumour-cell invasion and migration: Diversity and escape mechanisms. *Nat. Rev. Cancer* **2003**, *3*, 362–374. [CrossRef] [PubMed]
23. Zatulovskiy, E.; Tyson, R.; Bretschneider, T.; Kay, R.R. Bleb-driven chemotaxis of Dictyostelium cells. *J. Cell Biol.* **2014**, *204*, 1027–1044. [CrossRef]
24. Tyson, R.A.; Zatulovskiy, E.; Kay, R.R.; Bretschneider, T. How blebs and pseudopods cooperate during chemotaxis. *Proc. Nat. Acad. Sci. USA* **2014**, *111*, 11703–11708. [CrossRef]

25. Bergert, M.; Chandradoss, S.D.; Desai, R.A.; Paluch, E. Cell mechanics control rapid transitions between blebs and lamellipodia during migration. *Proc. Nat. Acad. Sci. USA* **2012**, *109*, 14434–14439. [CrossRef]
26. Charras, G.T.; Paluch, E. Blebs lead the way: How to migrate without lamellipodia. *Nat. Revs Molec Cell Biol.* **2008**, *9*, 730–736. [CrossRef]
27. Barry, N.P.; Bretscher, M.S. Dictyostelium amoebae and neutrophils can swim. *Proc. Nat. Acad. Sci. USA* **2010**, *107*, 11376–11380. [CrossRef]
28. Wang, Q.; Othmer, H.G. Computational analysis of amoeboid swimming at low Reynolds number. *J. Math. Biol.* **2015**, *72*, 1893–1926. [CrossRef]
29. Howe, J.D.; Barry, N.P.; Bretscher, M.S. How do amoebae swim and crawl? *PLoS ONE* **2013**, *8*, e74382. [CrossRef]
30. Renkawitz, J.; Sixt, M. Mechanisms of force generation and force transmission during interstitial leukocyte migration. *EMBO Rep.* **2010**, *11*, 744–750. [CrossRef]
31. Van Haastert, P.J.M. Amoeboid Cells Use Protrusions for Walking, Gliding and Swimming. *PLoS ONE* **2011**, *6*, e27532. [CrossRef] [PubMed]
32. Parkinson, J.S. Bacterial chemotaxis: A new player in response regulator dephosphorylation. *J. Bacteriol.* **2003**, *185*, 1492–1494. [CrossRef] [PubMed]
33. Xin, X.; Othmer, H.G. A Trimer of Dimers- Based Model for the Chemotactic Signal Transduction Network in Bacterial Chemotaxis. *Bull. Math. Biol.* **2012**, *74*, 2339–2382. [CrossRef] [PubMed]
34. Sasaki, A.T.; Janetopoulos, C.; Lee, S.; Charest, P.G.; Takeda, K.; Sundheimer, L.W.; Meili, R.; Devreotes, P.N.; Firtel, R.A. G protein-independent Ras/PI3K/F-actin circuit regulates basic cell motility. *J. Cell Biol.* **2007**, *178*, 185–191. [CrossRef] [PubMed]
35. Wessels, D.; Soll, D.; Knecht, D.; Loomis, W.; Lozanne, A.D.; Spudich, J. Cell Motility and Chemotaxis in *Dictyostelium* Amebae Lacking Myosin Heavy Chain. *Dev. Biol.* **1988**, *128*, 164–177. [CrossRef]
36. Condeelis, J.; Segall, J.E. Intravital imaging of cell movement in tumours. *Nat. Rev. Cancer* **2003**, *3*, 921–930. [CrossRef]
37. Li, L.; Nørrelykke, S.F.; Cox, E.C. Persistent cell motion in the absence of external signals: A search strategy for eukaryotic cells. *PLoS ONE* **2008**, *3*, e2093. [CrossRef]
38. Devreotes, P.; Horwitz, A.R. Signaling networks that regulate cell migration. *Cold Spring Harb. Perspect. Biol.* **2015**, *7*, a005959. [CrossRef]
39. Bretschneider, T.; Othmer, H.G.; Weijer, C.J. Progress and perspectives in signal transduction, actin dynamics, and movement at the cell and tissue level: Lessons from Dictyostelium. *Interface Focus* **2016**, *6*, 20160047. [CrossRef]
40. Wennerberg, K.; Rossman, K.L.; Der, C.J. The Ras superfamily at a glance. *J. Cell Sci.* **2005**, *118*, 843–846. [CrossRef]
41. Etienne-Manneville, S. Cdc42–the centre of polarity. *J. Cell Sci.* **2004**, *117*, 1291. [CrossRef]
42. Sanz-Moreno, V.; Marshall, C.J. The plasticity of cytoskeletal dynamics underlying neoplastic cell migration. *Curr. Opin. Cell Biol.* **2010**, *22*, 690–696. [CrossRef]
43. Charest, P.G.; Firtel, R.A. Big roles for small GTPases in the control of directed cell movement. *Biochem. J.* **2007**, *401*, 377–390. [CrossRef]
44. Katoh, K.; Kano, Y.; Amano, M.; Onishi, H.; Kaibuchi, K.; Fujiwara, K. Rho-kinase-mediated contraction of isolated stress fibers. *J. Cell Biol.* **2001**, *153*, 569–583. [CrossRef]
45. Kolsch, V.; Charest, P.G.; Firtel, R.A. The regulation of cell motility and chemotaxis by phospholipid signaling. *J. Cell Sci.* **2008**, *121*, 551–559. [CrossRef]
46. King, J.S.; Insall, R.H. Chemotaxis: Finding the way forward with Dictyostelium. *Trends Cell Biol.* **2009**, *19*, 523–530. [CrossRef]
47. Othmer, H.G.; Schaap, P. Oscillatory cAMP signaling in the development of *Dictyostelium discoideum*. *CMTS Theor. Biol.* **1998**, *5*, 175–282.
48. Chung, C.Y.; Firtel, R.A. PAKa, a putative PAK family member, is required for cytokinesis and the regulation of the cytoskeleton in Dictyostelium discoideum cells during chemotaxis. *J. Cell Biol.* **1999**, *147*, 559–576. [CrossRef]
49. Jilkine, A.; Edelstein-Keshet, L. A comparison of mathematical models for polarization of single eukaryotic cells in response to guided cues. *PLoS Comp. Biol.* **2011**, *7*, e1001121. [CrossRef]

50. Allard, J.; Mogilner, A. Traveling waves in actin dynamics and cell motility. *Curr. Opin. Cell Biol.* **2012**, *25*, 1–9. [CrossRef]
51. Sept, D.; Carlsson, A.E. Modeling large-scale dynamic processes in the cell: Polarization, waves, and division. *Q. Rev. Biophys.* **2014**, *47*, 221. [CrossRef] [PubMed]
52. Condeelis, J.; Bresnick, A.; Demma, M.; Dharmawardhane, S.; Eddy, R.; Hall, A.L.; Sauterer, R.; Warren, V. Mechanisms of amoeboid chemotaxis: An evaluation of the cortical expansion model. *Dev. Genet.* **1990**, *11*, 333–340. [CrossRef] [PubMed]
53. Postma, M.; Roelofs, J.; Goedhart, J.; Gadella, T.W.; Visser, A.J.; Haastert, P.J.V. Uniform cAMP stimulation of Dictyostelium cells induces localized patches of signal transduction and pseudopodia. *Mol. Biol. Cell* **2003**, *14*, 5019–5027. [CrossRef]
54. Chen, L.; Janetopoulos, C.; Huang, Y.E.; Iijima, M.; Borleis, J.; Devreotes, P.N. Two phases of actin polymerization display different dependencies on PI(3,4,5)P3 accumulation and have unique roles during chemotaxis. *Mol. Biol. Cell* **2003**, *14*, 5028–5037. [CrossRef]
55. Yan, J.; Mihaylov, V.; Xu, X.; Brzostowski, J.A.; Li, H.; Liu, L.; Veenstra, T.D.; Parent, C.A.; Jin, T. A $G\beta\gamma$ Effector, ElmoE, Transduces GPCR Signaling to the Actin Network during Chemotaxis. *Dev. Cell* **2012**, *22*, 92–103. [CrossRef] [PubMed]
56. Varnum, B.; Edwards, K.B.; Soll, D.R. *Dictyostelium* Amebae Alter Motil. Differ. Response Increasing Versus Decreasing Temporal Gradients cAMP. *J. Cell Biol.* **1985**, *101*, 1–5. [CrossRef]
57. Hall, A.L.; Schlein, A.; Condeelis, J. Relationship of pseudopod extension to chemotactic hormone-induced actin polymerization in amoeboid cells. *J. Cell Biol.* **1988**, *37*, 285–299. [CrossRef]
58. Wessels, D.; Murray, J.; Soll, D.R. Behavior of *Dictyostelium* amoebae is regulated primarily by the temporal dynamic of the natural cAMP wave. *Cell Motil. Cytoskelet.* **1992**, *23*, 145–156. [CrossRef]
59. Swanson, J.; Taylor, D.L. Local and spatially coordinated movements in Dictyostelium Discoideum Amoedae Chemotaxis. *Cell* **1982**, *28*, 225–232. [CrossRef]
60. Chung, C.Y.; Funamoto, S.; Firtel, R.A. Signaling pathways controlling cell polarity and chemotaxis. *Trends Biochem. Sci.* **2001**, *26*, 557–566. [CrossRef]
61. Takeda, K.; Sasaki, A.T.; Ha, H.; Seung, H.A.; Firtel, R.A. Role of phosphatidylinositol 3-kinases in chemotaxis in Dictyostelium. *J. Biol. Chem.* **2007**, *282*, 11874–11884. [CrossRef]
62. Iijima, M.; Huang, Y.E.; Devreotes, P. Temporal and spatial regulation of chemotaxis. *Dev. Cell* **2002**, *3*, 469–478. [CrossRef]
63. Parent, C.A.; Devreotes, P.N. A cell's sense of direction. *Science* **1999**, *284*, 765–770. [CrossRef] [PubMed]
64. Janetopoulos, C.; Jin, T.; Devreotes, P. Receptor-mediated activation of heterotrimeric G-proteins in living cells. *Science* **2001**, *291*, 2408–2411. [CrossRef]
65. Janetopoulos, C.; Firtel, R.A. Directional sensing during chemotaxis. *FEBS Lett.* **2008**, *582*, 2075–2085. [CrossRef]
66. Fets, L.; Nichols, J.M.; Kay, R.R. A PIP5 kinase essential for efficient chemotactic signaling. *Curr. Biol.* **2014**, *24*, 415–421. [CrossRef] [PubMed]
67. Killich, T.; Plath, P.J.; Haß, E.C.; Xiang, W.; Bultmann, H.; Rensing, L.; Vicker, M.G. Cell movement and shape are non-random and determined by intracellular, oscillatory rotating waves in Dictyostelium amoebae. *Biosystems* **1994**, *33*, 75–87. [CrossRef]
68. Killich, T.; Plath, P.J.; Wei, X.; Bultmann, H.; Rensing, L.; Vicker, M.G. The locomotion, shape and pseudopodial dynamics of unstimulated Dictyostelium cells are not random. *J. Cell Sci.* **1993**, *106*, 1005–1013.
69. Vicker, M.G. Eukaryotic cell locomotion depends on the propagation of self-organized reaction–diffusion waves and oscillations of actin filament assembly. *Exp. Cell Res.* **2002**, *275*, 54–66. [CrossRef]
70. Vicker, M.G. Reaction-diffusion waves of actin filament polymerization/depolymerization in Dictyostelium pseudopodium extension and cell locomotion. *Biophys. Chem.* **2000**, *84*, 87–98. [CrossRef]
71. Bretschneider, T.; Anderson, K.; Ecke, M.; Müller-Taubenberger, A.; Schroth-Diez, B.; Ishikawa-Ankerhold, H.C.; Gerisch, G. The three-dimensional dynamics of actin waves, a model of cytoskeletal self-organization. *Biophys. J.* **2009**, *96*, 2888–2900. [CrossRef]
72. Schroth-Diez, B.; Gerwig, S.; Ecke, M.; Hegerl, R.; Diez, S.; Gerisch, G. Propagating waves separate two states of actin organization in living cells. *HFSP J.* **2009**, *3*, 412–427. [CrossRef]
73. Arai, Y.; Shibata, T.; Matsuoka, S.; Sato, M.J.; Yanagida, T.; Ueda, M. Self-organization of the phosphatidylinositol lipids signaling system for random cell migration. *Proc. Natl. Acad. Sci. USA* **2010**, *107*, 12399–12404. [CrossRef]

74. Xiong, Y.; Huang, C.H.; Iglesias, P.A.; Devreotes, P.N. Cells navigate with a local-excitation, global-inhibition-biased excitable network. *Proc. Nat. Acad. Sci. USA* **2010**, *107*, 17079. [CrossRef]
75. Nishikawa, M.; Hörning, M.; Ueda, M.; Shibata, T. Excitable signal transduction induces both spontaneous and directional cell asymmetries in the phosphatidylinositol lipid signaling system for eukaryotic chemotaxis. *Biophys. J.* **2014**, *106*, 723–734. [CrossRef] [PubMed]
76. Masters, T.A.; Sheetz, M.P.; Gauthier, N.C. F-actin waves, actin cortex disassembly and focal exocytosis driven by actin-phosphoinositide positive feedback. *Cytoskeleton* **2016**, *73*, 180–196. [CrossRef] [PubMed]
77. Miao, Y.; Bhattacharya, S.; Edwards, M.; Cai, H.; Inoue, T.; Iglesias, P.A.; Devreotes, P.N. Altering the threshold of an excitable signal transduction network changes cell migratory modes. *Nat. Cell Biol.* **2017**, *19*, 329–340. [CrossRef]
78. Inagaki, N.; Katsuno, H. Actin waves: Origin of cell polarization and migration? *Trends Cell Biol.* **2017**, *27*, 515–526. [CrossRef]
79. Gerisch, G. Self-organizing actin waves that simulate phagocytic cup structures. *PMC Biophys.* **2010**, *3*, 7. [CrossRef]
80. Dai, J.; Ting-Beall, H.P.; Hochmuth, R.M.; Sheetz, M.P.; Titus, M.A. Myosin I contributes to the generation of resting cortical tension. *Biophys. J.* **1999**, *77*, 1168–1176. [CrossRef]
81. Brzeska, H.; Pridham, K.; Chery, G.; Titus, M.A.; Korn, E.D. The association of myosin IB with actin waves in Dictyostelium requires both the plasma membrane-binding site and actin-binding region in the myosin tail. *PLoS ONE* **2014**, *9*, e94306. [CrossRef] [PubMed]
82. Cai, L.; Marshall, T.W.; Uetrecht, A.C.; Schafer, D.A.; Bear, J.E. Coronin 1B coordinates Arp2/3 complex and cofilin activities at the leading edge. *Cell* **2007**, *128*, 915–929. [CrossRef] [PubMed]
83. Ecke, M.; Prassler, J.; Tanribil, P.; Müller-Taubenberger, A.; Körber, S.; Faix, J.; Gerisch, G. Formins specify membrane patterns generated by propagating actin waves. *Mol. Biol. Cell* **2020**. [CrossRef] [PubMed]
84. Othmer, H.G. Nonlinear wave propagation in reacting systems. *J. Math. Biol.* **1976**, *2*, 133–163. [CrossRef]
85. Mori, Y.; Jilkine, A.; Edelstein-Keshet, L. Wave-pinning and cell polarity from a bistable reaction-diffusion system. *Biophys. J.* **2008**, *94*, 3684–3697. [CrossRef] [PubMed]
86. van Haastert, P.J.; Keizer-Gunnink, I.; Kortholt, A. Coupled excitable Ras and F-actin activation mediates spontaneous pseudopod formation and directed cell movement. *Mol. Biol. Cell* **2017**, *28*, 922–934. [CrossRef] [PubMed]
87. de León, M.A.A.P.; Othmer, H.G. A phosphoinositide-based model of actin waves in frustrated phagocytosis. *PLoS Comput. Biol.* **2020**, submitted.
88. Fukushima, S.; Matsuoka, S.; Ueda, M. Excitable dynamics of Ras triggers spontaneous symmetry breaking of PIP3 signaling in motile cells. *J. Cell Sci.* **2019**, *132*, jcs224121. [CrossRef]
89. Shibata, T.; Nishikawa, M.; Matsuoka, S.; Ueda, M. Modeling the self-organized phosphatidylinositol lipid signaling system in chemotactic cells using quantitative image analysis. *J. Cell Sci.* **2012**, *125*, 5138–5150. [CrossRef]
90. Taniguchi, D.; Ishihara, S.; Oonuki, T.; Honda-Kitahara, M.; Kaneko, K.; Sawai, S. Phase geometries of two-dimensional excitable waves govern self-organized morphodynamics of amoeboid cells. *Proc. Natl. Acad. Sci. USA* **2013**, *110*, 5016–5021. [CrossRef]
91. Vazquez, F.; Matsuoka, S.; Sellers, W.R.; Yanagida, T.; Ueda, M.; Devreotes, P.N. Tumor suppressor PTEN acts through dynamic interaction with the plasma membrane. *Proc. Natl. Acad. Sci. USA* **2006**, *103*, 3633–3638. [CrossRef]
92. Gerisch, G.; Schroth-Diez, B.; Müller-Taubenberger, A.; Ecke, M. PIP3 waves and PTEN dynamics in the emergence of cell polarity. *Biophys. J.* **2012**, *103*, 1170–1178. [CrossRef] [PubMed]
93. Charest, P.G.; Firtel, R.A. Feedback signaling controls leading-edge formation during chemotaxis. *Curr. Opin. Genet. Dev.* **2006**, *16*, 339–347. [CrossRef] [PubMed]
94. Sasaki, A.T.; Chun, C.; Takeda, K.; Firtel, R.A. Localized Ras signaling at the leading edge regulates PI3K, cell polarity, and directional cell movement. *J. Cell Biol.* **2004**, *167*, 505–518. [CrossRef]
95. Tuosto, L.; Capuano, C.; Muscolini, M.; Santoni, A.; Galandrini, R. The multifaceted role of PIP2 in leukocyte biology. *Cell. Mol. Life Sci.* **2015**, *72*, 4461–4474. [CrossRef]
96. Funamoto, S.; Meili, R.; Lee, S.; Parry, L.; Firtel, R.A. Spatial and temporal regulation of 3-phosphoinositides by PI 3-kinase and PTEN mediates chemotaxis. *Cell* **2002**, *109*, 611–623. [CrossRef]

97. Iijima, M.; Devreotes, P. Tumor suppressor PTEN mediates sensing of chemoattractant gradients. *Cell* **2002**, *109*, 599–610. [CrossRef]
98. Song, M.S.; Salmena, L.; Pandolfi, P.P. The functions and regulation of the PTEN tumour suppressor. *Nat. Rev. Mol. Cell Biol.* **2012**, *13*, 283–296. [CrossRef]
99. Das, S.; Dixon, J.E.; Cho, W. Membrane-binding and activation mechanism of PTEN. *Proc. Natl. Acad. Sci. USA* **2003**, *100*, 7491–7496. [CrossRef]
100. Lee, J.O.; Yang, H.; Georgescu, M.M.; Cristofano, A.D.; Maehama, T.; Shi, Y.; Dixon, J.E.; Pandolfi, P.; Pavletich, N.P. Crystal structure of the PTEN tumor suppressor: Implications for its phosphoinositide phosphatase activity and membrane association. *Cell* **1999**, *99*, 323–334. [CrossRef]
101. Lumb, C.N.; Sansom, M.S. Defining the membrane-associated state of the PTEN tumor suppressor protein. *Biophys. J.* **2013**, *104*, 613–621. [CrossRef]
102. Nguyen, H.N.; Yang, J.M.; Miyamoto, T.; Itoh, K.; Rho, E.; Zhang, Q.; Inoue, T.; Devreotes, P.N.; Sesaki, H.; Iijima, M. Opening the conformation is a master switch for the dual localization and phosphatase activity of PTEN. *Sci. Rep.* **2015**, *5*. [CrossRef]
103. Rahdar, M.; Inoue, T.; Meyer, T.; Zhang, J.; Vazquez, F.; Devreotes, P.N. A phosphorylation-dependent intramolecular interaction regulates the membrane association and activity of the tumor suppressor PTEN. *Proc. Natl. Acad. Sci. USA* **2009**, *106*, 480–485. [CrossRef]
104. Nguyen, H.N.; Yang, J.M.; Afkari, Y.; Park, B.H.; Sesaki, H.; Devreotes, P.N.; Iijima, M. Engineering ePTEN, an enhanced PTEN with increased tumor suppressor activities. *Proc. Natl. Acad. Sci. USA* **2014**, *111*, E2684–E2693. [CrossRef]
105. Walker, S.M.; Leslie, N.R.; Perera, N.M.; Batty, I.H.; Downes, C.P. The tumour-suppressor function of PTEN requires an N-terminal lipid-binding motif. *Biochem. J.* **2004**, *379*, 301–307. [CrossRef]
106. Denning, G.; Jean-Joseph, B.; Prince, C.; Durden, D.; Vogt, P. A short N-terminal sequence of PTEN controls cytoplasmic localization and is required for suppression of cell growth. *Oncogene* **2007**, *26*, 3930–3940. [CrossRef]
107. Khamviwath, V.; Hu, J.; Othmer, H.G. A continuum model of actin waves in *D Ictyostelium discoideum*. *PLoS ONE* **2013**, *8*, e64272. [CrossRef]
108. Hoeller, O.; Toettcher, J.E.; Cai, H.; Sun, Y.; Huang, C.H.; Freyre, M.; Zhao, M.; Devreotes, P.N.; Weiner, O.D. Gβ regulates coupling between actin oscillators for cell polarity and directional migration. *PLoS Biol.* **2016**, *14*. [CrossRef]
109. Rosales, C.; Uribe-Querol, E. Phagocytosis: A fundamental process in immunity. *Biomed Res. Int.* **2017**, *2017*. [CrossRef]
110. Campanale, J.P.; Sun, T.Y.; Montell, D.J. Development and dynamics of cell polarity at a glance. *J. Cell Sci.* **2017**, *130*, 1201–1207. [CrossRef]
111. Goryachev, A.B.; Pokhilko, A.V. Dynamics of Cdc42 network embodies a Turing-type mechanism of yeast cell polarity. *FEBS Lett.* **2008**, *582*, 1437–1443. [CrossRef]
112. Muller, N.; Piel, M.; Calvez, V.; Voituriez, R.; Goncalves-Sá, J.; Guo, C.L.; Jiang, X.; Murray, A.; Meunier, N. A predictive model for yeast cell polarization in pheromone gradients. *PLoS Comp. Biol.* **2016**, *12*. [CrossRef] [PubMed]
113. Chiou, J.G.; Balasubramanian, M.K.; Lew, D.J. Cell polarity in yeast. *Annu. Rev. Cell Dev. Biol.* **2017**, *33*, 77–101. [CrossRef]
114. Chiou, J.G.; Ramirez, S.A.; Elston, T.C.; Witelski, T.P.; Schaeffer, D.G.; Lew, D.J. Principles that govern competition or co-existence in Rho-GTPase driven polarization. *PLoS Comp. Biol.* **2018**, *14*, e1006095. [CrossRef]
115. Vendel, K.J.; Tschirpke, S.; Shamsi, F.; Dogterom, M.; Laan, L. Minimal in vitro systems shed light on cell polarity. *J. Cell Sci.* **2019**, *132*, jcs217554. [CrossRef] [PubMed]
116. Woods, B.; Lew, D.J. Polarity establishment by Cdc42: Key roles for positive feedback and differential mobility. *Small GTPases* **2019**, *10*, 130–137. [CrossRef] [PubMed]
117. Sartorel, E.; Ünlü, C.; Jose, M.; Massoni-Laporte, A.; Meca, J.; Sibarita, J.B.; McCusker, D. Phosphatidylserine and GTPase activation control Cdc42 nanoclustering to counter dissipative diffusion. *Mol. Biol. Cell* **2018**, *29*, 1299–1310. [CrossRef]

118. Meca, J.; Massoni-Laporte, A.; Martinez, D.; Sartorel, E.; Loquet, A.; Habenstein, B.; McCusker, D. Avidity-driven polarity establishment via multivalent lipid–GTPase module interactions. *EMBO J.* **2019**, *38*, e99652. [CrossRef]
119. Hu, J.; Kang, H.W.W.; Othmer, H.G. Stochastic analysis of reaction–diffusion processes. *Bull. Math. Biol.* **2013**, 1–41. [CrossRef]
120. Turner, M.S.; Sens, P.; Socci, N.D. Nonequilibrium raftlike membrane domains under continuous recycling. *Phys. Rev. Lett.* **2005**, *95*, 168301. [CrossRef]
121. Richardson, G.; Cummings, L.J.; Harris, H.; O'Shea, P. Toward a mathematical model of the assembly and disassembly of membrane microdomains: Comparison with experimental models. *Biophys. J.* **2007**, *92*, 4145–4156. [CrossRef]
122. Altschuler, S.J.; Angenent, S.B.; Wang, Y.; Wu, L.F. On the spontaneous emergence of cell polarity. *Nature* **2008**, *454*, 886–889. [CrossRef]
123. Houk, A.R.; Jilkine, A.; Mejean, C.O.; Boltyanskiy, R.; Dufresne, E.R.; Angenent, S.B.; Altschuler, S.J.; Wu, L.F.; Weiner, O.D. Membrane tension maintains cell polarity by confining signals to the leading edge during neutrophil migration. *Cell* **2012**, *148*, 175–188. [CrossRef]
124. Marco, E.; Wedlich-Soldner, R.; Li, R.; Altschuler, S.J.; Wu, L.F. Endocytosis optimizes the dynamic localization of membrane proteins that regulate cortical polarity. *Cell* **2007**, *129*, 411–422. [CrossRef]
125. Meinhardt, H. Orientation of chemotactic cells and growth cones: Models and mechanisms. *J. Cell Sci.* **1999**, *17*, 2867–2874.
126. Tang, M.; Wang, M.; Shi, C.; Iglesias, P.A.; Devreotes, P.N.; Huang, C.H. Evolutionarily conserved coupling of adaptive and excitable networks mediates eukaryotic chemotaxis. *Nat. Commun.* **2014**, *5*. [CrossRef]
127. Umulis, D.; O'Connor, M.B.; Othmer, H.G. Robustness of Embryonic Spatial Patterning in *Dros. melanogater*. *Curr. Top Dev. Biol.* **2008**, *81*, 65–111.
128. Otsuji, M.; Ishihara, S.; Co, C.; Kaibuchi, K.; Mochizuki, A.; Kuroda, S. A mass conserved reaction—Diffusion system captures properties of cell polarity. *PLoS Comp. Biol.* **2007**, *3*. [CrossRef]
129. Mori, Y.; Jilkine, A.; Edelstein-Keshet, L. Asymptotic and bifurcation analysis of wave-pinning in a reaction-diffusion model for cell polarization. *SIAM J. Appl. Math.* **2011**, *71*, 1401–1427. [CrossRef]
130. Edelstein-Keshet, L.; Holmes, W.R.; Zajac, M.; Dutot, M. From simple to detailed models for cell polarization. *Philos. Trans. R. Soc. B* **2013**, *368*, 20130003. [CrossRef] [PubMed]
131. Dallon, J.C.; Othmer, H.G. A discrete cell model with adaptive signalling for aggregation of *Dictyostelium discoideum*. *Phil. Trans. R. Soc. Lond.* **1997**, *B352*, 391–417. [CrossRef] [PubMed]
132. Dallon, J.C.; Othmer, H.G. A continuum analysis of the chemotactic signal seen by *Dictyostelium discoideum*. *J. Theor. Biol.* **1998**, *194*, 461–483. [CrossRef] [PubMed]
133. Cheng, Y.; Othmer, H.G. A model for direction sensing in *Dictyostelium discoideum*: Ras activity and symmetry breaking driven by a $G_{\beta\gamma}$- mediated, $G_{\alpha 2}$-Ric8—Dependent signal transduction network. *PLoS Comp. Biol.* **2016**. [CrossRef]
134. Takeda, K.; Shao, D.; Adler, M.; Charest, P.G.; Loomis, W.F.; Levine, H.; Groisman, A.; Rappel, W.J.; Firtel, R.A. Incoherent feedforward control governs adaptation of activated Ras in a eukaryotic chemotaxis pathway. *Sci. Signal* **2012**, *5*, ra2. [CrossRef]
135. Kortholt, A.; Keizer-Gunnink, I.; Kataria, R.; Haastert, P.J.M. Ras activation and symmetry breaking during Dictyostelium chemotaxis. *J. Cell Sci.* **2013**, *126*, 4502–4513. [CrossRef]
136. Kataria, R.; Xu, X.; Fusetti, F.; Keizer-Gunnink, I.; Jin, T.; van Haastert, P.J.; Kortholt, A. Dictyostelium Ric8 is a nonreceptor guanine exchange factor for heterotrimeric G proteins and is important for development and chemotaxis. *Proc. Nat. Acad. Sci. USA* **2013**, *110*, 6424–6429. [CrossRef]
137. Jin, T.; Zhang, N.; Long, Y.; Parent, C.A.; Devreotes, P.N. Localization of the G protein $\beta\gamma$ complex in living cells during chemotaxis. *Science* **2000**, *287*, 1034–1036. [CrossRef]
138. Postma, M.; van Haastert, P.J. Mathematics of experimentally generated chemoattractant gradients. In *Chemotaxis*; Springer: Berlin, Germany, 2009; pp. 473–488.
139. Nichols, J.M.; Veltman, D.; Kay, R.R. Chemotaxis of a model organism: Progress with Dictyostelium. *Curr. Opin. Cell Biol.* **2015**, *36*, 7–12. [CrossRef]
140. Hohmann, T.; Dehghani, F. The cytoskeleton—A complex interacting meshwork. *Cells* **2019**, *8*, 362. [CrossRef]

141. Artemenko, Y.; Axiotakis, L.; Borleis, J.; Iglesias, P.A.; Devreotes, P.N. Chemical and mechanical stimuli act on common signal transduction and cytoskeletal networks. *Proc. Nat. Acad. Sci. USA* **2016**, *113*, E7500–E7509. [CrossRef]
142. Meili, R.; Ellsworth, C.; Lee, S.; Reddy, T.B.; Ma, H.; Firtel, R.A. Chemoattractant-mediated transient activation and membrane localization of Akt/PKB is required for efficient chemotaxis to cAMP in Dictyostelium. *EMBO J.* **1999**, *18*, 2092–2105. [CrossRef] [PubMed]
143. Xu, X.; Meier-Schellersheim, M.; Yan, J.; Jin, T. Locally controlled inhibitory mechanisms are involved in eukaryotic GPCR-mediated chemosensing. *J. Cell Biol.* **2007**, *178*, 141–153. [CrossRef]
144. Lebensohn, A.M.; Kirschner, M.W. Activation of the WAVE complex by coincident signals controls actin assembly. *Mol. Cell* **2009**, *36*, 512–524. [CrossRef]
145. Veltman, D.M.; King, J.S.; Machesky, L.M.; Insall, R.H. SCAR knockouts in Dictyostelium: WASP assumes SCAR's position and upstream regulators in pseudopods. *J. Cell Biol.* **2012**, *198*, 501–508. [CrossRef] [PubMed]
146. Filić, V.; Marinović, M.; Faix, J.; Weber, I. The IQGAP-related protein DGAP1 mediates signaling to the actin cytoskeleton as an effector and a sequestrator of Rac1 GTPases. *Cell. Mol. Life Sci.* **2014**, *71*, 2775–2785. [CrossRef] [PubMed]
147. Matsuoka, S.; Ueda, M. Mutual inhibition between PTEN and PIP3 generates bistability for polarity in motile cells. *Nat. Commun.* **2018**, *9*, 1–15. [CrossRef]
148. Jeon, T.J.; Lee, D.J.; Merlot, S.; Weeks, G.; Firtel, R.A. Rap1 controls cell adhesion and cell motility through the regulation of myosin II. *J. Cell Biol.* **2007**, *176*, 1021–1033. [CrossRef]
149. Parent, C.A. Making all the right moves: Chemotaxis in neutrophils and Dictyostelium. *Curr. Opin. Cell Biol.* **2004**, *16*, 4–13. [CrossRef]
150. Pramanik, M.K.; Iijima, M.; Iwadate, Y.; Yumura, S. PTEN is a mechanosensing signal transducer for myosin II localization in *Dictyostelium* cells. *Genes Cells* **2009**, *14*, 821–834. [CrossRef]
151. Yoshioka, D.; Fukushima, S.; Koteishi, H.; Okuno, D.; Ide, T.; Matsuoka, S.; Ueda, M. Single-molecule imaging of PI (4,5) P2 and PTEN in vitro reveals a positive feedback mechanism for PTEN membrane binding. *Commun. Biol.* **2020**, *3*, 1–14. [CrossRef]
152. Bosgraaf, L.; Van Haastert, P.J.M. The regulation of myosin II in *Dictyostelium*. *Eur. J. Cell Biol.* **2006**, *85*, 969–979. [CrossRef]
153. Sugiyama, T.; Pramanik, M.K.; Yumura, S. Microtubule-Mediated Inositol Lipid Signaling Plays Critical Roles in Regulation of Blebbing. *PLoS ONE* **2015**, *10*, e0137032. [CrossRef]
154. Fernandez-Gonzalez, R.; Simoes, S.M.; Röper, J.C.; Eaton, S.; Zallen, J.A. Myosin II dynamics are regulated by tension in intercalating cells. *Dev. Cell* **2009**, *17*, 736–743. [CrossRef]
155. Amato, C.; Thomason, P.A.; Davidson, A.J.; Swaminathan, K.; Ismail, S.; Machesky, L.M.; Insall, R.H. WASP restricts active Rac to maintain cells' front-rear polarization. *Curr. Biol.* **2019**, *29*, 4169–4182. [CrossRef]
156. Byrne, K.M.; Monsefi, N.; Dawson, J.C.; Degasperi, A.; Bukowski-Wills, J.C.; Volinsky, N.; Dobrzyński, M.; Birtwistle, M.R.; Tsyganov, M.A.; Kiyatkin, A.; et al. Bistability in the Rac1, PAK, and RhoA signaling network drives actin cytoskeleton dynamics and cell motility switches. *Cell Syst.* **2016**, *2*, 38–48. [CrossRef]
157. Sidani, M.; Wessels, D.; Mouneimne, G.; Ghosh, M.; Goswami, S.; Sarmiento, C.; Wang, W.; Kuhl, S.; El-Sibai, M.; Backer, J.M.; et al. Cofilin determines the migration behavior and turning frequency of metastatic cancer cells. *J. Cell Biol.* **2007**, *179*, 777–791. [CrossRef]
158. Funk, J.; Merino, F.; Venkova, L.; Heydenreich, L.; Kierfeld, J.; Vargas, P.; Raunser, S.; Piel, M.; Bieling, P. Profilin and formin constitute a pacemaker system for robust actin filament growth. *Elife* **2019**, *8*. [CrossRef]
159. Carlier, M.F.; Shekhar, S. Global treadmilling coordinates actin turnover and controls the size of actin networks. *Nat. Rev. Mol. Cell Biol.* **2017**, *18*, 389–401. [CrossRef]
160. Bieling, P.; Weichsel, J.; McGorty, R.; Jreij, P.; Huang, B.; Fletcher, D.A.; Mullins, R.D. Force feedback controls motor activity and mechanical properties of self-assembling branched actin networks. *Cell* **2016**, *164*, 115–127. [CrossRef]
161. Gerisch, G. Chemotaxis in *Dictyostelium*. *Annu. Rev. Physiol.* **1982**, *44*, 535–552. [CrossRef]
162. Faix, J.; Weber, I. A dual role model for active Rac1 in cell migration. *Small GTPases* **2013**, *4*, 110–115. [CrossRef]
163. Yamazaki, D.; Itoh, T.; Miki, H.; Takenawa, T. srGAP1 regulates lamellipodial dynamics and cell migratory behavior by modulating Rac1 activity. *Mol. Biol. Cell* **2013**, *24*, 3393–3405. [CrossRef]

164. Aspenström, P. BAR domain proteins regulate Rho GTPase signaling. *Small GTPases* **2014**, *5*, e972854. [CrossRef]
165. Dalous, J.; Burghardt, E.; Müller-Taubenberger, A.; Bruckert, F.; Gerisch, G.; Bretschneider, T. Reversal of cell polarity and actin-myosin cytoskeleton reorganization under mechanical and chemical stimulation. *Biophys. J.* **2008**, *94*, 1063–1074. [CrossRef]
166. Ruprecht, V.; Wieser, S.; Callan-Jones, A.; Smutny, M.; Morita, H.; Sako, K.; Barone, V.; Ritsch-Marte, M.; Sixt, M.; Voituriez, R.; et al. Cortical contractility triggers a stochastic switch to fast amoeboid cell motility. *Cell* **2015**, *160*, 673–685. [CrossRef]
167. Liu, Y.J.; Berre, M.L.; Lautenschlaeger, F.; Maiuri, P.; Callan-Jones, A.; Heuzé, M.; Takaki, T.; Voituriez, R.; Piel, M. Confinement and low adhesion induce fast amoeboid migration of slow mesenchymal cells. *Cell* **2015**, *160*, 659–672. [CrossRef]
168. Callan-Jones, A.; Ruprecht, V.; Wieser, S.; Heisenberg, C.P.; Voituriez, R. Cortical Flow-Driven Shapes of Nonadherent Cells. *Phys. Rev. Lett.* **2016**, *116*, 028102. [CrossRef]
169. Keren, K.; Pincus, Z.; Allen, G.M.; Barnhart, E.L.; Marriott, G.; Mogilner, A.; Theriot, J.A. Mechanism of shape determination in motile cells. *Nature* **2008**, *453*, 475–481. [CrossRef]
170. Stinner, B.; Bretschneider, T. Mathematical modelling in cell migration: Tackling biochemistry in changing geometries. *Biochem. Soc. Trans.* **2020**, *48*, 419–428. [CrossRef]
171. Traynor, D.; Milne, J.L.; Insall, R.H.; Kay, R.R. Ca^{2+} signalling is not required for chemotaxis in *Dictyostelium*. *EMBO J.* **2000**, *19*, 4846–4854. [CrossRef]
172. Lusche, D.F.; Wessels, D.; Soll, D.R. The effects of extracellular calcium on motility, pseudopod and uropod formation, chemotaxis, and the cortical localization of myosin II in *Dictyostelium discoideum*. *Cell Motil. Cytoskelet.* **2009**, *66*, 567–587. [CrossRef]
173. Melchionda, M.; Pittman, J.K.; Mayor, R.; Patel, S. Ca^{2+}/H^+ exchange by acidic organelles regulates cell migration in vivo. *J. Cell Biol.* **2016**, *212*, 803–813. [CrossRef]
174. Tsai, F.C.; Seki, A.; Yang, H.W.; Hayer, A.; Carrasco, S.; Malmersjö, S.; Meyer, T. A polarized Ca^{2+}, diacylglycerol and STIM1 signalling system regulates directed cell migration. *Nat. Cell Biol.* **2014**, *16*, 133–144. [CrossRef]
175. Evans, J.H.; Falke, J.J. Ca^{2+} influx is an essential component of the positive-feedback loop that maintains leading-edge structure and activity in macrophages. *Proc. Nat. Acad. Sci. USA* **2007**, *104*, 16176–16181. [CrossRef]
176. Wei, C.; Wang, X.; Chen, M.; Ouyang, K.; Song, L.S.; Cheng, H. Calcium flickers steer cell migration. *Nature* **2009**, *457*, 901–905. [CrossRef]
177. Valeyev, N.V.; Kim, J.S.; Heslop-Harrison, J.P.; Postlethwaite, I.; Kotov, N.V.; Bates, D.G. Computational modelling suggests dynamic interactions between Ca^{2+}, IP3 and G protein-coupled modules are key to robust *Dictyostelium* aggregation. *Mol. BioSyst.* **2009**, *5*, 612–628. [CrossRef]
178. Catacuzzeno, L.; Franciolini, F. Role of KCa3.1 channels in modulating Ca^{2+} oscillations during glioblastoma cell migration and invasion. *Int. J. Mol. Sci.* **2018**, *19*, 2970. [CrossRef]
179. Cuddapah, V.A.; Robel, S.; Watkins, S.; Sontheimer, H. A neurocentric perspective on glioma invasion. *Nat. Rev. Neurosci.* **2014**, *15*, 455–465. [CrossRef]
180. Bominaar, A.A.; Haastert, P.J.V. Phospholipase C in *Dictyostelium discoideum*. Identification of stimulatory and inhibitory surface receptors and G-proteins. *Biochem. J.* **1994**, *297*, 189–193. [CrossRef]
181. Gresset, A.; Sondek, J.; Harden, T.K. The phospholipase C isozymes and their regulation. In *Phosphoinositides I: Enzymes of Synthesis and Degradation*; Springer: Berlin, Germany, 2012; pp. 61–94.
182. Kortholt, A.; King, J.S.; Keizer-Gunnink, I.; Harwood, A.J.; Haastert, P.J.M.V. Phospholipase C regulation of phosphatidylinositol 3,4,5-trisphosphate-mediated chemotaxis. *Mol. Biol. Cell* **2007**, *18*, 4772–4779. [CrossRef]
183. Falcke, M. Reading the patterns in living cells—the physics of Ca^{2+} signaling. *Adv. Phys.* **2004**, *53*, 255–440. [CrossRef]
184. Cohen, R.; Torres, A.; Ma, H.T.; Holowka, D.; Baird, B. Ca^{2+} waves initiate antigen-stimulated Ca^{2+} responses in mast cells. *J. Immunol.* **2009**, *183*, 6478–6488. [CrossRef]

185. Lechleiter, J.; Girard, S.; Peralta, E.; Clapham, D. Spiral Calcium wave propagation and annihilation in *Xenopus laevis* Oocytes. *Science* **1991**, *252*, 123–126. [CrossRef]
186. Hu, J.; Khamviwath, V.; Othmer, H. A Stochastic Model for Actin Waves in Eukaryotic Cells. *bioRxiv* **2019**. [CrossRef]

Communication

Unilateral Cleavage Furrows in Multinucleate Cells

Julia Bindl [1], Eszter Sarolta Molnar [1], Mary Ecke [1], Jana Prassler [1], Annette Müller-Taubenberger [2] and Günther Gerisch [1,*]

1 Max Planck Institute of Biochemistry, Am Klopferspitz 18, D-82152 Martinsried, Germany; j.bindl@campus.lmu.de (J.B.); eszter.molnar@campus.lmu.de (E.S.M.); ecke@biochem.mpg.de (M.E.); prassler@biochem.mpg.de (J.P.)

2 LMU Munich, Department of Cell Biology (Anatomy III), Biomedical Center, D-82152 Planegg-Martinsried, Germany; amueller@bmc.med.lmu.de

* Correspondence: gerisch@biochem.mpg.de; Tel.: +49-898-578-2326

Received: 28 May 2020; Accepted: 16 June 2020; Published: 18 June 2020

Abstract: Multinucleate cells can be produced in *Dictyostelium* by electric pulse-induced fusion. In these cells, unilateral cleavage furrows are formed at spaces between areas that are controlled by aster microtubules. A peculiarity of unilateral cleavage furrows is their propensity to join laterally with other furrows into rings to form constrictions. This means cytokinesis is biphasic in multinucleate cells, the final abscission of daughter cells being independent of the initial direction of furrow progression. Myosin-II and the actin filament cross-linking protein cortexillin accumulate in unilateral furrows, as they do in the normal cleavage furrows of mononucleate cells. In a myosin-II-null background, multinucleate or mononucleate cells were produced by cultivation either in suspension or on an adhesive substrate. Myosin-II is not essential for cytokinesis either in mononucleate or in multinucleate cells but stabilizes and confines the position of the cleavage furrows. In fused wild-type cells, unilateral furrows ingress with an average velocity of 1.7 $\mu m \times min^{-1}$, with no appreciable decrease of velocity in the course of ingression. In multinucleate myosin-II-null cells, some of the furrows stop growing, thus leaving space for the extensive broadening of the few remaining furrows.

Keywords: cell fusion; cortexillin; cytokinesis; *Dictyostelium*; myosin

1. Introduction

Mitotic cell division is typically mediated by a contractile ring that, after segregation of the chromosomes, forms a cleavage furrow to separate the daughter cells. Double-headed myosin II has been shown to be a common driver of ring constriction in cells as divergent as *Schizosaccharomyces pombe*, sea urchin blastomeres, and mammalian cells [1,2]. Constriction of the ring is based on the interaction of the bipolar filaments of myosin-II with anti-parallel actin filaments that are linked to the membrane [3].

Nevertheless, there are exceptional modes of cleavage. First, in a variety of parasitic protozoans, cytokinesis is performed without the participation of myosin-II [4]. Second, mitotic cleavages can be accomplished by furrows that ingress laterally from one side of the cleavage region. The formation of unilateral furrows raises two principal questions. The first one is: how do these furrows progress to separate the daughter cells? The second question is: can mitotic cells alternate between cytokinesis by normal, ring-shaped cleavage furrows and by unilateral ones? This would suggest that the machinery responsible for cytokinesis is flexible enough to perform the constriction of a cytokinetic ring as well as the unilateral ingression of a furrow. Here, we used mono- and multinucleate cells of the eukaryotic microorganism *Dictyostelium discoideum* to address these questions.

Unilateral furrows are of general interest since they are formed under various conditions in cells other than the multinucleate *Dictyostelium* cells. In *Physarum polycephalum*, unilateral furrowing occurs during transition from the mononucleate amoebal state to the multinucleate plasmodium [5]. This transition is accomplished by the switch from astral mitosis connected with cytokinetic furrowing to anastral mitosis, with an intranuclear spindle that omits cytokinesis. The formation of incomplete cleavage furrows as intermediates between complete cytokinesis and lack of it is favored by the presence of additional microtubule-organizing centers, which unilaterally prevent the formation of a furrow.

Similarly, intracellular particles can asymmetrically block the formation of a cleavage furrow in HeLa cells infected with *Chlamydia*; the bacteria causing these human cells to form a unilateral furrow. At the side close to the inclusion of *Chlamydia* particles, the accumulation of RhoA became attenuated [6], and consequently the accumulation of anillin and the assembly of myosin-II were inhibited. The asymmetric accumulation of RhoA was due to the truncated localization of a RhoGEF, Ect2. In line with these observations, a unilateral furrow could be experimentally induced in HeLa cells using optogenetics to locally activate Ect2 in one area of the cleavage plane [7].

Unilateral furrows with microfilaments decorating the cleavage zone in an "arcuate manner" are formed in cnidarian eggs of the genus *Aequorea* [8]. These furrows begin at the animal pole and ingress toward the vegetal pole. In amphibian embryos, isolated blastomeres of the outermost layer, called "superficial cells", form unilateral furrows beginning at their adhesive basal surface and progressing toward their non-adhesive apical surface [9]. A unilateral furrow is also formed at the basal region of *Echinarachnius* (sand dollar) eggs that are forced into a conical shape [10].

Related to unilateral furrowing is the ingression of furrows between nuclei during blastoderm formation in early insect embryogenesis [11]. The subsequent constriction of a ring at the blastoderm–yolk interface occurs in two steps: a first slow phase and a second faster phase. Only the first phase is myosin-II-dependent [12].

During meiosis II in mouse oocytes, a unilateral furrow initiates polar body formation [13]. After turning of the spindle, this furrow is converted into a bilateral one or a contractile ring (reviewed by Uraji et al. [14]). Finally, there are unilateral cleavage furrows formed in electrofused mammalian cells, as reported for PtK1 cells by Savoian et al. [15].

In summary, there are three conditions under which unilateral cleavage furrows have been observed: (1) during plasmodium formation as an intermediate state between mitosis with and without cytokinesis, (2) in cells with an asymmetric architecture perpendicular to the direction of the mitotic spindle or with an asymmetry caused by the lateral location of an obstructing structure, and (3) in multinucleate cells that are too large to be cleaved by a circular furrow.

One point to be clarified in the context of unilateral furrowing is the role of myosin-II. Cytokinesis in *D. discoideum* is not exceptional, in the sense that myosin-II accumulates in the furrow region [16] and cytokinesis is impaired in myosin-II-null mutants [17–19]. However, there are two features that are remarkable. First, cytokinesis is strongly inhibited only if cells are cultivated in shaken suspension. When attached to an adhesive substrate surface, the mutant cells are capable of performing cytokinesis, albeit less efficiently than wild-type cells: this means they are forming a cleavage furrow linked to mitosis [20]. Second, in multinucleate myosin-II-null cells grown in shaken suspension, nuclei divide synchronously, but cytokinesis is impaired. When brought into contact with an adhesive substrate surface, these large cells form multiple cleavage furrows ingressing from their border [21].

A non-motor protein required for cytokinesis in *D. discoideum* is cortexillin that causes anti-parallel bundling of actin filaments [22]. Three isoforms, cortexillin I to III, form preferentially heterodimers [23,24]. Cells lacking both cortexillins I and II show severely impaired cytokinesis [22]. Cortexillin accumulates in the cleavage furrow [24–26] and has been proposed to interact there with myosin-II in a mechanosensory control system of contractility [27]. We have used GFP-cortexillin I to visualize cleavage furrows in myosin-II-null cells, where cortexillin accumulates at higher levels than in wild-type cells [28].

In order to study unilateral furrowing in cells other than myosin-II-null cells, we have produced multinucleate wild-type cells of *D. discoideum* through electric pulse-induced cell fusion [29]. We show that these fused cells are optimally suited to study cytokinesis by the ingression of unilateral furrows and compare them with multinucleate myosin-II-null cells.

2. Materials and Methods

2.1. Cell Strains and Culture Conditions

Fluorescent proteins were expressed in the AX2-214 strain of *D. discoideum* [30] or in the HS2205 strain derived from it. In HS2205, myosin-II heavy chain has been deleted [19]. In the AX2-214 strain, GFP-myosin-II [31] together with mRFPM-α-tubulin [32], GFP-α-tubulin [21] together with mRFP1-histone 2B, or mRFPM-LimEΔ [32] together with GFP-α-tubulin were expressed. In the HS2205 strain, GFP-cortexillin I [28] was expressed together with mRFPM-histone 2B.

2.2. Design of mRFP-Histone 2B Vectors

For the expression of histone 2B C-terminally of mRFP, two vectors were constructed, one conferring resistance to blasticidin, the other to hygromycin. For the blasticidin vector, the coding sequence of the *D. discoideum* histone variant H2Bv3 (DDB0231622|DDB_G0286509) was cloned into the *Eco*RI-site 3′ of mRFP1 [33] and expressed under control of an actin-15 promoter, using a pDEX-based vector conferring resistance to blasticidin [34].

To construct an expression vector with a hygromycin selection marker [35], the cassette A15P-mRFPmars-A8T consisting of actin-15 promoter, mRFPmars coding region [32], and actin-8 terminator was cloned between the *Sma*I and *Sph*I sites of the multiple cloning site of the pGEM7-based plasmid pHygTm(plus)/pG7 (a kind gift of Jeff Williams and Masashi Fukuzawa, University of Dundee). Subsequently, the coding sequence of histone H2Bv3 was inserted into the *Eco*RI site 3′ of mRFPmars.

2.3. Culture Conditions and Sample Preparation for Confocal Microscopy

Cells were cultivated in Petri dishes containing nutrient medium [36] supplemented with 10 μg/mL of blasticidin S (Gibco, Life Technologies Corporation, Grand Island, NY, USA), 10 μg/mL of geneticin (Sigma-Aldrich, St. Louis, MO, USA), or 33 μg/mL hygromycin B (EMD Millipore Corp., Billerica, MA, USA) at 21 ± 2 °C.

For imaging, cells rinsed off the Petri dish were transferred to an HCl-cleaned cover-glass bottom dish (FluoroDish, WPI INC., Sarasota, FL, USA) and kept for 1 to 2 h in LoFlo medium (ForMedium Ltd., Norfolk, UK). The rate of mitosis could be increased by incubating the cells for about 20 h at 4 °C in Petri dishes with nutrient medium and subsequently bringing them to room temperature before transfer to LoFlo medium.

Large cells with wild-type AX2-214 background were produced by electric pulse-induced fusion as described by Gerisch et al. [29]. Myosin-II-null cells were cultivated in shaken suspension in nutrient medium for about 36 h to get large multinucleate cells [20]. Multinucleate cells were transferred onto HCl-cleaned cover-glass bottom dishes and incubated in LoFlo medium for about 1 h before imaging was started. Where indicated in the figure legends, cells were overlaid with a thin agarose sheet [37,38]. The velocity of unilateral furrow propagation was measured beginning at 1 μm of ingression.

2.4. Confocal Image Acquisition and Data Processing

For confocal images, a Zeiss LSM 780 microscope equipped with a Plan-Apo 63x/NA 1.46 oil immersion objective was used (Zeiss AG, Oberkochen, Germany). Images were processed using the image-processing package Fiji (http://Fiji.sc/Fiji) developed by Schindelin et al. [39] on the basis of ImageJ (http://imagej.nih.gov/ij). For bleach correction of the red channel, the total mean grey level for each image of a series was measured to calculate the percentage of bleaching, and accordingly, the brightness of the red channel was linearly enhanced.

For 3D rendering and animation, the images were first deconvolved with the software of Huygens Essential, version 18.04 (Scientific Volume Imaging b.v., Hilversum, The Netherlands) and then animated and displayed in UCSF Chimera, version 1.14 (https://www.cgl.ucsf.edu/chimera) [40].

3. Results

3.1. Mitosis in Wild-Type and Myosin-II-Null Cells

As a reference for mitosis, we will first show spindle dynamics and chromosome segregation in a wild-type cell containing a single dividing nucleus (Figure 1A and Supplementary Video S1). This cell expressed GFP-α-tubulin (green) as a constituent of the mitotic apparatus together with mRFP-histone 2B to label chromosomes (red). During metaphase (0-s frame), the two centrosomes stayed at a distance of only 1 to 2 µm from each other. Accordingly, the connecting spindle was short; subsequently it elongated, before disrupting in the middle. At anaphase, the two sets of daughter chromosomes immediately followed the separating centrosomes, which means that centromere-associated microtubules remained short. During the entire process, aster microtubules connected the centrosomes with the polar regions of the cell cortex. This example is representative of the dynamics of the mitotic apparatus in mononucleate and multinucleate wild-type or myosin-II-null cells (Table 1). Not seen in Figure 1 is the nuclear membrane, which separates the centrosomes from the intranuclear spindle in the semi-closed mitosis of *Dictyostelium* [41,42].

Table 1. Mitotic spindles in wild-type and myosin-II-null cells.

Mononucleate Cells	**Wild-Type**	**Myosin-II-Null**
Velocity of spindle elongation	44 ± 15 s.d. nm/s	56 nm/s
Maximal spindle length	11.8 ± 1.3 s.d. µm	14.0 µm
Number of measured cells	7	1 for velocity, 3 for spindle length
Multinucleate Cells	**Wild-Type**	**Myosin-II-Null**
Velocity of spindle elongation	46 ± 11 s.d. nm/s	51 ± 14 s.d. nm/s
Maximal spindle length	14.9 ± 2.1 s.d. µm	13.9 ± 1.4 s.d. µm
Number of measured spindles	17 in 5 cells	18 in 6 cells

The velocity of elongation was measured between 3 and 10 µm of spindle length. The spindle length between the center of the centrosomes was determined in the frame before the spindle was disrupted.

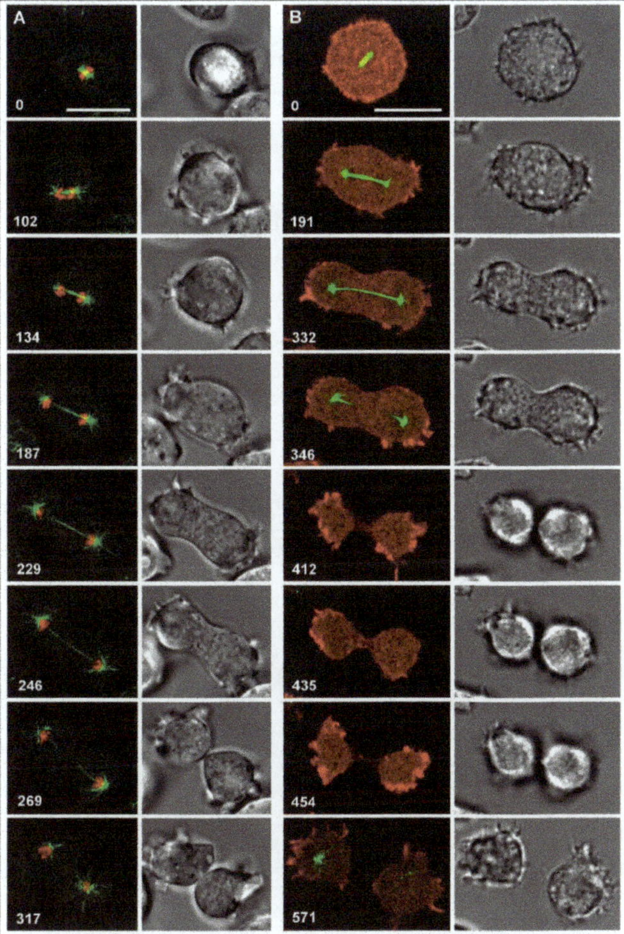

Figure 1. Mitosis and cytokinesis in wild-type cells of *Dictyostelium discoideum*. The left panels in the time series of (**A**) and (**B**) show confocal dual-color fluorescence images, the right panels show DIC bright-field images. Time after the first frame of each series is indicated in seconds. Scale bars, 10 µm. (**A**) A cell expressing GFP-α-tubulin as a label for the mitotic apparatus (green) and mRFP-histone 2B to visualize the chromosomes (red). The elongated spindle is disrupted between the 246 s and 269 s frames. The 317 s frame shows in the right cell radial microtubules connecting the centrosome with the cell cortex. (**B**) A cell expressing GFP-α-tubulin (green) and mRFP-LimEΔ as a label for filamentous actin (red). Fluorescence images are primarily focused on the spindle or on polar protrusions; for the 435 s frame, the focus was changed to the cleavage furrow, where little actin had accumulated. The cell was flattened by agarose overlay. The same sequence is shown in Supplementary Video S1. For a 3D display showing polar protrusions attached to the substrate surface, see Supplementary Video S2.

3.2. *Cytokinesis in Mononucleate Wild-Type and Myosin-II-Null Cells*

Upon disruption of the spindle, the cleavage furrow ingressed in the midplane of wild-type cells, as shown in Figure 1A. The furrow separated the daughter cells except for a thin tubular bridge that was finally disrupted by a dynamin A-dependent mechanism [43]. Filamentous actin accumulated most prominently in rounded protrusions at the polar regions of the dividing cell and only faintly in the cleavage furrow (Figure 1B, Supplementary Video S1 and Supplementary Video S2).

Mononucleate myosin-II-null cells that completed cytokinesis, accumulated cortexillin in the cleavage furrow (Figure 2A and Supplementary Video S3, top). Instability of furrow positioning was indicated by those cells in which the furrow, initiated as usual in the middle of the cell, slipped to one side, such that it would separate a binucleate portion from an anucleate one (Figure 2B and Supplementary Video S3, bottom). In that way, the furrow became located on top of aster microtubules, which did not support its further ingression. The anucleate portion was rather retracted, resulting in failure of cytokinesis.

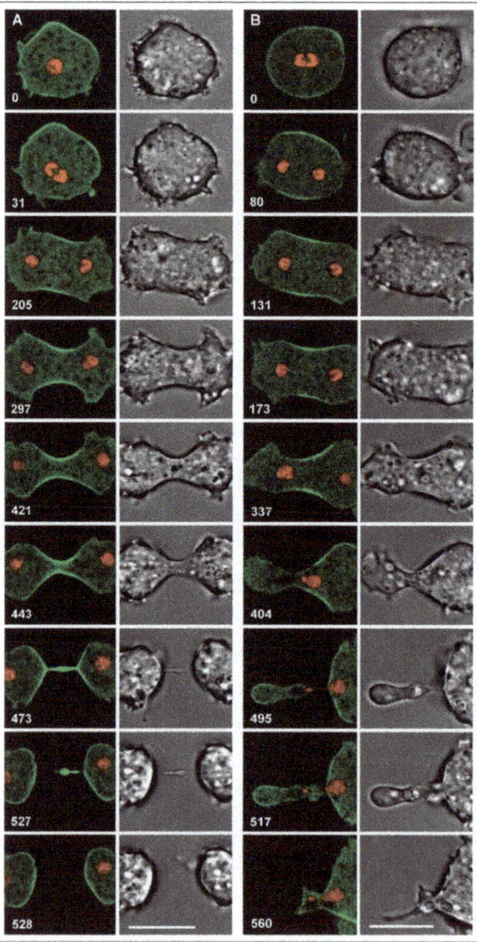

Figure 2. Successful and failing cytokinesis in mononucleate myosin-II-null cells. Cells attached to a glass surface are shown in confocal fluorescence (left panels) and bright field images (right panels). The cells expressed GFP-cortexillin I (green), together with mRFP-histone 2B as a label of the chromosomes. Time is indicated in seconds after the first frame. Scale bars, 10 μm. The entire sequences of (**A**) and (**B**) are shown in Supplementary Video S3. (**A**) Successful cell division showing cortexillin accumulating in the cleavage furrow that divides the cell between the two daughter nuclei. (**B**) Unsuccessful cytokinesis, which begins like the successful one with the accumulation of cortexillin in the midzone. However, subsequently, the furrow becomes asymmetric, and the left nucleus slips toward the right, the now anucleate part of the cell becoming integrated into the binucleate one. This cell was gently compressed by an agar overlay, so that the upper and lower surfaces were brought into parallel planes, and the dividing nuclei were kept in focus.

3.3. Unilateral Cleavage Furrows in Multinucleate Wild-Type Cells

In all kinds of the multinucleate cells tested, that is, in fused wild-type cells, myosin-II-null cells pre-grown in suspension, and cortexillin I- and II-null cells, the nuclei divide synchronously. To label unilateral cleavage furrows in multinucleate cells, we alternatively used two proteins that are involved in cleavage furrow formation accompanying mitosis, i.e., the two-headed motor protein myosin-II [31] and cortexillin, an anti-parallel bundler of actin filaments [22].

The division of large wild-type cells produced by electric pulse-induced fusion is exemplified in Figure 3 and Supplementary Video S4, showing a cell that expressed mRFP-α-tubulin to label the mitotic apparatus together with GFP-myosin-II heavy chains to visualize the accumulation of myosin in unilateral furrows. The mitotic apparatus showed the docking of aster microtubules to the cell cortex, as previously observed in myosin-II-null cells [20]. Docking resulted in bending of the spindle (159 s frame) and in the induction of protrusions where furrow formation was inhibited. Again, in accord with previous findings on myosin-II-null cells [21], furrows ingressed at spaces not occupied by microtubule asters, independent of whether or not these spaces were bridged by a spindle.

The unilateral furrows in wild-type cells ingressed with an average velocity of 1.7 $\mu m \times min^{-1}$, with no appreciable decrease of the velocity in the course of ingression (Figure 4A,B). They widened during ingression and tended to cooperate with neighboring furrows to form constricting rings that separated mono- or oligo-nucleate portions from the multinucleate cell mass (Figure 4C). These cases are of interest because here the final constriction was not a continuation of the initial furrow ingression but proceeded laterally between two furrows.

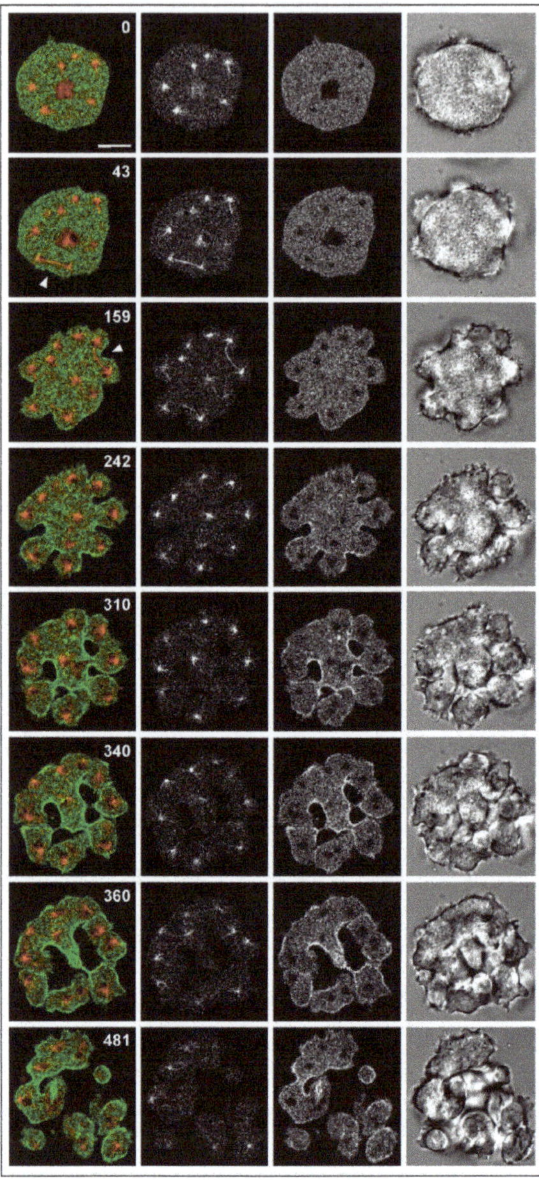

Figure 3. Unilateral furrows in a multinucleate wild-type cell produced by electric pulse-induced fusion. The cells fused expressed mRFP-α-tubulin (red) and GFP-myosin-II heavy chains (green in the merged panels). Confocal fluorescence images show the labels merged (left) but also separated (second row: α-tubulin; third row: myosin-II). On the right, DIC bright-field images are shown. In the merged panels, a straight spindle before furrow ingression and a spindle bent after the onset of furrowing are indicated by arrowheads. In the middle of the large cell, an interphase cell was apparently entrapped by accident (seen in the 0 and 43 s frames). Time is indicated in seconds after the first frame. Scale bar, 10 μm. The entire time series from which these images were taken is covered in Supplementary Video S4.

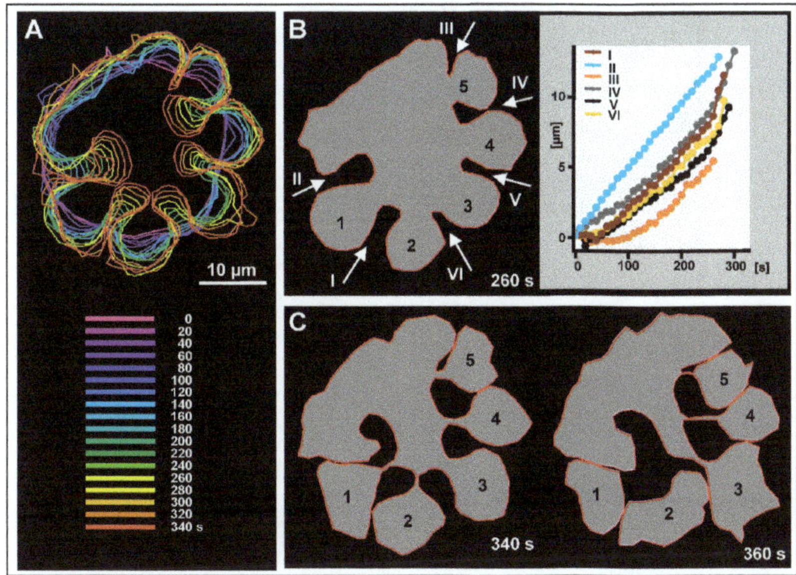

Figure 4. Progression of cleavage furrows in the multinucleate wild-type cell of Figure 3. (**A**) Color-coded cell boundaries within the 0 to 340 s time span derived from the confocal fluorescence images of myosin-II. (**B**) Left panel: cell boundary at the 260 s time point, showing unilateral cleavage furrows I to VI and incipient daughter cells 1 to 5. Right panel: progression of the six cleavage furrows, measured in the direction indicated by the arrows in the left panel. (**C**) Comparison of cell boundaries at the 340 s and 360 s time points. Within the time span of 20 s, the daughter cells 1 and 2 became disconnected, while separation of the incipient daughter cells 3 to 5 proceeded. In all these cases, abscission occurred obliquely to the initial furrowing. The drawings in (**B**) and (**C**) show the cell shapes in the confocal plane, with additional information on connectivity gathered from the bright-field images.

3.4. Division of Multinucleate Myosin-II-Null Cells

Multinucleate myosin-II-null cells can divide by unilateral furrows to which cortexillin is localized. However, these mutant cells appeared to inefficiently restrict the expansion of these furrows: while some furrows strongly expanded before the daughter cells were separated, other furrows stopped growing. An extreme example is shown in Figure 5 and Supplementary Video S5, with two major and three minor furrows. The multinucleate cell shown in Figure 6 and Supplementary Video S6 also formed a long furrow, the only one which continued to propagate (indicated as furrow 1 by the arrowhead). In cooperation with furrow 3, the broad furrow 1 gave rise to a daughter cell, while interaction with furrow 2 failed. All three nuclei in the incipient daughter cell slipped through the gap into the major part of the cell (774 s and 936 s frames). The anucleate remnant directed protrusions toward the major part (996 s frame) and reintegrated, similar to the anucleate portion of the mitotic cell in Figure 2B.

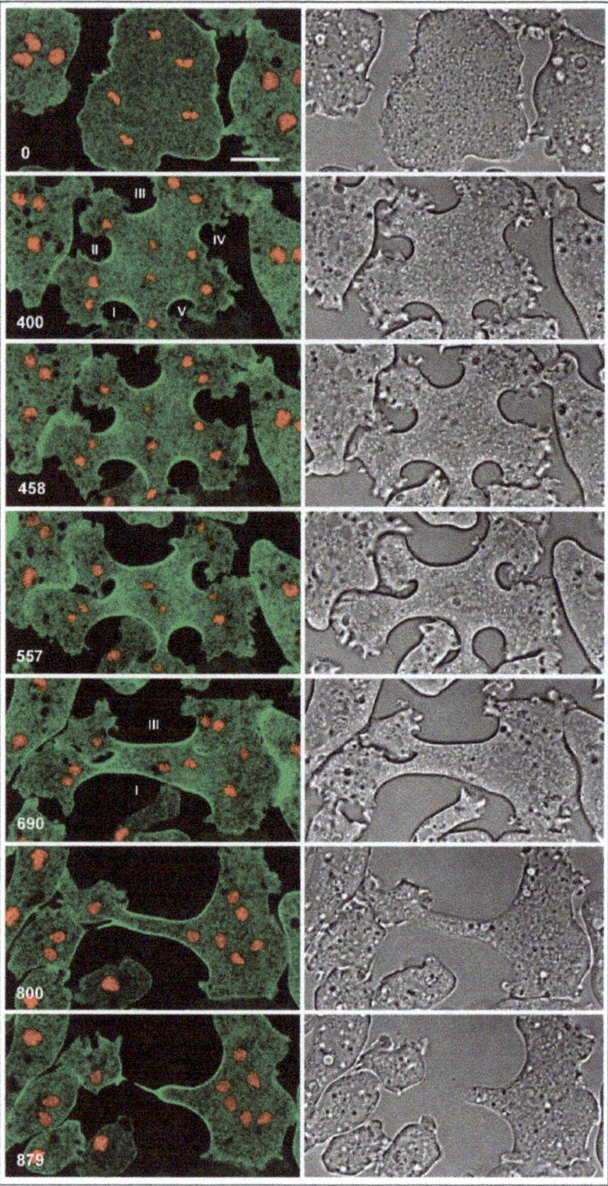

Figure 5. Synchronous mitosis and division of a multinucleate myosin-II-null cell. Like the cells in Figure 2, this cell expressed GFP-cortexillin I (green) and mRFP-histone 2B (red). Initially, five unilateral furrows formed that were enriched in cortexillin (400 s to 557 s frames). Later on, two long furrows prevailed (690 s and 800 s frames), and a final abscission occurred between them (879 s frame). The cell was flattened by agarose overlay. Time is indicated in seconds after the first frame. Scale bar, 10 μm. The entire sequence is shown in Supplementary Video S5.

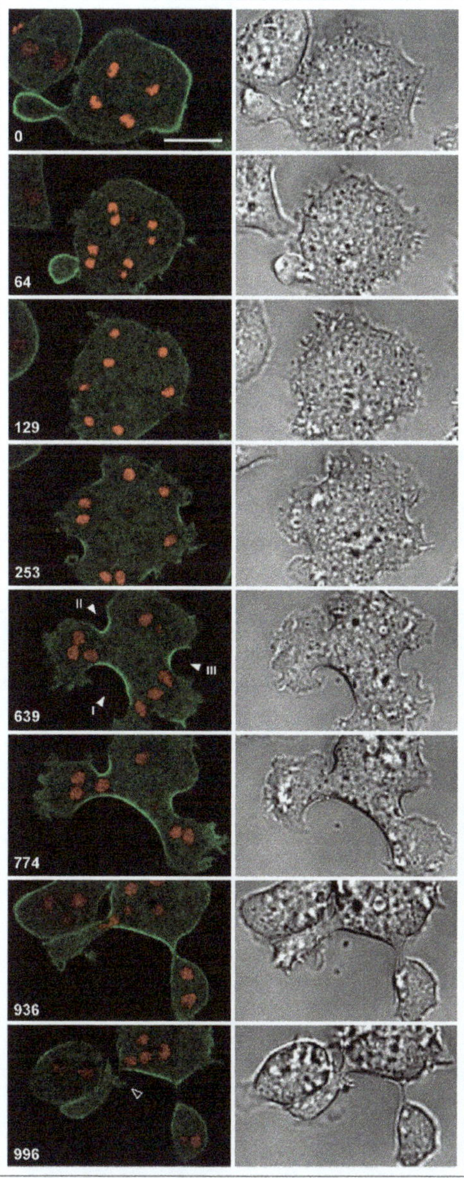

Figure 6. A multinucleate myosin-II-null cell undergoing successful and unavailing cytokinesis. The cell is labeled, like the cell in Figure 5, for cortexillin I (green) and for the chromosomes (red). Three unilateral furrows are indicated by arrowheads in the 639 s frame. Between furrow I and III, abscission of a daughter cell proceeds, whereas between furrows I and II three nuclei slip from the left portion into the major cell body (774 s and 936 s frames). Consequently, the anucleate part is united with the major cell body, forming protrusions toward the latter (open arrowhead in the 996 s frame). Time is indicated in seconds after the first frame. Scale bar, 10 µm. The same sequence is shown in Supplementary Video S6.

Figure 7 provides quantitative data on furrow progression and arrest in the cells shown in Figures 5 and 6. Figure 7A refers to the cell in Figure 5, in which furrows 2 and 4 stopped growing, while furrows 1 and 3 progressed. Furrow 5 showed a hybrid behavior: it first stopped and subsequently united with furrow 1 and resumed growing. Figure 7B, which refers to the cell in Figure 6, shows a clear distinction between furrow I that progressed continuously and furrows II and III that completely stopped after an initial phase of incision. These data indicate that in the absence of myosin-II, unilateral furrows can progress with almost the same velocity as in wild-type background and that two furrows can join to separate daughter cells. However, often furrows come to an early arrest, and the separation of daughter cells may fail.

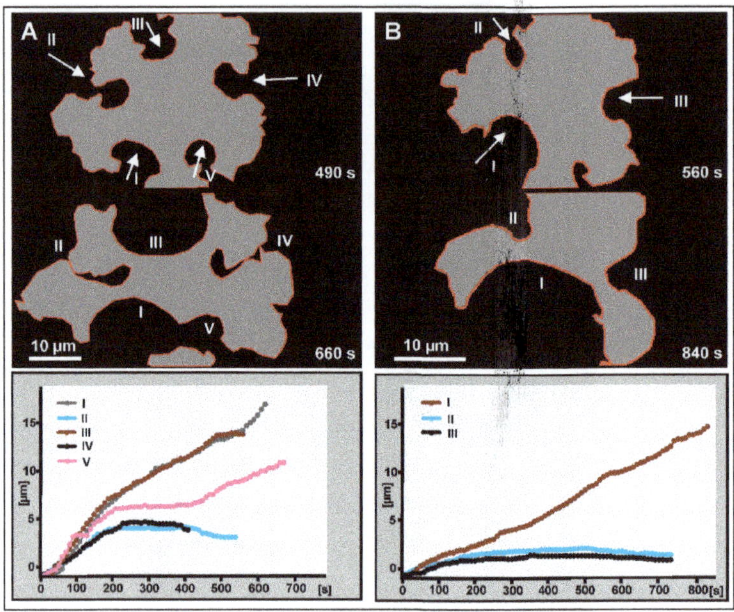

Figure 7. Progression and arrest of unilateral furrows in the myosin-II-null cells shown in the time series of Figures 5 and 6. (**A**) The cell shown in Figure 5 and in Supplementary Video S5 forms two progressing furrows (I and III) and two furrows that stop growing (II and IV). Furrow V enters a phase of arrest before it merges with the broad furrow I and continues propagating as part of a joint furrow. (**B**) The cell shown in Figure 6, with furrow I continuously progressing, and furrows II and III coming to a halt. The top and middle panels show the cell boundaries at earlier and later stages of furrow incision. The indicated times correspond to the time scales in Figures 5 and 6, respectively. Zero-time in the curves is set to the beginning of furrow ingression.

A summary of the data obtained for myosin-II-null cells is provided in Table 2, showing that altogether, half of the mutant cells succeeded in completing cytokinesis when attached to an adhesive substrate surface. We wish to add that the high proportion of failures held only for myosin-II-null cells grown axenically in nutrient medium. Cells fed with bacteria performed cytokinesis with higher efficiency, as previously reported [28].

Table 2. Mitosis in myosin-II-null cells attached to a glass surface with successful or unsuccessful cytokinesis.

Pre-Culture	Dividing Nuclei	Completed	Failed
(1) On substrate	Mono-nucleate	3	4 [1]
	Multi-nucleate	2 [1]	1 [1]
(2) Shaken culture	Mono-nucleate	2	1
	Multi-nucleate	20	9
(3) Electric pulse-fused	Mono-nucleate	2	0
	Multi-nucleate	8	3
Total		37	18

The cells were pre-cultured under one of three different conditions: (1) in a plastic Petri dish; (2) in shaken suspension; (3) in Petri dishes and subsequently fused. With three exceptions, cell division was observed without application of an agar overlay. The number of cells which were gently compressed by an overlay is provided in squared brackets.

4. Discussion

In *D. discoideum*, myosin-II-null as well as wild-type cells provide the possibility of studying the division of mononucleate and multinucleate cells and thus of normal and unilateral cleavage furrow formation in an identical genetic background. The inability of myosin-II-null cells to perform cytokinesis in suspension and the support of their mitotic division by an adhesive substrate made it easy to compare mononucleate and multinucleate cells in the absence of myosin-II.

Both wild-type and myosin-II-null cells showed how two unilateral furrows join to form a ring that separates a daughter cell from the multinucleate cell body, but the fused wild-type cells were superior because of the higher efficiency of furrow formation (Figure 3). In myosin-II-null cells, this ring constricted by a myosin-II independent mechanism, though frequently failed to complete cytokinesis (Figure 2). Together, the data obtained in mononucleate and multinucleate myosin-II-null cells are in accord with previous findings indicating that myosin-II stabilizes the position of the cleavage furrow [26,44], preventing its lateral sliding (Figure 2B) or expansion (Figures 5 and 6).

The sites of unilateral cleavage furrows are negatively controlled by positioning of the microtubule asters that emanate from the centrosomes during mitosis. The aster microtubules induce the cell cortex to ruffle, thus preventing the formation of a furrow. The cells are cleaved at spaces between asters, even if the flanking asters are not connected by a spindle [21]. The ingression of cleavage furrows between centrosomes that are not connected by a spindle relates cytokinesis in multinucleate *Dictyostelium* cells to furrow formation in sand dollar eggs [45]. However, the argument that these furrows are induced by aster microtubules [46] is probably not applicable to *Dictyostelium*. The asters may only act in inducing actin-rich protrusions, which correspond to polar regions in mononucleate cells ([20,21] and Figure 1B of the present paper). In multinucleate cells, there is no recognizable microtubule structure that may act as a source of signals for furrow ingression in the spindle-free inter-centrosomal spaces.

A protein important for the formation of cleavage furrows in *Dictyostelium* is cortexillin, which bundles actin-filaments [22]. Cortexillin may serve a function similar to that of anillin, another actin-bundling protein [47,48], which is missing in *Dictyostelium*. In various other cells, anillin is recruited to the furrow region [49] to fulfil multiple functions in cytokinesis [50]. In vitro, anillin can be shown to act independently of myosin in the constriction of an actin ring [51].

Cytokinesis in *Dictyostelium* cells is extremely adaptable since, depending on environmental conditions, different mechanisms are exploited on the route to complete the separation of the daughter cells [52]. The unilateral furrows studied here indicate that for initiation of a cleavage furrow, no contractile ring is required. The accumulation of myosin-II and cortexillin in unilateral furrows formed in multinucleate wild-type cells identifies two molecular players in the ingression of these furrows, which are established constituents of the machinery involved in constricting the ring-shaped furrows in mononucleate cells.

5. Conclusions

A peculiar feature of cytokinesis in multinucleate cells of *Dictyostelium* is the variable geometric relationship between furrow ingression and the subsequent ring contraction that finally results in the abscission of daughter cells (Figure 3). Ring formation may just be a continuation of the initial ingression events. This is the case when two furrows meet from opposite sides (Figure 8A). However, different from cytokinesis in mononucleate cells, the rings may also be formed laterally with respect of the initial ingression (Figure 8B), thus separating the entire process of cleavage into two distinct phases.

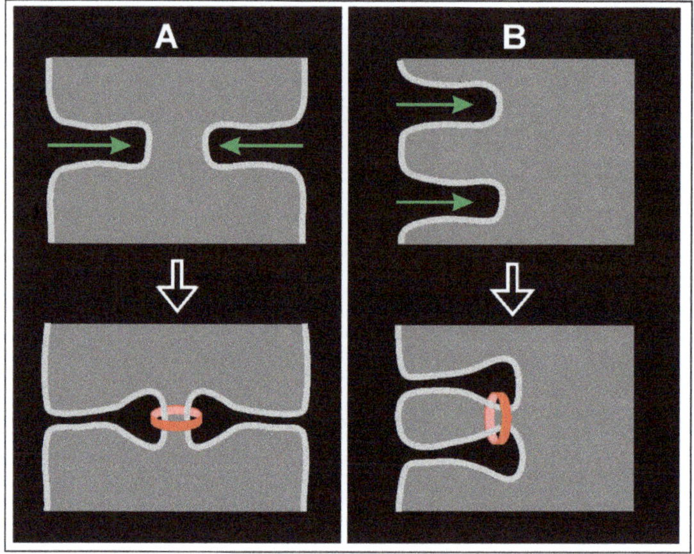

Figure 8. Diagram of cytokinesis in multinucleate cells, illustrating the variable spatial relationship of initial furrow ingression (green arrows) and final abscission (red ring). (**A**) Two furrows ingressing from opposite sides, and separation proceeding along the furrow directions. (**B**) Two furrows ingressing in parallel directions, and abscission occurring obliquely to these directions.

Supplementary Materials: The following are available online at http://www.mdpi.com/2073-4409/9/6/1493/s1, Video S1: Mitosis and cytokinesis in a wild-type cell of *Dictyostelium discoideum*, Video S2: Animation of three stages of cell division of a 3D-rendered wild-type cell of *D. discoideum*, Video S3: Successful (top) and failing (bottom) cytokinesis in myosin-II-null cells, Video S4: A multinucleate wild-type cell produced by electric pulse-induced fusion, Video S5: Synchronous mitosis and division of a multinucleate myosin-II-null cell, Video S6: A multinucleate myosin-II-null cell.

Author Contributions: J.B., E.S.M., J.P., M.E., conducted the experiments and analyzed the data, A.M.-T. provided vectors, G.G. designed the project and wrote the paper. All authors have read and agreed to the published version of the manuscript.

Funding: This work was supported by funds of the Max Planck Society to G.G.

Acknowledgments: We thank Martin Spitaler and his team at the Imaging Facility of the Max Planck Institute of Biochemistry for cooperation and providing the software Huygens Essential, Giulia Lang and Elisa Adani for contributing data, and Petra Fey and dictyBase for providing information [53].

Conflicts of Interest: The authors declare no competing or financial interests.

References

1. Henson, J.H.; Ditzler, C.E.; Germain, A.; Irwin, P.M.; Vogt, E.T.; Yang, S.; Wu, X.; Shuster, C.B. The ultrastructural organization of actin and myosin II filaments in the contractile ring: New support for an old model of cytokinesis. *Mol. Biol. Cell* **2017**, *28*, 613–623. [CrossRef] [PubMed]
2. Pollard, T.D.; O'Shaughnessy, B. Molecular mechanism of cytokinesis. *Annu. Rev. Biochem.* **2019**, *88*, 661–689. [CrossRef] [PubMed]
3. Swulius, M.T.; Nguyen, L.T.; Ladinsky, M.S.; Ortega, D.R.; Aich, S.; Mishra, M.; Jensen, G.J. Structure of the fission yeast actomyosin ring during constriction. *Proc. Natl. Acad. Sci. USA* **2018**, *115*, E1455. [CrossRef] [PubMed]
4. Hammarton, T.C. Who needs a contractile actomyosin ring? The plethora of alternative ways to divide a protozoan parasite. *Front. Cell. Infect. Microbiol.* **2019**, *9*, 397. [CrossRef]
5. Solnica-Krezel, L.; Burland, T.G.; Dove, W.F. Variable pathways for developmental changes of mitosis and cytokinesis in Physarum polycephalum. *J. Cell Biol.* **1991**, *113*, 591–604. [CrossRef]
6. Sun, H.S.; Wilde, A.; Harrison, R.E. Chlamydia trachomatis inclusions induce asymmetric cleavage furrow formation and ingression failure in host cells. *Mol. Cell Biol.* **2011**, *31*, 5011. [CrossRef]
7. Kotýnková, K.; Su, K.-C.; West, S.C.; Petronczki, M. Plasma membrane association but not midzone recruitment of RhoGEF ECT2 is essential for cytokinesis. *Cell Rep.* **2016**, *17*, 2672–2686. [CrossRef]
8. Szollosi, D. Cortical cytoplasmic filaments of cleaving eggs: A structural element corresponding to the contractile ring. *J. Cell Biol.* **1970**, *44*, 192–209. [CrossRef]
9. Roberson, M.; Armstrong, J.; Armstrong, P. Adhesive and non-adhesive membrane domains of amphibian embryo cells. *J. Cell Sci.* **1980**, *44*, 19.
10. Rappaport, R.; Rappaport, B.N. Cleavage in Conical Sand Dollar Eggs. *Dev. Biol.* **1994**, *164*, 258–266. [CrossRef]
11. Schejter, E.D.; Wieschaus, E. Functional elements of the cytoskeleton in the early Drosophila embryo. *Annu. Rev. Cell Biol.* **1993**, *9*, 67–99. [CrossRef] [PubMed]
12. Xue, Z.; Sokac, A.M. Back-to-back mechanisms drive actomyosin ring closure during Drosophila embryo cleavage. *J. Cell Biol.* **2016**, *215*, 335–344. [CrossRef] [PubMed]
13. Wang, Q.; Racowsky, C.; Deng, M. Mechanism of the chromosome-induced polar body extrusion in mouse eggs. *Cell Div.* **2011**, *6*, 17. [CrossRef] [PubMed]
14. Uraji, J.; Scheffler, K.; Schuh, M. Functions of actin in mouse oocytes at a glance. *J. Cell Sci.* **2018**, *131*, jcs218099. [CrossRef] [PubMed]
15. Savoian, M.S.; Khodjakov, A.; Rieder, C.L. Unilateral and wandering furrows during mitosis in vertebrates: Implications for the mechanism of cytokinesis. *Cell Biol. Int.* **1999**, *23*, 805–812. [CrossRef]
16. Fukui, Y.; Inoue, S. Cell division in Dictyostelium with special emphasis on actomyosin organization in cytokinesis. *Cell Motil. Cytoskel.* **1991**, *18*, 41–54. [CrossRef]
17. De Lozanne, A.; Spudich, J.A. Disruption of the Dictyostelium myosin heavy chain gene by homologous recombination. *Science* **1987**, *236*, 1086–1091. [CrossRef]
18. Knecht, D.A.; Loomis, W.F. Antisense RNA inactivation of myosin heavy chain gene expression in Dictyostelium discoideum. *Science* **1987**, *236*, 1081–1085. [CrossRef]
19. Manstein, D.J.; Titus, M.A.; De Lozanne, A.; Spudich, J.A. Gene replacement in Dictyostelium: Generation of myosin null mutants. *EMBO J.* **1989**, *8*, 923–932. [CrossRef]
20. Neujahr, R.; Heizer, C.; Gerisch, G. Myosin II-independent processes in mitotic cells of Dictyostelium discoideum: Redistribution of the nuclei, re-arrangement of the actin system and formation of the cleavage furrow. *J. Cell Sci.* **1997**, *110*, 123.
21. Neujahr, R.; Albrecht, R.; Köhler, J.; Matzner, M.; Schwartz, J.M.; Westphal, M.; Gerisch, G. Microtubule-mediated centrosome motility and the positioning of cleavage furrows in multinucleate myosin II-null cells. *J. Cell Sci.* **1998**, *111*, 1227. [PubMed]
22. Faix, J.; Steinmetz, M.; Boves, H.; Kammerer, R.A.; Lottspeich, F.; Mintert, U.; Murphy, J.; Stock, A.; Aebi, U.; Gerisch, G. Cortexillins, major determinants of cell shape and size, are actin-bundling proteins with a parallel coiled-coil tail. *Cell* **1996**, *86*, 631–642. [CrossRef]

23. Faix, J.; Weber, I.; Mintert, U.; Köhler, J.; Lottspeich, F.; Marriott, G. Recruitment of cortexillin into the cleavage furrow is controlled by Rac1 and IQGAP-related proteins. *EMBO J.* **2001**, *20*, 3705–3715. [CrossRef] [PubMed]
24. Liu, X.; Shu, S.; Yu, S.; Lee, D.-Y.; Piszczek, G.; Gucek, M.; Wang, G.; Korn, E.D. Biochemical and biological properties of cortexillin III, a component of Dictyostelium DGAP1–cortexillin complexes. *Mol. Biol. Cell* **2014**, *25*, 2026–2038. [CrossRef] [PubMed]
25. Gerisch, G.; Weber, I. Cytokinesis without myosin II. *Curr. Opin. Cell Biol.* **2000**, *12*, 126–132. [CrossRef]
26. Weber, I.; Neujahr, R.; Du, A.; Köhler, J.; Faix, J.; Gerisch, G. Two-step positioning of a cleavage furrow by cortexillin and myosin II. *Curr. Biol.* **2000**, *10*, 501–506. [CrossRef]
27. Srivastava, V.; Iglesias, P.A.; Robinson, D.N. Cytokinesis: Robust cell shape regulation. *Semin. Cell Dev. Biol.* **2016**, *53*, 39–44. [CrossRef]
28. Weber, I.; Gerisch, G.; Heizer, C.; Murphy, J.; Badelt, K.; Stock, A.; Schwartz, J.M.; Faix, J. Cytokinesis mediated through the recruitment of cortexillins into the cleavage furrow. *EMBO J.* **1999**, *18*, 586–594. [CrossRef]
29. Gerisch, G.; Ecke, M.; Neujahr, R.; Prassler, J.; Stengl, A.; Hoffmann, M.; Schwarz, U.S.; Neumann, E. Membrane and actin reorganization in electropulse-induced cell fusion. *J. Cell Sci.* **2013**, *126 Pt 9*, 2069–2078. [CrossRef]
30. Rossier, C.; Gerisch, G.; Malchow, D. Action of a slowly hydrolysable cyclic AMP analogue on developing cells of Dictyostelium discoideum. *J. Cell Sci.* **1979**, *35*, 321.
31. Robinson, D.N.; Cavet, G.; Warrick, H.M.; Spudich, J.A. Quantitation of the distribution and flux of myosin-II during cytokinesis. *BMC Cell Biol.* **2002**, *3*, 4. [CrossRef]
32. Fischer, M.; Haase, I.; Simmeth, E.; Gerisch, G.; Müller-Taubenberger, A. A brilliant monomeric red fluorescent protein to visualize cytoskeleton dynamics in Dictyostelium. *FEBS Lett.* **2004**, *577*, 227–232. [CrossRef]
33. Campbell, R.E.; Tour, O.; Palmer, A.E.; Steinbach, P.A.; Baird, G.S.; Zacharias, D.A.; Tsien, R.Y. A monomeric red fluorescent protein. *Proc. Natl. Acad. Sci. USA* **2002**, *99*, 7877. [CrossRef]
34. Müller-Taubenberger, A. Application of fluorescent protein tags as reporters in live-cell imaging studies. In *Dictyostelium Discoideum Protocols*; Eichinger, L., Rivero, F., Eds.; Humana Press: Totowa, NJ, USA, 2006; Volume 346, pp. 229–246.
35. Gritz, L.; Davies, J. Plasmid-encoded hygromycin B resistance: The sequence of hygromycin B phosphotransferase gene and its expression in Escherichia coli and Saccharomyces cerevisiae. *Gene* **1983**, *25*, 179–188. [CrossRef]
36. Malchow, D.; Nägele, B.; Schwarz, H.; Gerisch, G. Membrane-bound cyclic AMP phosphodiesterase in chemotactically responding cells of Dictyostelium discoideum. *Eur. J. Biochem.* **1972**, *28*, 136–142. [CrossRef]
37. Fukui, Y.; Yumura, S.; Yumura, T.K. Agar-overlay immunofluorescence: High-resolution studies of cytoskeletal components and their changes during chemotaxis. *Methods Cell Biol.* **1987**, *28*, 347–356.
38. Samereier, M.; Meyer, I.; Koonce, M.P.; Gräf, R. Live cell-Imaging techniques for analyses of microtubules in Dictyostelium. In *Methods in Cell Biology*; Cassimeris, L., Tran, P., Eds.; Academic Press: Cambridge, MA, USA, 2010; Volume 97, pp. 341–357.
39. Schindelin, J.; Arganda-Carreras, I.; Frise, E.; Kaynig, V.; Longair, M.; Pietzsch, T.; Preibisch, S.; Rueden, C.; Saalfeld, S.; Schmid, B. Fiji: An open-source platform for biological-image analysis. *Nat. Methods* **2012**, *9*, 676–682. [CrossRef]
40. Pettersen, E.F.; Goddard, T.D.; Huang, C.C.; Couch, G.S.; Greenblatt, D.M.; Meng, E.C.; Ferrin, T.E. UCSF Chimera—A visualization system for exploratory research and analysis. *J. Comput. Chem.* **2004**, *25*, 1605–1612. [CrossRef] [PubMed]
41. Leo, M.; Santino, D.; Tikhonenko, I.; Magidson, V.; Khodjakov, A.; Koonce, M.P. Rules of engagement: Centrosome–nuclear connections in a closed mitotic system. *Biol. Open* **2012**, *1*, 1111. [CrossRef]
42. McIntosh, J.R.; Roos, U.P.; Neighbors, B.; McDonald, K.L. Architecture of the microtubule component of mitotic spindles from Dictyostelium discoideum. *J. Cell Sci.* **1985**, *75*, 93.
43. Wienke, D.C.; Knetsch, M.L.W.; Neuhaus, E.M.; Reedy, M.C.; Manstein, D.J. Disruption of a dynamin homologue affects endocytosis, organelle morphology, and cytokinesis in Dictyostelium discoideum. *Mol. Biol. Cell* **1999**, *10*, 225–243. [CrossRef]
44. Neujahr, R.; Heizer, C.; Albrecht, R.; Ecke, M.; Schwartz, J.-M.; Weber, I.; Gerisch, G. Three-dimensional patterns and redistribution of myosin II and actin in mitotic Dictyostelium cells. *J. Cell Biol.* **1997**, *139*, 1793. [CrossRef]

45. Rappaport, R. Experiments concerning the cleavage stimulus in sand dollar eggs. *J. Exp. Zool.* **1961**, *148*, 81–89. [CrossRef]
46. Rappaport, R. Cytokinesis in Animal Cells. In *Biomechanics of Active Movement and Deformation of Cells*; Akkaş, N., Ed.; Springer: Berlin/Heidelberg, Germany, 1990; Volume H42, pp. 1–34.
47. Field, C.M.; Alberts, B.M. Anillin, a contractile ring protein that cycles from the nucleus to the cell cortex. *J. Cell Biol.* **1995**, *131*, 165–178. [CrossRef]
48. Jananji, S.; Risi, C.; Lindamulage, I.K.S.; Picard, L.-P.; Van Sciver, R.; Laflamme, G.; Albaghjati, A.; Hickson, G.R.X.; Kwok, B.H.; Galkin, V.E. Multimodal and polymorphic interactions between anillin and actin: Their implications for cytokinesis. *J. Mol. Biol.* **2017**, *429*, 715–731. [CrossRef]
49. Piekny, A.J.; Glotzer, M. Anillin is a scaffold protein that links RhoA, Actin, and Myosin during cytokinesis. *Curr. Biol.* **2008**, *18*, 30–36. [CrossRef]
50. Piekny, A.J.; Maddox, A.S. The myriad roles of Anillin during cytokinesis. *Semin. Cell Dev. Biol.* **2010**, *21*, 881–891. [CrossRef]
51. Kučera, O.; Janda, D.; Siahaan, V.; Dijkstra, S.H.; Pilátová, E.; Zatecka, E.; Diez, S.; Braun, M.; Lansky, Z. Anillin propels myosin-independent constriction of actin rings. *bioRxiv* **2020**. [CrossRef]
52. Taira, R.; Yumura, S. A novel mode of cytokinesis without cell-substratum adhesion. *Sci. Rep.* **2017**, *7*, 17694. [CrossRef]
53. Fey, P.; Dodson, R.J.; Basu, S.; Chisholm, R.L. One stop shop for everything Dictyostelium: DictyBase and the Dicty Stock Center in 2012. In *Methods Mol. Biol.*; Eichinger, L., Rivero, F., Eds.; Humana Press: Totowa, NJ, USA, 2013; pp. 59–92.

Article

Flow Induced Symmetry Breaking in a Conceptual Polarity Model

Manon C. Wigbers †, Fridtjof Brauns †, Ching Yee Leung † and Erwin Frey *

Arnold Sommerfeld Center for Theoretical Physics (ASC) and Center for NanoScience (CeNS), Department of Physics, Ludwig-Maximilians-Universität München, Theresienstraße 37, D–80333 München, Germany; m.wigbers@physik.uni-muenchen.de (M.C.W.); fridtjof.brauns@physik.uni-muenchen.de (F.B.); c.leung@physik.uni-muenchen.de (C.Y.L.)

* Correspondence: frey@lmu.de

† These authors contributed equally to this work.

Received: 19 May 2020; Accepted: 19 June 2020; Published: 23 June 2020

Abstract: Important cellular processes, such as cell motility and cell division, are coordinated by cell polarity, which is determined by the non-uniform distribution of certain proteins. Such protein patterns form via an interplay of protein reactions and protein transport. Since Turing's seminal work, the formation of protein patterns resulting from the interplay between reactions and diffusive transport has been widely studied. Over the last few years, increasing evidence shows that also advective transport, resulting from cytosolic and cortical flows, is present in many cells. However, it remains unclear how and whether these flows contribute to protein-pattern formation. To address this question, we use a minimal model that conserves the total protein mass to characterize the effects of cytosolic flow on pattern formation. Combining a linear stability analysis with numerical simulations, we find that membrane-bound protein patterns propagate against the direction of cytoplasmic flow with a speed that is maximal for intermediate flow speed. We show that the mechanism underlying this pattern propagation relies on a higher protein influx on the upstream side of the pattern compared to the downstream side. Furthermore, we find that cytosolic flow can change the membrane pattern qualitatively from a peak pattern to a mesa pattern. Finally, our study shows that a non-uniform flow profile can induce pattern formation by triggering a regional lateral instability.

Keywords: symmetry breaking; cytoplasmic flow; phase-space analysis; pattern formation

---

1. Introduction

Many biological processes rely on the spatiotemporal organization of proteins. Arguably one of the most elementary forms of such organization is cell polarization—the formation of a "cap" or spot of high protein concentration that determines a direction. Such a polarity axis then coordinates downstream processes including motility [1,2], cell division [3], and directional growth [4]. Cell polarization is an example for symmetry breaking [5], as the orientational symmetry of the initially homogeneous protein distribution is broken by the formation of the polar cap.

Intracellular protein patterns arise from the interplay between protein interactions (chemical reactions) and protein transport. Diffusion in the cytosol serves as the most elementary means of transport. Pattern formation resulting from the interplay of reactions and diffusion has been widely studied since Turing's seminal work [6]. In addition to diffusion, proteins can be transported by fluid flows in the cytoplasm [7–9] and along cytoskeletal structures (vesicle trafficking, cortical contractions) driven by molecular motors [10–12]. These processes lead to advective transport of proteins.

Recently, it has been shown experimentally that advective transport (caused by cortical flows) induces polarization of the PAR system in the *C. elegans* embryo [13–15]. Furthermore, in vitro studies with the MinDE system of *E. coli*, reconstituted in microfluidic chambers, have shown that the flow of the bulk fluid has a strong effect on the protein patterns that form on the membrane [16,17]. Increasing evidence shows that cortical and cytosolic flows (also called "cytoplasmic streaming") are present in many cells [18–23]. In addition, cortical contractions can drive cell-shape deformations [24], inducing flows in the incompressible cytosol [8,25]. However, the role of flows for protein-pattern formation remains elusive. This motivates to study the role of advective flow from a conceptual perspective, with a minimal model. The insights thus gained will help to understand the basic, principal effects of advective flow on pattern formation and reveal the underlying elementary mechanisms.

The basis of our study is a paradigmatic class of models for cell polarization that describe a single protein species which has a membrane-bound state and a cytosolic state. Such two-component mass-conserving reaction–diffusion (2cMcRD) systems serve as conceptual models for cell polarization [7,26–31]. Specifically, they have been used to model Cdc42 polarization in budding yeast [32] and PAR-protein polarity [33]. 2cMcRD systems generically exhibit both spontaneous and stimulus-induced polarization [5,31,33]. In the former case, a spatially uniform steady state is unstable against small spatial perturbations ("Turing instability" [6]). Adjacent to the parameter regime of this lateral instability, a sufficiently strong, localized stimulus (e.g., an external signal) can induce the formation of a pattern starting from a stable spatially uniform state. The steady state patterns that form in two-component McRD systems are generally stationary (there are no traveling or standing waves). Moreover, the final stationary pattern has no characteristic wavelength. Instead, the peaks that grow initially from the fastest growing mode ("most unstable wavelength") compete for mass until only a single peak remains ("winner takes all") [30,34,35]. The location of this peak can be controlled by external stimuli (e.g., spatial gradients in the reaction rates) [34,36].

Recently, a theoretical framework, termed local equilibria theory, has been developed to study these phenomena using a geometric analysis in the phase plane of the protein concentrations [31,37]. With this framework one can gain insight into the mechanisms underlying the dynamics of McRD systems both in the linear and in the strongly nonlinear regime, thereby bridging the gap between these two regimes.

Here, we show that cytosolic flow in two-component systems always induces upstream propagation of the membrane-bound pattern. In other words, the peak moves against the cytosolic flow direction. This propagation is driven by a higher protein influx on the upstream side of the membrane-concentration peak compared to its downstream side. Using this insight, we are able to explain why the propagation speed becomes maximal at intermediate flow speeds and vanishes when the rate of advective transport becomes fast compared to the rate of diffusive transport or compared to the reaction rates. We first study a uniform flow profile using periodic boundaries. This effectively represents a circular flow, which is observed in plant cells (where this phenomenon is called cytoplasmic streaming or cyclosis) [38]. It also represents an in vitro system in a laterally large microfluidic chamber. We then study the effect of a spatially non-uniform flow profile in a system with reflective boundaries, as a minimal system for flows close to the membrane [7,13,15], e.g., in the actin cortex. We show that a non-uniform flow profile redistributes the protein mass, which can trigger a regional lateral instability and thereby induce pattern formation from a stable homogeneous steady state.

The remainder of the paper is structured as follows. We first introduce the model in Section 2. We then perform a linear stability analysis in Section 3 to show how spatially uniform cytosolic flow influences the dynamics close to a homogeneous steady state. In Section 4, we use numerical simulations to study the fully nonlinear long-term behavior of the system. Next, we show that upon increasing the cytosolic flow velocity, the pattern can qualitatively change from a mesa pattern to a peak pattern in Section 5. Finally, in Section 6, we study how a spatially non-uniform cytosolic flow can trigger a regional lateral instability and thus induce pattern formation. Implications of our findings and links to earlier literature are briefly discussed at the end of each section. We conclude with a brief outlook section.

2. Model

We consider a spatially one-dimensional system of length L. The proteins can cycle between a membrane-bound state (concentration $m(x,t)$) and a cytosolic state (concentration $c(x,t)$), and diffuse with diffusion constants D_m and D_c, respectively (Figure 1). In cells, the diffusion constant on the membrane is typically much smaller than the diffusion constant in the cytosol. In the cytosol, the proteins are assumed to be advected with a speed $v_f(x)$, as indicated by the blue arrow in Figure 1. Thus, the reaction-diffusion-advection equations for the cytosolic density and membrane density read

$$\partial_t c + \partial_x(v_f c) = D_c \partial_x^2 c - f(m,c), \tag{1a}$$

$$\partial_t m = D_m \partial_x^2 m + f(m,c), \tag{1b}$$

with either periodic or reflective boundary conditions. The nonlinear function $f(m,c)$ describes the reaction kinetics of the system. Attachment–detachment kinetics can generically be written in the form

$$f(m,c) = a(m)c - d(m)m, \tag{2}$$

where $a(m) > 0$ and $d(m) > 0$ denote the rate of attachment from the cytosol to the membrane and detachment from the membrane to the cytosol, respectively. The dynamics given by Equation (1) conserve the average total density

$$\bar{n} = \frac{1}{L}\int_0^L \mathrm{d}x\, n(x,t). \tag{3}$$

Here, we introduced the local total density $n(x,t) := m(x,t) + c(x,t)$.

For illustration purposes, we will use a specific realization of the reaction kinetics [31],

$$a(m) = k_{\mathrm{on}} + k_{\mathrm{fb}} m \quad \text{and} \quad d(m) = \frac{k_{\mathrm{off}}}{K_{\mathrm{D}} + m}, \tag{4}$$

describing attachment with a rate k_{on}, self-recruitment with a rate k_{fb}, and enzyme-driven detachment with a rate k_{off} and the Michaelis–Menten constant K_{D}, respectively. However, our results do not depend on the specific choice of the reaction kinetics. Unless stated otherwise, we use the parameters: $k_{\mathrm{on}} = 1\,\mathrm{s}^{-1}, k_{\mathrm{fb}} = 1\,\mu\mathrm{m}\,\mathrm{s}^{-1}, k_{\mathrm{off}} = 2\,\mathrm{s}^{-1}, K_{\mathrm{D}} = 1\,\mu\mathrm{m}^{-1}, \bar{n} = 5\,\mu\mathrm{m}^{-1}, D_m = 0.01\,\mu\mathrm{m}^2/\mathrm{s}, D_c = 10\,\mu\mathrm{m}^2/\mathrm{s}$.

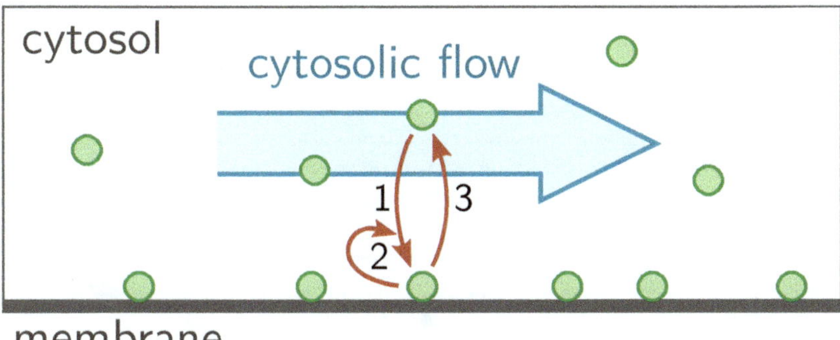

Figure 1. One-dimensional two-component system with cytosolic flow into the positive x-direction. The reaction kinetics include **(1)** attachment, **(2)** self-recruitment, and **(3)** enzyme-driven detachment.

3. Linear Stability Analysis

3.1. Linearized Dynamics and Basic Results

To study how cytosolic flow affects the formation of protein patterns, we first consider a spatially uniform flow profile (i.e., constant $v_f(x) = v_f$) and perform a linear stability analysis of a spatially homogeneous steady state $\mathbf{u}^* = (c^*, m^*)$:

$$f(m^*, c^*) = 0, \quad m^* + c^* = \bar{n}. \tag{5}$$

Following the standard procedure, we linearize the dynamics for small perturbations $\mathbf{u}(x,t) = (c(x,t), m(x,t)) = \mathbf{u}^* + \delta\mathbf{u}(x,t)$ around the homogeneous steady state. Expanding $\delta\mathbf{u}(x,t)$ in exponentially growing (or decaying) Fourier modes $\delta\mathbf{u} = \hat{\mathbf{u}}_q e^{\sigma t} e^{iqx}$ leads to the eigenvalue problem

$$\mathcal{J}\hat{\mathbf{u}}_q = \sigma\hat{\mathbf{u}}_q, \tag{6}$$

with the Jacobian

$$\mathcal{J} = \begin{pmatrix} -D_c q^2 - iv_f q - f_c & -f_m \\ f_c & -D_m q^2 + f_m \end{pmatrix},$$

where $f_c = \partial_c f|_{\mathbf{u}^*}$ and $f_m = \partial_m f|_{\mathbf{u}^*}$ encode the linearized reaction kinetics. Note that for reaction kinetics of the form Equation (2), $f_c = a(m) > 0$ and we consider this case in the following.

For each mode with wavenumber q, there are two eigenvalues $\sigma_{1,2}(q)$. The case $q = 0$ corresponds to spatially homogeneous perturbations, where the two eigenvalues are given by $\sigma_1 = f_m - f_c$ and $\sigma_2 = 0$ [31]. Here, we restrict our analysis to homogeneously stable states ($\sigma_1 < 0$). The second eigenvalue ($\sigma_2 = 0$) corresponds to perturbations that change the average mass $\bar{n}$ and therefore shift the homogeneous steady state $\mathbf{u}^*(\bar{n})$ along the nullcline $f = 0$. As a result that these perturbations break mass-conservation, they are not relevant for the stability of a closed system as considered here. The modes $q > 0$ determine the stability of the system against spatially inhomogeneous perturbations (lateral stability). The eigenvalue with the larger real part determines the stability and will be denoted by $\sigma(q)$, suppressing the index.

A typical dispersion relation with a band of unstable modes is shown in Figure 2A. The real part (solid line), indicating the mode's growth rate, has a band of unstable modes $[0, q_{\max}]$ where $\operatorname{Re}\sigma(q) > 0$. The fastest growing mode q^* determines the wavelength λ of the pattern that initially grows, triggered by a small, random perturbation of the spatially homogeneous steady state. For $v_f = 0$, the imaginary part of $\sigma(q)$ vanishes, for locally stable steady states ($\sigma(0) \leq 0$) [31]. However, in the presence of flow, the imaginary part of $\sigma(q)$ is non-zero (dashed line in Figure 2A), which implies a propagation of each mode with the phase velocity $v_{\text{phase}}(q) = -\operatorname{Im}\sigma(q)/q$. This means that a mode q not only grows over time (orange arrows in Figure 2B), but also propagates as indicated by the pink arrows in Figure 2B. Further below, in Section 3.4, we will show that $\operatorname{Im}\sigma(q)$ always has the same sign as the flow velocity v_f, such that all modes propagate against the flow direction.

To gain physical insight into the mechanisms underlying the growth and propagation of perturbations (modes) we will first give an intuitive explanation of a lateral instability in McRD systems, building on the concepts of local equilibria theory [31,37]. We then provide a more detailed analysis in the limits of long wavelength as well as fast and slow flow.

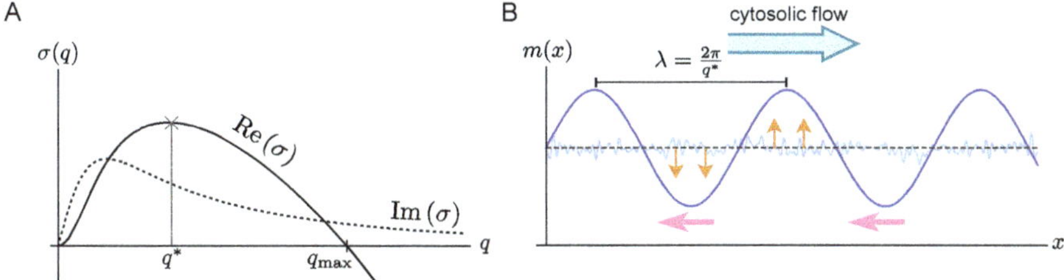

Figure 2. (**A**) Sketch of real (solid) and imaginary (dotted) part of a typical dispersion relation with a band $[0, q_{max}]$ of unstable modes. (**B**) The initial dynamics of a spatially homogeneous state with a small random perturbation (blue thin line). The direction of cytosolic flow is indicated by a blue arrow. The typical wavelength (λ) of the initial pattern is determined by the fastest growing mode q^* and the phase velocity is determined by the value of the imaginary part of dispersion relation at the fastest growing mode ($v_{phase} = -\mathrm{Im}\sigma(q^*)/q^*$). The growth of the pattern is indicated by orange arrows, while the traveling direction is indicated by pink arrows.

3.2. *Intuition for the Flow-Driven Instability and Upstream Propagation of the Unstable Mode*

Lateral instability in McRD systems can be understood as a mass-redistribution instability [31]. Let us briefly recap the mechanism underlying this instability for a system without flow. To this end, we first discuss the effect of reactions and diffusion separately, and explain how these effects together drive the mass-redistribution instability. We then explain how this instability is affected by cytosolic flow.

Consider a spatially homogeneous steady state, perturbed by a slight redistribution of the local total density $n(x,t)$. The dashed orange line in Figure 3A shows such a perturbation where the membrane concentration (Figure 3A top) is slightly perturbed in a sinusoidal fashion. In phase space this is represented by a density distribution that slightly deviates from the spatially homogeneous steady state (marked by the orange dashed line). Here, the open star and open circle mark the minimum and maximum of the local total density, respectively. The local total density determines the local reactive equilibrium concentrations $m^*(n)$ and $c^*(n)$ (cf. Equation (5), replacing the average mass $\bar{n}$ by the local mass $n(x,t)$). In phase space (Figure 3A bottom) these local equilibria can be read off from the intersections (marked by black circles) of the reactive subspaces $n(x,t) = m(x,t) + c(x,t)$ (gray solid lines) and the reactive nullcline (black solid lines). A slight redistribution of the local total density shifts the reactive equilibria, leading to reactive flows towards these shifted equilibria (red and green arrows in Figure 3A). Thus, the reactive equilibria, and thereby the reactive flows, are encoded in the shape of the reactive nullcline in phase space. If the nullcline slope is negative, increasing the total density leads to a decreasing equilibrium cytosolic concentration and therefore to attachment (green arrows in Figure 3A). Conversely, in regions of lower total density, the equilibrium cytosolic concentration increases via detachment (red arrows in Figure 3A). Hence, regions of high total density become self-organized attachment zones and regions of low total density become self-organized detachment zones [37] (green and red areas in Figure 3 top and middle).

These attachment and detachment zones act as sinks and sources for diffusive mass-transport on the membrane and in the cytosol: The attachment zone acts as a cytosolic sink and membrane source, and the detachment zone acts as a cytosolic source and a membrane sink (blue arrows in Figure 3B). As diffusion in the cytosol is much faster than in the membrane, mass is transported faster in the cytosol than on the membrane, as indicated by the size of the blue arrows in Figure 3B top and middle. This leads to net mass transport from the detachment zone to the attachment zone. As the local total density increases in the attachment zone, it facilitates further attachment and thereby the growth of the pattern on the membrane. In short, the mechanism underlying the mass-redistribution instability is a cascade of attachment–detachment kinetics (Figure 3A) and net mass-transport towards attachment zones (Figure 3B).

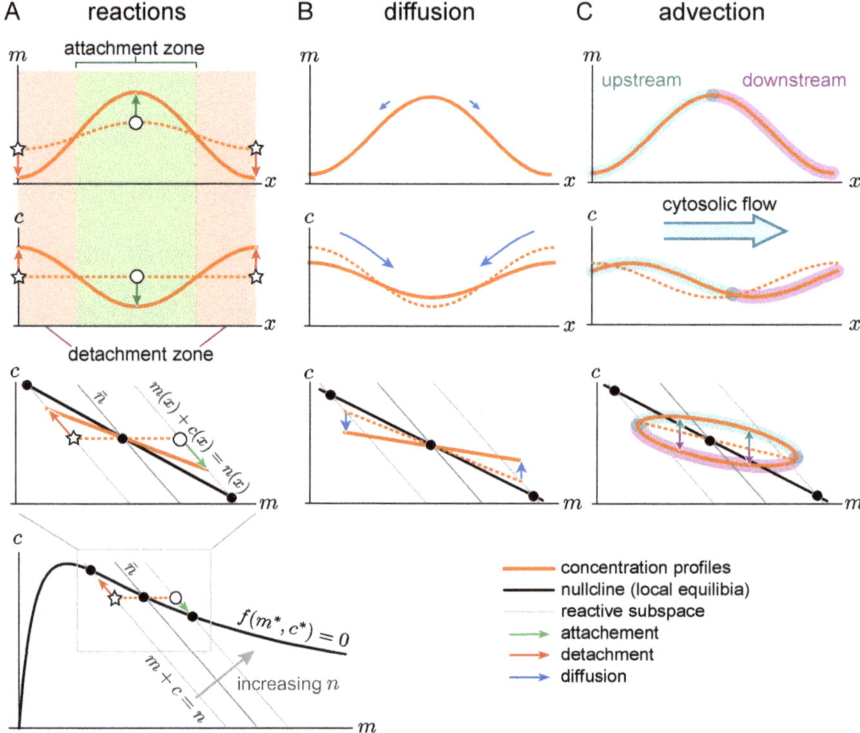

Figure 3. Sketch of the initial dynamics of an laterally unstable spatially homogeneous steady state. The role of reactions (**A**), diffusion (**B**), and advection (**C**) for a mass-redistribution instability are presented for the membrane (top) and cytosolic (middle) concentration profiles and in phase space (bottom). (**A**) A small perturbation of the spatially homogeneous membrane concentration (orange dashed lines in top panel) leads to a spatially varying local total density $n(x)$, with a larger total density at the maximum of the membrane profile (open circle) and a smaller total density at the minimum (open star). These local variations in total density lead to attachment zones (green region) and detachment zones (red region). The reactive flow, indicated by the red and green arrows, points along the reactive subspace (gray lines) in phase space towards the shifted local equilibria (black circles). These reactive flows lead to the solid orange density profiles after a small amount of time. (**B**) Faster diffusion in the cytosol compared to the membrane (indicated by the large and small blue arrows in the middle and top panel, respectively), lead to net mass transport from the detachment zone to the attachment zone. Again, dashed and solid lines indicate the state before and after a short time interval of diffusive transport. (**C**) Cytosolic flow shifts the cytosolic concentration with respect to the membrane concentration (orange dashed to orange solid lines), increasing the cytosolic concentration on the upstream side of the pattern and decreasing the cytosolic concentration on the downstream side. In phase space, the trajectory of this density profile forms a 'loop'.

How does cytosolic fluid flow affect the mass-redistribution instability? Cytosolic flow transports proteins advectively. This advective transport shifts the cytosolic density profile downstream relative to the membrane density profile (dashed to solid orange line in Figure 3C middle). This shift leads to an increase of the cytosolic density on the upstream (cyan) side of the membrane peak and a decrease on the downstream (magenta) side, in Figure 3C (middle), respectively. In phase space, this asymmetry is reflected as a 'loop' shape of the phase space trajectory that corresponds to the real space pattern (Figure 3C bottom). The higher cytosolic density on the upstream side increases attachment relative to the downstream side. This leads to a propagation of the membrane concentration profile in the upstream direction.

3.3. Long Wavelength Limit

To complement this intuitive picture we consider the long wavelength limit $q \to 0$. (In principle, the dispersion relation can be easily obtained in closed form using the formula for eigenvalues of 2×2 matrices: $\sigma_{1,2} = \frac{1}{2}\,\mathrm{tr}\,\mathcal{J} \mp \frac{1}{2}\sqrt{(\mathrm{tr}\,\mathcal{J})^2 - 4\det\mathcal{J}}$ where $\mathrm{tr}\,\mathcal{J}$ and $\det\mathcal{J}$ are the Jacobian's trace and determinant, respectively. As a result that the resulting expression is rather lengthy, we do not write it out it explicitly here.) In this limit, the dispersion relation expanded to second order in q reads

$$\sigma(q) \approx -\frac{1}{1+s_{\mathrm{nc}}}\left[i s_{\mathrm{nc}} v_{\mathrm{f}} q + (D_m + s_{\mathrm{nc}} D_c) q^2 + \frac{s_{\mathrm{nc}}\, v_{\mathrm{f}}^2}{f_c (1+s_{\mathrm{nc}})^2} q^2 \right], \tag{7}$$

where $s_{\mathrm{nc}} = -f_m/f_c$ is the slope of the reactive nullcline. The imaginary part $\mathrm{Im}\,\sigma(q)$ is linear in q to lowest order, implying a phase velocity $v_{\mathrm{phase}} = v_{\mathrm{f}} s_{\mathrm{nc}}/(1+s_{\mathrm{nc}})$ that is independent of the wavelength. The growth rate $\mathrm{Re}\,\sigma(q)$ is quadratic in q to lowest order. If this quadratic term is positive, there is a band of unstable modes (Homogeneous stability implies that the nullcline slope s_{nc} is larger than -1 [31], such that the prefactor $(1+s_{\mathrm{nc}})^{-1}$ is positive.). Hence, the criterion for a mass-redistribution instability can be expressed in terms of the nullcline slope [31]

$$s_{\mathrm{nc}} < -\frac{D_m}{D_c}\left[1 + \frac{v_{\mathrm{f}}^2}{(1+s_{\mathrm{nc}})^2 D_c f_c}\right]^{-1}. \tag{8}$$

In the absence of flow, $v_{\mathrm{f}} = 0$, we recover the slope criterion $s_{\mathrm{nc}} < -D_m/D_c$ for a mass-redistribution instability driven by cytosolic diffusion [31]. We find that flow always increases the range of instability since the second term in the square brackets monotonically increases with flow speed $|v_{\mathrm{f}}|$. Furthermore, the instability criterion becomes independent of the diffusion constants in the limit of fast flow ($|v_{\mathrm{f}}| \gg \sqrt{D_c f_c}$). The criterion for the (flow-driven) mass-redistribution instability then simply becomes $s_{\mathrm{nc}} < 0$, independently of the ratio of the diffusion constants. This has the interesting consequence that, for sufficiently fast flow, a mass-redistribution instability can be driven solely via cytoplasmic flow, independent of diffusion.

3.4. Limits of Slow and fast Flow

To analyze the effect of flow for wavelengths away from the long wavelength limit, it is instructive to consider the limit cases of slow and fast flow speed.

We first consider a limit where advective transport $(q v_{\mathrm{f}})^{-1}$ is slow compared either to the chemical reactions or to diffusive transport. To lowest order in v_{f}, the dispersion relation is given by (see Appendix A)

$$\sigma(q) \approx \sigma^{(0)}(q) + i\frac{v_{\mathrm{f}} q}{2} A(q), \tag{9}$$

where the zeroth order term, $\sigma^{(0)}(q)$, is the dispersion relation in the absence of flow, which has no imaginary part [31] (cf. Equation (A1)). The function $A(q)$ is positive for all laterally unstable modes ($\mathrm{Re}\,\sigma(q) > 0$). Equation (9) shows that to lowest order (linear in v_{f}) the effect of cytosolic flow is to induce propagation of the modes with the phase velocity $v_{\mathrm{phase}}(q) = -\mathrm{Im}\,\sigma(q)/q \approx -v_{\mathrm{f}} A(q)$. Since $A(q) > 0$ for laterally unstable modes, all growing perturbations propagate against the direction of the flow (as illustrated in Figure 2B).

In the limit of fast flow (compared either to reactions or to cytosolic transport) we find that the dispersion relation (given by the eigenvalue problem Equation (6)) reduces to

$$\sigma(q) \approx f_m - D_m q^2 + i\frac{f_c f_m}{v_{\mathrm{f}} q} \tag{10}$$

for non-zero wavenumbers. The real part of the dispersion relation in this fast flow limit becomes identical to the dispersion relation in the limit of fast diffusion [31]. In both limits, cytosolic transport becomes (near) instantaneous. In particular, in the limit of fast flow, advective transport completely dominates over diffusive transport in the cytosol such that the dispersion relation becomes independent of the cytosol diffusion constant D_c.

From the imaginary part of $\sigma(q)$, we obtain the phase velocity $v_{\text{phase}} = -f_c f_m / (v_f q^2)$. In other words, an increase in cytosolic flow leads to a decrease of the phase velocity. This is opposite to the slow flow limit discussed above, where the phase velocity increased linearly with the flow speed.

To rationalize these findings, we recall the propagation mechanism as discussed above. There, we argued that a phase shift between the membrane and the cytosol pattern is responsible for the pattern propagation, as it leads to an asymmetry in the attachment–detachment balance upstream and downstream. This phase shift increases with the flow velocity and eventually saturateslat $\pi/4$ (The phase shift can be read off from the real and imaginary parts of the eigenvectors in the linear stability analysis.). On the other hand, the cytosol concentration gradients become shallower the faster the flow. To understand why this is, imagine a small volume element in the cytosol being advected with the flow. The faster the flow, the less time it has to interact with each point on the membrane it passes. Therefore, for faster advective flow, the attachment–detachment flux at the membrane is effectively diluted over a larger cytosolic volume. This leads to a flattening of the cytosolic concentration profile (see Supplementary Materials Movie 2), and therefore a reduction in the upstream–downstream asymmetry of attachment. As a result, in the limit of fast flow, the pattern propagates slower the faster the flow, whereas, in the limit of slow flow, the pattern propagates faster the faster the flow. Thus, comparing these two limits, we learn that the phase velocity reaches a maximum at intermediate flow speeds.

3.5. Summary and Discussion of Linear Stability

Let us briefly summarize our main findings from linear stability analysis. We found that the leading order effect of cytosolic flow is to induce upstream propagation of patterns. This propagation is driven by the faster resupply of protein mass on the upstream side of the pattern compared to the downstream side. A similar effect was previously found for vegetation patterns which move uphill because nutrients are transported downhill by water flow [39]. Even though these systems are not strictly mass conserving, their pattern propagation underlies the same principle: The nutrient uptake in regions of high vegetation density creates a nutrient sink which is resupplied asymmetrically due to the downhill flow of water and nutrients.

Moreover, we used a phase-space analysis to explain how flow extends the range of parameters where patterns emerge spontaneously, i.e., where the homogeneous steady state is laterally unstable. This was previously shown mathematically for general two-component reaction–diffusion systems (not restricted to mass-conserving ones) [39,40]. Our analysis in the long wavelength limit explains the physical mechanism of this instability for mass-conserving systems: The flow-driven instability is a mass-redistribution instability, driven by a self-amplifying cascade of (flow-driven) mass transport and the self-organized formation of attachment and detachment zones (shifting reactive equilibria). This shows that the instability mechanism is identical to the mass-redistribution instability that underlies pattern formation in systems without flow (i.e., where only diffusion drives mass transport) [31]. For these systems, the instability strictly requires $D_c > D_m$. In contrast, we find that for sufficiently fast flow, there can be a mass-redistribution instability even in the absence of cytosolic diffusion ($D_c = 0$). While the case $D_c = 0$ is not physiologically relevant in the context of intracellular pattern formation, it may be relevant for the formation of vegetation patterns on sloped terrain [41], where c and m are the soil-nutrient concentration and plant biomass density, respectively. In conclusion, advective flow can fully replace diffusion as the mass-transport mechanism driving the mass-redistribution instability.

4. Pattern Propagation in the Nonlinear Regime

So far we have analyzed how cytosolic flow affects the dynamics of the system in the vicinity of a homogeneous steady state, using linear stability analysis. However, patterns generically do not saturate at small amplitudes but continue to grow into the strongly nonlinear regime [31] (see Supplementary Materials Movie 1 for an example in which a small perturbation of the homogeneous steady state evolves into a large amplitude pattern in the presence of flow).

To study the long time behavior (steady state) far away from the spatially homogeneous steady state, we performed finite element simulations in Mathematica [42]. To interpret the results of these numerical simulations, we will use local equilibria theory, building on the phase-space analysis introduced in Refs. [31,37].

Figure 4A shows the space-time plot (kymograph) of a system where there is initially no flow ($t < t_0$), such that the system is in a stationary state with a single peak. For such a stationary steady state, diffusive fluxes on the membrane and in the cytosol have to balance exactly. This diffusive flux balance imposes the constraint that in the (m, c)-phase plane, the trajectory corresponding to the pattern lies on a straight line with slope $-D_m/D_c$, called 'flux-balance subspace' (FBS) [31] (see light blue line in Figure 4C). At the plateaus and inflection points of the pattern, the net diffusive flow vanishes and attachment and detachment are balanced, i.e., the system is locally in reactive equilibrium ($f = 0$). Hence, plateaus and inflection points of the spatial concentration profile correspond to intersection points between the reactive nullcline and the FBS in the (m, c)-phase plane (blue and green points in Figure 4C). At the first intersection point (blue), the nullcline slope is larger than the FBS slope. Thus, by the slope criterion $s_{nc} < -D_m/D_c$ for lateral instability, this point corresponds to a laterally stable state in the spatial domain—i.e., a plateau. Following a spatial perturbation, the concentrations will relax back towards the flat plateau.

At the second intersection point (green point in Figure 4C), the nullcline slope is more negative than the FBS slope, indicating a laterally unstable state. This state corresponds to the inflection point of the pattern and the lateral instability there can be thought of as "spanning" the interfacial region of the pattern that connects the two plateaus. An in-depth analysis of stationary patterns based on these geometric relations in phase space can be found in Ref. [31]. Here we ask how the phase portrait changes in the presence of flow.

At time $t = t_0$, a constant cytosolic flow in the positive x-direction is switched on. Consistent with the expectation from linear stability analysis, we find that the peak propagates against the flow direction in the negative x-direction (solid lines in Figure 4A). The diffusive fluxes no longer balance for this propagating steady state, such that the phase-space trajectory is no longer embedded in the FBS. Instead, as advective flow shifts the cytosol concentration profile relative to the membrane profile, the phase-space trajectory becomes a loop (Figure 4C). On the upstream side of the peak, the cytosolic density is increased, such that net attachment—which is proportional to the cytosolic density—is increased relative to net detachment. Conversely, the reactive balance is shifted towards detachment on the downstream side. As a result of the reactive flow is approximately proportional to the distance from the reactive nullcline in phase space, the asymmetry between net attachment and detachment on the upstream and downstream side of the peak can be estimated by the area enclosed by the loop-shaped trajectory in phase space.

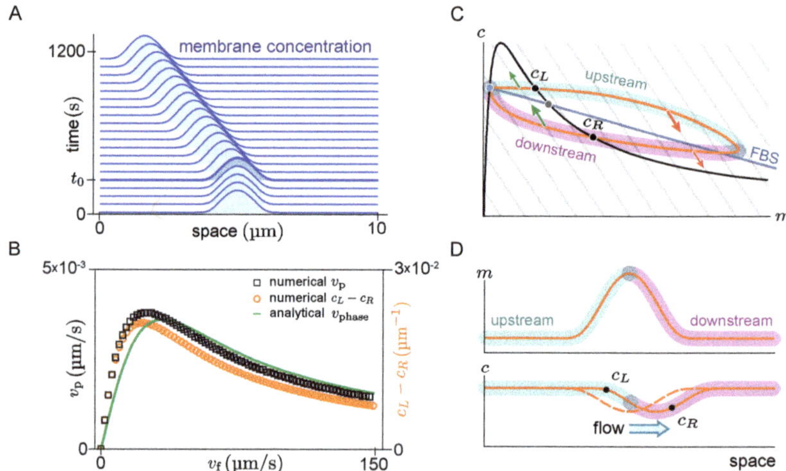

Figure 4. Pattern dynamics far from the spatially homogeneous steady state. (**A**) Time evolution of the membrane-bound protein concentration. At time $t_0 = 240$ s a constant cytosolic flow with velocity $v_f = 20\,\mu m/s$ towards the right is switched on (cf. Supplementary Materials Movie 3). (**B**) Relation between the peak speed (v_p) and flow speed (v_f). Results from finite element simulations (black open squares) are compared to the phase velocity of the mode q_{max} obtained from linear stability analysis (green solid line) and to an approximation (orange open circles) of the area enclosed by the density distribution trajectory in phase space (area enclosed by the 'loop' in **D**). (The domain size, $L = 10\,\mu m$, is chosen large enough compared to the peak width such that boundary effects are negligible.) (**C**) A schematic of the phase portrait corresponding to the pattern in **D**. The density distribution in the absence of flow is embedded in the flux-balance subspace (FBS) (blue straight line). In the presence of flow, the density distribution trajectory forms a 'loop' in phase space. The upstream and downstream side of the pattern are highlighted in cyan and magenta, respectively. Red and green arrows indicate the direction of the reactive flow in the attachment and detachment zones, respectively. At intersection points of the density distribution with the nullcline (c_L and c_R) the system is at its local reactive equilibrium. (**D**) Sketch of the membrane (orange solid line, top) and cytosolic (orange dashed line, bottom) concentration profiles for a stationary pattern in the absence of cytosolic flow. Flow shifts the cytosol profile downstream (orange solid line, bottom).

To test whether the attachment–detachment asymmetry explains the propagation speed of the peak, we estimate the enclosed area in phase space by the difference in cytosolic concentrations at the points c_L and c_R (black dots in Figure 4C,D) where the loop intersects the reactive nullcline ($f = 0$ black line Figure 4C). At these points, the system is in a local reactive equilibrium. Indeed, we find that the propagation speed of the pattern obtained from numerical simulations (black open squares in Figure 4B) is well approximated by the difference in cytosolic density ($v_p \propto c_L - c_R$) for all flow speeds (orange open circles in Figure 4B). Furthermore, in the limit of slow and fast flow, the peak propagation speed is well approximated by the propagation speed of the unstable traveling mode with the longest wavelength, as obtained from linear stability analysis (The phase velocity depends on the mode's wavelength. The relevant length scale for the peak's propagation is its width, which is approximately given by $2\pi/q_{max}$ at the pattern's inflection point [31]. Thus, we infer the peak propagation speed from $-\operatorname{Im}\sigma(q_{max})/q_{max}$ at the inflection point of the stationary peak.). For small flow speeds, the pattern's propagation speed v_p increases linearly with v_f (cf. Equation (7)) and for large flow speeds the pattern speed is proportional to $1/v_f$ (cf. Equation (10)).

In summary, we found that the peak propagation speed in the slow and fast flow limits is well described by the propagation speed of the linearly unstable mode with the longest wavelength (i.e., the right edge of the band of unstable modes q_{max}). Moreover, we approximated the asymmetry of protein attachment by the area enclosed by the density distribution in phase space, and found that this is proportional to the peak speed for all flow speeds.

5. Flow-Induced Transition from Mesa to Peak Patterns

So far we have studied the propagation of patterns in response to cytosolic flow. Next, we will show how cytosolic flow can also drive the transition between qualitatively different pattern types. We distinguish two pattern types exhibited by McRD systems, peaks, and mesas [30,31]. Mesa patterns are composed of plateaus (low density and high density) connected by interfaces, while a peak can be pictured as two interfaces concatenated directly (cf. Figure 5A). Mesa patterns form if protein attachment saturates in regions of high total density, forming a plateau there. As we argued above, the low- and high-density plateaus correspond to laterally stable steady states, marked—in the phase plane—by intersection points between the FBS and the reactive nullcline where the nullcline slope is larger than the FBS slope. Peaks form if the attachment rate does not saturate at high density, i.e., if the third intersection point between nullcline and FBS is not reached [31]. Thus, while the amplitude of mesa patterns is determined by the attachment–detachment balance in the two plateaus, the amplitude (maximum concentration) of a peak is determined by the total mass available in the system [31].

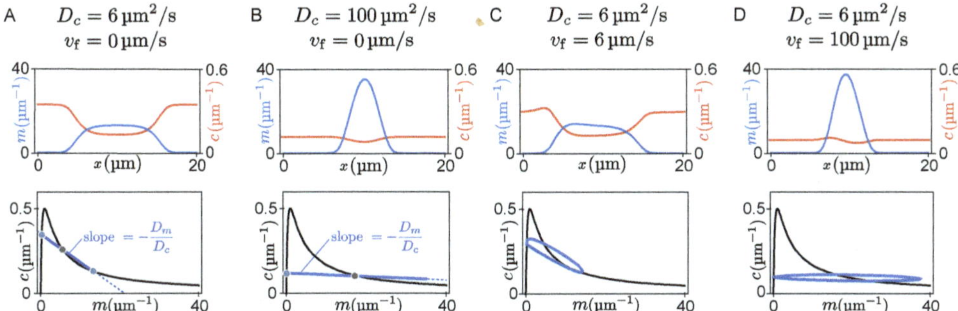

Figure 5. Demonstration of the transition from a mesa pattern to a peak pattern. Each panel shows a snapshot from finite element simulations in steady state. Top concentration profiles in real space; bottom: corresponding trajectory (blue solid line) in phase space. (**A**) Mesa pattern in the case of slow cytosol diffusion and no flow. The two plateaus (blue dots) and the inflection point (gray dot) of the pattern correspond to the intersection points of the FBS (blue dashed line) with the reactive nullcline (black line). (**B**) For fast cytosol diffusion, the third intersection point between FBS and nullcline lies at much higher membrane concentration such that it no longer limits the pattern amplitude. Therefore, a peak forms whose amplitude is limited by the total protein mass in the system. (**C**) Slow flow only slightly deforms the mesa pattern, compare to (**A**). Fast cytosolic flow leads to formation of a peak pattern (**D**), similarly to fast diffusion. Parameters: $\bar{n} = 7\,\mu m^{-1}$, $D_m = 0.1\,\mu m^2/s$, and $L = 20\,\mu m$.

How does protein transport affect whether a peak or a mesa forms? As we argued above, a peak pattern forms if protein attachment in regions of high density does not saturate. In general, this will happen if attachment to the membrane depletes proteins from the cytosol slower than lateral transport can resupply proteins (see Figure 5A). Let us first recap the situation without flow, where proteins are resupplied by diffusion from the detachment zone to the attachment zone across the pattern's interface with width ℓ_{int}. Thus, a peak pattern forms if the rate of transport by cytosolic diffusion is faster than the attachment rate ($D_c/\ell_{int}^2 \gg \tau_{react}^{-1}$). Further using that the interface width is given by a balance of membrane diffusion and local reactions ($\ell_{int}^2 \sim \tau_{react} D_m$), we obtain the condition $D_c \gg D_m$ for the formation of peak patterns.

In terms of phase space geometry, this means that the slope $-D_m/D_c$ of the flux-balance subspace in phase space must be sufficiently shallow. For a steep slope $-D_m/D_c$ of the FBS, the membrane concentration saturates at the point where the FBS intersects with the reactive nullcline blue dots in

Figure 5A. There, attachment and detachment balance such that a mesa forms (Figure 5A). For faster cytosol diffusion, the flux-balance subspace is shallower such that the third FBS-NC intersection point shifts to higher densities. Thus, for sufficiently fast cytosol diffusion a peak forms (Figure 5B).

Adding slow cytosolic flow does not significantly contribute to the resupply of the cytosolic sink (i.e., attachment zone) and therefore does not alter the pattern type (Figure 5C). In contrast, when cytosolic protein transport (by advection and/or diffusion) is fast compared to the reaction kinetics, the cytosolic sink gets resupplied quickly, leading to a flattening of the cytosolic concentration profile. Accordingly, the density distribution in phase space approaches a horizontal line, both for fast cytosolic diffusion (Figure 5B) and for fast cytosolic flow (Figure 5D). As a consequence, the point where the density distribution meets the nullcline shifts towards larger membrane concentrations, resulting in an increasing amplitude of the mesa pattern. Eventually, when the amplitude of the pattern can not grow any further due to limiting total mass, a peak pattern forms (Figure 5B,D). Hence, an increased flow velocity can cause a transition from a mesa pattern to a peak pattern (see Supplementary Materials Movie 4).

In summary, we found that cytosolic flow can qualitatively change the membrane-bound protein pattern from a small-amplitude, wide mesa pattern to a large-amplitude, narrow peak pattern. In cells, such flows could therefore promote the precise positioning of polarity patterns on the membrane. Furthermore, we hypothesize that flow can contribute to the selection of a single peak by accelerating the coarsening dynamics of the pattern via two distinct mechanisms. First, flow accelerates protein transport that drives coarsening. Second, as peak patterns coarsen faster than mesa patterns [30,43], flow can accelerate coarsening via the flow-driven mesa-to-peak transition. Such fast coarsening may be important for the selection of a single polarity axis, e.g., a single budding site in *S. cerevisiae* [4], for axon formation in neurons [44], and to establish a distinct front and back in motile cells [2,45].

6. Flow-Induced Pattern Formation

So far we have studied how a uniform flow profile affects pattern formation on a domain with periodic boundary conditions, representing circular flows along the cell membrane and bulk flows in microfluidic in vitro setups. However, flows in the vicinity of the membrane can be non-uniform. For example, one (or more) components of the pattern forming system may be embedded in the cell cortex [13,15,46] which is a contractile medium driven by myosin-motor activity. Furthermore, the incompressible cytosol can flow in the direction normal to the membrane, such that the 3D flow field of the cytosol is perceived as a compressible flow along the membrane [9]. In this section, we will discuss how such non-uniform, uni-directional flows lead to pattern formation.

A non-uniform flow transports the proteins at different speeds along the membrane. Starting from a spatially homogeneous initial state, such a non-uniform flow leads to a redistribution of mass. It has been demonstrated in previous work that this non-uniform flow can induce pattern formation even if the homogeneous steady state is laterally stable (i.e., there is no spontaneous pattern formation) [7,13,15]. Based on numerical simulations, a transition from flow-guided to self-organized dynamics has been reported [15]. However, the physical mechanism underlying this transition, and what determines the transition point have remained unclear.

We address this question using the two-component model, which serves a conceptual model that mimics the qualitative behavior of the more complex PAR system [33]. While flow in the PAR system is governed by the myosin concentration, we assume a stationary parabolic flow profile that vanishes at the system boundaries (Figure 6A, top). We use a one-dimensional domain with no-flux boundary conditions that correspond to the symmetry axis of a rotationally symmetric flow profile. In the following, we describe the flow-induced dynamics starting from a spatially homogeneous steady state to the final polarity pattern observed in numerical simulations (see Supplementary Materials Movie 5). Figure 6 visualizes these dynamics in real space (A) and in the (m, c)-phase plane (B). To relate our findings to the previous study Ref. [15], we also visualize the dynamics in an abstract representation of the state space (comprising all concentration profiles) used in this previous study. In this state space, steady states are points and the time evolution of the system is a trajectory (thick blue/orange line in Figure 6C).

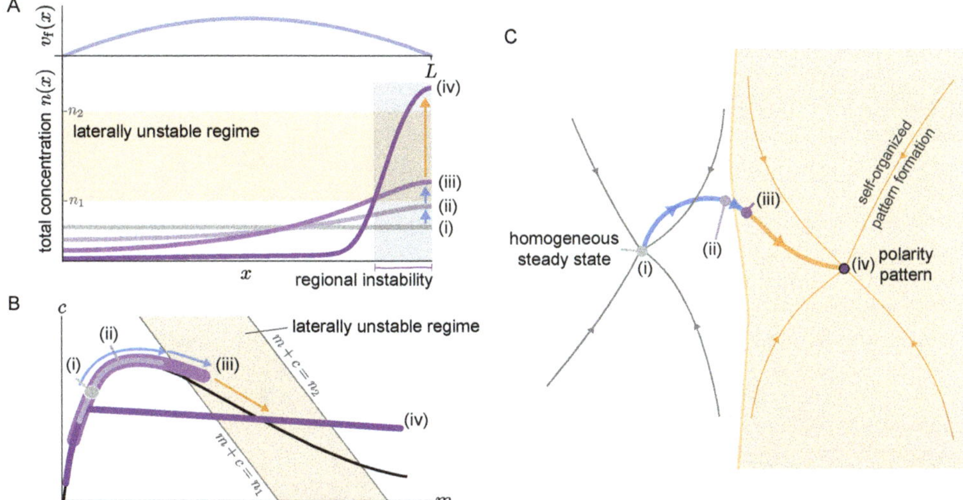

Figure 6. Flow-driven protein mass accumulation can induce pattern formation by triggering a regional lateral instability. **(A)** Top: quadratic flow velocity profile: $v_f(x)/v_{\max} = 1 - 4\,(x/L - 1/2)^2$. Bottom: illustration of the total density profiles at different time points starting from a homogeneous steady state (*i*) to the final pattern (*iv*); see Movie 5. Mass redistribution due to the non-uniform flow velocity drives mass towards the right hand side of the system, as indicated by the blue arrows. The range of total densities shaded in orange indicates the laterally unstable regime determined by linear stability analysis. Once the total density reaches this regime locally, a regional lateral instability is triggered resulting in the self-organized formation of a peak (orange arrow). **(B)** Sketch of the phase space representation corresponding to the profiles shown in A. Note that the concentrations are slaved to the reactive nullcline (black line) until the regional lateral instability is triggered. **(C)** Schematic representation of the state space of concentration patterns in a case where both the homogeneous steady state and a stationary polarity pattern are stable. Thin trajectories indicate the dynamics in the absence of flow and the pattern's basin of attraction is shaded in orange. The thick trajectory connecting both steady states shows the flow-induced dynamics, corresponding to the sequence of states (*i*)–(*iv*) shown in **A** and **B**.

Starting from the homogeneous steady state (*i*), the non-uniform advective flow redistributes mass in the cytosol (*ii*). Due to this redistribution of mass, the local reactive equilibria shift as we have seen repeatedly here and in earlier studies of mass-conserving systems [31,47]. In fact, as long as the gradients of both the membrane and cytosol profiles are shallow, the concentrations remain close to the local equilibria, as evidenced by the density distribution in phase space spreading along the reactive nullcline (see profile (*ii*) in Figure 6A,B). As long as there is no laterally unstable region, the mass accumulation is limited by the counteracting diffusive flow in the cytosol. Eventually, the region where mass accumulates (here the right edge of the domain) enters the laterally unstable regime (see profile *iii*). In this laterally unstable region, cytosol diffusion will enhance the accumulation of mass via the mass-redistribution instability, until it is limited by the much slower membrane diffusion. In the phase plane (Figure 6B), the laterally unstable regime corresponds to the range of total densities $\bar{n}$ where the nullcline slope has a steeper negative value than the flux-balance subspace slope ($s_{\mathrm{nc}} < -D_m/D_c$) (More precisely, the size of the laterally unstable region must be larger than the shortest unstable mode (corresponding to the right edge of the band of unstable modes in the dispersion relation (Figure 2A))). The mass-redistribution instability in this region, based on the self-organized formation of attachment and detachment zones (cf. Section 3.2) will lead to the formation of a polarity pattern there (*iv*). Thus, the onset of a regional lateral instability marks the transition from flow-guided dynamics to self-organized dynamics.

In the abstract state space visualization (Figure 6C) the area shaded in orange indicates the polarity pattern's basin of attraction comprising all states (concentration profiles) where a spatial region in the system is laterally unstable. In the absence of flow, states that do not exhibit such a laterally unstable region return to the homogeneous steady state (thin gray lines). Non-uniform cytosolic flow induces mass-redistribution, that can drive an initially homogeneous system (i) into the polarity pattern's basin of attraction. From there on, self-organized pattern formation takes over, leading to the formation of a polarity pattern (iv), essentially independently of the advective flow (orange trajectory). In future work, it would be interesting to make the abstract state space representation, Figure 6C, more quantitative. For example, one could try to estimate the minimal flow velocity required to drive the system past the separatrix, i.e., into the basin of attraction of the polarity pattern. A promising approach is to use the fact that prior to the onset of regional lateral instability, the concentrations are slaved to the local equilibria that depend on the local total density. Thus, one can obtain an approximate, closed equation for the flow-driven evolution of the total density, similar to the "adiabatic scaffolding approximation" made in [31]. Solving this equation would provide a criterion for when the total density exceeds the critical density for lateral instability in some spatial region that initiates the self-organized formation of a polarity pattern there.

Similar pattern forming mechanisms based on a regional instability have previously been shown to also underlie stimulus-induced pattern formation following a sufficiently strong initial perturbation [31] and peak formation at a domain edge where the reaction kinetics abruptly change [36]. Thus, an overarching principle for stimulus-induced pattern formation emerges: To trigger (polarity) pattern formation, the stimulus, be it advective flow or heterogeneous reaction kinetics, has to redistribute protein mass in a way such that a regional (lateral) instability is triggered.

It remains to be discussed what happens once the cytoplasmic flow is switched off after the polarity pattern has formed. In general, the polarity pattern will persist (see Supplementary Materials Movie 5), since it is maintained by self-organized attachment and detachment zones, largely independent of the flow. However, as long as there is flow, the average mass on the right hand side of the system (downstream of the flow) is higher than on the left hand side. Hence, flow can maintain a polarity pattern even if the average mass in the system as a whole is too low to sustain polarity patterns in the absence of flow (see bifurcation analysis in Ref. [31]). If this is the case, the peak disappears once the flow is switched off (see Supplementary Materials Movie 6).

In summary, the redistribution of the protein mass is key to induce (polarity) pattern formation starting from a stable homogeneous state.

7. Conclusions and Outlook

Inside cells, proteins are transported via diffusion and fluid flows, which, in combination with reactions, can lead to the formation of protein patterns on the cell membrane. To characterize the role fluid flows play in pattern formation, we studied the effect of flow on the formation of a polarity pattern, using a generic two-component model. We found that flow leads to propagation of the polarity pattern against the flow direction with a speed that is maximal for intermediate flow speeds, i.e., when the rate of advective transport is comparable to either the reaction rates or to the rate diffusive transport in the cytosol. Using a phase-space analysis, we showed that the propagation of the pattern is driven by an asymmetric influx of protein mass to a self-organized protein-attachment zone. As a consequence, attachment is stronger on the upstream side of the pattern compared to the downstream side, leading to upstream propagation of the membrane bound pattern. Furthermore, we have shown that flow can qualitatively change the pattern from a wide mesa pattern (connecting two plateaus) to a narrow peak pattern. Finally, we have presented a phase-space analysis to elucidate the interplay between flow-guided dynamics and self-organized pattern formation. This interplay was previously studied numerically in the context of PAR-protein polarization [13,15]. Our analysis reveals the underlying cause for the transition from flow-guided to self-organized dynamics: the regional onset of a mass-redistribution instability.

We discussed implications of our results and links to earlier literature at the end of each section. Here, we conclude with a brief outlook. We expect that the insights obtained from the minimal two-component model studied here generalize to systems with more components and multiple protein species. For example, in vitro studies of the reconstituted MinDE system of *E. coli* show that MinD and MinE spontaneously form dynamic membrane-bound patterns, including spiral waves [48] and quasi-stationary patterns [49]. These patterns emerge from the competition of MinD self-recruitment and MinE-mediated detachment of MinD [50,51]. In the presence of a bulk flow, the traveling waves were found to propagate upstream [16]. Our analysis based on a simple conceptual model suggests that this upstream propagation is caused by a larger influx of the self-recruiting MinD on the upstream flanks compared to the downstream flanks of the traveling waves. However, the bulk flow also increases the resupply of MinE on the upstream flanks. As MinE mediates the detachment of MinD and therefore effectively antagonizes MinD's self-recruitment, this may drive the membrane-bound patterns to propagate downstream instead of upstream. Which one of the two processes dominates—MinD-induced upstream propagation or MinE-induced downstream propagation—likely depends on the details of their interactions. This interplay will be the subject of future work.

A different route of generalization is to consider advective flows that depend on the protein concentrations. In cells, such coupling arises, for instance, from myosin-driven cortex contractions [15,52] and shape deformations [8,25]. Myosin-motors, in turn, may be advected by the flow and their activity is controlled by signaling proteins such as GTPases and kinases [53]. This can give rise to feedback loops between flow and protein patterns. Previous studies show that such feedback loops can give rise to mechano-chemical instabilities [54], drive pulsatile (standing-wave) patterns [55,56] or cause the breakup of traveling waves [57]. We expect that our analysis based on phase-space geometry can provide insight into the mechanisms underlying these phenomena.

Supplementary Materials: The following are available online at http://www.mdpi.com/2073-4409/9/6/1524/s1. Movie Descriptions:

1. Growth of a pattern from a homogeneous steady state in the presence of flow. Top: concentration profiles in space; bottom: corresponding density distribution in the phase space. (Parameters: $D_m = 0.1\,\mu m^2/s$, $\bar{n} = 3\,\mu m^{-1}$, $L = 20\,\mu m$ and $v_f = 20\,\mu m/s$.)
2. Simulation with adiabatically increasing flow speed from $v_f = 0\,\mu m/s$ to $v_f = 100\,\mu m/s$. Note the flattening of the cytosolic concentration profile as the flow speed increases. (Fixed parameters as for Movie 1.)
3. Corresponds to the space-time plot in Figure 4A.
4. Pattern transformation from a mesa pattern to a peak pattern as flow speed is adiabatically increased from $v_f = 0$ to $v_f = 45\,\mu m/s$. (Fixed parameters as for Movie 1.)
5. Pattern formation triggered by mass-redistribution due to a spatially non-uniform flow (parabolic flow profile shown in Figure 6A). After the flow is switched off at $t = 200\,s$, the pattern is maintained. (Parameters: $D_m = 0.1\,\mu m^2/s$, $L = 30\,\mu m$, $v_{max} = 1\,\mu m/s$, and $\bar{n} = 1\,\mu m^{-1}$.)
6. As Movie 5, but with lower average mass, $\bar{n} = 0.8\,\mu m^{-1}$. This mass is not sufficient to maintain a stationary peak in the absence of flow. Therefore, the peak disappears after the flow is switched off ($t > 200\,s$).

Author Contributions: All authors designed and carried out the research; M.C.W., F.B., and E.F. wrote the paper; C.Y.L. visualized the findings. All authors have read and agreed to the published version of the manuscript.

Funding: This work was funded by the Deutsche Forschungsgemeinschaft (DFG, German Research Foundation)—Project-ID 201269156—Collaborative Research Center (SFB) 1032 – Project B2. M.C.W. and C.Y.L. are supported by a DFG fellowship through the Graduate School of Quantitative Biosciences Munich (QBM). M.C.W. acknowledges the Joachim Herz Stiftung for support.

Conflicts of Interest: The authors declare no conflict of interest.

Abbreviations

The following abbreviations are used in this manuscript:

McRD	mass-conserving reaction–diffusion
2cMcRD	two-component mass-conserving reaction–diffusion
FBS	flux-balance subspace

Appendix A. Limit of Slow Flow and Timescale Comparison

The dispersion relation in the absence of flow ($v_\mathrm{f} = 0$) reads

$$\sigma^{(0)} = -\frac{1}{2}\left[(D_m + D_c)\,q^2 + f_c - f_m\right] + \frac{B(q)}{2A(q)}, \tag{A1}$$

with $A(q) = \left[1 - 4f_c f_m / B(q)^2\right]^{-1/2} - 1$ and $B(q) = f_m + f_c + (D_c - D_m)q^2$. To find the effect of slow flow, we first need to identify the relevant timescales such that we can define a dimensionless small parameter to expand in. As a result that pattern formation is driven by transport in the cytosol (diffusive and advective) and attachment from the cytosol to the membrane, there are three relevant timescales: (*i*) The the rate of advective transport on length scale q^{-1} is given by qv_f; (*ii*) the rate of diffusive transport on that scale, given by $D_c q^2$; and (*iii*) the attachment rate $f_c = a(m)$ (cf. Equation (2)). To compare these timescales, we form two dimensionless numbers: the Peclét number $\mathrm{Pe} = v_\mathrm{f}/(D_c q)$ and the Damköhler number $\mathrm{Da} = f_c/(v_\mathrm{f} q)$. Flow can either be slow compared to reactions ($\mathrm{Da} \gg 1$) or slow compared to diffusion ($\mathrm{Pe} \ll 1$). In both cases, expanding the the dispersion relation $\sigma(q)$ to first order yields

$$\sigma(q) = \sigma^{(0)}(q) + i\frac{v_\mathrm{f} q}{2}A(q) + \mathcal{O}(\varepsilon^2), \tag{A2}$$

where $\varepsilon = \min(\mathrm{Pe}, \mathrm{Da}^{-1})$. By elementary algebra using the assumptions $D_c > D_m$ and $f_c > 0$ made above, it follows that $A(q)$ is positive when $s_\mathrm{nc} < 0$. As Equation (8) in Section 3.3 shows, the condition $s_\mathrm{nc} < 0$ is necessary for a band of unstable modes to exist. Therefore, $A(q)$ is positive for all unstable modes.

References

1. Weiner, O.D. Regulation of Cell Polarity during Eukaryotic Chemotaxis: The Chemotactic Compass. *Curr. Opin. Cell Biol.* **2002**, *14*, 196–202. [CrossRef]
2. Keilberg, D.; Søgaard-Andersen, L. Regulation of Bacterial Cell Polarity by Small GTPases. *Biochemistry* **2014**, *53*, 1899–1907. [CrossRef] [PubMed]
3. Bi, E.; Park, H.O. Cell Polarization and Cytokinesis in Budding Yeast. *Genetics* **2012**, *191*, 347–387. [CrossRef] [PubMed]
4. Chiou, J.G.; Balasubramanian, M.K.; Lew, D.J. Cell Polarity in Yeast. *Annu. Rev. Cell Dev. Biol.* **2017**, *33*, 77–101. [CrossRef]
5. Goryachev, A.B.; Leda, M. Many Roads to Symmetry Breaking: Molecular Mechanisms and Theoretical Models of Yeast Cell Polarity. *Mol. Biol. Cell* **2017**, *28*, 370–380. [CrossRef]
6. Turing, A.M. The Chemical Basis of Morphogenesis. *Philos. Trans. R. Soc. Lond. B Ser. Biol. Sci.* **1952**, *237*, 37–72. [CrossRef]
7. Diegmiller, R.; Montanelli, H.; Muratov, C.B.; Shvartsman, S.Y. Spherical Caps in Cell Polarization. *Biophys. J.* **2018**, *115*, 26–30. [CrossRef]
8. Klughammer, N.; Bischof, J.; Schnellbächer, N.D.; Callegari, A.; Lénárt, P.; Schwarz, U. Cytoplasmic Flows in Starfish Oocytes Are Fully Determined by Cortical Contractions. *PLoS Comput. Biol.* **2018**, *14*, e1006588. [CrossRef]
9. Mittasch, M.; Gross, P.; Nestler, M.; Fritsch, A.W.; Iserman, C.; Kar, M.; Munder, M.; Voigt, A.; Alberti, S.; Grill, S.W.; et al. Non-Invasive Perturbations of Intracellular Flow Reveal Physical Principles of Cell Organization. *Nat. Cell Biol.* **2018**, *20*, 344–351. [CrossRef]
10. Hawkins, R.J.; Bénichou, O.; Piel, M.; Voituriez, R. Rebuilding Cytoskeleton Roads: Active-Transport-Induced Polarization of Cells. *Phys. Rev. E* **2009**, *80*, 040903. [CrossRef]

11. Calvez, V.; Lepoutre, T.; Meunier, N.; Muller, N. Non-Linear Analysis of a Model for Yeast Cell Communication. *ESAIM Math. Model. Numer. Anal.* **2020**, *54*, 619–648. [CrossRef]
12. Yochelis, A.; Ebrahim, S.; Millis, B.; Cui, R.; Kachar, B.; Naoz, M.; Gov, N.S. Self-Organization of Waves and Pulse Trains by Molecular Motors in Cellular Protrusions. *Sci. Rep.* **2015**, *5*, 13521. [CrossRef] [PubMed]
13. Goehring, N.W.; Trong, P.K.; Bois, J.S.; Chowdhury, D.; Nicola, E.M.; Hyman, A.A.; Grill, S.W. Polarization of PAR Proteins by Advective Triggering of a Pattern-Forming System. *Science* **2011**, *334*, 1137–1141. [CrossRef] [PubMed]
14. Illukkumbura, R.; Bland, T.; Goehring, N.W. Patterning and Polarization of Cells by Intracellular Flows. *Curr. Opin. Cell Biol.* **2020**, *62*, 123–134. [CrossRef]
15. Gross, P.; Kumar, K.V.; Goehring, N.W.; Bois, J.S.; Hoege, C.; Jülicher, F.; Grill, S.W. Guiding Self-Organized Pattern Formation in Cell Polarity Establishment. *Nat. Phys.* **2019**, *15*, 293–300. [CrossRef]
16. Ivanov, V.; Mizuuchi, K. Multiple Modes of Interconverting Dynamic Pattern Formation by Bacterial Cell Division Proteins. *Proc. Natl. Acad. Sci. USA* **2010**, *107*, 8071–8078. [CrossRef]
17. Vecchiarelli, A.G.; Li, M.; Mizuuchi, M.; Mizuuchi, K. Differential Affinities of MinD and MinE to Anionic Phospholipid Influence Min Patterning Dynamics in vitro: Flow and Lipid Composition Effects on Min Patterning. *Mol. Microbiol.* **2014**, *93*, 453–463. [CrossRef]
18. Grill, S.W.; Gönczy, P.; Stelzer, E.H.K.; Hyman, A.A. Polarity Controls Forces Governing Asymmetric Spindle Positioning in the Caenorhabditis Elegans Embryo. *Nature* **2001**, *409*, 630–633. [CrossRef]
19. Munro, E.; Nance, J.; Priess, J.R. Cortical Flows Powered by Asymmetrical Contraction Transport PAR Proteins to Establish and Maintain Anterior-Posterior Polarity in the Early C. Elegans Embryo. *Dev. Cell* **2004**, *7*, 413–424. [CrossRef]
20. Hecht, I.; Rappel, W.J.; Levine, H. Determining the Scale of the Bicoid Morphogen Gradient. *Proc. Natl. Acad. Sci. USA* **2009**, *106*, 1710–1715. [CrossRef]
21. Mayer, M.; Depken, M.; Bois, J.S.; Jülicher, F.; Grill, S.W. Anisotropies in Cortical Tension Reveal the Physical Basis of Polarizing Cortical Flows. *Nature* **2010**, *467*, 617–621. [CrossRef] [PubMed]
22. Goldstein, R.E.; van de Meent, J.W. A Physical Perspective on Cytoplasmic Streaming. *Interface Focus* **2015**, *5*, 20150030. [CrossRef] [PubMed]
23. Mogilner, A.; Manhart, A. Intracellular Fluid Mechanics: Coupling Cytoplasmic Flow with Active Cytoskeletal Gel. *Annu. Rev. Fluid Mech.* **2018**, *50*, 347–370. [CrossRef]
24. Bischof, J.; Brand, C.A.; Somogyi, K.; Májer, I.; Thome, S.; Mori, M.; Schwarz, U.S.; Lénárt, P. A Cdk1 Gradient Guides Surface Contraction Waves in Oocytes. *Nat. Commun.* **2017**, *8*, 1–10. [CrossRef] [PubMed]
25. Koslover, E.F.; Chan, C.K.; Theriot, J.A. Cytoplasmic Flow and Mixing Due to Deformation of Motile Cells. *Biophys. J.* **2017**, *113*, 2077–2087. [CrossRef]
26. Mori, Y.; Jilkine, A.; Edelstein-Keshet, L. Wave-Pinning and Cell Polarity from a Bistable Reaction-Diffusion System. *Biophys. J.* **2008**, *94*, 3684–3697. [CrossRef]
27. Jilkine, A.; Edelstein-Keshet, L. A Comparison of Mathematical Models for Polarization of Single Eukaryotic Cells in Response to Guided Cues. *PLoS Comput. Biol.* **2011**, *7*, e1001121. [CrossRef]
28. Klünder, B.; Freisinger, T.; Wedlich-Söldner, R.; Frey, E. GDI-Mediated Cell Polarization in Yeast Provides Precise Spatial and Temporal Control of Cdc42 Signaling. *PLoS Comput. Biol.* **2013**, *9*, e1003396. [CrossRef]
29. Freisinger, T.; Klünder, B.; Johnson, J.; Müller, N.; Pichler, G.; Beck, G.; Costanzo, M.; Boone, C.; Cerione, R.A.; Frey, E.; et al. Establishment of a Robust Single Axis of Cell Polarity by Coupling Multiple Positive Feedback Loops. *Nat. Commun.* **2013**, *4*, 1807. [CrossRef]
30. Chiou, J.G.; Ramirez, S.A.; Elston, T.C.; Witelski, T.P.; Schaeffer, D.G.; Lew, D.J. Principles That Govern Competition or Co-Existence in Rho-GTPase Driven Polarization. *PLoS Comput. Biol.* **2018**, *14*, e1006095. [CrossRef]
31. Brauns, F.; Halatek, J.; Frey, E. Phase-Space Geometry of Reaction–Diffusion Dynamics. *arXiv* **2018**, arXiv:1812.08684.
32. Goryachev, A.B.; Pokhilko, A.V. Dynamics of Cdc42 Network Embodies a Turing-Type Mechanism of Yeast Cell Polarity. *FEBS Lett.* **2008**, *582*, 1437–1443. [CrossRef] [PubMed]
33. Trong, P.K.; Nicola, E.M.; Goehring, N.W.; Kumar, K.V.; Grill, S.W. Parameter-Space Topology of Models for Cell Polarity. *New J. Phys.* **2014**, *16*, 065009. [CrossRef]
34. Otsuji, M.; Ishihara, S.; Co, C.; Kaibuchi, K.; Mochizuki, A.; Kuroda, S. A Mass Conserved Reaction–Diffusion System Captures Properties of Cell Polarity. *PLoS Comput. Biol.* **2007**, *3*, e108. [CrossRef]

35. Ishihara, S.; Otsuji, M.; Mochizuki, A. Transient and Steady State of Mass-Conserved Reaction-Diffusion Systems. *Phys. Rev. E* **2007**, *75*, 015203. [CrossRef]
36. Wigbers, M.C.; Brauns, F.; Hermann, T.; Frey, E. Pattern Localization to a Domain Edge. *Phys. Rev. E* **2020**, *101*, 022414. [CrossRef]
37. Halatek, J.; Brauns, F.; Frey, E. Self-Organization Principles of Intracellular Pattern Formation. *Philos. Trans. R. Soc. B Biol. Sci.* **2018**, *373*, 20170107. [CrossRef]
38. Allen, N.S.; Allen, R.D. Cytoplasmic Streaming in Green Plants. *Annu. Rev. Biophys. Bioeng.* **1978**, *7*, 497–526. [CrossRef]
39. Siero, E.; Doelman, A.; Eppinga, M.; Rademacher, J.D.; Rietkerk, M.; Siteur, K. Striped Pattern Selection by Advective Reaction-Diffusion Systems: Resilience of Banded Vegetation on Slopes. *Chaos Interdiscip. J. Nonlinear Sci.* **2015**, *25*, 036411. [CrossRef]
40. Perumpanani, A.J.; Sherratt, J.A.; Maini, P.K. Phase Differences in Reaction-Diffusion-Advection Systems and Applications to Morphogenesis. *IMA J. Appl. Math.* **1995**, *55*, 19–33. [CrossRef]
41. Samuelson, R.; Singer, Z.; Weinburd, J.; Scheel, A. Advection and Autocatalysis as Organizing Principles for Banded Vegetation Patterns. *J. Nonlinear Sci.* **2019**, *29*, 255–285. [CrossRef]
42. Wolfram Research, Inc. *Mathematica*; Wolfram Research, Inc.: Champaign, IL, USA, 2019.
43. Brauns, F.; Weyer, H.; Halatek, J.; Yoon, J.; Frey, E. Coarsening in (Nearly) Mass-Conserving Two-Component Reaction Diffusion Systems. *arXiv* **2020**, arXiv:2005.01495.
44. Fivaz, M.; Bandara, S.; Inoue, T.; Meyer, T. Robust Neuronal Symmetry Breaking by Ras-Triggered Local Positive Feedback. *Curr. Biol.* **2008**, *18*, 44–50. [CrossRef] [PubMed]
45. Wang, Y.; Ku, C.J.; Zhang, E.R.; Artyukhin, A.B.; Weiner, O.D.; Wu, L.F.; Altschuler, S.J. Identifying Network Motifs That Buffer Front-to-Back Signaling in Polarized Neutrophils. *Cell Rep.* **2013**, *3*, 1607–1616. [CrossRef]
46. Wang, S.C.; Low, T.Y.F.; Nishimura, Y.; Gole, L.; Yu, W.; Motegi, F. Cortical Forces and CDC-42 Control Clustering of PAR Proteins for Caenorhabditis Elegans Embryonic Polarization. *Nat. Cell Biol.* **2017**, *19*, 988–995. [CrossRef]
47. Halatek, J.; Frey, E. Rethinking Pattern Formation in Reaction–Diffusion Systems. *Nat. Phys.* **2018**, *14*, 507–514. [CrossRef]
48. Loose, M.; Fischer-Friedrich, E.; Ries, J.; Kruse, K.; Schwille, P. Spatial Regulators for Bacterial Cell Division Self-Organize into Surface Waves in Vitro. *Science* **2008**, *320*, 789–792. [CrossRef]
49. Glock, P.; Ramm, B.; Heermann, T.; Kretschmer, S.; Schweizer, J.; Mücksch, J.; Alagöz, G.; Schwille, P. Stationary Patterns in a Two-Protein Reaction-Diffusion System. *ACS Synth. Biol.* **2019**, *8*, 148–157. [CrossRef]
50. Huang, K.C.; Meir, Y.; Wingreen, N.S. Dynamic Structures in Escherichia Coli: Spontaneous Formation of MinE Rings and MinD Polar Zones. *Proc. Natl. Acad. Sci. USA* **2003**, *100*, 12724–12728. [CrossRef]
51. Halatek, J.; Frey, E. Highly Canalized MinD Transfer and MinE Sequestration Explain the Origin of Robust MinCDE-Protein Dynamics. *Cell Rep.* **2012**, *1*, 741–752. [CrossRef]
52. Deneke, V.E.; Puliafito, A.; Krueger, D.; Narla, A.V.; De Simone, A.; Primo, L.; Vergassola, M.; De Renzis, S.; Di Talia, S. Self-Organized Nuclear Positioning Synchronizes the Cell Cycle in Drosophila Embryos. *Cell* **2019**, *177*, 925–941.e17. [CrossRef]
53. Iden, S.; Collard, J.G. Crosstalk between Small GTPases and Polarity Proteins in Cell Polarization. *Nat. Rev. Mol. Cell Biol.* **2008**, *9*, 846–859. [CrossRef] [PubMed]
54. Bois, J.S.; Jülicher, F.; Grill, S.W. Pattern Formation in Active Fluids. *Phys. Rev. Lett.* **2011**, *106*, 028103. [CrossRef] [PubMed]
55. Radszuweit, M.; Alonso, S.; Engel, H.; Bär, M. Intracellular Mechanochemical Waves in an Active Poroelastic Model. *Phys. Rev. Lett.* **2013**, *110*, 138102. [CrossRef] [PubMed]

56. Kumar, K.V.; Bois, J.S.; Jülicher, F.; Grill, S.W. Pulsatory Patterns in Active Fluids. *Phys. Rev. Lett.* **2014**, *112*. [CrossRef]

57. Nakagaki, T.; Yamada, H.; Ito, M. Reaction–Diffusion–Advection Model for Pattern Formation of Rhythmic Contraction in a Giant Amoeboid Cell of thePhysarumPlasmodium. *J. Theor. Biol.* **1999**, *197*, 497–506. [CrossRef] [PubMed]

Perspective

Why a Large-Scale Mode Can Be Essential for Understanding Intracellular Actin Waves

Carsten Beta [1], Nir S. Gov [2] and Arik Yochelis [3,4,*]

[1] Institute of Physics and Astronomy, University of Potsdam, 14476 Potsdam, Germany; beta@uni-potsdam.de

[2] Department of Chemical and Biological Physics, Weizmann Institute of Science, Rehovot 76100, Israel; nir.gov@weizmann.ac.il

[3] Department of Solar Energy and Environmental Physics, Blaustein Institutes for Desert Research (BIDR), Ben-Gurion University of the Negev, Sede Boqer Campus, Midreshet Ben-Gurion 8499000, Israel

[4] Department of Physics, Ben-Gurion University of the Negev, Be'er Sheva 8410501, Israel

* Correspondence: yochelis@bgu.ac.il

Received: 16 April 2020; Accepted: 18 June 2020; Published: 23 June 2020

Abstract: During the last decade, intracellular actin waves have attracted much attention due to their essential role in various cellular functions, ranging from motility to cytokinesis. Experimental methods have advanced significantly and can capture the dynamics of actin waves over a large range of spatio-temporal scales. However, the corresponding coarse-grained theory mostly avoids the full complexity of this multi-scale phenomenon. In this perspective, we focus on a minimal continuum model of activator–inhibitor type and highlight the qualitative role of mass conservation, which is typically overlooked. Specifically, our interest is to connect between the mathematical mechanisms of pattern formation in the presence of a large-scale mode, due to mass conservation, and distinct behaviors of actin waves.

Keywords: nonlinear waves; actin polymerization; bifurcation theory; mass conservation; spatial localization; pattern formation; activator–inhibitor models

---

1. Introduction

Biological pattern formation refers to the emergence of complex spatiotemporal variations in living systems that are typically far from thermodynamic equilibrium [1–3]. Even though these systems can differ in composition and scales, they share many similarities and generic phenomena that are observed in a wide variety of natural settings, such as stationary periodic patterns of pigments on animal skins, spiral waves in biological cells and cardiac arrhythmia, or swarming phenomena in bacterial colonies and in flocks of birds or fish. The theoretical study of biological pattern formation can be roughly divided into two time periods: (i) The second half of the twentieth century, following the seminal works by Turing on morphogenesis [4] and by Hodgkin and Huxley (HH) on action potentials in the giant squid axon [5], and (ii) the beginning of the twenty-first century, where an ever-increasing amount of quantitative biological data provided the basis for more detailed mechanistic models of biological systems.

During the first period, theoretical studies were largely limited to a few prototypical reaction–diffusion (RD) or activator–inhibitor (AI) models [6], such as the FitzHugh–Nagumo (FHN) [7,8], Gierer–Meinhardt [9], and Keller–Segel equations [10]. Based on the relative simplicity of these models, e.g., the FHN model as compared to the HH equations, and their relation with models of inanimate matter, e.g., the Swift-Hohenberg model of thermal fluid convection [11] and the Gray–Scott model of chemical reactions [12–14], several pattern formation methodologies have been advanced [1], such as weakly nonlinear and singular-perturbation methods. These models provided deep insights into universal aspects of pattern formation phenomena and generic relations to applications were substantiated, such as frequency locking and spiral waves in the cardiac system. The second time period has manifested a gradual shift of research interests towards specific detailed biological and medical systems [15,16], including, for example, micron-scale intracellular waves, the development of tissues and organs, sound discrimination in the auditory system, and pathologies such as cancer metastasis. In particular, systems in these contexts are generally described by elaborate, system-specific models that are less amenable to mathematical analysis than earlier toy models of pattern formation.

Consequently, despite the common pattern formation thread that connects these different biological and medical applications, it became difficult to navigate through the vast number of distinct models and approaches, particularly in cases where technical jargon makes it difficult to adopt cross-disciplinary integration between different communities including biophysics, computational biology, mathematical biology, biochemistry, dynamical systems, and numerical analysis. Even though the formulation of complete models for biological systems is currently unrealistic, uncovering partial mechanisms that drive pattern formation phenomena remains of utmost importance for understanding functional aspects of living systems and for developing technological and medical applications, such as drugs or implants. Moreover, mechanistic studies of pattern forming systems are also fertile sources of new mathematical questions that advance the development of analytical and numerical methods [2,17–25], which, in turn, contribute new insights into the original applications.

In this perspective, we will focus on intracellular actin waves (IAW), a topic that recently gained much interest not only in the biological context but also as an inspiring showcase of active matter. More specifically, we are interested in IAW that are affected by a large-scale mode—a situation that arises due to the conservation of actin monomers (over the time-scale of the IAW phenomenon). We note that phenomena such as Ca^{2+} waves are, in general, beyond the scope of this perspective as they involve the transport of ions between the cell interior and the extracellular space (which acts as an infinite reservoir) [26–28], unless conservation can be accounted for [29]. Moreover, we emphasize that we aim to provide a perspective and not a comprehensive review, as such reviews are already available, see, e.g., in [30–36].

The perspective is organized as follows. In Section 2, we introduce the rich phenomenology of IAW and the modeling aspects that are associated with mass conservation. Then, in Section 3 we present the theoretical aspects of a large scale mode in the context of physicochemical settings and also indicate its significance to IAW as a consequence of the conservation of actin monomers on the time scales at which many actin-driven processes operate. Finally, in Section 4 we discuss why incorporation of mass conservation is a plausible qualitative step in unfolding the robustness of IAW mechanisms, and in Section 5 we conclude by emphasizing the theoretical strategies for modeling and control of wave persistence as a potential roadmap toward applications in synthetic biology.

2. Intracellular Actin Waves

The functions of many cells are tied to their ability to dynamically change their shape, mostly via the spatiotemporal organization of their actin cytoskeleton. Examples of this include diverse cell types, such as human neutrophils, fish keratocytes, or the social amoeba *Dictyostelium discoideum*. Among the most prominent dynamical patterns in the actin cytoskeleton are IAW, which have attracted much attention over the past decade [31]. These waves are assumed to play a role in several essential cellular functions, among them cell locomotion, cytokinesis, and phagocytic uptake of extracellular matter. Many competing models at different levels of complexity have been developed to describe cortical actin waves, mostly relying on coupled nonlinear AI equations. Even though intracellular actin waves involve a large number of interacting molecular species as well as multiple local and global interactions, prototypical AI models have been shown to capture many features of the overall dynamics. However, important effects due to mass conservation constraints have been hitherto largely neglected.

2.1. Phenomenology from Experiments

Actin waves are characterized by propagating cytoskeletal regions that are enriched in filamentous actin and actin-related proteins. Depending on the cell type, IAW may differ in their biochemical composition and dynamics, including different wave morphologies and propagation speeds. One of the earliest examples of IAW was reported from cultured neurons that show the propagation of fin-like actin-filled membrane protrusions along their axon [37]. They were found to depend on actin polymerization and have been associated with neural polarization [38,39]. Similar fin-like actin waves also emerge in non-neural cell types when cultured on thin fibers [40]. Moreover, adherent cells that are attached to flat substrates may display traveling wave-like protrusions of their cell shape. They are particularly prominent when moving laterally along the cell border, such as in mouse embryonic fibroblasts [41] or at the leading edge of fish keratocytes [42].

Traveling actin waves have also been observed at the dorsal and ventral sides of adherent cells. In neutrophils, small dynamic wave fragments emerge that organize cell polarity and leading edge formation [43]. Larger ring-shaped waves were found to travel across the substrate-attached bottom membrane of *D. discoideum* cells [44]. They enclose a region that is structurally distinct from the cortical area outside the actin ring [45,46], and their dynamics often shows rotating spiral cores and mutual annihilation upon collision [47,48], but they could not be initiated by external receptor stimuli [49]. While understanding the rich dynamics of IAW is challenging on its own right, there are prominent applications and functional properties that stimulate further studies of IAW in different contexts:

Motility. Recently, clear evidence was reported that actin waves directly impact the motility of immune cells, see Figure 1A,B. In particular, dendritic cells, which move in an amoeboid fashion and search the human body for pathogens, display a random walk pattern that can switch between diffusive and persistent states of motion, a direct consequence of the intracellular actin wave dynamics [50];

Cell division. In oocytes and embryonic cells of frog and echinoderms, excitable waves of Rho activity in conjunction with actin polymerization waves were observed shortly after anaphase onset, providing an explanation for the sensitivity of the cell cortex to signals generated by the mitotic spindle [51]. Similarly, in metaphase mast cells, concentric target and spiral waves of Cdc42 and of the F-BAR protein FBP17 were found to set the site of cell division in a size-dependent manner [52]. IAW can also act as the force-generating element that directly drives the division process in a contractile ring-independent form of cytofission. This was observed in *D. discoideum* cells beyond a critical size, where waves that collide with the cell border not only induce strong deformations of the cell shape but also trigger the division into smaller daughter cells—a cell cycle-independent form of wave-mediated cytofission, see Figure 1C [53].

Macropinocytosis. While functional roles in phagocytosis and motility have been proposed [54,55], recent genetic studies suggest a relation to macropinocytosis [56]. This is supported by similarities between the basal actin waves and circular dorsal ruffles (CDR) [57,58]. The latter also adopt a ring-shaped structure but meander across the apical membrane, where they induce membrane ruffles that were related to the formation of macropinocytic cups [59].

Cancer. Macropinocytosis has been also identified as an important mechanism of nutrient uptake in tumor cells [60]. Specifically, the inability of cells to undergo efficient macropinocytosis, e.g., thorough disordered IAW behavior or suppressed activity via pinning of IAW to cell boundaries [58], has been associated with cancerous phenotypes [61,62].

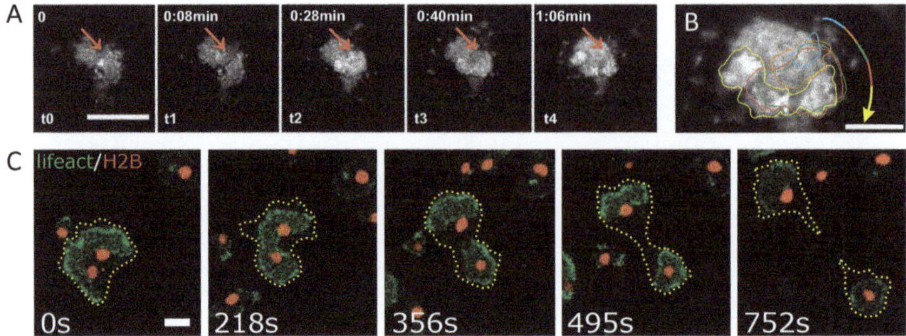

Figure 1. Examples of intracellular actin waves. (**A**) Actin wave nucleation and propagation in a migrating immature dendritic cell. Red arrows indicate the origin of the wave, scale bar 20 µm. (**B**) Overlay of contours of representative actin waves shown in panel (**A**) during propagation. (**C**) Wave-mediated binary cytofission in a *Dictyostelium discoideum* cell, scale bar 10 µm. An actin wave in a cell with two nuclei becomes unstable and splits into two independent segments that move in opposite directions and induce a cytofission event. Panels (**A**) and (**B**) are reproduced from [50] and panel (**C**) is reproduced from [53]. Copyright 2020 National Academy of Sciences.

Despite intense studies over the past years, the molecular details of IAW mechanisms remain largely unclear and most likely vary between different cell types.

2.2. Modeling Approaches of Actin Waves

Following the numerous experimental observations of IAW in different cell types and during different cellular functions, many model equations have been proposed to describe this phenomenon. Here, we will briefly describe the main types and features of theoretical models that have been employed while referring the reader to [31,34,35,63,64] for more details.

The growth of the cortical actin network within IAW is a complex dynamical process that involves many components that perform a coordinated set of functions, giving rise to the formation of a three-dimensional network of actin filaments that propagates along the cell membrane. This process involves the activation of actin-associated proteins some of them membrane bound, that initiate the nucleation of actin polymerization, branching of actin filaments, cross-linking, and bundling, as well as severing and depolymerization. There are very few theoretical models that attempt to give a molecular-scale description of the IAW phenomenon, where all of these processes are described. One example for such a model that attempts to describe the waves at the scale of the individual actin filaments is given in [65]. Although providing detailed pictures of the actin network, it is difficult and time-consuming to use such modeling to extract understanding regarding the large-scale dynamics of the IAW. Such modeling efforts could in the future include more molecular components [66,67], on larger length and time scales, and provide a platform for theoretical advances in this field, that works in conjunction with filament-scale experimental data [46].

As the IAW have characteristic spatial scales in the range of hundreds of nanometers, propagate over tens of microns, and persist over hours, it is natural to describe them using coarse-grained models that avoid prescribing the molecular-scale details of the actin network. As will be shown, many of these models agree with some qualitative or even quantitative features of the observed IAW in cells. It is therefore difficult at present to reach a consensus regarding the validity of these models. Comparisons between such models is also complicated as they often include different components, and it is unclear if and which one of those components play a fundamental role in the emergence of IAW or can be neglected, otherwise.

Among the coarse-grained models, we can find a small class of models that contain biophysical elements, such as forces and/or the membrane shape, which play a key role in the mechanism that drives the propagation of the IAW. One example is well demonstrated by Gholami et al. [68], who show that the dynamics of the actin polymerization/depolymerization drive the oscillatory propagation of waves. When actin filaments polymerize against the cell membrane, they exert a protrusive pressure on the membrane, which pushes the membrane forward and the actin network backwards. The interplay between the rate of actin polymerization and the rate at which the actin filaments are cross-linked into a stable gel-like network, determines if the cortical actin is stable or whether it exhibits an unstable oscillatory regime.

Another group of biophysics-based models contains curved membrane proteins that nucleate the cortical actin polymerization [69–73]. In these models, the curved proteins flow/adsorb to the membrane regions that have a curvature similar to their intrinsic shape, and their concentration is therefore affected by the membrane deformations that are induced by the forces exerted by the actin cytoskeleton. These forces include the protrusive force of actin polymerization, as well as contractile forces due to myosin-II mediated contractility. Recently, also models combining an RD kinetics coupled to mechanical properties through the impact of curved actin nucleators and/or membrane shape and tension were introduced [42,74]. Other models combine the RD dynamics with a physical effect, such that the directed or random lateral actin polymerization can physically drive the treadmilling of the IAW components along the membrane [75]. The advantage of the biophysical class of models is that they can naturally account for the observed effects of physical parameters on the IAW, such as membrane tension [68,74] or the contractile forces of myosin-II motors [76].

In many cases, however, RD equations that include both positive and negative feedback loops, are sufficient to demonstrate the formation of propagating waves, fronts, or pulses. These models exhibit different levels of complexity and different numbers of components. In the simplest cases, generic activator–inhibitor models of FHN-type were proposed. In particular, they were used together with a local-excitation, global-inhibition (LEGI) mechanism to account for the response of the receptor-mediated signaling pathway and the downstream actin cytoskeleton to external cues [77–79]. Other basic RD models describe the actin dynamics, including the monomeric and filamentous species, and one form of an actin activator, using the filamentous actin itself as a source of negative [65] or positive [80] feedback. More complex models include different numbers of activators of actin polymerization, inhibitors, and their complex network of interactions [55,81,82]. Yet, in general, RD equations are not subjected to the conservation of mass constraint although often some of the components are conserved, for example, when they represent two different forms of the same protein [83]. In other cases, the actin is conserved as it is converted from monomeric to filamentous forms and back, see for example [58,84]. In what follows, we address the qualitative role of conservation, which is reflected by the existence of a large scale mode, on the dynamics of IAW, using as much as possible generic principles, i.e., extracting conclusions that are qualitatively independent of the specific molecular details that are included in the model.

3. Actin Dynamics as a Constrained Continuous Medium: Implications and Applications

The phenomenology of dissipative waves can be conveniently demonstrated through a dynamical systems approach via prototypical models, such as FHN. As summarized above, many variants of such activator–inhibitor models have been used to describe different aspects of cytoskeletal dynamics and in particular the formation of actin waves. Although these are heuristic models, they are analytically tractable and thus allow for fundamental insights into spatiotemporal behavior, which cannot be obtained through the analysis of more realistic multivariable equation sets. Propagating waves are traditionally classified into three universality classes [1,2,6,85]:

Oscillatory dynamics, which represent traveling waves that develop via a Hopf instability of a uniform steady state.

Excitability, corresponding to supra–threshold solitary waves (pulses) that propagate on top of a linearly stable uniform steady state.

Bistability, which describes traveling domain walls or fronts, i.e., an interface that connects two linearly stable uniform steady states.

While the mathematical mechanisms are distinct, the emerging patterns can show similar characteristics, for example, all classes may display the formation of spiral waves [2,85]. Consequently, comparisons to experimental observations can often only be qualitative, making insights uncertain. Moreover, it is not always clear whether the simplified models comprise the minimal set of qualitative ingredients, e.g., interactions (local vs. non-local), spatial coupling, essential degrees of freedom and feedback loops, finite domain effects, or existence of conserved observable(s). In a broader context, IAW can be classified as AI type media [31,86], although unlike the typical RD media, the number of actin monomers is conserved over the time scales of wave dynamics. As such, mass conservation is an inherent constraint of the modeling framework [35,58,86,87], which is generically reflected by coupling to a large scale mode in the dispersion relation, as illustrated in Figure 2.

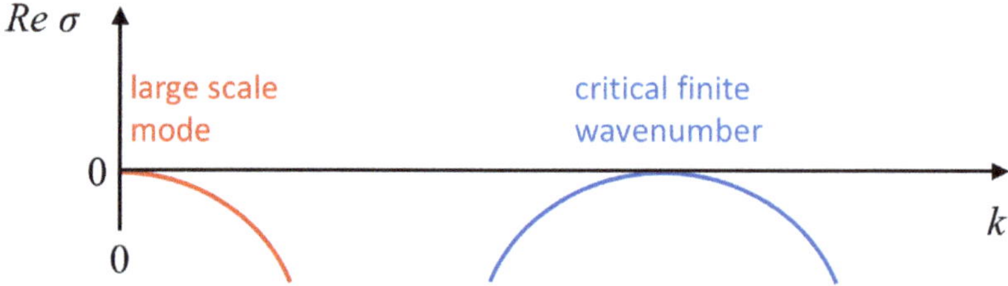

Figure 2. Schematic representation of a dispersion relation obtained from infinitesimal periodic perturbations, proportional to $\exp(\sigma t + ikx)$, about a uniform steady state; $Re[\sigma]$ is the perturbation growth rate and k its wavenumber. The right-hand part of the dispersion relation represents the onset of an instability of a finite wavenumber type (often also referred to as Turing instability), while the left-hand part reflects a conserved quantity and stays always neutral; both parts are model-independent. The curves may connect as typically occurs in systems such as (1) or belong to different curves, such as for (5). The imaginary part of σ corresponds to stationary nonuniform patterns if zero, and otherwise describes time-dependent solutions.

3.1. Conservation in Physicochemical Systems

It is convenient to first consider total conservation of an observable, described by the continuity equation

$$\frac{\partial u}{\partial t} = \nabla \cdot \left[M(u) \, \nabla \frac{\delta F(u)}{\delta u} \right], \qquad (1)$$

where u is a scalar field, M is a mobility function, and F is a free energy. If the free energy contains an intrinsic length scale, like in the phase field crystal model or in wetting, stationary periodic and localized patterns may emerge [88–95]. The mutual aspect is coupling between the large-scale mode ($k = 0$), which is model independent and remains always neutral due to conservation (also known as the Goldstone mode), and the pattern forming instability of finite wavenumber (Turing) type [96], as shown in Figure 2. The impact of the conserved quantity has been analyzed mostly via a weakly nonlinear reduction to a set of two amplitude equations: One is the complex Ginzburg–Landau equation for the finite wave-length mode, while the other is for the neutral large scale mode [97–106]. Both super- and sub-critical bifurcation types have been studied, and showed that indeed inclusion of the large-scale mode may qualitatively change the nature of the solution in terms of organization and stability [107,108] (and references therein). For example, in the absence of the large-scale mode, spatially localized solutions form in coexistence with a periodic (Turing-type) solution and are organized in a vertical snakes-and-ladders homoclinic structure. In the presence of a large-scale mode, these solutions can also form outside the existence region of periodic solutions and only partially overlap, i.e., the homoclinic snaking structure becomes slanted [101,109].

In fact, similar asymptotic intuition and analysis methods apply also if the observable has a velocity-like behavior (often Galilean invariance) [110], obeying the symmetry $x \to -x$ and $u \to -u$. Such behavior arises in systems that are being driven out of equilibrium, such as convection [111–114], propagation of flames [115,116], surface waves [117–119], and electro-diffusion in ion channels [120,121]. In such cases, leading order approximations show that the dynamics can still be enslaved to an oscillatory (Hopf) finite wavenumber mode and a large scale mode [122–126]. While many fundamental advances have been made in understanding the coupling between the complex Ginzburg–Landau equation and the large scale mode, e.g., in terms of stability of periodic and solitary waves in one space dimension and dynamics of spiral waves in two-dimensional systems, several pattern formation issues remain open [127]. Consequently, as over the time scales on which IAW occur the system is far from equilibrium, it is natural to assume that a large scale mode due to mass conservation alters the pattern formation mechanism, even without explicit flux conservation.

3.2. Activator–Inhibitor Patterns with Conservation

In general, AI systems are modeled in a similar fashion as chemical reactions [6,15,16,36,128–131], which are not limited by supply of new substrates into the reactor:

$$\frac{\partial u}{\partial t} = f(u,v) + D_u \nabla^2 u, \tag{2a}$$

$$\frac{\partial v}{\partial t} = g(u,v) + D_v \nabla^2 v, \tag{2b}$$

where u is the activator that typically contains an autocatalytic or enzymatic kinetic term and a diffusion constant D_u, and v is an inhibitor that rapidly diffuses with a diffusion constant D_v, where typically $D_v \gg D_u$. Note that we do not discuss transport by cross-diffusion here. As intracellular processes often take place on very different time scales, effective mass conservation may arise, for example, in cases where protein synthesis and/or degradation occurs much slower than a particular biochemical reaction of interest. Conservation in AI models is associated with a local conservation of mass

$$\int_\Omega [u(\mathbf{x},t) + v(\mathbf{x},t)]\, d\mathbf{x} = \text{constant}, \tag{3}$$

where Ω is the physical domain, or by defining in (2)

$$g(u,v) = -f(u,v). \tag{4}$$

Linear stability analysis about uniform solutions leads to dispersion relations that contain the persistent neutral (large scale) mode, as shown in Figure 2. As in the case of Equation (1), also Equation (4) supports multiplicity of uniform solutions since u depends on an arbitrarily chosen value of v (or vise versa), and this degenerate degree of freedom appears as the $k = 0$ mode. This constraint plays effectively the role of a chemical potential. However, in the pattern forming case, where an additional bifurcation is present (e.g., a Turing bifurcation), the dispersion relations may contain both the neutral mode at $k = 0$ and another at a finite wavenumber. In this formulation, models for cell polarity [132] and molecular motors [133] had inspired several mathematical works in the context of existence and emergence of stationary [134–138] and time-dependent [139–141] patterns.

However, as has been described in Section 2.2, IAW are multicomponent processes and essentially comprise involve a large number activators and inhibitors. Moreover, in such an AI network, not all the components obey conservation [86,142], namely, to Equations (2) and (4) can be added at least one additional non-conserved observable w,

$$\frac{\partial u}{\partial t} = f(u,v,w) + D_u \nabla^2 u, \quad (5a)$$

$$\frac{\partial v}{\partial t} = -f(u,v,w) + D_v \nabla^2 v, \quad (5b)$$

$$\frac{\partial w}{\partial t} = h(u,v,w) + D_w \nabla^2 w, \quad (5c)$$

where h can be either a linear or a nonlinear functional and essentially does not have to include transport of w via diffusion; these details are naturally determined by the characteristics of the biological system. Equation (5) thus reflects only a partial conservation and has been employed to study the emergence of IAW in the context of CDR [58], where a variety of complex behaviors have been observed experimentally, ranging from distinct types of propagating fronts to apparently chaotic spiral waves.

4. Discussion and Example

The complex pattern formation exhibited by CDR raises the question about the modeling strategy, specifically, with respect to the minimal set of equations and the necessity of a conserved quantity. As has already been indicated in Section 2.2, there are many ways to model IAW but all of them are prone to subjective interpretations.

In the absence of a clear physical intuition, as IAW are far from equilibrium phenomena, dynamical systems offer an efficient platform for creating an appropriate qualitative framework. More specifically, the study of bifurcations may provide the minimal qualitative set of constraints, exactly as phase-transitions allow us to classify many types of physical phenomena. On the other hand, bifurcation analysis can also be a tedious task as there may be many local and global bifurcations that coexist in a given parameter range (as an example we refer the reader to a systematic extension of excitable media by Champneys et al. [143]). Nevertheless, utilizing recent advances in nonlinear perturbations [83,144] and numerical path continuation methods [145–149] it might be possible to navigate between coexisting bifurcations and a multiplicity of emerging stable and unstable solutions [144,150]. Next, we turn to conservation and ask whether it may prescribe a fundamental and robust qualitative change, as compared to typical local RD modeling in the absence of conserved quantities. To exemplify this case, we exploit a reduced CDR model (of Equations (5) type), which has been used to examine solitary wave collisions in the context of IAW [151]. In the reduced CDR model, the conserved AI system of Equations (5) is replaced by the conservation of the actin monomers, as they are converted from the monomeric to the filamentous form (and back) while the IAW propagates.

Observation of solitary waves dates back to John S. Russell (1834), yet only after the work in Zabusky and Kruskal [152] were solitary waves distinguished by their collision properties [153,154]: *solitons* if after a collision of two pulses, two pulses emerge (particle-like identity), and *dissipative solitons* or *excitable pulses* if they are annihilated. Solitons are often being discussed in the context of conservative media, which mathematically means exploiting the integrable nature of the governing model equations [107,155], while excitable pulses often arise in RD type systems. Although collisions of solitons may involve high spatiotemporal complexity, the outcome of two colliding solitons remains unchanged (i.e., elastic particle-like dynamics) [156,157]. On the other hand, the annihilation of excitable pulses after the collision is recognized as paramount for electrophysiological function, i.e., it would be impossible to maintain directionality, and thus rhythmic behavior, under the reflection of action potentials [158].

Importantly, collision of pulses implies merging of the pulses in space, i.e, through the formation of a *collision zone*. This behaviour is distinct from the *interaction* between excitable pulses that is due to repulsion and can exhibit dynamics that may resemble a solitonic behavior [159,160]. Moreover, more complex scattering scenarios have been observed in generic RD models such as, for example, the Gray–Scott model [161,162]. Note that there exists a vast literature on the latter topic that we do not intend to review in total here. Taken together, the distinction between solitons and excitable pulses is important for numerous applications.

Yochelis et al. [151] showed that the minimal IAW model in one space dimension, in the setting of Equations (5), may indeed support rich and robust spatiotemporal dynamics following pulse collisions, in contrast to IAW models which do not contain explicit mass conservation [28,55,81,163,164]: annihilation, reflection, and "birth" of new pulses after reflection, as shown in Figure 3. In a broader RD context, where similar aspects have been also observed, these dynamics do not require special properties, such as non-locality [165–168], cross-diffusion [169], and heterogeneity [170–172]. Moreover, the phenomenon is robust and occurs over a wide range of parameter values, whereas for a typical RD model without mass conservation, such as FHN, somewhat similar dynamics of propagating pulses are observed only in a narrow range near the onset of an oscillatory Hopf bifurcation about a uniform steady state [173,174]. The distinction between the FHN model and a system of Equations (5) type can be elaborated by geometrical intuition, as pulses are of large amplitude and thus cannot be unfolded using weakly nonlinear analysis such as in Section 3. Argentina et al. [173] showed that in the FHN model a manifold construction about the collision state of two pulses ("collision droplet", Figure 4A) can explain why a Hopf bifurcation may impact the collision zone and thus generate crossover of pulses (soliton-like behavior). A similar geometric picture shows that mass conservation in Equations (5) changes the nature of the collision zone by addition of a generic two-dimensional neutral manifold (Figure 4B), relating the pulse crossover behavior to a localized unstable mode and does not require any Hopf bifurcation of the uniform state [173,174]. In other words, for the colliding pulses to avoid annihilation, there has to be a mechanism for recovery—a spontaneous regrowth of the fields after the collision. In the FHN model [173], the proximity to the oscillatory onset can re-initiate the pulses. In the case of actin conservation, the colliding pulses first disintegrate the polymerized actin, thereby releasing a large local pool of monomers. If these monomers do not diffuse too fast, they are available to re-initiate the pulses by polymerization. For more details, we refer the reader to the work in [151].

Figure 3. Space–time plots showing (from left to right) annihilation, reflection/crossover, and "birth" of new pulses following collision (a behavior that resembles backfiring), respectively, as obtained from direct numerical integration of the minimal CDR model equations [151] that have the same structure as Equations (5). No-flux boundary conditions were used. From left to right the amount of actin monomers increases (see details in [151]). The dark shaded color indicates higher values of filamentous actin in the IAW. Reprinted figure with permission from the work in [151] Copyright 2020 by the American Physical Society.

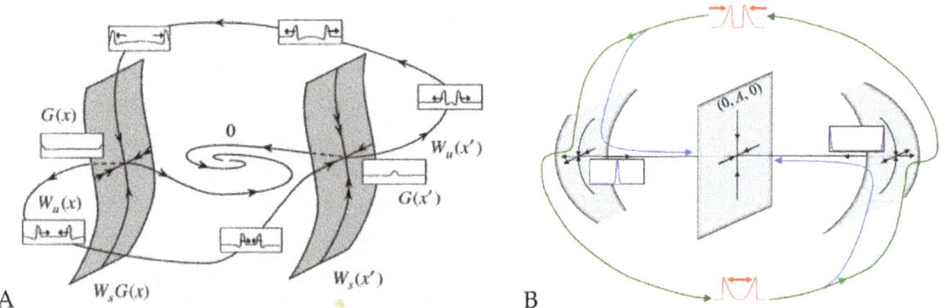

Figure 4. Excitable solitons, geometric schematics of the dynamics during collision of two pulses. (**A**) FitzHugh–Nagumo model and (**B**) an reaction–diffusion model with mass conservation, of Equation (5) type. (**A**) Reprinted from Publication [173], with permission from Elsevier and (**B**) from [151], Copyright 2020 by the American Physical Society.

Note that the nucleation of new pulses after a collision should not be confused with the well-known scenario of backfiring, an instability that appears when a localized propagating pulse becomes unstable and splits into two new counterpropagating pulses that, upon collision, annihilate [175]. Backfiring has been observed in a wide range of model systems [166,176,177], and also in recent experiments of CO electrooxidation on Pt [178]. However, in contrast to backfiring, the nucleation of new pulses that we addressed here and that is shown in Figure 3 always requires a preceding collision event and thus has to be distinguished from the classical backfiring scenario.

5. Conclusions

The case of the reduced CDR model discussed above provides a glimpse of the profound impact of mass conservation on the dynamics. In conventional FHN-type AI models without mass conservation, colliding pulses typically annihilate upon collision. Here, a soliton-like crossover occurs only under special conditions, e.g., near a Hopf point, and thus requires fine-tuning of the parameters. In contrast, if mass conservation is taken into account, propagating pulses robustly exhibit rich collision scenarios over a wide range of parameters, including crossover and formation of new pulses following a collision. Even though this has only been demonstrated for a simple toy model, the universal nature of the underlying bifurcations suggests that similar behavior will be observed also in more detailed, high-dimensional models of IAW, provided that mass conservation is included, e.g., for mechanochemical waves under the conservation of calcium [29].

The impact of mass conservation on pattern formation in biological systems has recently attracted increasing attention, in particular in the context of well-controlled, confined systems such as the bacterial Min protein oscillator [87]. However, many biological systems involve multiple components, not all of which are conserved, so that the consequences of strict mass conservation as implied by Equations (2) and (4) are often relaxed and require a more general view. This is provided, in the simplest case, by adding a third dynamical variable to the system that is coupled to the conserved quantities but does not obey mass conservation itself, see Equations (5). It demonstrates that a large scale mode is a key feature that mass conservation introduces to the system and that triggers specific dynamical properties, such as soliton-like crossover of pulses and the collision-induced birth of new pulses in a wide range of parameters. Equations (5), and its resulting dynamics, can serve as motivation for future studies of the synthesis between classical AI models and models with complete mass conservation (such as those used in the context of the Min and Par systems [179]).

Similar to neural systems, where annihilation of colliding pulses is essential to maintain directionality of information transport, we conjecture that also in the case of IAW, the crossover of colliding pulses, which is favored due to the mass conservation constraint, plays an important functional role. This may be particularly true when sustained wave activity is a key requirement for proper cell functions, as, for example, in cases where cell locomotion or nutrient uptake depend on IAW (see Section 2.1). For traditional excitable pulses that annihilate upon collision, wave activity is likely to get extinguished regularly, thus hampering cellular activities that rely on persistent IAW. In contrast, soliton-like crossover and collision-induced nucleation of new pulses that are robust properties of a mass-conserved system may ensure prolonged wave activity even in the absence of actively triggered pulse nucleation or local heterogeneities that may serve as pacemakers. Moreover, cells may also actively exploit shifts between parameter regimes of pulse annihilation and soliton-like behavior to control their level of IAW activity, as shown in Figure 3.

Finally, the study of simplified models to elucidate generic properties of IAW patterns may also prove useful for the future design of synthetic cellular systems. A current focus of bottom-up approaches in synthetic biology is to introduce artificial cytoskeletal structures into membrane vesicles, thus assembling the essential building blocks of a primitive cell [180,181]. The next logical step along this line of research will be to endow the artificial cytoskeletal components with the simple pattern forming properties that may ultimately serve as a basis for essential cellular functions, such as motility and cytokinesis. This requires a thorough understanding of the key properties that are necessary to reconstitute the desired wave patterns in a minimal model system. We thus expect that an understanding of the essential bifurcations and instabilities that govern the dynamics of IAW will provide a useful guideline for the future design of artificial cell cortices.

Author Contributions: Conceptualization and writing—original draft preparation, A.Y.; writing—review and editing, C.B., N.S.G., A.Y. All authors have read and agreed to the published version of the manuscript.

Funding: N.S.G. is the incumbent of the Lee and William Abramowitz Professorial Chair of Biophysics and thanks the support by the Israel Science Foundation (Grant No. 1459/17).

Conflicts of Interest: The authors declare no conflicts of interest.

Abbreviations

The following abbreviations are used in this manuscript:

HH	Hodgkin–Huxley
RD	Reaction–diffusion
AI	Activator–inhibitor
FHN	FitzHugh–Nagumo
IAW	Intracellular actin waves
CDR	Circular dorsal ruffles

References

1. Cross, M.C.; Hohenberg, P.C. Pattern formation outside of equilibrium. *Rev. Mod. Phys.* **1993**, *65*, 851–1112. [CrossRef]
2. Pismen, L. *Patterns and Interfaces in Dissipative Dynamics*; Springer: Berlin, Germany, 2006.
3. Cross, M.; Greenside, H. *Pattern Formation and Dynamics in Nonequilibrium Systems*; Cambridge University Press: Cambridge, UK, 2009.
4. Turing, A. The chemical basis of morphogenesis. *Philos. Trans. R. Soc. B* **1952**, *237*, 37–72.
5. Hodgkin, A.L.; Huxley, A.F. A quantitative description of membrane current and its application to conduction and excitation in nerve. *J. Physiol.* **1952**, *117*, 500. [CrossRef]
6. Murray, J.D. *Mathematical Biology. II Spatial Models and Biomedical Applications {Interdisciplinary Applied Mathematics V. 18}*; Springer: New York, NY, USA, 2001.
7. FitzHugh, R. Impulses and physiological states in theoretical models of nerve membrane. *Biophys. J.* **1961**, *1*, 445. [CrossRef]
8. Nagumo, J.; Arimoto, S.; Yoshizawa, S. An active pulse transmission line simulating nerve axon. *Proc. IRE* **1962**, *50*, 2061–2070. [CrossRef]
9. Gierer, A.; Meinhardt, H. A theory of biological pattern formation. *Kybernetik* **1972**, *12*, 30–39. [CrossRef]
10. Keller, E.F.; Segel, L.A. Model for chemotaxis. *J. Theor. Biol.* **1971**, *30*, 225–234. [CrossRef]

11. Swift, J.; Hohenberg, P.C. Hydrodynamic fluctuations at the convective instability. *Phys. Rev. A* **1977**, *15*, 319. [CrossRef]
12. Gray, P.; Scott, S. Autocatalytic reactions in the isothermal, continuous stirred tank reactor: Isolas and other forms of multistability. *Chem. Eng. Sci.* **1983**, *38*, 29–43. [CrossRef]
13. Gray, P.; Scott, S. Autocatalytic reactions in the isothermal, continuous stirred tank reactor: Oscillations and instabilities in the system A+ 2B $\rightarrow$ 3B; B $\rightarrow$ C. *Chem. Eng. Sci.* **1984**, *39*, 1087–1097. [CrossRef]
14. Pearson, J.E. Complex patterns in a simple system. *Science* **1993**, *261*, 189–192. [CrossRef] [PubMed]
15. Keener, J.; Sneyd, J. *Mathematical Physiology. Part I: Cellular Physiology*; Springer: New York, NY, USA, 1998.
16. Keener, J.; Sneyd, J. *Mathematical Physiology. Part II: Systems Physiology*; Springer Science+Business Media: New York, NY, USA, 2008.
17. Tyson, J.J.; Keener, J.P. Singular perturbation theory of traveling waves in excitable media (a review). *Phys. D Nonlinear Phenom.* **1988**, *32*, 327–361. [CrossRef]
18. Golubitsky, M.; Stewart, I. *The Symmetry Perspective*; Birkhäuser Verlag: Basel, Switzerland, 2002.
19. Hoyle, R. *Pattern Formation*; Cambridge University Press: Cambridge, UK, 2006.
20. Chow, S.N.; Hale, J.K. *Methods of Bifurcation Theory. Grundlehren der Mathematischen Wissenschaften [Fundamental Principles of Mathematical Science]*; Springer: New York, NY, USA, 1982; Volume 251.
21. Collet, P.; Eckmann, J.P. *Instabilities and Fronts in Extended Systems*; Princeton University Press: Princeton, NJ, USA, 1990.
22. Kuznetsov, Y.A. *Elements of Applied Bifurcation Theory*, 3rd ed.; Applied Mathematical Sciences; Springer: New York, NY, USA, 2004; Volume 112.
23. Mei, Z. *Numerical Bifurcation Analysis for Reaction-Diffusion Equations*; Springer: Berlin, Germany, 2000.
24. Pavliotis, G.; Stuart, A. *Multiscale Methods: Averaging and Homogenization*; Springer Science & Business Media: Berlin, Germany, 2008.
25. Schneider, G.; Uecker, H. *Nonlinear PDE—A Dynamical Systems Approach*; Graduate Studies Mathematics; AMS: Premstätten, Austria, 2017; Volume 182.
26. Dupont, G. Modeling the intracellular organization of calcium signaling. *Wiley Interdiscip. Rev. Syst. Biol. Med.* **2014**, *6*, 227–237. [CrossRef] [PubMed]
27. Falcke, M. Deterministic and stochastic models of intracellular Ca2+ waves. *New J. Phys.* **2003**, *5*, 96. [CrossRef]
28. Kulawiak, D.A.; Löber, J.; Bär, M.; Engel, H. Active poroelastic two-phase model for the motion of physarum microplasmodia. *PLoS ONE* **2019**, *14*, e0217447. [CrossRef]
29. Radszuweit, M.; Alonso, S.; Engel, H.; Bär, M. Intracellular mechanochemical waves in an active poroelastic model. *Phys. Rev. Lett.* **2013**, *110*, 138102. [CrossRef]
30. Onsum, M.D.; Rao, C.V. Calling heads from tails: The role of mathematical modeling in understanding cell polarization. *Curr. Opin. Cell Biol.* **2009**, *21*, 74–81. [CrossRef]
31. Allard, J.; Mogilner, A. Traveling waves in actin dynamics and cell motility. *Curr. Opin. Cell Biol.* **2013**, *25*, 107–115. [CrossRef]
32. Blanchoin, L.; Boujemaa-Paterski, R.; Sykes, C.; Plastino, J. Actin Dynamics, Architecture, and Mechanics in Cell Motility. *Physiol. Rev.* **2014**, *94*, 235–263. [CrossRef]
33. Inagaki, N.; Katsuno, H. Actin waves: Origin of cell polarization and migration? *Trends Cell Biol.* **2017**, *27*, 515–526. [CrossRef]
34. Beta, C.; Kruse, K. Intracellular oscillations and waves. *Annu. Rev. Condens. Matter Phys.* **2017**, *8*, 239–264. [CrossRef]
35. Halatek, J.; Brauns, F.; Frey, E. Self-organization principles of intracellular pattern formation. *Philos. Trans. R. Soc. B Biol. Sci.* **2018**, *373*, 20170107. [CrossRef] [PubMed]
36. Deneke, V.E.; Di Talia, S. Chemical waves in cell and developmental biology. *J. Cell Biol.* **2018**, *217*, 1193–1204. [CrossRef] [PubMed]
37. Ruthel, G.; Banker, G. Actin-dependent anterograde movement of growth-cone-like structures along growing hippocampal axons: A novel form of axonal transport? *Cell Motil. Cytoskelet.* **1998**, *40*, 160–173. [CrossRef]
38. Toriyama, M.; Shimada, T.; Kim, K.B.; Mitsuba, M.; Nomura, E.; Katsuta, K.; Sakumura, Y.; Roepstorff, P.; Inagaki, N. Shootin1: A protein involved in the organization of an asymmetric signal for neuronal polarization. *J. Cell Biol.* **2006**, *175*, 147–157. [CrossRef] [PubMed]

39. Tomba, C.; Braïni, C.; Bugnicourt, G.; Cohen, F.; Friedrich, B.M.; Gov, N.S.; Villard, C. Geometrical determinants of neuronal actin waves. *Front. Cell. Neurosci.* **2017**, *11*, 86. [CrossRef]
40. Guetta-Terrier, C.; Monzo, P.; Zhu, J.; Long, H.; Venkatraman, L.; Zhou, Y.; Wang, P.; Chew, S.Y.; Mogilner, A.; Ladoux, B.; et al. Protrusive waves guide 3D cell migration along nanofibers. *J. Cell Biol.* **2015**, *211*, 683–701. [CrossRef]
41. Döbereiner, H.G.; Dubin-Thaler, B.J.; Hofman, J.M.; Xenias, H.S.; Sims, T.N.; Giannone, G.; Dustin, M.L.; Wiggins, C.H.; Sheetz, M.P. Lateral Membrane Waves Constitute a Universal Dynamic Pattern of Motile Cells. *Phys. Rev. Lett.* **2006**, *97*, 038102. [CrossRef]
42. Barnhart, E.L.; Allard, J.; Lou, S.S.; Theriot, J.A.; Mogilner, A. Adhesion-dependent wave generation in crawling cells. *Curr. Biol.* **2017**, *27*, 27–38. [CrossRef]
43. Weiner, O.D.; Marganski, W.A.; Wu, L.F.; Altschuler, S.J.; Kirschner, M.W. An Actin-Based Wave Generator Organizes Cell Motility. *PLoS Biol.* **2007**, *5*, e221. [CrossRef]
44. Gerisch, G.; Bretschneider, T.; Müller-Taubenberger, A.; Simmeth, E.; Ecke, M.; Diez, S.; Anderson, K. Mobile Actin Clusters and Traveling Waves in Cells Recovering from Actin Depolymerization. *Biophys. J.* **2004**, *87*, 3493–3503. [CrossRef] [PubMed]
45. Schroth-Diez, B.; Gerwig, S.; Ecke, M.; Hegerl, R.; Diez, S.; Gerisch, G. Propagating waves separate two states of actin organization in living cells. *HFSP J.* **2009**, *3*, 412–427. [CrossRef] [PubMed]
46. Jasnin, M.; Beck, F.; Ecke, M.; Fukuda, Y.; Martinez-Sanchez, A.; Baumeister, W.; Gerisch, G. The architecture of traveling actin waves revealed by cryo-electron tomography. *Structure* **2019**, *27*, 1211–1223. [CrossRef] [PubMed]
47. Gerhardt, M.; Ecke, M.; Walz, M.; Stengl, A.; Beta, C.; Gerisch, G. Actin and PIP3 waves in giant cells reveal the inherent length scale of an excited state. *J. Cell Sci.* **2014**, *127*, 4507–4517. [CrossRef] [PubMed]
48. Miao, Y.; Bhattacharya, S.; Edwards, M.; Cai, H.; Inoue, T.; Iglesias, P.A.; Devreotes, P.N. Altering the threshold of an excitable signal transduction network changes cell migratory modes. *Nat. Cell Biol.* **2017**, *19*, 329–340. [CrossRef] [PubMed]
49. Gerhardt, M.; Walz, M.; Beta, C. Signaling in chemotactic amoebae remains spatially confined to stimulated membrane regions. *J. Cell Sci.* **2014**, *127*, 5115–5125. [CrossRef]
50. Stankevicins, L.; Ecker, N.; Terriac, E.; Maiuri, P.; Schoppmeyer, R.; Vargas, P.; Lennon-Duménil, A.M.; Piel, M.; Qu, B.; Hoth, M.; et al. Deterministic actin waves as generators of cell polarization cues. *Proc. Natl. Acad. Sci. USA* **2020**, *117*, 826–835. [CrossRef]
51. Bement, W.M.; Leda, M.; Moe, A.M.; Kita, A.M.; Larson, M.E.; Golding, A.E.; Pfeuti, C.; Su, K.C.; Miller, A.L.; Goryachev, A.B.; et al. Activator-inhibitor coupling between Rho signalling and actin assembly makes the cell cortex an excitable medium. *Nat. Cell Biol.* **2015**, *17*, 1471–1483. [CrossRef]
52. Xiao, S.; Tong, C.; Yang, Y.; Wu, M. Mitotic Cortical Waves Predict Future Division Sites by Encoding Positional and Size Information. *Dev. Cell* **2017**, *43*, 493–506.e3. [CrossRef]
53. Flemming, S.; Font, F.; Alonso, S.; Beta, C. How cortical waves drive fission of motile cells. *Proc. Natl. Acad. Sci. USA* **2020**, *117*, 6330–6338. [CrossRef]
54. Gerisch, G.; Ecke, M.; Schroth-Diez, B.; Gerwig, S.; Engel, U.; Maddera, L.; Clarke, M. Self-organizing actin waves as planar phagocytic cup structures. *Cell Adhes. Migr.* **2009**, *3*, 373–382. [CrossRef] [PubMed]
55. Miao, Y.; Bhattacharya, S.; Banerjee, T.; Abubaker-Sharif, B.; Long, Y.; Inoue, T.; Iglesias, P.A.; Devreotes, P.N. Wave patterns organize cellular protrusions and control cortical dynamics. *Mol. Syst. Biol.* **2019**, *15*, e8585. [CrossRef] [PubMed]
56. Veltman, D.M.; Williams, T.D.; Bloomfield, G.; Chen, B.C.; Betzig, E.; Insall, R.H.; Kay, R.R. A plasma membrane template for macropinocytic cups. *eLife* **2016**, *5*, e20085. [CrossRef] [PubMed]
57. Bernitt, E.; Koh, C.G.; Gov, N.; Döbereiner, H.G. Dynamics of Actin Waves on Patterned Substrates: A Quantitative Analysis of Circular Dorsal Ruffles. *PLoS ONE* **2015**, *10*, e0115857. [CrossRef] [PubMed]
58. Bernitt, E.; Döbereiner, H.G.; Gov, N.S.; Yochelis, A. Fronts and waves of actin polymerization in a bistability-based mechanism of circular dorsal ruffles. *Nat. Commun.* **2017**, *8*, 15863. [CrossRef] [PubMed]
59. Buccione, R.; Orth, J.D.; McNiven, M.A. Foot and mouth: Podosomes, invadopodia and circular dorsal ruffles. *Nat. Rev. Mol. Cell Biol.* **2004**, *5*, 647–657. [CrossRef]
60. Commisso, C.; Davidson, S.M.; Soydaner-Azeloglu, R.G.; Parker, S.J.; Kamphorst, J.J.; Hackett, S.; Grabocka, E.; Nofal, M.; Drebin, J.A.; Thompson, C.B.; et al. Macropinocytosis of protein is an amino acid supply route in Ras-transformed cells. *Nature* **2013**, *497*, 633–637. [CrossRef]

61. Itoh, T.; Hasegawa, J. Mechanistic insights into the regulation of circular dorsal ruffle formation. *J. Biochem.* **2012**, *153*, 21–29. [CrossRef]
62. Hoon, J.L.; Wong, W.K.; Koh, C.G. Functions and Regulation of Circular Dorsal Ruffles. *Mol. Cell. Biol.* **2012**, *32*, 4246–57. [CrossRef]
63. Ryan, G.L.; Watanabe, N.; Vavylonis, D. A review of models of fluctuating protrusion and retraction patterns at the leading edge of motile cells. *Cytoskeleton* **2012**, *69*, 195–206. [CrossRef]
64. Sept, D.; Carlsson, A.E. Modeling large-scale dynamic processes in the cell: Polarization, waves, and division. *Q. Rev. Biophys.* **2014**, *47*, 221–248. [CrossRef] [PubMed]
65. Carlsson, A.E. Dendritic actin filament nucleation causes traveling waves and patches. *Phys. Rev. Lett.* **2010**, *104*, 228102. [CrossRef] [PubMed]
66. Huber, F.; Käs, J.; Stuhrmann, B. Growing actin networks form lamellipodium and lamellum by self-assembly. *Biophys. J.* **2008**, *95*, 5508–5523. [CrossRef] [PubMed]
67. Khamviwath, V.; Hu, J.; Othmer, H.G. A continuum model of actin waves in Dictyostelium discoideum. *PLoS ONE* **2013**, *8*, e64272. [CrossRef] [PubMed]
68. Gholami, A.; Enculescu, M.; Falcke, M. Membrane waves driven by forces from actin filaments. *New J. Phys.* **2012**, *14*, 115002. [CrossRef]
69. Gov, N.S.; Gopinathan, A. Dynamics of membranes driven by actin polymerization. *Biophys. J.* **2006**, *90*, 454–469. [CrossRef]
70. Shlomovitz, R.; Gov, N. Membrane waves driven by actin and myosin. *Phys. Rev. Lett.* **2007**, *98*, 168103. [CrossRef]
71. Veksler, A.; Gov, N.S. Calcium-actin waves and oscillations of cellular membranes. *Biophys. J.* **2009**, *97*, 1558–1568. [CrossRef]
72. Peleg, B.; Disanza, A.; Scita, G.; Gov, N. Propagating cell-membrane waves driven by curved activators of actin polymerization. *PLoS ONE* **2011**, *6*, e18635. [CrossRef]
73. Naoz, M.; Gov, N.S. Cell-Substrate Patterns Driven by Curvature-Sensitive Actin Polymerization: Waves and Podosomes. *Cells* **2020**, *9*, 782. [CrossRef]
74. Wu, Z.; Su, M.; Tong, C.; Wu, M.; Liu, J. Membrane shape-mediated wave propagation of cortical protein dynamics. *Nat. Commun.* **2018**, *9*, 1–12. [CrossRef] [PubMed]
75. Katsuno, H.; Toriyama, M.; Hosokawa, Y.; Mizuno, K.; Ikeda, K.; Sakumura, Y.; Inagaki, N. Actin migration driven by directional assembly and disassembly of membrane-anchored actin filaments. *Cell Rep.* **2015**, *12*, 648–660. [CrossRef] [PubMed]
76. Chen, C.H.; Tsai, F.C.; Wang, C.C.; Lee, C.H. Three-dimensional characterization of active membrane waves on living cells. *Phys. Rev. Lett.* **2009**, *103*, 238101. [CrossRef]
77. Beta, C.; Amselem, G.; Bodenschatz, E. A bistable mechanism for directional sensing. *New J. Phys.* **2008**, *10*, 083015. [CrossRef]
78. Xiong, Y.; Huang, C.H.; Iglesias, P.A.; Devreotes, P.N. Cells navigate with a local-excitation, global-inhibition-biased excitable network. *Proc. Natl. Acad. Sci. USA* **2010**, *107*, 17079–17086. [CrossRef]
79. Devreotes, P.N.; Bhattacharya, S.; Edwards, M.; Iglesias, P.A.; Lampert, T.; Miao, Y. Excitable Signal Transduction Networks in Directed Cell Migration. *Annu. Rev. Cell Dev. Biol.* **2017**, *33*, 103–125. [CrossRef]
80. Sambeth, R.; Baumgaertner, A. Autocatalytic polymerization generates persistent random walk of crawling cells. *Phys. Rev. Lett.* **2001**, *86*, 5196. [CrossRef]
81. Dreher, A.; Aranson, I.S.; Kruse, K. Spiral actin-polymerization waves can generate amoeboidal cell crawling. *New J. Phys.* **2014**, *16*, 055007. [CrossRef]
82. Cao, Y.; Ghabache, E.; Rappel, W.J. Plasticity of cell migration resulting from mechanochemical coupling. *eLife* **2019**, *8*, e48478. [CrossRef]
83. Mata, M.A.; Dutot, M.; Edelstein-Keshet, L.; Holmes, W.R. A model for intracellular actin waves explored by nonlinear local perturbation analysis. *J. Theo. Biol.* **2013**, *334*, 149–161. [CrossRef]
84. Wasnik, V.; Mukhopadhyay, R. Modeling the dynamics of dendritic actin waves in living cells. *Phys. Rev. E* **2014**, *90*, 052707. [CrossRef] [PubMed]
85. Meron, E. *Nonlinear Physics of Ecosystems*; CRC Press: Boca Raton, FL, USA, 2015.
86. Jilkine, A.; Edelstein-Keshet, L. A comparison of mathematical models for polarization of single eukaryotic cells in response to guided cues. *PLoS Comput. Biol.* **2011**, *7*, e1001121. [CrossRef] [PubMed]

87. Halatek, J.; Frey, E. Rethinking pattern formation in reaction–diffusion systems. *Nat. Phys.* **2018**, *14*, 507–514. [CrossRef]
88. Golovin, A.; Davis, S.H.; Voorhees, P.W. Self-organization of quantum dots in epitaxially strained solid films. *Phys. Rev. E* **2003**, *68*, 056203. [CrossRef] [PubMed]
89. Ziebert, F.; Zimmermann, W. Pattern formation driven by nematic ordering of assembling biopolymers. *Phys. Rev. E* **2004**, *70*, 022902. [CrossRef] [PubMed]
90. Golovin, A.; Levine, M.; Savina, T.; Davis, S.H. Faceting instability in the presence of wetting interactions: A mechanism for the formation of quantum dots. *Phys. Rev. B* **2004**, *70*, 235342. [CrossRef]
91. Weliwita, J.; Rucklidge, A.; Tobias, S. Skew-varicose instability in two-dimensional generalized Swift-Hohenberg equations. *Phys. Rev. E* **2011**, *84*, 036201. [CrossRef]
92. Robbins, M.J.; Archer, A.J.; Thiele, U.; Knobloch, E. Modeling the structure of liquids and crystals using one-and two-component modified phase-field crystal models. *Phys. Rev. E* **2012**, *85*, 061408. [CrossRef]
93. Thiele, U.; Frohoff-Hülsmann, T.; Engelnkemper, S.; Knobloch, E.; Archer, A.J. First order phase transitions and the thermodynamic limit. *New J. Phys.* **2019**, *21*, 123021. [CrossRef]
94. Barker, B.; Jung, S.; Zumbrun, K. Turing patterns in parabolic systems of conservation laws and numerically observed stability of periodic waves. *Phys. D Nonlinear Phenom.* **2018**, *367*, 11–18. [CrossRef]
95. Hilder, B. Modulating traveling fronts for the Swift-Hohenberg equation in the case of an additional conservation law. *arXiv* **2018**, arXiv:1811.12178.
96. Tribelsky, M.I.; Velarde, M.G. Short-wavelength instability in systems with slow long-wavelength dynamics. *Phys. Rev. E* **1996**, *54*, 4973. [CrossRef]
97. Matthews, P.; Cox, S.M. Pattern formation with a conservation law. *Nonlinearity* **2000**, *13*, 1293. [CrossRef]
98. Cox, S.; Matthews, P. Instability and localisation of patterns due to a conserved quantity. *Phys. D Nonlinear Phenom.* **2003**, *175*, 196–219. [CrossRef]
99. Cox, S.M. The envelope of a one-dimensional pattern in the presence of a conserved quantity. *Phys. Lett. A* **2004**, *333*, 91–101. [CrossRef]
100. Shiwa, Y. Hydrodynamic coarsening in striped pattern formation with a conservation law. *Phys. Rev. E* **2005**, *72*, 016204. [CrossRef]
101. Dawes, J.H. Localized pattern formation with a large-scale mode: Slanted snaking. *SIAM J. Appl. Dyn. Syst.* **2008**, *7*, 186–206. [CrossRef]
102. Ohnogi, H.; Shiwa, Y. Instability of spatially periodic patterns due to a zero mode in the phase-field crystal equation. *Phys. D Nonlinear Phenom.* **2008**, *237*, 3046–3052. [CrossRef]
103. Golovin, A.; Kanevsky, Y.; Nepomnyashchy, A. Feedback control of subcritical Turing instability with zero mode. *Phys. Rev. E* **2009**, *79*, 046218. [CrossRef]
104. Huang, Z.F.; Elder, K.; Provatas, N. Phase-field-crystal dynamics for binary systems: Derivation from dynamical density functional theory, amplitude equation formalism, and applications to alloy heterostructures. *Phys. Rev. E* **2010**, *82*, 021605. [CrossRef]
105. Kanevsky, Y.; Nepomnyashchy, A. Patterns and Waves Generated by a Subcritical Instability in Systems with a Conservation Law under the Action of a Global Feedback Control. *Math. Model. Nat. Phenom.* **2011**, *6*, 188–208. [CrossRef]
106. Thiele, U.; Archer, A.J.; Robbins, M.J.; Gomez, H.; Knobloch, E. Localized states in the conserved Swift-Hohenberg equation with cubic nonlinearity. *Phys. Rev. E* **2013**, *87*, 042915. [CrossRef]
107. Knobloch, E. Localized structures and front propagation in systems with a conservation law. *IMA J. Appl. Math.* **2016**, *81*, 457–487. [CrossRef]
108. Schneider, G.; Zimmermann, D. The Turing Instability in Case of an Additional Conservation Law—Dynamics Near the Eckhaus Boundary and Open Questions. In *International Conference on Patterns of Dynamics*; Springer: Berlin, Germany, 2016; pp. 28–43.
109. Firth, W.; Columbo, L.; Maggipinto, T. On homoclinic snaking in optical systems. *Chaos Interdiscip. J. Nonlinear Sci.* **2007**, *17*, 037115. [CrossRef]
110. Winterbottom, D.M. Pattern Formation with a Conservation Law. Ph.D. Thesis, University of Nottingham, Nottingham, UK, 2006.
111. Matthews, P.; Rucklidge, A. Travelling and standing waves in magnetoconvection. *Proc. R. Soc. Lond. Ser. A Math. Phys. Sci.* **1993**, *441*, 649–658.
112. Cox, S.; Matthews, P. Instability of rotating convection. *J. Fluid Mech.* **2000**, *403*, 153–172. [CrossRef]

113. Cox, S.M.; Matthews, P.; Pollicott, S. Swift-Hohenberg model for magnetoconvection. *Phys. Rev. E* **2004**, *69*, 066314. [CrossRef] [PubMed]
114. Jacono, D.L.; Bergeon, A.; Knobloch, E. Magnetohydrodynamic convectons. *J. Fluid Mech.* **2011**, *687*, 595–605. [CrossRef]
115. Golovin, A.; Nepomnyashchy, A.; Matkowsky, B.J. Traveling and spiral waves for sequential flames with translation symmetry: Coupled CGL–Burgers equations. *Phys. D Nonlinear Phenom.* **2001**, *160*, 1–28. [CrossRef]
116. Golovin, A.; Matkowsky, B.J.; Nepomnyashchy, A. A complex Swift–Hohenberg equation coupled to the Goldstone mode in the nonlinear dynamics of flames. *Phys. D Nonlinear Phenom.* **2003**, *179*, 183–210. [CrossRef]
117. Tsimring, L.S.; Aranson, I.S. Localized and cellular patterns in a vibrated granular layer. *Phys. Rev. Lett.* **1997**, *79*, 213. [CrossRef]
118. Snezhko, A.; Aranson, I.; Kwok, W.K. Surface wave assisted self-assembly of multidomain magnetic structures. *Phys. Rev. Lett.* **2006**, *96*, 078701. [CrossRef] [PubMed]
119. Pradenas, B.; Araya, I.; Clerc, M.G.; Falcón, C.; Gandhi, P.; Knobloch, E. Slanted snaking of localized Faraday waves. *Phys. Rev. Fluids* **2017**, *2*, 064401. [CrossRef]
120. Kramer, S.C.; Kree, R. Pattern formation of ion channels with state-dependent charges and diffusion constants in fluid membranes. *Phys. Rev. E* **2002**, *65*, 051920. [CrossRef]
121. Peter, R.; Zimmermann, W. Traveling ion channel density waves affected by a conservation law. *Phys. Rev. E* **2006**, *74*, 016206. [CrossRef]
122. Coullet, P.; Fauve, S. Propagative phase dynamics for systems with Galilean invariance. *Phys. Rev. Lett.* **1985**, *55*, 2857. [CrossRef]
123. Riecke, H. Solitary waves under the influence of a long-wave mode. *Phys. D Nonlinear Phenom.* **1996**, *92*, 69–94. [CrossRef]
124. Ipsen, M.; Sørensen, P.G. Finite wavelength instabilities in a slow mode coupled complex Ginzburg-Landau equation. *Phys. Rev. Lett.* **2000**, *84*, 2389. [CrossRef]
125. Winterbottom, D.; Matthews, P.; Cox, S.M. Oscillatory pattern formation with a conserved quantity. *Nonlinearity* **2005**, *18*, 1031. [CrossRef]
126. Hek, G.; Valkhoff, N. Pulses in a complex Ginzburg–Landau system: Persistence under coupling with slow diffusion. *Phys. D Nonlinear Phenom.* **2007**, *232*, 62–85. [CrossRef]
127. Nepomnyashchy, A.; Shklyaev, S. Longwave oscillatory patterns in liquids: Outside the world of the complex Ginzburg–Landau equation. *J. Phys. A Math. Theor.* **2016**, *49*, 053001. [CrossRef]
128. Koch, A.; Meinhardt, H. Biological pattern formation: From basic mechanisms to complex structures. *Rev. Mod. Phys.* **1994**, *66*, 1481. [CrossRef]
129. Maini, P.; Painter, K.; Chau, H.P. Spatial pattern formation in chemical and biological systems. *J. Chem. Soc. Faraday Trans.* **1997**, *93*, 3601–3610. [CrossRef]
130. Volpert, V.; Petrovskii, S. Reaction–diffusion waves in biology. *Phys. Life Rev.* **2009**, *6*, 267–310. [CrossRef]
131. Baker, R.E.; Gaffney, E.; Maini, P. Partial differential equations for self-organization in cellular and developmental biology. *Nonlinearity* **2008**, *21*, R251. [CrossRef]
132. Otsuji, M.; Ishihara, S.; Co, C.; Kaibuchi, K.; Mochizuki, A.; Kuroda, S. A mass conserved reaction–diffusion system captures properties of cell polarity. *PLoS Comput. Biol.* **2007**, *3*, e108. [CrossRef]
133. Yochelis, A.; Ebrahim, S.; Millis, B.; Cui, R.; Kachar, B.; Naoz, M.; Gov, N.S. Self-organization of waves and pulse trains by molecular motors in cellular protrusions. *Sci. Rep.* **2015**, *5*, 13521. [CrossRef] [PubMed]
134. Ishihara, S.; Otsuji, M.; Mochizuki, A. Transient and steady state of mass-conserved reaction-diffusion systems. *Phys. Rev. E* **2007**, *75*, 015203. [CrossRef] [PubMed]
135. Morita, Y.; Ogawa, T. Stability and bifurcation of nonconstant solutions to a reaction–diffusion system with conservation of mass. *Nonlinearity* **2010**, *23*, 1387. [CrossRef]
136. Chern, J.L.; Morita, Y.; Shieh, T.T. Asymptotic behavior of equilibrium states of reaction–diffusion systems with mass conservation. *J. Differ. Equ.* **2018**, *264*, 550–574. [CrossRef]
137. Kuwamura, M.; Seirin-Lee, S.; Ei, S.i. Dynamics of localized unimodal patterns in reaction-diffusion systems for cell polarization by extracellular signaling. *SIAM J. Appl. Math.* **2018**, *78*, 3238–3257. [CrossRef]
138. Ei, S.I.; Tzeng, S.Y. Spike solutions for a mass conservation reaction-diffusion system. *Dis. Contin. Dyn. Syst.-A* **2020**, *40*, 3357. [CrossRef]

139. Sakamoto, T.O. Hopf bifurcation in a reaction–diffusion system with conservation of mass. *Nonlinearity* **2013**, *26*, 2027. [CrossRef]
140. Yochelis, A.; Bar-On, T.; Gov, N.S. Reaction–diffusion–advection approach to spatially localized treadmilling aggregates of molecular motors. *Physica D* **2016**, *318*, 84–90. [CrossRef]
141. Zmurchok, C.; Small, T.; Ward, M.J.; Edelstein-Keshet, L. Application of quasi-steady-state methods to nonlinear models of intracellular transport by molecular motors. *Bull. Math. Biol.* **2017**, *79*, 1923–1978. [CrossRef]
142. Goryachev, A.B.; Pokhilko, A.V. Dynamics of Cdc42 network embodies a Turing-type mechanism of yeast cell polarity. *FEBS Lett.* **2008**, *582*, 1437–1443. [CrossRef]
143. Champneys, A.R.; Kirk, V.; Knobloch, E.; Oldeman, B.E.; Sneyd, J. When Shil'nikov meets Hopf in excitable systems. *SIAM J. Appl. Dyn. Syst.* **2007**, *6*, 663–693. [CrossRef]
144. Yochelis, A.; Knobloch, E.; Xie, Y.; Qu, Z.; Garfinkel, A. Generation of finite wave trains in excitable media. *Europhys. Lett.* **2008**, *83*, 64005. [CrossRef]
145. Doedel, E.; Champneys, A.R.; Fairgrieve, T.F.; Kuznetsov, Y.A.; Sandstede, B.; Wang, X. AUTO: Continuation and Bifurcation Software for Ordinary Differential Equations (with HomCont). Available online: http://cmvl.cs.concordia.ca/auto/ (accessed on 22 June 2020).
146. Sherratt, J.A. Numerical continuation methods for studying periodic travelling wave (wavetrain) solutions of partial differential equations. *Appl. Math. Comput.* **2012**, *218*, 4684–4694. [CrossRef]
147. Uecker, H.; Wetzel, D.; Rademacher, J.D. pde2path-A Matlab package for continuation and bifurcation in 2D elliptic systems. *Numer. Math. Theory Methods Appl.* **2014**, *7*, 58–106. [CrossRef]
148. Bindel, D.; Friedman, M.; Govaerts, W.; Hughes, J.; Kuznetsov, Y. Numerical computation of bifurcations in large equilibrium systems in Matlab. *J. Comput. Appl. Math.* **2014**, *261*, 232–248. [CrossRef]
149. Dankowicz, H.; Schilder, F. *Recipes for continuation*; SIAM: Philadelphia, PA, USA, 2013.
150. Yochelis, A.; Knobloch, E.; Köpf, M.H. Origin of finite pulse trains: Homoclinic snaking in excitable media. *Phys. Rev. E* **2015**, *91*, 032924. [CrossRef] [PubMed]
151. Yochelis, A.; Beta, C.; Gov, N.S. Excitable solitons: Annihilation, crossover, and nucleation of pulses in mass-conserving activator–inhibitor media. *Phys. Rev. E* **2020**, *101*, 022213. [CrossRef]
152. Zabusky, N.J.; Kruskal, M.D. Interaction of "solitons" in a collisionless plasma and the recurrence of initial states. *Phys. Rev. Lett.* **1965**, *15*, 240. [CrossRef]
153. Scott, A.C.; Chu, F.; McLaughlin, D.W. The soliton: A new concept in applied science. *Proc. IEEE* **1973**, *61*, 1443–1483. [CrossRef]
154. Scott, A.C. The electrophysics of a nerve fiber. *Rev. Mod. Phys.* **1975**, *47*, 487. [CrossRef]
155. Knobloch, E. Spatial localization in dissipative systems. *Annu. Rev. Condens. Matter Phys.* **2015**, *6*, 325–359. [CrossRef]
156. Ablowitz, M.J.; Baldwin, D.E. Nonlinear shallow ocean-wave soliton interactions on flat beaches. *Phys. Rev. E* **2012**, *86*, 036305. [CrossRef] [PubMed]
157. Santiago-Rosanne, M.; Vignes-Adler, M.; Velarde, M.G. Dissolution of a drop on a liquid surface leading to surface waves and interfacial turbulence. *J. Colloid Interface Sci.* **1997**, *191*, 65–80. [CrossRef]
158. Alonso, S.; Bär, M.; Echebarria, B. Nonlinear physics of electrical wave propagation in the heart: A review. *Rep. Prog. Phys.* **2016**, *79*, 096601. [CrossRef]
159. Petrov, V.; Scott, S.K.; Showalter, K. Excitability, wave reflection, and wave splitting in a cubic autocatalysis reaction-diffusion system. *Philos. Trans. R. Soc. London. Ser. A Phys. Eng. Sci.* **1994**, *347*, 631–642.
160. Argentina, M.; Coullet, P.; Mahadevan, L. Colliding Waves in a Model Excitable Medium: Preservation, Annihilation, and Bifurcation. *Phys. Rev. Lett.* **1997**, *79*, 2803–2806. [CrossRef]
161. Nishiura, Y.; Teramoto, T.; Ueda, K.I. Scattering and separators in dissipative systems. *Phys. Rev. E* **2003**, *67*, 056210. [CrossRef]
162. Nishiura, Y.; Teramoto, T.; Ueda, K.I. Dynamic transitions through scattors in dissipative systems. *Chaos* **2003**, *13*, 962–972. [CrossRef]
163. Whitelam, S.; Bretschneider, T.; Burroughs, N.J. Transformation from spots to waves in a model of actin pattern formation. *Phys. Rev. Lett.* **2009**, *102*, 198103. [CrossRef]
164. Alonso, S.; Stange, M.; Beta, C. Modeling random crawling, membrane deformation and intracellular polarity of motile amoeboid cells. *PLoS ONE* **2018**, *13*, e0201977. [CrossRef] [PubMed]

165. Krischer, K.; Mikhailov, A. Bifurcation to traveling spots in reaction-diffusion systems. *Phys. Rev. Lett.* **1994**, *73*, 3165. [CrossRef] [PubMed]
166. Bär, M.; Hildebrand, M.; Eiswirth, M.; Falcke, M.; Engel, H.; Neufeld, M. Chemical turbulence and standing waves in a surface reaction model: The influence of global coupling and wave instabilities. *Chaos* **1994**, *4*, 499–508. [CrossRef] [PubMed]
167. Mimura, M.; Kawaguchi, S. Collision of travelling waves in a reaction-diffusion system with global coupling effect. *SIAM J. Appl. Math.* **1998**, *59*, 920–941. [CrossRef]
168. Coombes, S.; Owen, M. Exotic dynamics in a firing rate model of neural tissue. In Proceedings of the Fluids and Waves: Recent Trends in Applied Analysis: Research Conference, the Universtiy of Memphis, Memphis, TN, USA, 11–13 May 2006; Volume 440, p. 123.
169. Tsyganov, M.; Brindley, J.; Holden, A.; Biktashev, V. Quasisoliton interaction of pursuit-evasion waves in a predator-prey system. *Phys. Rev. Lett.* **2003**, *91*, 218102. [CrossRef]
170. Bär, M.; Eiswirth, M.; Rotermund, H.H.; Ertl, G. Solitary-wave phenomena in an excitable surface reaction. *Phys. Rev. Lett.* **1992**, *69*, 945. [CrossRef] [PubMed]
171. Nishiura, Y.; Teramoto, T.; Yuan, X.; Ueda, K.I. Dynamics of traveling pulses in heterogeneous media. *Chaos* **2007**, *17*, 037104. [CrossRef] [PubMed]
172. Yuan, X.; Teramoto, T.; Nishiura, Y. Heterogeneity-induced defect bifurcation and pulse dynamics for a three-component reaction-diffusion system. *Phys. Rev. E* **2007**, *75*, 036220. [CrossRef]
173. Argentina, M.; Coullet, P.; Krinsky, V. Head-on collisions of waves in an excitable FitzHugh–Nagumo system: a transition from wave annihilation to classical wave behavior. *J. Theor. Biol.* **2000**, *205*, 47–52. [CrossRef] [PubMed]
174. Bordyugov, G.; Engel, H. Anomalous pulse interaction in dissipative media. *Chaos* **2008**, *18*, 026104. [CrossRef] [PubMed]
175. Argentina, M.; Rudzick, O.; Velarde, M.G. On the back-firing instability. *Chaos* **2004**, *14*, 777–783. [CrossRef] [PubMed]
176. Zimmermann, M.G.; Firle, S.O.; Natiello, M.A.; Hildebrand, M.; Eiswirth, M.; Bär, M.; Bangia, A.K.; Kevrekidis, I.G. Pulse bifurcation and transition to spatiotemporal chaos in an excitable reaction-diffusion model. *Phys. D Nonlinear Phenom.* **1997**, *110*, 92–104. [CrossRef]
177. Nishiura, Y.; Ueyama, D. Spatio-temporal chaos for the Gray–Scott model. *Phys. D: Nonlinear Phenom.* **2001**, *150*, 137–162. [CrossRef]
178. Bauer, P.R.; Bonnefont, A.; Krischer, K. Dissipative solitons and backfiring in the electrooxidation of CO on Pt. *Sci. Rep.* **2015**, *5*, 1–8. [CrossRef]
179. Geßele, R.; Halatek, J.; Würthner, L.; Frey, E. Geometric cues stabilise long-axis polarisation of PAR protein patterns in C. elegans. *Nat. Commun.* **2020**, *11*, 1–12. [CrossRef]
180. Siton-Mendelson, O.; Bernheim-Groswasser, A. Toward the reconstitution of synthetic cell motility. *Cell Adhes. Migr.* **2016**, *10*, 461–474. [CrossRef] [PubMed]
181. Schwille, P.; Spatz, J.; Landfester, K.; Bodenschatz, E.; Herminghaus, S.; Sourjik, V.; Erb, T.J.; Bastiaens, P.; Lipowsky, R.; Hyman, A.; et al. MaxSynBio: Avenues Towards Creating Cells from the Bottom Up. *Angew. Chem. Int. Ed.* **2018**, *57*, 13382–13392. [CrossRef] [PubMed]

Perspective

Size-Regulated Symmetry Breaking in Reaction-Diffusion Models of Developmental Transitions

Jake Cornwall Scoones [1,†], Deb Sankar Banerjee [2,†] and Shiladitya Banerjee [2,*]

1 Division of Biology and Biological Engineering, California Institute of Technology, Pasadena, CA 91125, USA; jakecs@caltech.edu
2 Department of Physics, Carnegie Mellon University, Pittsburgh, PA 15213, USA; debsankb@andrew.cmu.edu
* Correspondence: shiladtb@andrew.cmu.edu
† These authors contributed equally to this work.

Received: 21 April 2020; Accepted: 6 July 2020; Published: 9 July 2020

Abstract: The development of multicellular organisms proceeds through a series of morphogenetic and cell-state transitions, transforming homogeneous zygotes into complex adults by a process of self-organisation. Many of these transitions are achieved by spontaneous symmetry breaking mechanisms, allowing cells and tissues to acquire pattern and polarity by virtue of local interactions without an upstream supply of information. The combined work of theory and experiment has elucidated how these systems break symmetry during developmental transitions. Given that such transitions are multiple and their temporal ordering is crucial, an equally important question is how these developmental transitions are coordinated in time. Using a minimal mass-conserved substrate-depletion model for symmetry breaking as our case study, we elucidate mechanisms by which cells and tissues can couple reaction–diffusion-driven symmetry breaking to the timing of developmental transitions, arguing that the dependence of patterning mode on system size may be a generic principle by which developing organisms measure time. By analysing different regimes of our model, simulated on growing domains, we elaborate three distinct behaviours, allowing for clock-, timer- or switch-like dynamics. Relating these behaviours to experimentally documented case studies of developmental timing, we provide a minimal conceptual framework to interrogate how developing organisms coordinate developmental transitions.

Keywords: symmetry breaking; pattern formation; reaction-diffusion; developmental transitions

---

1. Introduction

In developmental systems, it is important for the mechanistic constituents to "know" about the size of the living system as a whole [1,2]. This is most apparent in developmental transitions, which in many cases only proceed when cells or tissues have reached a critical size. Such control strategies allow living systems to couple developmental time to their size and geometry. What is the physical basis of these phenomena?

One can envisage two broad classes of size-control mechanisms: The first proposes size-dependent transitions are under external regulation: in developing tissues, this could be manifested as a cell-intrinsic clock, whereby a transition is achieved after a pre-defined time interval [3], or a gradient-based mechanism, whereby a distance critically far from the source of signalling molecules triggers a transition [4]. Alternatively, size-regulation could be an emergent property of collective decision-making, akin to quorum sensing [5–7]: communication between mechanistic constituents allows the system to sense its size. A priori, these control mechanisms have several conceptual benefits: synchrony in a transition is more robust, given decisions are made collectively, and such decision-making does not rely on a subset of constituents (e.g., source cells in gradient generation), instead being decentralised [1,8]. Can we find examples of emergent size-regulation from collective decision-making in living systems?

Many developmental transitions couple size- and temporal-control to a change in polarity regime: at a critical size, the system may spontaneously break symmetry to polarise, depolarise, bipolarise, or even radically change its pattern. We hypothesise that size regulation of such developmental transitions are an emergent property of the many mechanisms of polarisation. This view provides a framework for understanding developmental time [9], placing a critical emphasis on system size.

Here, we outline a minimal reaction–diffusion model for size-dependent polarisation in developing systems, arguing that the underlying regulatory motifs can be understood via a substrate-depletion feedback motif coupled with the growth of the system. We then overview cases of size-dependent symmetry breaking across scales, focusing on biochemical systems. We consider size-dependent decision-making within individual cells through to analogous processes in developing multicellular systems, proposing that our minimal model can help unify these divergent processes within a common theoretical framework. We conclude by speculating on the role of these mechanisms in coupling size-dependent transitions to developmental time.

2. Reaction–Diffusion as a Framework to Understand Size-Regulated Symmetry-Breaking

Pattern forming reaction–diffusion (RD) systems [10] are widely used to characterise the complex networks of molecular and cellular interactions that underlie biological symmetry breaking. In these systems, pattern formation can arise spontaneously driven by feedback motifs between diffusible molecules (intracellular polarity proteins or extracellular morphogens). Since Turing's insight in 1952, many different RD motifs have been proposed as the physical basis for the emergence of developmental patterns across scales [11]—from polarity establishment at the scale of a single cell [12] to pattern formation on the scale of a whole organism [11,13].

Perhaps the most famous phrasing of an RD system is the activator–inhibitor circuit, originally developed by Gierer and Meinhardt [14]. In activator–inhibitor systems, an activator molecule promotes its own production as well as the production of its fast-diffusing inhibitor that suppresses autocatalytic production of the activator. This motif has remarkable explanatory power across contexts, being used to describe the spontaneous establishment of hair follicle spacing [15], left-right asymmetry establishment in vertebrates [16], skeletal patterns in growing limbs [17–19], as well as pole-to-pole oscillation of Min proteins during bacterial cell division [20], and self-organisation of Rho GTPases in the animal cell cortex [21,22]. Substrate depletion models can also yield the spontaneous emergence of periodic patterns [23,24]. In these models, the activator consumes its own substrate to promote its autocatalytic production, leading to out-of-phase patterning of the activator and the substrate molecules (Figure 1a). For example, a substrate-depletion model was been used to explain lung branching, explaining out of phase patterns of gene expression between Shh (the activator) and FGF (the substrate) [25].

Substrate-depletion models are particularly relevant in the study of intracellular pattern formation and polarity establishment as feedback can be phrased in a mass-conserved manner. In such models, patterns emerge by the redistribution of polarity proteins [26,27]: proteins that form a polarity patch engage in self-recruitment, acting as activators, but this positive feedback is limited by a finite pool of (typically cytoplasmic) subunits. Variants on mass-conserved substrate depletion models have been used to understand PAR polarity establishment in the *C. elegans* embryo [28–30] and Cdc42 polarisation in *S. cerevisiae* [31–33]. Further, such models exhibit dynamic regimes, for example helping to explain oscillations in the *E. coli* Min-protein system [34–36].

While the mechanistic constituents and precise feedback architectures of RD mechanisms differ, many rely on a central motif of local activation and long-range inhibition [37]. This concept has acted as an important heuristic in framing models of biological pattern formation, but importantly unifies diverse RD systems within a common mathematical framework. Specifically, recent theoretical models have demonstrated that most RD models of pattern formation can be approximated by the same mathematical formulation: the Swift–Hohenberg equation [38]. Strikingly, other mechanisms of periodic pattern formation that rely on cell movement [39] or mechanical instabilities [40] also rely on local-activation and long-range inhibition [38,41]. Therefore, many dynamical features of these models are applicable across systems and length scales.

In this perspective, we restrict our focus to RD models for biochemical pattern formation, elucidating the biological significance of a common dynamical feature shared across many motifs: the role of a critical system size for symmetry breaking [23,42,43]. To demonstrate this, we develop a mathematical framework for a mass-conserving RD system with a feedback motif similar to activator–substrate models for cell polarisation (Figure 1a).

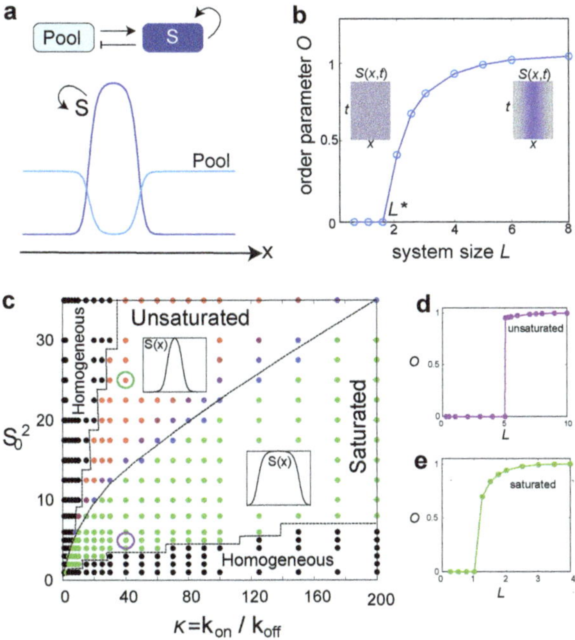

Figure 1. Size-regulated symmetry breaking in activator–substrate model. (**a**) Pattern formation in a model of positive feedback coupled to a finite constituent pool. (**b**) Patterns form above a critical system size (L^*), corresponding to the largest mode where the homogeneous state becomes unstable and the system breaks symmetry. The order parameter $\mathcal{O}$ for symmetry breaking is defined as $\mathcal{O}(L) = \int_0^L |dS/dx| / \left[\max_L \int_0^L |dS/dx|\right]$, where $\mathcal{O}$ is zero for a homogeneous state and $\mathcal{O} = 1$ for a symmetry broken patterned state. All parameters other than system size was kept constant in this analysis and initial perturbations were so chosen that N/L is constant, where $N = \int P(x,t) + S(x,t)\,dx$ is the total pool size. Parameters: $D_P = 1$, $D_S = 0.05$, $k_{\text{on}}/k_{\text{off}} = 20$, $S_0^2 = 10$ and $N/L = 1.5$. (**c**) Phase diagram in the plane of autocatalytic activity κ and the Hill saturation parameter S_0^2, showing three different phases: homogeneous state (black), symmetry broken saturated state (green), and symmetry-broken unsaturated state (red). Colormap (green to red) denotes the average value of the reaction rate $F_{av} = \int F(x)dx$, computed in the high density region, with $F_{av} = 0$ in the saturated state and $F_{av} \neq 0$ in the unsaturated state. $F_{av} = 0.01$ (blue points) defines the crossover value from the saturated state to the unsaturated state. Parameter values are the same as (**b**) except for $D_S = 0.01$ and $L = 3$. Inset: Profiles of $S(x)$. (**d–e**) Order parameter $\mathcal{O}$ for the unsaturated (**d**) and the saturated (**e**) regimes, showing that the symmetry of the homogeneous state is broken beyond a critical system size. We used periodic boundary conditions for all numerical simulations, unless otherwise specified. For initial conditions, we assumed a homogeneous $P(x)$ and a sinusoidal $S(x)$ profile of large wavelength.

3. A Minimal Model for Size-Regulated Symmetry-Breaking

To analyse the role of system size on the timing of symmetry breaking in a biochemical system, we consider a minimal model for a mass-conserved substrate-depletion system. Specifically, we model the spatiotemporal dynamics of a regulatory structure S in a living system of size L and coupled to a finite pool of building blocks. Let $P(\mathbf{x}, t)$ denote the concentration of building blocks in the subunit pool at location $\mathbf{x}$ at time t, and $S(\mathbf{x}, t)$ is the concentration of building blocks incorporated in the regulatory structure. S increases in amount by depleting the subunit pool P, and S can undergo dissociation into P (Figure 1a). The coupled dynamics of S and P are given by

$$\partial_t S = D_s \nabla^2 S + k_{\text{on}} P f(S) - k_{\text{off}} S \, , \tag{1}$$

$$\partial_t P = D_p \nabla^2 P - k_{\text{on}} P f(S) + k_{\text{off}} S \, , \tag{2}$$

where D_p and D_s are the diffusion constants of the subunits and the structure S ($D_s \ll D_p$), k_{on} parameterises the association rate of subunits to S, and k_{off} is the constant rate of disassembly of S. The function f defines the size-dependence of the autocatalytic production rate of S. We assume the functional form $f(S) = S^n/(S_0^n + S^n)$, where the constant $n > 1$ controls the strength of cooperative assembly of S. The total amount of P and S remains conserved at all times, i.e., $N = \int P(x,t) + S(x,t)\, dx =$ constant, and is assumed to scale linearly with the system size L.

As the structure S grows by locally depleting the pool P, localized patterns of S will exist in low density regions of P as long as $D_P \gg D_S$ (Figure 1a). The symmetry of the homogeneous state breaks and patterns appear above a critical domain size, making this transition size-dependent (Figure 1b). The critical size can be obtained from linear stability analysis as

$$L^* = \left(\frac{2 D_S D_P}{D_S \partial_P F + D_P \partial_S F} \right)^{\frac{1}{2}} \tag{3}$$

where $F = k_{\text{on}} P f(S) - k_{\text{off}} S$, and the derivatives are evaluated at the homogeneous steady state. This property of size dependent symmetry-breaking can serve as a decision-making rule to enact developmental state transitions when the system size reaches a critical value L^*. As the system size gets larger more discrete structures will emerge (Figure 2a). Such sequential pattern formation may regulate size-dependent developmental transitions, as we discuss later.

The S-P model introduced above (for $n = 2$) is similar to previously studied mass-conserved RD models such as wave-pinning [26] and the Turing-like autocatalytic model [31]. These activator–substrate models exhibit two distinct dynamic regimes [44]: the wave-pinning regime, characterised by wide mesa-like patterns and saturated subunit association kinetics and the Turing regime that yields narrow concentration peaks by virtue of competition between structures. The Turing regime operates below saturation ($S \ll S_0$), where a winner-take-all competition between structures asymptotically results in a single concentration peak [31,45,46]. By contrast, in the regime above saturation ($S \gg S_0$), the structures can co-exist for very long timescales, with the timescale of coarsening determined by parameters such as the diffusion coefficients or the reaction fluxes [47]. The S-P model exhibits both the saturated and the unsaturated regimes that can be obtained by tuning the strength of autocatalytic activity ($\kappa = k_{\text{on}}/k_{\text{off}}$) and the Hill saturation parameter S_0^2 (Figure 1c). Both these dynamical regimes exhibit size-dependent symmetry breaking (Figure 1d,e), as well as sequential pattern formation for increasing system sizes (Figure 2). In the latter case, we assume that the subunit density is constant for increasing L, $\int_0^L (S + P)\, dx \propto L$, as macromolecular contents often scale with cell size [48]. For appropriate choice of model parameters in the unsaturated regime, an increase in total subunit pool size coupled with local depletion of the subunit pool gives rise to coexisting peaks of the same height over biologically relevant timescales, much shorter than the long timescale of structure coarsening.

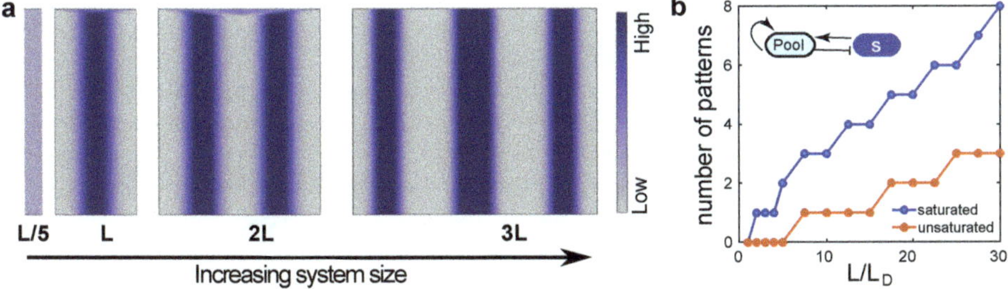

Figure 2. Sequential pattern formation in growing domains. (**a**) Kymographs of $S(x,t)$ for increasing system size. Figures show spontaneous pattern formation via symmetry breaking of the homogeneous state above a critical system size L^*. As we simulate larger systems (Equations (1) and (2)), multiple patterns emerge in a size-dependent manner. Simulations were done as described in Figure 1b. (**b**) Number of patterns as a function of system size, for both the saturated and unsaturated regimes of the S-P model. Parameter values for (**a**,**b**) are the same as Figure 1b,c, respectively. Small amplitude random (uniform distribution) initial conditions for S and P were used for (**b**).

4. Critical Size for Polarisation Can Be Utilised to Enact Cell State Transitions

Our minimal model demonstrates how a positive-feedback motif coupled to features of system size, such as a limiting cytoplasmic pool, can yield size-regulated symmetry breaking of regulatory structures. In this section we explore biological realisations of our model around the bifurcation from unpolarised to polarised, arguing that cells may utilise these size-dependent properties to coordinate state transitions.

4.1. Cell Size Dependent Transition from Asymmetric to Symmetric Division in the Early C. elegans Embryo

The polarisation of the early *C. elegans* embryo has become a paradigm in biological symmetry breaking [49]. Anterior-posterior (AP) polarity in *C. elegans* is established before the first cell division [50]. Polarity establishment is achieved by the segregation of two groups of partitioning-defective (PAR) proteins to the anterior versus posterior [28–30]. Initially anterior PARs (aPARs) cover the entire membrane of the egg, but upon fertilisation at the posterior, serving as the symmetry breaking cue [51], aPARs segregate anteriorly, and posterior PARs (pPARs) posteriorly. Segregated PARs coordinate polarised division, whereby the division plane is set by the boundary of the two PAR domains [52,53].

PARs bind the membrane from a common, finite cytoplasmic pool and diffuse freely. Symmetry breaking is achieved by phosphorylation-dependent mutual inhibition between aPARs and pPARs [54,55], patterning the cell membrane into two polarisation domains. This double-negative feedback structure reinforces biases in the localisation of PARs and thus plays a similar role as the positive feedback motif in our minimal model. Indeed explicit substrate-depletion models yield phenomenologically identical results [56].

The boundary between PAR domains is regulated by the relative diffusivities of aPARs and pPARs, as well as the relative off rates. Boundary length is set by $L_D = \sqrt{D/k_{\text{off}}}$, where D is the aPAR diffusion constant and k_{off} is the aPAR dissociation rate from the membrane. Therefore, provided diffusion and dissociation rates are independent of system size, L_D will also be independent of cell size. Modelling confirms that this holds true regardless of structural differences in the model [56]. This length-scale thus sets a minimum cell size that can sustain polarised PAR domains (grey region in Figure 3a): below this critical size, diffusion overwhelms the capacity for PAR segregation, resulting in homogeneous localisation of either aPARs or pPARs (pink and blue regions of Figure 3a)

The physical critical size limit may be used by developing *C. elegans* embryos to coordinate a developmental transition. Sequential divisions of generating the first three posterior cells (P1-3) are asymmetric, each generating two daughters of different fates via the segregation of germline determinants. This pattern shifts to a symmetric mode by the third division of P4 (Figure 3a), generating the two founding cells of the germline lineage (Z2/Z3) [57]. Divisions are fast, meaning cell volume declines progressively, falling beneath the theoretical critical cell size for polarisation by P4 [56]. By quantifying division symmetry using 3D reconstructions of PAR distributions, Habatsch et al. [56] found that the polarisation regime shifts from asymmetric to symmetric by P4. The timing of this regime shift can be changed by reducing embryo size through genetic (ima3 RNAi) or physical (laser mediated extrusion) perturbations. Thus, a reduction in cell size coordinates a developmental transition in *C. elegans* embryos from asymmetric to symmetric cell division.

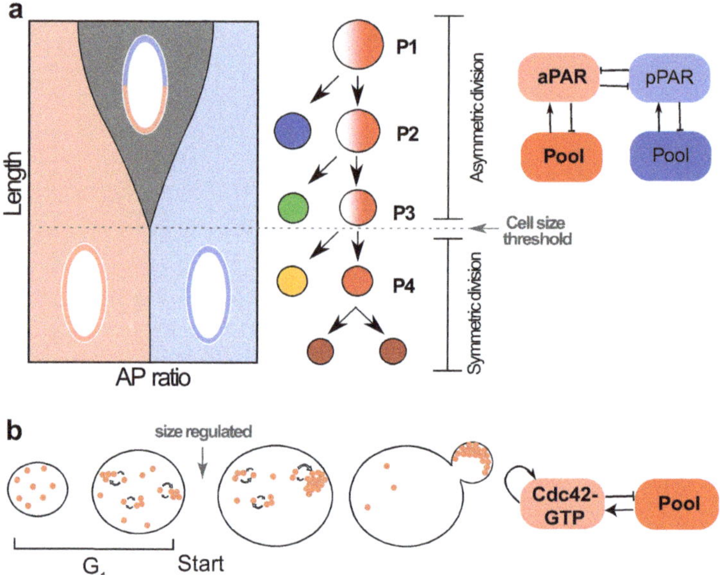

Figure 3. Size-regulated symmetry breaking in single cells. (**a**) A phase-diagram for the PAR system, considering polarisation state as a function of the circumferential length of the embryo ("Length"), and the ratio of aPAR to pPAR pool size ("AP ratio"). The diagram demonstrates a bipolar state (grey region) becomes unstable below a critical circumferential embryo length (pink and blue regions). Schematics for each of the three states are overlayed, with aPARs denoted in pink and pPARs denoted in blue. This bifurcation point quantitatively matches the critical size for which dividing P-cells in the early *C. elegans* embryo transition from asymmetric to symmetric division. Figure adapted from the work in [56]. Adjacent is the feedback motif that drives pattern formation. (**b**) In the budding yeast (*S. cerevisiae*), cell cycle commitment to Start is linked to the localisation of Cdc42 effectors at the presumptive bud site [58]. Cdc42 polarity establishment is related to the duration of the G_1 phase of the cell cycle, which ends at a critical cell size [59]. Models have shown that a growth process with positive feedback leads to Cdc42 polarisation at a single site [31].

4.2. Size-Dependent Polarity Establishment in Budding Yeast

Cell cycle commitment to budding in *S. cerevisiae* follows from Cdc42 polarity establishment at the presumptive budding site (Figure 3b). The small Rho GTPase Cdc42-GTP forms a polarity patch to mark the bud location [60,61]. This polarity pattern emerges from an autocatalytic positive feedback via Bem1 on the clustering of slowly diffusing membrane bound Cdc42-GTP [12], while the cytosolic Cdc42-GDP diffuses fast (Figure 3b). Polarity establishment in the system can be captured by mass-conserved substrate-depletion model, with a slow-diffusing activator and a fast-diffusing substrate [26,31,45]. With appropriate choice of parameters, activator-substrate models would predict the formation of a polarised cluster beyond a critical cell size [44] (Figure 1). Thus, the establishment of Cdc42 polarisation can be linked to a critical cell size, consistent with models of critical cell size threshold at the termination of G_1 phase of the cell cycle [59,62]. Molecular rewiring experiments have shown that when the Bem1 is tweaked to diffuse very slowly, multiple Cdc42 polarity patches are formed [63]. As the onset of pattern formation depends on L/L_D, with L_D the diffusion length, slowing down diffusion is equivalent to increasing the system size, so multiple patterns emerge in accordance with predictions from increasing domain length in our minimal RD model (Figure 2a,b).

5. Sequential Pattern Formation and Polarisation Can Be Coordinated by a Growing Domain

Across scales of biological organisation, structures of a bipolar or iterative nature are abundant, and their development is often sequential. While clock-and-wavefront models have successfully explained sequential patterning of axial segmentation in vertebrates [64] and recently also some invertebrates [65], sequential pattern formation can also arise from collective decision-making. Our minimal mass-conserved substrate-depletion model illustrates how an RD mechanism can give rise to sequential periodic patterning: smaller domain sizes can sustain only a single pattern via an instability of the homogeneous state, whereas domain growth provides sufficient space to accommodate multiple patterns (Figure 2a,b). Here, we implicitly assume that the RD system relaxes much faster than the timescale of domain growth, and that subunit concentration remains unchanged. When domain growth rate is comparable to the reaction rate, different dynamic patterns emerge as discussed in Section 6. In spite of differences in their mechanistic bases, sequential patterning through domain growth is common among many RD models (activator–inhibitor and substrate-depletion) [23]. In the following, we expand our focus beyond just mass-conserved models, to illustrate how local-activation and long-range inhibition can explain how growth can couple developmental tempo to state transitions.

5.1. *Neuronal Sequential Bipolarisation Coordinated by Membrane Growth*

Neuronal polarisation is critical for brain development. Polarisation commences as soon as neurones complete their final division, by a process of neurite formation and selection. In vitro studies have suggested neurones acquire a bipolar phenotype, generating a leading neurite, key in guiding migration, and a trailing neurite which later acquires axonal fate [66,67]. Bipolarisation in vitro is achieved stochastically, whereby the position of the first neurite is seemingly random, with the second being positioned opposite to the first [68]. How are these patterns coordinated?

Menchón et al. [69] proposed an activator–inhibitor Turing model for cell polarisation. They argued that the necessary feedback architecture for a Turing instability is manifest in developing neurones: integral membrane proteins (the polarisation cue) undergo cooperative self-recruitment, i.e., local activation, and also recruit more diffusive endocytosis modulators which facilitate their removal, i.e., long-range inhibition (Figure 4a). Indeed, in the right parameter regime in a finite domain, simulations suggest neurones can spontaneously break symmetry. The polarity regime is critically dependent on membrane size: a subcritical size prohibits symmetry breaking (like in *C. elegans*); an intermediate size allows for a single polarity axis and larger sizes allow for a bipolar phenotype (Figure 4a). Sequential and "mirrored" polarisation can be achieved by membrane growth. A growing domain leads to a time-dependent bifurcation, whereby the cell transitions from a unipolar to bipolar stability regime. The "mirroring" of the second neurite on the first can be rationalised in terms of the feedback circuit: the region of membrane furthest from the first neurite will display the lowest concentration of inhibitor. Neurones thus coordinate the developmental timing of bipolarity through a size-dependent process.

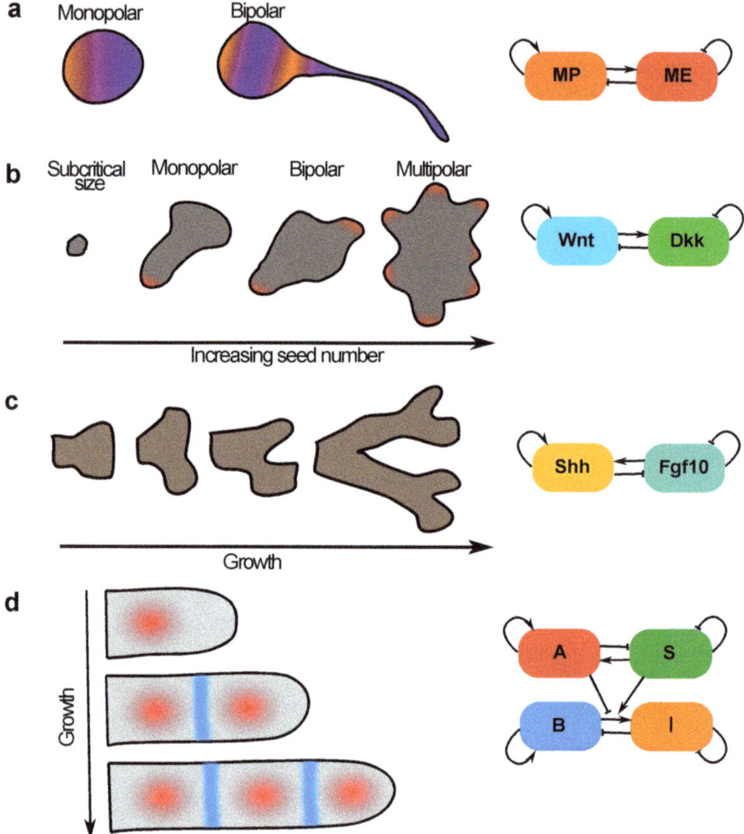

Figure 4. Sequential pattern formation in developmental systems. (**a**) In in vitro cultured neurones, polarity arises sequentially. A second polarity axis is formed after cell growth, and its orientation is "mirrored" off the first. A putative symmetry breaking circuit is presented adjacently [69], considering a membrane-protein (MP) activator coupled to modulators of endocytosis (ME), representing the effects of small GTPases. (**b**) Gastruloids polarise and elongate only when initialised with a critical number of cells [70]. For seed numbers beyond this initial bifurcation value, gastruloids can self-organise more axes. T/Brachyury expression is localised to the protrusion in monopolarised gastruloids, and is speculated to also be localised to further protrusions in multipolar variants. A potential feedback circuit is drawn adjacently, which remains to be investigated. (**c**) An activator–substrate model for lung branching [25], based on autocatalytic production of the signalling molecule Shh (activator) at the lung bud tip, via consumption of the substrate molecule Fgf10. (**d**) A dot-stripe mechanism is proposed to pattern the joints of developing digits: a Turing-like dot-forming system specifies the positions of bones and orients through repression a Turing-like stripe-forming system to specify joints. Modelled on a growing domain, sequential joint specification emerges, with joints forming near the developing tip. A coupled Turing scheme is described adjacent, considering a dot-forming substrate-depletion module (A,S) coupled to a stripe-forming activator–inhibitor module (B,I).

5.2. Size-Dependent Sequential Patterning in Mammalian Development—Insights from Gastruloids

Unlike in *C. elegans*, establishment of anterior–posterior polarity in the epiblast of mammalian embryos occurs well after the first cell division, an axis that lays the ground plan for the commitment of germ layers during gastrulation. In mice, AP symmetry breaking has long thought to be coordinated by the positioning of extra-embryonic cues to the posterior and hence specifying the future primitive streak [71]. This view of sequential polarity hand-off has been thrown into question in recent years by several in vitro systems, suggesting that epiblast has the capacity to break symmetry spontaneously in the absence of extra-embryonic cues [70,72–75]. While the precise genetic constituents of this symmetry breaking are under contention [76,77], several that argue some form of reaction–diffusion system is at play, citing for example the co-expression of morphogens with their extra-cellular antagonists, e.g., Wnt and its antagonist Dkk [78,79].

Consistent with this hypothesis, in vitro systems display size-dependence in symmetry breaking capacity and patterning modality [70,75]. This is seen in gastruloids, small aggregates of embryonic stem cells (ESCs) that can spontaneously break symmetry [70], axially elongating and displaying polarised expression of primitive streak marker T/Brachyury. In refining their protocol for generating gastruloids, van den Brink et al. [70] found that seeding microwells with different numbers of ESCs yielded qualitatively different phenomenology (Figure 4b): critically small ($\leq$200 cells) aggregates could not break symmetry, aggregates of intermediate size ($\sim$300–400 cells) could subsume a unipolar state, aggregates of double that size ($\sim$800 cells) displayed two oppositely positioned poles, and critically large aggregates ($>$1600 cells) generated many poles. These results are consistent with a Turing-like system controlling polarity, whereby critically small domains cannot sustain an instability, whereas increasingly large domains can maintain progressively more "peaks". This is furthered by the observation that bipolar gastruloids polarise sequentially, with the second pole seemingly emerging after growth, protruding from the opposite edge of the structure.

5.3. Sequential Patterning of Phalanges in Developing Digits Is Coordinated by Coupling Patterning to Growth

An analogous mechanism may explain the sequential specification of joints in developing digits of tetrapod limbs. Digit patterning and growth are concomitant, with joints being laid down sequentially as progenitor cells are added to the distal tip. Guided by gene expression patterns, as well as mutant phenotypes, joint patterning has been proposed to be governed by a coupled Turing system [80] (Figure 4d). An activator–substrate system specifies the positions of bones (phalanges) by prescribing a series of "dots" of gene expression, which represses a second activator–inhibitor system to specify joints as "stripes" of gene expression at alternate positions. Simulations on a static domain recapitulate both wild type and mutant expression patterns, but patterning occurs simultaneously across the entire digit. However, simulated on a growing domain, adding new cells distally, leads to a shift in dynamics in favour of sequential patterning. Therefore, here too, the coordination of developmental timing and growth may be an emergent property of patterning by collective decision-making.

6. Regulating Pattern Size and Lifetime in Growing Systems

Pattern-forming systems that align with activator–substrate or activator–inhibitor motifs are able to undergo sequential transitions in pattern concomitant with domain growth (Figures 2 and 4). As domain growth continues, patterns undergo further bifurcations to establish periodicity. While these motifs allow irreversible transitions in pattern, it may be desirable for systems to sense intermediate sizes, and for these transitions to be reversible. One can conceive of a timer-like set-up in a biphasic scheme: in the assembly stage, symmetry is broken at a critical size, and in the proceeding dissolution stage, patterns are lost at some larger size. To investigate the emergence of timer-like behaviour, we couple an activator–substrate system to a growing domain (Figure 5a).

Specifically, we consider isotropic growth of the domain [81] (all parts of the domain grow in a similar fashion) and the subunit pool density P grows homogeneously with a constant rate. Due to domain growth, both the system size $L(t)$ and the total amount of building block pool, $N(t) = \int (S+P)dx$, are time-dependent. The growth of the system introduces local flow and dilution of both S and P [81]. The coupled dynamics of S and P are given by (Figure 5a)

$$\partial_t S + \frac{\dot{r}}{r}(x\partial_x S + S) = D_s \partial_x^2 S + k_{\mathrm{on}} P f(S) - k_{\mathrm{off}} S \,, \tag{4}$$

$$\partial_t P + \frac{\dot{r}}{r}(x\partial_x P + P) = D_p \partial_x^2 P - k_{\text{on}} P f(S) + k_{\text{off}} S + G\,, \tag{5}$$

where G is the growth rate of the subunit pool and the domain growth function $r(t)$ is defined as $L(t) = L(0)r(t)$, where $L(0)$ is the initial system size. We write the assembly rate function as $f(S) = \kappa_0 + S^n/(S^n + S_0^n)$, where κ_0 defines the size-independent rate of assembly of S. Motivated by exponentially growing cells and tissues, we specifically consider the case of exponential growth in system size, such that $r(t) = e^{\alpha t}$ with α the growth rate. As the macromolecular composition of cells scales with the cell size, we assume that the total amount of building blocks, $\int P dx$, grows at a rate proportional to system size L, resulting in a constant rate of growth G. While the formation of patterns occurs beyond a critical system size $L > L^*$, the stability of the pattern depends on the interplay between the rates of growth-induced dilution, synthesis of the subunit pool P, and autocatalysis of S.

6.1. Case 1: Transient Polarity Pattern Due to Growth-Induced Dilution

When the subunit pool grows at a rate much slower than the rate of system growth $\tilde{G} \ll \alpha$ (where $\tilde{G} = GL(0)$), the structure is formed transiently and dissolves after a critical time T^c. The initial slow growth of the system size allows the formation of pattern beyond a critical size L^*. However, as dL/dt increases rapidly (due to exponential growth) and becomes much faster than the rate of pool synthesis, the subunit density starts decreasing. Below a critical density of subunits, the pattern dissolves and the system reaches a homogeneous state (Figure 5b, left). Transient structure formation has been observed in slime mould [82] and during mammalian development [83], and modelled using stochastic RD systems [84]. Here, we argue that system growth can also induce such transient polarity formation.

The patterned state makes a transition to a homogeneous state at a time T^c when the system size reaches L^c. The critical time for the transition to the homogeneous state, T^c, is determined by the parameters of the feedback motif ($k_{\text{on}}/k_{\text{off}}$, κ_0) and the growth rates α and $\tilde{G}$ (Figure 5c). The lifetime of the pattern, T^c, and the system size at transition to the homogeneous state, L^c, can be tuned independently of each other by modulating κ_0, α and $\tilde{G}$ (Figure 5d). Controlling the lifetime of polarity patterns is essential for regulating developmental transitions, and further experimental studies are essential to uncover such control mechanisms.

6.2. Case 2: Pattern Scaling Due to Proportional Growth of System Size and the Subunit Pool

When the subunit pool grows at a rate comparable to the rate of growth of system size, $\tilde{G} \sim \alpha$, the subunits can reach a homeostatic density in time. As a result, the patterns formed during growth do not dissolve. Strong autocatalytic growth prevents delocalisation of the early pattern and prevents the possibility of period doubling as seen in Schnakenberg kinetics and Gierer and Meinhardt model [81]. This leads to a dynamic pattern scaling behaviour where the size of the pattern scales with the size of the system (Figure 5b, middle). This mechanism of scaling is notably different from the morphogen gradient scaling [85]. Here, pattern scaling is a consequence of system growth where the polarity pattern that does not change qualitatively with system size.

6.3. Case 3: Pattern Splitting

When the rate of autocatalysis is sufficiently high, $\kappa_0 \neq 0$, and the subunit pool and system size grow at similar rates, $\tilde{G} \gtrsim \alpha$, dynamic pattern splitting can emerge in the context of growth (Figure 5b, right). Here, slower positive feedback weakens the long-range inhibition arising as the consequence of pool depletion, allowing new peaks to emerge, in contrast to case 2. Adaptive benefits of pattern splitting may be multiple, for example, allowing for the emergence of sequential patterning, and in maintaining relative stasis in local patterns upon domain extension.

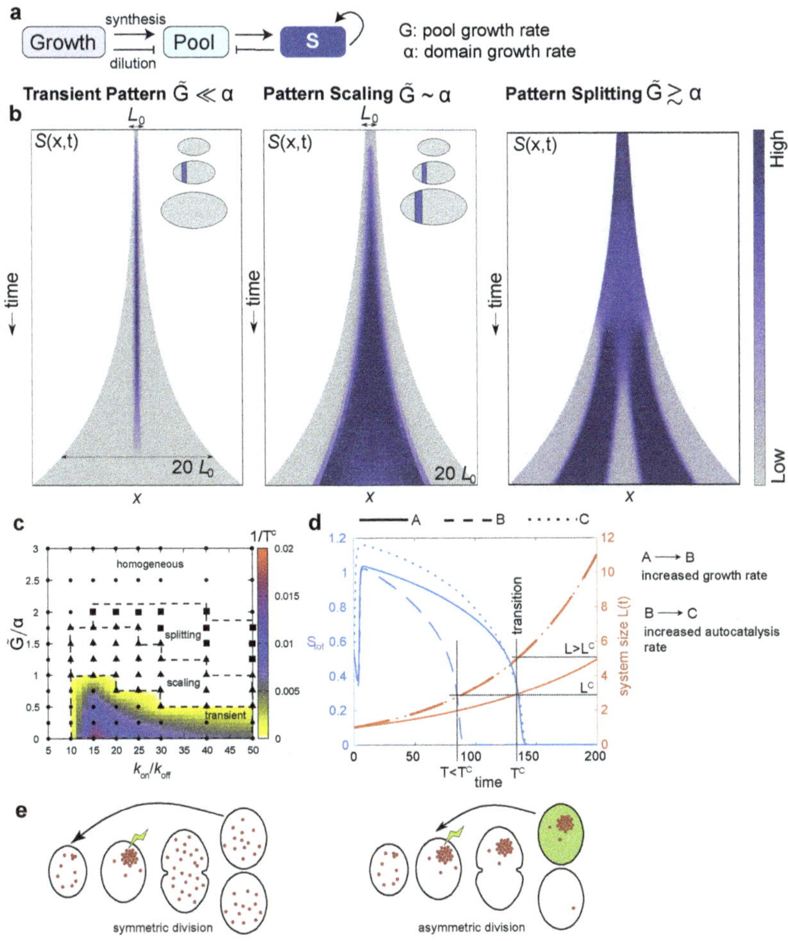

Figure 5. Pattern scaling, splitting and transient pattern formation in growing domains. (**a**) Feedback motif for an activator–substrate system coupled to a growing domain. (**b**) (Left) When subunits are produced at a rate slower than the rate of domain growth, growth-induced dilution leads to transient pattern establishment. (Middle) When the production of subunit pool occurs at a rate comparable to system size growth, the pattern formed grows in proportion to system size, exhibiting a dynamic scaling behaviour. This is different from sequential pattern formation as the polarity is preserved during growth. (Right) In the case of strong autocatalysis of S, the pattern spontaneously splits. (**c**) Phase diagram for pattern formation as functions of pool growth rate relative to the system, $\tilde{G}/\alpha$, and k_{on}/k_{off}. Colormap denotes the inverse of the pattern lifetime, $1/T^c$. (**d**) Time evolution of structure size $S_{tot} = \int_0^L S dx$ (blue) and system size L (red) for the case of transient pattern formation. The lifetime of the pattern T^c is coupled to the system size at transition to the homogeneous state, L^c. They can be tuned independently of each other, for example, by changing growth rate α (case B) where only the transition time changes, or by changing autocatalysis rate (case C) where T^c remains the same but transition happens at a different system size. (**e**) Tunability of pattern lifetime can be utilised as a control mechanism for symmetric and asymmetric cell division (in terms of polarity protein content). When the pattern is transient (left) the dissolution of the structure will make the daughter cells symmetric in fate, containing the same amount of polarity proteins. If the pattern persists (right) then the division will lead to asymmetric fate inheritance. Parameter values: $D_P = 1$, $D_S = 0.005$, $\kappa_0 = 0.85$, $S_0^2 = 10$, $\alpha = 0.01$, $L(0) = 1$, and total pool density $\int(P+S)dx/L = 2$, with G and k_{on}/k_{off} variable.

7. Using Growth as a Timer: Transient Symmetry Breaking at Intermediate Size

Analysis of our minimal model shows that a timer-like control of pattern is an emergent property of symmetry-breaking systems that couple pool size to system volume. We speculate that transient symmetry breaking may serve as a control strategy to mediate shifts between symmetric and asymmetric cell division in stem cell homeostasis. In particular, the biphasic nature of transient symmetry breaking scheme can help rationalise the equal distribution of determinants in the face of unequal nature of cell division. Suppose this system underlies the establishment of cell polarity required for asymmetric division. A critical cell size would allow polarity to be established, preventing precocious cell division. After the cell size timer has elapsed, and the cell enters the dissolution phase, polarity proteins will return to the fast-mixing pool (Figure 5e, left). If cell polarity can have some effect on differential daughter cell fate, such a scheme would dissolve any bias prior to division.

The significance of a mechanism like this can be understood in cases where cell lineages undergo switches between symmetric and asymmetric cell division. Cases 1 and 2 of our model are identical besides the relative rates of pool synthesis, system size growth and autocatalysis. Therefore, tuning the coupling between pool production rate and system size can regulate transitions between reversible polarisation (case 1) and irreversible polarisation (case 2), wherein polarity is maintained through a cell division event, maintaining a bias in determinants and seeding differential daughter cell fate (Figure 5e, right). In situations where such switches between symmetric and asymmetric division are dynamic, regulating the extent of growth-induced dilution to effect switches between transient versus irreversible polarisation may be an optimal control strategy.

Given the switch between reversible and irreversible symmetry breaking in our minimal model is governed by a single parameter change, it is tempting to speculate that such a mechanism may be responsible for the mixed modes of stem cell proliferation, which in many systems are seemingly stochastic. Under such a scenario, the switch between these division modes may be noisy at the level of individual cell decisions, given the requirement of being poised near the transition point. However, given fate decisions and patterns of cell division are known to be influenced by signals emanating from stem cells or their progeny in many systems, such a model would confer plasticity and robustness in stem cell homeostasis at the level of the population. We stress that this mechanism remains a theoretical prediction and are intrigued as to whether such a control strategy is indeed utilised in nature.

8. Overcoming Size Constraints: Scaling Patterns in Growing Systems

Not all biological systems that display symmetry breaking also show size-dependent pattern formation. Indeed, it may be adaptive for systems to canalise their patterning mode irrespective of size. This is a feature of case 2 of our minimal model (Figure 5b, middle), which features commensurate growth of pool size and system size, leading to scale invariance upon domain growth: if the system breaks symmetry to form a single structure at a smaller size, upon isotropic growth, the system maintains a single structure which grows in proportion to the domain as a whole. This motif utilises the symmetry breaking capacity of reaction–diffusion systems but subverts the feature of intrinsic wavelengths characteristic of traditional Turing circuits. In this section, we delineate two potential modes of scale-invariant symmetry breaking systems—one with history dependence, and one without—and argue that such systems display adaptive features in certain contexts.

8.1. Autocatalysis as a Mechanism to Preserve Patterns in the Face of Growth

Patterning in developing systems is almost invariably proceeded by growth, which is often proportional to the initial pattern. Traditional morphogen gradient hypotheses [86] have implicitly assumed that the tissue is initially patterned when it is small and subsequently undergoes growth, facilitating proportional extension of the pattern. This two-phase model of patterning, where cell fates are assigned during an initial patterning phase, face the challenge of noise: while growth can lock in the lower positional error entailed by patterning in small fields of cells, this error cannot be reduced through growth. Accordingly, small errors in boundary position can be amplified upon growth, demanding the read-out of positional information at early stages is exquisitely tuned. If, however, fates are assigned in a self-organised manner, as in symmetry-breaking the systems we overview in our minimal model, this hard limit on noise in boundary positioning can be surpassed. Provided the symmetry breaking system scales with domain size, absolute noise in boundary position if anything reduces with growth; self-organised systems such as these continually refine boundaries throughout growth, rather than amplifying noise in initial specification.

Case 2 of our minimal model allows for pattern scaling via proportional growth of pool and domain size, and autocatalysis, which in effect instils history-dependence in pattern formation, thus helping to preserve proportions. Therefore, the pattern generated at small domain sizes is preserved upon elongation. Given the diffusion length scale shortens with respect to relative domain size upon growth, boundaries sharpen over time. In the context of a developing field of cells, this autocatalysis could represent positive feedback in master transcription factors or indeed epigenetic changes, which allow cells and their progeny to remember past states. Therefore, such a model may help provide alternative mechanisms for scaling of patterns with growth, whereby initial stages establish the crude pattern (e.g., number and position of structures), which is in turn refined over time. The hallmark of actively scaling processes such as these is the reduction of noise in boundary positioning, i.e., violating the data-processing inequality [87].

8.2. Expander-Coupled Systems Can Scale Patterns to Domain Size Irrespective of History

While symmetry-breaking schemes incorporating autocatalysis show benefits of maintaining patterns with growth, certain systems may require patterning to be scale-invariant without the requirement of time-dependence. This is exemplified in regenerating systems, which are able to regrow organs or entire organisms in the correct proportions, in spite of drastically different starting sizes. Recent theoretical work has advanced understandings of how scale-invariant symmetry breaking could operate. Werner et al. [88] proposed that a third component is required, analogous to expanders in morphogen gradient scaling, which dynamically modulates patterning wavelength as a function of system size by tuning levels of pattern forming molecules. This model demonstrated time-independent scaling across several orders of magnitude differences in domain size. While the model is based on a traditional activator-inhibitor model, the scheme is generalisable to other modes of expander-mediated modulation and other symmetry breaking motifs such as substrate depletion.

9. Discussion

In this perspective, we presented a minimal model for symmetry breaking to serve as a unifying framework to understand pattern formation in the context of timing and growth. We argue that systems that display positive feedback in activator recruitment, drawn from a limiting pool, can yield spontaneous symmetry breaking. This basic scheme is mathematically akin to other RD mechanisms including activator–inhibitor or substrate-depletion motifs, all relying on a common logic of local activation and long-range inhibition. Thus, the insights gleaned from the phenomenological behaviour of this system is applicable to diverse systems.

Across the cases of the minimal model we consider, we observe a hard size limit on pattern formation: below a critical size, diffusive dispersion overwhelms the capacity to break symmetry. Given developing systems across scales typically display patterning and growth occurring in unison, if this critical size is within biologically meaningful length scales, such behaviour can elicit qualitative changes in patterning: growth above a critical size leads to a bifurcation, whereby the system transitions from unpolarised to polarised. Viewing growth as a control parameter of the system that increases system size at a predictable rate, developmental systems can utilise this bifurcation to enact developmental transitions at the right place and time. As a generic by-product of symmetry breaking systems, we predict that this time-keeping mechanism may be more abundant than anticipated. We note that this feature is the most generic among RD models of pattern formation, and among the different cases of our minimal model: the diffusion length-scale sets a physical limit on pattern formation.

Beyond this first bifurcation, our mass-conserved RD model predicts different dynamic behaviours depending on the regulatory motifs. These include sequential pattern formation, transient pattern formation, pattern scaling, and pattern splitting in growing systems. In line with the well-established literature on domain size in Turing patterns [23], our model predicts clock-like sequential pattern formation (Figures 2 and 4): as the system grows larger, given patterning wavelength is intrinsic to the system, the domain can accommodate multiple structures. An important dynamical consequence of this is temporal ordering: growth elicits consecutive bifurcations, resulting in sequential patterning, shown to be instrumental in neuronal cell (bi)polarity [69], and joint patterning in digits [80]. Alternatively, growth-induced pool dilution can drive systems back towards an unpolarised state, allowing for transient pattern formation at intermediate size (Figure 5). Thus, an alteration in growth regulation can yield timer-like dynamics, which we hypothesise may be important in orchestrating switches between asymmetric and symmetric stem cell division modes. A qualitatively different behaviour upon continued growth is scale invariance, whereby the proportions of the pattern are maintained upon domain elongation. Scale-invariant systems show switch-like dynamics, becoming time-independent after the first bifurcation. We argue that such behaviour could allow patterned tissues to maintain proportions upon proliferation, where self-organisation continually refines boundary position instead of stretching noise in initial specification.

Our reaction–diffusion framework for understanding developmental time in terms of size-dependent symmetry breaking is generalisable beyond the systems that couple increases in size to developmental transitions via biochemical circuits. First, decreases in system size can also be utilised by developmental systems to temporal transitions. For example, the transition from asymmetric to symmetric division in the P-lineage of *C. elegans* can be understood in terms of sequential reductions in cell volume pushing the system over the critical cell size threshold for polarisation. Second, the organising principle of Turing-like pattern formation—local-activation and long-range inhibition—extends beyond systems based solely on chemical cross-talk [38]: pattern formation can emerge from cell–cell interactions or mechanical instabilities [89–91]. While we restricted our focus in this paper to biochemical systems, future work should attempt to unify these results with mechanically driven size-dependent symmetry breaking. Indeed, we may see strong parallels in how nature utilises chemical or mechanical instabilities to regulate the timing of developmental transitions. We hope that our proposed strategies for time-keeping in natural living systems can also provide inspirations for engineering of synthetic circuits with tunable dynamics.

Author Contributions: Conceptualization, J.C.S., D.S.B. and S.B.; methodology, D.S.B.; validation, J.C.S., D.S.B. and S.B.; formal analysis, D.S.B.; investigation, J.C.S., D.S.B. and S.B.; resources, S.B.; data curation, J.C.S., and D.S.B.; writing—original draft preparation, J.C.S., D.S.B. and S.B.; writing—review and editing, J.C.S., D.S.B. and S.B.; visualization, J.C.S., D.S.B. and S.B.; supervision, S.B.; project administration, S.B.; funding acquisition, S.B. All authors have read and agreed to the published version of the manuscript.

Funding: SB acknowledges funding from Royal Society University Research Fellowship URF/R1/180187, Human Frontiers Science Program (HFSP) grant number RGY0073/2018, and Carnegie Mellon startup funds.

Acknowledgments: The authors thank Andrew Goryachev and the anonymous reviewers for many useful comments and suggestions.

Conflicts of Interest: The authors declare no conflicts of interest. The funders had no role in the design of the study; in the collection, analyses or interpretation of data; in the writing of the manuscript; or in the decision to publish the results.

References

1. Meinhardt, H. Models of biological pattern formation: From elementary steps to the organization of embryonic axes. *Curr. Top. Dev. Biol.* **2008**, *81*, 1–63. [PubMed]
2. Green, J.B.; Sharpe, J. Positional information and reaction-diffusion: Two big ideas in developmental biology combine. *Development* **2015**, *142*, 1203–1211. [CrossRef] [PubMed]

3. Saiz-Lopez, P.; Chinnaiya, K.; Campa, V.M.; Delgado, I.; Ros, M.A.; Towers, M. An intrinsic timer specifies distal structures of the vertebrate limb. *Nat. Commun.* **2015**, *6*, 1–9. [CrossRef] [PubMed]
4. Briscoe, J.; Small, S. Morphogen rules: Design principles of gradient-mediated embryo patterning. *Development* **2015**, *142*, 3996–4009. [CrossRef] [PubMed]
5. Waters, C.M.; Bassler, B.L. Quorum sensing: Cell-to-cell communication in bacteria. *Annu. Rev. Cell Dev. Biol.* **2005**, *21*, 319–346. [CrossRef] [PubMed]
6. Jörg, D.J.; Kitadate, Y.; Yoshida, S.; Simons, B.D. Competition for Stem Cell Fate Determinants as a Mechanism for Tissue Homeostasis. *arXiv* **2019**, arXiv:1901.03903.
7. Altschuler, S.J.; Angenent, S.B.; Wang, Y.; Wu, L.F. On the spontaneous emergence of cell polarity. *Nature* **2008**, *454*, 886–889. [CrossRef]
8. Chara, O.; Tanaka, E.M.; Brusch, L. Mathematical modeling of regenerative processes. In *Current Topics in Developmental Biology*; Elsevier: Amsterdam, The Netherlands, 2014; Volume 108, pp. 283–317.
9. Ebisuya, M.; Briscoe, J. What does time mean in development? *Development* **2018**, *145*, dev164368. [CrossRef] [PubMed]
10. Turing, A.M. The chemical basis of morphogenesis. *Phil. Trans. R. Soc. Lond.* **1952**, *237*, 37–72.
11. Kondo, S.; Miura, T. Reaction-Diffusion Model as a Framework for Understanding Biological Pattern Formation. *Science* **1999**, *329*, 1616–1620. [CrossRef]
12. Butty, A.; Perrinjaquet, N.; Petit, A.; Jaquenoud, M.; Segall, J.; Hofmann, K.; Zwahlen, C.; Peter, M. A positive feedback loop stabilizes the guanine-nucleotide exchange factor Cdc24 at sites of polarization. *EMBO J.* **2002**, *21*, 1565–1576. [CrossRef]
13. Kauffman, S.A. Pattern formation in the Drosophila embryo. *Phil. Trans. R. Soc. Lond. B* **1981**, *295*, 567–594.
14. Meinhardt, A.; Gierer, J. Applications of a theory of biological pattern formation based on lateral inhibition. *J. Cell Sci.* **1974**, *15*, 321–346. [PubMed]
15. Sick, S.; Reinker, S.; Timmer, J.; Schlake, T. WNT and DKK determine hair follicle spacing through a reaction-diffusion mechanism. *Science* **2006**, *314*, 1447–1450. [CrossRef] [PubMed]
16. Nakamura, T.; Mine, N.; Nakaguchi, E.; Mochizuki, A.; Yamamoto, M.; Yashiro, K.; Meno, C.; Hamada, H. Generation of robust left-right asymmetry in the mouse embryo requires a self-enhancement and lateral-inhibition system. *Dev. Cell* **2006**, *11*, 495–504. [CrossRef] [PubMed]
17. Newman, S.A.; Frisch, H.L. Dynamics of skeletal pattern formation in developing chick limb. *Science* **1979**, *205*, 662–668. [CrossRef] [PubMed]
18. Miura, T.; Shiota, K. TGFbeta2 acts as an "activator" molecule in reaction-diffusion model and is involved in cell sorting phenomenon in mouse limb micromass culture. *Dev. Dyn.* **2000**, *217*, 241–249. [CrossRef]
19. Raspopovic, J.; Marcon, L.; Russo, L.; Sharpe, J. Digit patterning is controlled by a Bmp-Sox9-Wnt Turing network modulated by morphogen gradients. *Science* **2014**, *345*, 566–570. [CrossRef]
20. Meinhardt, H.; Boer, P.A.J.d. Pattern formation in Escherichia coli: A model for the pole-to-pole oscillations of Min proteins and the localization of the division site. *Proc. Natl. Acad. Sci. USA* **2001**, *98*, 14202–14207. [CrossRef] [PubMed]
21. Bement, W.M.; Leda, M.; Moe, A.M.; Kita, A.M.; Larson, M.E.; Golding, A.E.; Pfeuti, C.; Su, K.C.; Miller, A.L.; Goryachev, A.B.; et al. Activator–inhibitor coupling between Rho signalling and actin assembly makes the cell cortex an excitable medium. *Nat. Cell Biol.* **2015**, *17*, 1471–1483. [CrossRef]
22. Michaux, J.B.; Robin, F.B.; McFadden, W.M.; Munro, E.M. Excitable RhoA dynamics drive pulsed contractions in the early C. elegans embryo. *J. Cell Biol.* **2018**, *217*, 4230–4252. [CrossRef] [PubMed]
23. Koch, A.; Meinhardt, H. Biological pattern formation: From basic mechanisms to complex structures. *Rev. Mod. Phys.* **1994**, *66*, 1481. [CrossRef]
24. Marcon, L.; Sharpe, J. Turing patterns in development: What about the horse part? *Curr. Opin. Genet. Dev.* **2012**, *22*, 578–584. [CrossRef] [PubMed]
25. Menshykau, D.; Kraemer, C.; Iber, D. Branch mode selection during early lung development. *PLoS Comput. Biol.* **2012**, *8*, e1002377. [CrossRef] [PubMed]
26. Mori, Y.; Jilkine, A.; Edelstein-Keshet, A. Wave-Pinning and Cell Polarity from a Bistable Reaction-Diffusion System. *Biophys J.* **2008**, *94*, 3684–3697. [CrossRef] [PubMed]
27. Halatek, J.; Brauns, F.; Frey, E. Self-organization principles of intracellular pattern formation. *Phil. Trans. R. Soc. Lond B* **2018**, *373*, 20170107. [CrossRef]

28. Munro, E.; Nance, J.; Priess, J.R. Cortical flows powered by asymmetrical contraction transport PAR proteins to establish and maintain anterior-posterior polarity in the early C. elegans embryo. *Dev. Cell* **2004**, *7*, 413–424. [CrossRef]
29. Goehring, N.W.; Trong, P.K.; Bois, J.S.; Chowdhury, D.; Nicola, E.M.; Hyman, A.A.; Grill, S.W. Polarization of PAR proteins by advective triggering of a pattern-forming system. *Science* **2011**, *334*, 1137–1141. [CrossRef]
30. Tostevin, F.; Howard, M. Modeling the establishment of PAR protein polarity in the one-cell *C. elegans* embryo. *Biophys. J.* **2008**, *95*, 4512–4522. [CrossRef]
31. Goryachev, A.B.; Pokhilko, A.V. Dynamics of Cdc42 network embodies a Turing-type mechanism of yeast cell polarity. *FEBS Lett.* **2008**, *582*, 1437–1443. [CrossRef]
32. Savage, N.S.; Layton, A.T.; Lew, D.J. Mechanistic mathematical model of polarity in yeast. *Mol. Biol. Cell* **2012**, *23*, 1998–2013. [CrossRef] [PubMed]
33. Goryachev, A.B.; Leda, M. Many roads to symmetry breaking: Molecular mechanisms and theoretical models of yeast cell polarity. *Mol. Biol. Cell* **2017**, *28*, 370–380. [CrossRef] [PubMed]
34. Howard, M.; Rutenberg, A.D.; de Vet, S. Dynamic compartmentalization of bacteria: Accurate division in *E. coli*. *Phys. Rev. Lett.* **2001**, *87*, 278102. [CrossRef] [PubMed]
35. Kruse, K. A dynamic model for determining the middle of Escherichia coli. *Biophys. J.* **2002**, *82*, 618–627. [CrossRef]
36. Huang, K.C.; Meir, Y.; Wingreen, N.S. Dynamic structures in Escherichia coli: Spontaneous formation of MinE rings and MinD polar zones. *Proc. Natl. Acad. Sci. USA* **2003**, *100*, 12724–12728. [CrossRef] [PubMed]
37. Gierer, A.; Meinhardt, H. A theory of biological pattern formation. *Kybernetik* **1972**, *12*, 30–39. [CrossRef] [PubMed]
38. Hiscock, T.W.; Megason, S.G. Mathematically guided approaches to distinguish models of periodic patterning. *Development* **2015**, *142*, 409–419. [CrossRef]
39. Frohnhöfer, H.G.; Krauss, J.; Maischein, H.M.; Nüsslein-Volhard, C. Iridophores and their interactions with other chromatophores are required for stripe formation in zebrafish. *Development* **2013**, *140*, 2997–3007. [CrossRef]
40. Murray, J.; Oster, G. Cell traction models for generating pattern and form in morphogenesis. *J. Math. Biol.* **1984**, *19*, 265–279. [CrossRef]
41. Howard, J.; Grill, S.W.; Bois, J.S. Turing's next steps: The mechanochemical basis of morphogenesis. *Nat. Rev. Mol. Cell Biol.* **2011**, *12*, 392–398. [CrossRef]
42. Granero, M.; Porati, A.; Zanacca, D. A bifurcation analysis of pattern formation in a diffusion governed morphogenetic field. *J. Math. Biol.* **1977**, *4*, 21–27. [CrossRef] [PubMed]
43. Zadorin, A.S.; Rondelez, Y.; Gines, G.; Dilhas, V.; Urtel, G.; Zambrano, A.; Galas, J.C.; Estevez-Torres, A. Synthesis and materialization of a reaction–diffusion French flag pattern. *Nat. Chem.* **2017**, *9*, 990–996. [CrossRef] [PubMed]
44. Chiou, J.G.; Ramirez, S.A.; Elston, T.C.; Witelski, T.P.; Schaeffer, D.G.; Lew, D.J. Principles that govern competition or co-existence in Rho-GTPase driven polarization. *PLoS Comput. Biol.* **2018**, *14*, e1006095. [CrossRef] [PubMed]
45. Otsuji, M.; Ishihara, S.; Co, C.; Kaibuchi, K.; Mochizuki, A.; Kuroda, S. A mass conserved reaction–diffusion system captures properties of cell polarity. *PLoS Comput. Biol.* **2007**, *3*, e108. [CrossRef]
46. Banerjee, D.S.; Banerjee, S. Size regulation of multiple organelles competing for a shared subunit pool. *bioRxiv* **2020**, bioRxiv:902783. [CrossRef]
47. Brauns, F.; Halatek, J.; Frey, E. Phase-space geometry of reaction–diffusion dynamics. *arXiv* **2018**, arXiv:1812.08684.
48. Goehring, N.W.; Hyman, A.A. Organelle growth control through limiting pools of cytoplasmic components. *Curr. Biol.* **2012**, *22*, R330–R339. [CrossRef]
49. Hoege, C.; Hyman, A.A. Principles of PAR polarity in Caenorhabditis elegans embryos. *Nat. Rev. Mol. Cell Biol.* **2013**, *14*, 315–322. [CrossRef]
50. Kemphues, K.J.; Priess, J.R.; Morton, D.G.; Cheng, N. Identification of genes required for cytoplasmic localization in early C. elegans embryos. *Cell* **1988**, *52*, 311–320. [CrossRef]
51. Goldstein, B.; Hird, S.N. Specification of the anteroposterior axis in Caenorhabditis elegans. *Development* **1996**, *122*, 1467–1474.

52. Nguyen-Ngoc, T.; Afshar, K.; Gönczy, P. Coupling of cortical dynein and Gα proteins mediates spindle positioning in Caenorhabditis elegans. *Nat. Cell Biol.* **2007**, *9*, 1294–1302. [CrossRef] [PubMed]
53. Gönczy, P. Mechanisms of asymmetric cell division: Flies and worms pave the way. *Nat. Rev. Mol. Cell Biol.* **2008**, *9*, 355–366. [CrossRef] [PubMed]
54. Motegi, F.; Zonies, S.; Hao, Y.; Cuenca, A.A.; Griffin, E.; Seydoux, G. Microtubules induce self-organization of polarized PAR domains in Caenorhabditis elegans zygotes. *Nat. Cell Biol.* **2011**, *13*, 1361–1367. [CrossRef] [PubMed]
55. Sailer, A.; Anneken, A.; Li, Y.; Lee, S.; Munro, E. Dynamic opposition of clustered proteins stabilizes cortical polarity in the C. elegans zygote. *Dev. Cell* **2015**, *35*, 131–142. [CrossRef] [PubMed]
56. Hubatsch, L.; Peglion, F.; Reich, J.D.; Rodrigues, N.T.; Hirani, N.; Illukkumbura, R.; Goehring, N.W. A cell-size threshold limits cell polarity and asymmetric division potential. *Nat. Phys.* **2019**, *15*, 1078–1085. [CrossRef] [PubMed]
57. Sulston, J.E.; Schierenberg, E.; White, J.G.; Thomson, J. The embryonic cell lineage of the nematode Caenorhabditis elegans. *Dev. Biol.* **1983**, *100*, 64–119. [CrossRef]
58. Chiou, J.g.; Balasubramanian, M.K.; Lew, D.J. Cell polarity in yeast. *Annu. Rev. Cell Dev. Biol.* **2017**, *33*, 77–101. [CrossRef] [PubMed]
59. Di Talia, S.; Skotheim, J.M.; Bean, J.M.; Siggia, E.D.; Cross, F.R. The effects of molecular noise and size control on variability in the budding yeast cell cycle. *Nature* **2007**, *448*, 947–951. [CrossRef] [PubMed]
60. Ozbudak, E.M.; Becskei, A.; Oudenaarden, A. A system of counteracting feedback loops regulates Cdc42p activity during spontaneous cell polarization. *Dev. Cell* **2005**, *9*, 565–571. [CrossRef] [PubMed]
61. Okada, S.; Leda, M.; Hanna, J.; Savage, N.S.; Bi, E.; Goryachev, A.B. Daughter Cell Identity Emerges from the Interplay of Cdc42, Septins, and Exocytosis. *Dev. Cell* **2013**, *26*, 148–161. [CrossRef]
62. Turner, J.J.; Ewald, J.C.; Skotheim, J.M. Cell size control in yeast. *Curr. Biol.* **2012**, *22*, R350–R359. [CrossRef] [PubMed]
63. Howell, A.S.; Savage, N.S.; Johnson, S.A.; Bose, I.; Wagner, A.W.; Zyla, T.R.; Nijhout, H.F.; Reed, M.C.; Goryachev, A.B.; Lew, D.J. Singularity in Polarization: Rewiring Yeast Cells to Make Two Buds. *Cell* **2009**, *139*, 731–743. [CrossRef] [PubMed]
64. Maroto, M.; Bone, R.A.; Dale, J.K. Somitogenesis. *Development* **2012**, *139*, 2453–2456. [CrossRef] [PubMed]
65. Sarrazin, A.F.; Peel, A.D.; Averof, M. A segmentation clock with two-segment periodicity in insects. *Science* **2012**, *336*, 338–341. [CrossRef]
66. LoTurco, J.J.; Bai, J. The multipolar stage and disruptions in neuronal migration. *Trends Neurosci.* **2006**, *29*, 407–413. [CrossRef]
67. Noctor, S.C.; Martínez-Cerdeño, V.; Ivic, L.; Kriegstein, A.R. Cortical neurons arise in symmetric and asymmetric division zones and migrate through specific phases. *Nat. Neurosci.* **2004**, *7*, 136–144. [CrossRef]
68. De Anda, F.C.; Gärtner, A.; Tsai, L.H.; Dotti, C.G. Pyramidal neuron polarity axis is defined at the bipolar stage. *J. Cell Sci.* **2008**, *121*, 178–185. [CrossRef]
69. Menchón, S.A.; Gärtner, A.; Román, P.; Dotti, C.G. Neuronal (bi) polarity as a self-organized process enhanced by growing membrane. *PLoS ONE* **2011**, *6*, e24190. [CrossRef] [PubMed]
70. Van den Brink, S.C.; Baillie-Johnson, P.; Balayo, T.; Hadjantonakis, A.K.; Nowotschin, S.; Turner, D.A.; Arias, A.M. Symmetry breaking, germ layer specification and axial organisation in aggregates of mouse embryonic stem cells. *Development* **2014**, *141*, 4231–4242. [CrossRef] [PubMed]
71. Arnold, S.J.; Robertson, E.J. Making a commitment: Cell lineage allocation and axis patterning in the early mouse embryo. *Nat. Rev. Mol. Cell Biol.* **2009**, *10*, 91–103. [CrossRef] [PubMed]
72. Harrison, S.E.; Sozen, B.; Christodoulou, N.; Kyprianou, C.; Zernicka-Goetz, M. Assembly of embryonic and extraembryonic stem cells to mimic embryogenesis in vitro. *Science* **2017**, *356*, eaal1810. [CrossRef] [PubMed]
73. Warmflash, A.; Sorre, B.; Etoc, F.; Siggia, E.D.; Brivanlou, A.H. A method to recapitulate early embryonic spatial patterning in human embryonic stem cells. *Nat. Methods* **2014**, *11*, 847. [CrossRef] [PubMed]
74. Ten Berge, D.; Koole, W.; Fuerer, C.; Fish, M.; Eroglu, E.; Nusse, R. Wnt signaling mediates self-organization and axis formation in embryoid bodies. *Cell Stem Cell* **2008**, *3*, 508–518. [CrossRef] [PubMed]
75. Sagy, N.; Slovin, S.; Allalouf, M.; Pour, M.; Savyon, G.; Boxman, J.; Nachman, I. Prediction and control of symmetry breaking in embryoid bodies by environment and signal integration. *Development* **2019**, *146*, dev181917. [CrossRef]

76. Turner, D.A.; Girgin, M.; Alonso-Crisostomo, L.; Trivedi, V.; Baillie-Johnson, P.; Glodowski, C.R.; Hayward, P.C.; Collignon, J.; Gustavsen, C.; Serup, P.; et al. Anteroposterior polarity and elongation in the absence of extra-embryonic tissues and of spatially localised signalling in gastruloids: mammalian embryonic organoids. *Development* **2017**, *144*, 3894–3906. [CrossRef] [PubMed]
77. Morgani, S.M.; Hadjantonakis, A.K. Signaling regulation during gastrulation: Insights from mouse embryos and in vitro systems. *eLife* **2019**, *137*, 391–431.
78. Juan, H.; Hamada, H. Roles of nodal-lefty regulatory loops in embryonic patterning of vertebrates. *Genes Cells* **2001**, *6*, 923–930. [CrossRef]
79. Shahbazi, M.N.; Siggia, E.D.; Zernicka-Goetz, M. Self-organization of stem cells into embryos: A window on early mammalian development. *Science* **2019**, *364*, 948–951. [CrossRef]
80. Cornwall Scoones, J.; Hiscock, T.W. A dot-stripe Turing model of joint patterning in the tetrapod limb. *Development* **2020**, *147*, dev183699. [CrossRef]
81. Crampin, E.J.; Gaffney, E.A.; Maini, P.K. Reaction and diffusion on growing domains: Scenarios for robust pattern formation. *Bull. Math. Biol.* **1999**, *61*, 1093–1120. [CrossRef]
82. Postma, M.; Roelofs, J.; Goedhart, J.; Gadella, T.W.; Visser, A.J.; Van Haastert, P.J. Uniform cAMP stimulation of Dictyostelium cells induces localized patches of signal transduction and pseudopodia. *Mol. Biol. Cell* **2003**, *14*, 5019–5027. [CrossRef] [PubMed]
83. Erzurumlu, R.S.; Jhaveri, S.; Benowitz, L.I. Transient patterns of GAP-43 expression during the formation of barrels in the rat somatosensory cortex. *J. Comp. Neurol.* **1990**, *292*, 443–456. [CrossRef] [PubMed]
84. Hecht, I.; Kessler, D.A.; Levine, H. Transient localized patterns in noise-driven reaction-diffusion systems. *Phys. Rev. Lett.* **2010**, *104*, 158301. [CrossRef] [PubMed]
85. Ben-Zvi, D.; Shilo, B.Z.; Barkai, N. Scaling of morphogen gradients. *Curr. Opin. Genet. Dev.* **2011**, *21*, 704–710. [CrossRef] [PubMed]
86. Wolpert, L. Positional information and the spatial pattern of cellular differentiation. *J. Theor. Biol.* **1969**, *25*, 1–47. [CrossRef]
87. Kinney, J.B.; Atwal, G.S. Equitability, mutual information, and the maximal information coefficient. *Proc. Natl. Acad. Sci. USA* **2014**, *111*, 3354–3359. [CrossRef]
88. Werner, S.; Stückemann, T.; Amigo, M.B.; Rink, J.C.; Jülicher, F.; Friedrich, B.M. Scaling and regeneration of self-organized patterns. *Phys. Rev. Lett.* **2015**, *114*, 138101. [CrossRef] [PubMed]
89. Mark, S.; Shlomovitz, R.; Gov, N.S.; Poujade, M.; Grasland-Mongrain, E.; Silberzan, P. Physical model of the dynamic instability in an expanding cell culture. *Biophys. J.* **2010**, *98*, 361–370. [CrossRef] [PubMed]
90. Banerjee, S.; Utuje, K.J.; Marchetti, M.C. Propagating stress waves during epithelial expansion. *Phys. Rev. Lett.* **2015**, *114*, 228101. [CrossRef] [PubMed]
91. Ravasio, A.; Le, A.P.; Saw, T.B.; Tarle, V.; Ong, H.T.; Bertocchi, C.; Mège, R.M.; Lim, C.T.; Gov, N.S.; Ladoux, B. Regulation of epithelial cell organization by tuning cell–substrate adhesion. *Integr. Biol.* **2015**, *7*, 1228–1241. [CrossRef]

Article

Fission Yeast Polarization: Modeling Cdc42 Oscillations, Symmetry Breaking, and Zones of Activation and Inhibition

Bita Khalili [1,†], Hailey D. Lovelace [1,2,†], David M. Rutkowski [1,†], Danielle Holz [1] and Dimitrios Vavylonis [1,*]

1 Department of Physics, Lehigh University, Bethlehem, PA 18015, USA; bitakhalili@protonmail.com (B.K.); hlovela@g.clemson.edu (H.D.L.); dmr518@lehigh.edu (D.M.R.); dah414@lehigh.edu (D.H.)
2 Department of Physics and Astronomy, Clemson University, Clemson, SC 29631, USA
* Correspondence: vavylonis@lehigh.edu; Tel.: +1-610-758-3724
† These authors contributed equally to this work.

Received: 17 June 2020; Accepted: 23 July 2020; Published: 24 July 2020

Abstract: Cells polarize for growth, motion, or mating through regulation of membrane-bound small GTPases between active GTP-bound and inactive GDP-bound forms. Activators (GEFs, GTP exchange factors) and inhibitors (GAPs, GTPase activating proteins) provide positive and negative feedbacks. We show that a reaction–diffusion model on a curved surface accounts for key features of polarization of model organism fission yeast. The model implements Cdc42 membrane diffusion using measured values for diffusion coefficients and dissociation rates and assumes a limiting GEF pool (proteins Gef1 and Scd1), as in prior models for budding yeast. The model includes two types of GAPs, one representing tip-localized GAPs, such as Rga3; and one representing side-localized GAPs, such as Rga4 and Rga6, that we assume switch between fast and slow diffusing states. After adjustment of unknown rate constants, the model reproduces active Cdc42 zones at cell tips and the pattern of GEF and GAP localization at cell tips and sides. The model reproduces observed tip-to-tip oscillations with periods of the order of several minutes, as well as asymmetric to symmetric oscillations transitions (corresponding to NETO "new end take off"), assuming the limiting GEF amount increases with cell size.

Keywords: cell polarization; mathematical model; fission yeast; reaction–diffusion model; small GTPases; Cdc42 oscillations

---

1. Introduction

The ability of cells to establish an axis for directed growth, motion, or mating relies on their ability to localize signaling proteins at the growing or leading edge of the cell. Such processes enable motile cells to migrate, epithelial cells to develop and maintain tissues, and neurons to grow axons and dendrites [1–3]. Cell polarization generally arises from symmetry-breaking formation of robust protein localization patterns along the cell membrane [4–6]. Small GTPases, such as Ras and Cdc42, play a central role in cell polarization by switching between active GTP-bound and inactive GDP-bound forms. A system of activators (GEFs) and inhibitors (GAPs) provide positive and negative feedbacks for small GTPase activation and inactivation [7–10]. Through self-organization, this results in the formation of membrane regions enriched in activated signaling proteins including Cdc42-GTP.

Extensive experimental and modeling studies in budding yeast, *S. cerevisiae*, have highlighted important mechanisms in polarization, namely formation of a stable patch along the cell membrane, the site of bud growth. These mechanisms are related to the process of Turing pattern formation [5,11–13]. Activated Cdc42 accumulates at a dominant patch where it forms slowly-diffusing aggregates (corresponding to a slowly diffusing local activation) that recruit the Cdc42 GEF, while at the same time depleting it from elsewhere in the cell (a type of global inhibition) [11]. The positive feedback through the "winner-take-all" mechanism is also enhanced by the actin system [14–18].

While polarization in budding yeast involves selection of a single growth site, some organisms can maintain multiple active sites. This includes rod-shaped fission yeast, *S. pombe*, a model organism for studies of cell shape. Many mutations perturb its normal tubular shape towards thinner, wider, round, T-shaped, banana-shaped or other shapes [19]. Studies of fission yeast cell polarity have highlighted several additional phenomena suggestive of a modeling approach:

(i) Fission yeast is able to maintain two stable sites of growth and Cdc42-GTP localization (at the two tips) rather than one. After cell division, a fission yeast cell begins monopolar growth from the old end inherited from the mother cell. The cell subsequently experiences new end take off (NETO) and enters a bipolar growth phase from both cell tips [20]. NETO transition has been described as a result of competition over a limiting component between the two tips that can reach saturation [21–24].

(ii) Cdc42 oscillatory and fluctuating states underlie the monopolar and bipolar growth states of fission yeast [22,25] (reminiscent of the Min protein oscillation system in bacteria [26,27]). During mating, Ras1/Cdc42 patch appearance and disappearance dynamics are also crucial for cells to find and polarize towards a mating partner [28]. These observations suggest that Cdc42 oscillations and fluctuations embody an exploratory mechanism, enabling cells to adapt their growth pattern in response to external and internal cues to maximize cell survival [22,29].

(iii) The two Cdc42 GEFs, Scd1 and Gef1, localize at cell tips, together with Cdc42-GTP [22,30–35]. By contrast, three known Cdc42 GAPs establish an intriguing pattern, with Rga4 [36–38] and Rga6 [39] decorating primarily the cell sides while Rga3 accumulates primarily at the cell tips [40]. Fluorescence recovery after photobleaching (FRAP) studies indicate different dynamics of Rga6 at cell tips as compared to cell sides: the percent recovery at the cell sides is smaller than the tips over the same time period [39].

(iv) Recent evidence suggests that localization of Cdc42-GTP and Ras1-GTP to cell tips occurs primarily through fast membrane diffusion of Cdc42-GDP and Ras1-GDP, converting to slowly-diffusing GTP forms at the cell tips [9,33,41]. While polarization is generally thought to require Cdc42-GDP extraction from the cell membrane through guanine nucleotide dissociation inhibitor (GDI), the effect of the fission yeast GDI Rdi1 is relatively small as cells are able to polarize in its absence [33]. Membrane diffusion coefficients and dissociation rates of Cdc42 have been previously estimated and can be incorporated into models [33,41].

In this short article, we focus on the broad dynamic and geometric features of the fission yeast polarization system to propose a reaction–diffusion model that can account for the polarity transitions and spatial pattern of Cdc42, its activators and inhibitors. The aim of this top-down approach is to (i) indicate the minimum level of complexity required to describe the broad features mentioned above, (ii) motivate experiments to measure unknown model parameters, and (iii) serve as a framework to more accurately incorporate missing biological mechanisms. Compared to previous models of fission yeast polarization [21–24,42–45], here we include GAP localization on the cell membrane, we use diffusion coefficients of Cdc42-GTP and Cdc42-GDP on the cell membrane estimated in experiments, and we implement the 3D geometry of the rod-shaped fission yeast.

To model GAP plasma membrane recruitment, we suggest a mechanism similar to our previous model of localized membrane recruitment of Gap1, the GAP of Ras1, at the exploratory Ras1 patch during fission yeast mating [41]. The effect of local Gap1 recruitment and diffusion around the mating patch was to restrict the zone of Ras1 activation and regulate the lifetime of the exploratory patch [41,46].

2. Model

We developed a system of partial differential equations that implements diffusion on a 3D curved surface representing a fission yeast plasma membrane. We used the same numerical methods as an earlier study for reaction–diffusion of Ras1 during cell mating [41,47]. In the simulations we compute the surface concentrations of Cdc42-GDP, C_D, and Cdc42-GTP, C_T through diffusion and reaction (Figure 1). The surface representing the plasma membrane has the shape of fission yeast cells with a cylindrical body of radius 2 μm capped by hemispherical tips at either end. We do not implement a reaction scheme for Ras1, so in this preliminary model C_T and C_D can be thought to represent the combined Ras1/Cdc42 polarity patch. The activation and deactivation of Cdc42 is regulated by the surface concentrations of GEFs, C_{GEF}, and two types of GAPs: GAP_I, C_{GAP_I}, and GAP_{II} that we assume exists in two distinct diffusive states, $C^{fast}_{GAP_{II}}$ and $C^{slow}_{GAP_{II}}$, to be described below. The equations describing Cdc42 dynamics are as follows:

$$\partial C_D/\partial t = D_D\Delta_S C_D + f^p_D + (k^n_1 + k^n_2 C_{GAP_I} + k^n_3 C^{fast}_{GAP_{II}} + k^n_8 C^{slow}_{GAP_{II}})C_T - k^p_0 e^{-\frac{s}{\lambda}} C_{GEF}C_D - r_D C_D - r_{noise}C_D, \quad (1)$$

$$\partial C_T/\partial t = D_T\Delta_S C_T + k^p_0 e^{-\frac{s}{\lambda}} C_{GEF}C_D - (k^n_1 + k^n_2 C_{GAP_I} + k^n_3 C^{fast}_{GAP_{II}} + k^n_8 C^{slow}_{GAP_{II}})C_T - r_T C_T + r_{noise}C_D. \quad (2)$$

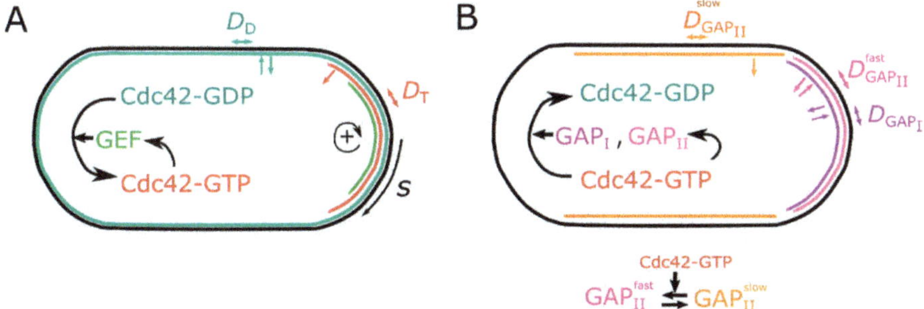

Figure 1. Modeling Cdc42 activation and regulator distribution. (**A**) Schematic illustrates GEF-mediated positive feedback at cell tips, with colored regions indicating zones of active Cdc42-GTP patch. The model equations are identical for both tips; however, they allow symmetry-breaking states and one dominant tip as shown. Parameter s indicates arc length distance from the nearest cell tip. Colored arrows indicate association, dissociation, and diffusion along cell membrane. Cdc42-GDP (teal) associates, dissociates, and diffuses on the plasma membrane with diffusion coefficient D_D. It converts to slowly-diffusing Cdc42-GTP (red, diffusion coefficient D_T) by GEF (green) that is recruited to the membrane by Cdc42-GTP in a nonlinear manner, establishing a positive feedback (+ arrow). (**B**) Schematic illustrates the negative feedback through GAPs. GAP_I (purple, diffusion coefficient D_{GAP_I}) and fast-diffusing GAP_{II} (pink, diffusion coefficient $D^{fast}_{GAP_{II}}$) are recruited to the membrane through Cdc42-GTP. Fast-diffusing GAP_{II} spontaneously converts to slow-diffusing GAP_{II} (orange, diffusion coefficient $D^{slow}_{GAP_{II}}$), while the reverse (slow to fast) is catalyzed by Cdc42-GTP. All GAPs catalyze hydrolysis of Cdc42-GTP.

Here Δ_S is the Laplace–Beltrami operator, and D here and below represents diffusion coefficients. Symbols k and r indicate reaction and membrane dissociation rate constants, respectively. Superscripts p and n indicate positive and negative feedback contributions, respectively. Constant rate f^p_D represents uniform association of Cdc42-GDP to the plasma membrane from a cytoplasmic pool, which we assume has a constant concentration. We used prior experimentally estimated diffusion coefficients and membrane dissociation constants r_D and r_T, which are assumed to implicitly include the effect of GDI Rdi1. In Equations (1) and (2), GAPs promote conversion of Cdc42-GTP to Cdc42-GDP and the reverse conversion is promoted by GEFs. Activation of Cdc42 by GEFs is biased to occur close to the cell tips, where it should be enhanced through microtubule-based tip delivery of Tea proteins, restriction of activation zone through the endoplasmic reticulum (ER), and possibly actin polymerization [32,48–50]. This is implemented through the exponential term where s is the arc length distance to the closest tip

(Figure 1) and parameter λ indicates the scale over which activation at cell tips is assumed to occur. The model allows for random activation of Cdc42-GDP with rate r_{noise}, implemented similarly as in [41].

Accumulation of GEFs in response to Cdc42-GTP is assumed to occur through an autocatalytic mechanism that has a functional form similar to the positive feedback proposed for *S. cerevisiae* polarization [11]. In this positive feedback mechanism, a finite amount of GEF in the system is assumed to be distributed in quasi-static equilibrium with higher proportions at sites with higher active Cdc42 concentration:

$$C_{\mathrm{GEF}} = k_1^p E_c C_{\mathrm{T}}/V + k_2^p E_c C_{\mathrm{T}}^2/V, \tag{3}$$

$$E_c = E_c{}^{\mathrm{tot}}/(1 + \int \left[k_1^p C_{\mathrm{T}}/V + k_2^p C_{\mathrm{T}}^2/V\right] da), \tag{4}$$

where E_c is the available number of GEF molecules in the cytoplasm, $E_c{}^{\mathrm{tot}}$ is the total number of GEF molecules in the cell, and V is the cell volume. In Equation (4), the integral is over the cell's surface area. The quasi-static approximation is introduced for simplicity, to avoid additional parameters related to a GEF concentration field in the model; earlier work has shown this approximation is valid in the limit of sufficiently fast GEF membrane dissociation rate [11,41]. Membrane-bound Cdc42 GEFs Scd1 and Gef1 are indeed localized at the cell tip, where they are expected to form complexes with Cdc42-GTP [22,30,31,35]. Within the simplifying quasi-static approximation, we are thus consistently assuming a GEF membrane diffusion coefficient similar to that of Cdc42-GTP. The nonlinear dependence in Equation (4) leads to a positive feedback strong enough to break symmetry and establish a Cdc42-GTP patch [5], and is supported by experiments showing recruitment of GEF Scd1 depends on scaffold protein Scd2, which itself depends on Cdc42-GTP [35].

In the model, we include a negative inhibitor that we designate $\mathrm{GAP_I}$, which accumulates at cell tips through Cdc42-GTP-mediated recruitment, and provides a nonlinear negative feedback able to generate fluctuations and oscillations. Including such a component is motivated by the observed tip localization of Rga3 [40], but we also bundle together all tip-localized inhibition mechanisms of Ras1 and Cdc42 through Gap1, Pak1 and actin. We assumed a functional form similar to the Ras1 GAP, Gap1, recruitment to the exploratory mating patch [41] and the negative feedback for Cdc42 oscillations in budding yeast [51]:

$$\partial C_{\mathrm{GAP_I}}/\partial t = D_{\mathrm{GAP_I}} \Delta_S C_{\mathrm{GAP_I}} + k_4^n C_{\mathrm{T}}^h/(k_{\mathrm{sat}}^h + C_{\mathrm{T}}^h) - r_{\mathrm{GAP_I}} C_{\mathrm{GAP_I}}\,. \tag{5}$$

The second term on the right hand side represents cooperative recruitment at small Cdc42-GTP concentrations, reaching a plateau for concentration above k_{sat}. We used a value $h = 2$ that was sufficient to provide delayed negative feedback needed for oscillations.

To generate a spatial pattern of inhibitors such as Rga4 and Rga6, which accumulate in "collar" or "corset" shapes around growing cell tips [36–39], we make the bold assumption that these inhibitors that we collectively call $\mathrm{GAP_{II}}$ are also recruited to the plasma membrane through Cdc42-GTP (similar to $\mathrm{GAP_I}$). We further assume that $\mathrm{GAP_{II}}$ proteins that diffuse away from the cell tip convert to slowly-diffusing forms, possibly through binding to each other, thus accumulating away from the active region. Support for such a differential mobility along the plasma membrane is the observation of larger FRAP recovery of Rga6 at cell tips compared to cells sides, and a pattern of FRAP recovery consistent with Rga6 membrane diffusion [39]. The dynamics of the fast and slow $\mathrm{GAP_{II}}$ components are described by:

$$\partial C_{\mathrm{GAP_{II}}}^{\mathrm{fast}}/\partial t = D_{\mathrm{GAP_{II}}}^{\mathrm{fast}} \Delta_S C_{\mathrm{GAP_{II}}}^{\mathrm{fast}} + k_5^n C_T - k_6^n C_{\mathrm{GAP_{II}}}^{\mathrm{fast}} + k_7^n C_{\mathrm{GAP_{II}}}^{\mathrm{slow}} C_T - r_{\mathrm{GAP_{II}}}^{\mathrm{fast}} C_{\mathrm{GAP_{II}}}^{\mathrm{fast}}\,, \tag{6}$$

$$\partial C_{\mathrm{GAP_{II}}}^{\mathrm{slow}}/\partial t = D_{\mathrm{GAP_{II}}}^{\mathrm{slow}} \Delta_S C_{\mathrm{GAP_{II}}}^{\mathrm{slow}} + k_6^n C_{\mathrm{GAP_{II}}}^{\mathrm{fast}} - k_7^n C_{\mathrm{GAP_{II}}}^{\mathrm{slow}} C_T - r_{\mathrm{GAP_{II}}}^{\mathrm{slow}} C_{\mathrm{GAP_{II}}}^{\mathrm{slow}}. \tag{7}$$

Both fast and slow forms can hydrolyze Cdc42-GTP, at different rates as shown in Equations (1) and (2). We also assumed that Cdc42-GTP can catalyze conversion of $C_{\mathrm{GAP_{II}}}^{\mathrm{slow}}$ and $C_{\mathrm{GAP_{II}}}^{\mathrm{fast}}$ through the k_7^n

terms in Equations (6) and (7) (since otherwise the distribution of $C^{slow}_{GAP_{II}}$ would be peaked at the tips instead of away from cell tips). Though we are not aware of experimental evidence in support of the latter assumption, this process might occur through release of slow $C^{slow}_{GAP_{II}}$ from a protein complex after binding to Cdc42-GTP.

For most of this study we kept cells at a fixed reference length of 8 μm, and other parameters as in Table 1. The area of each Voronoi cell used in the discretization of the surface area was between 0.017 to 0.046 μm^2. We used a simulation time step of 0.01 s and started the simulations from an unpolarized state, with Cdc42-GDP at the concentration it would have at steady state in the absence of activation ($C_D = j^p_D / r_D$) plus or minus small relative random fluctuations. We also initialize a smaller random C_T field and checked the evolution of the system over hundreds or thousands of seconds.

Table 1. List of parameters in reaction–diffusion equations, their reference value in the simulations, and their physical meaning in the model. Unless indicated, the values were estimated or adjusted to match experimental observations as mentioned in the main text.

Variable	Reference Value	Description
D_T	0.02 $\mu m^2/s$	Diffusion coefficient of Cdc42-GTP from [33]
D_D	0.2 $\mu m^2/s$	Diffusion coefficient of Cdc42-GDP from [33]
D_{GAP_I}	0.03 $\mu m^2/s$	Diffusion coefficient of GAP_I, estimated
$D^{fast}_{GAP_{II}}$	0.0625 $\mu m^2/s$	Diffusion coefficient of fast GAP_{II}, estimated
$D^{slow}_{GAP_{II}}$	0.005 $\mu m^2/s$	Diffusion coefficient of slow GAP_{II}, estimated
k^n_1	0.000625/s	Spontaneous rate Cdc42-GTP hydrolysis, adjusted
k^n_2	0.00325 $\mu m^2/s$	Rate constant of GAP_I-mediated Cdc42-GTP hydrolysis, adjusted
k^n_3	0.00125 $\mu m^2/s$	Rate constant of fast GAP_{II}-mediated Cdc42-GTP hydrolysis, adjusted
k^n_4	250/s	Rate constant of Cdc42-GTP-mediated GAP_I recruitment, adjusted
k^n_5	0.03/s	Rate constant of Cdc42-GTP-mediated GAP_{II} recruitment, adjusted
k^n_6	2/s	Rate of fast GAP_{II} conversion to slow form, adjusted
k^n_7	0.025 $\mu m^2/s$	Cdc42-GTP-mediated conversion of slow to fast GAP_{II}, adjusted
k^n_8	0.0005 $\mu m^2/s$	Rate constant of slow GAP_{II}-mediated Cdc42-GTP hydrolysis, adjusted
k_{sat}	600/μm^2	Saturating concentration of GAP_I negative feedback, adjusted
k^p_0	0.0025 $\mu m^2/s$	Rate constant of GEF-mediated Cdc42-GDP activations, adjusted
k^p_1	0.5 μm^3	Linear rate constant of GEF recruitment to Cdc42-GTP, adjusted
k^p_2	0.1 μm^5	Quadratic rate constant of GEF recruitment to Cdc42-GTP, adjusted
E_c^{tot}	250	Total pool of GEFs, estimated
j^p_D	2.4/s/μm^2	Flux of Cdc42-GDP from cytoplasm to membrane, estimated
r_T	0.005/s	Rate of Cdc42-GTP dissociation from membrane from [33]
r_D	0.03/s	Rate of Cdc42-GDP dissociation from membrane from [33]
r_{GAP_I}	0.01/s	Rate of GAP_I dissociation from membrane, adjusted
$r^{fast}_{GAP_{II}}$	0.0125/s	Rate of fast GAP_{II} dissociation from membrane, adjusted
$r^{slow}_{GAP_{II}}$	0.0025/s	Rate of slow GAP_{II} dissociation from membrane, adjusted
r_{noise}	0.0021/s	Rate of random Cdc42-GDP conversion to Cdc42-GTP, adjusted
λ	2.5 μm	GEF activation scale at cell tips, estimated
L	8 μm	Cell length

3. Results

3.1. Goals

We explored our model's ability to capture basic phenomenology of fission yeast Cdc42 polarization by adjusting the unknown rate constants while using reported estimates for the membrane diffusion coefficients and dissociation rates of Cdc42-GTP and Cdc42-GDP. The desired phenomena include: (i) ability of the system to exhibit asymmetric, symmetric, and oscillatory states; (ii) a pattern of polarity transitions as function of changing rate constants matching fission yeast polarity change with cell growth; (iii) oscillatory states with periods in the range of 4–6 min [22]; (iv) enhancement of Cdc42 concentration (combined Cdc42-GDP and Cdc42-GTP) by 2–3-fold at cell tips compared to cell sides [33]; (v) establishment of micron-scale active regions at cell tips [22,32,33,52]; (vi) accumulation of GAP_I at cell tips and GAP_{II} in collar/corset manner away from cell tips.

3.2. Dynamical States Observed in Parameter Scan

Through a systematic but non-exhaustive scan of unknown rate constants (see Table 1), we were able to show that the system of Equations (1)–(7) provides a mechanism with solutions that can describe, to different extents, all the desired phenomena mentioned above. The model also accounted for the geometric features of fission yeast cells through our implementation of reaction–diffusion equations on a curved surface.

For a large range of model parameter values, the system converged to stationary solutions that were either asymmetric (MPS, monopolar stable) or symmetric (BPS, bipolar stable), as shown in Figure 2. Because both tips have identical rate constants, the MPS states represent symmetry-breaking states. The zones of activation were always found at one or both of the tips since we bias Cdc42 activation to the cell tip region.

For the MPS and BPS examples in Figure 2, the concentration of Cdc42 at activated cell tips is enhanced by a factor of 2–3 compared to cell sides: This is due to accumulation of Cdc42-GTP at cells tips, consistent with prior experiments [33]. The concentration profile of Cdc42-GDP exhibits a small dip closer to the active cell tip: This reflects the diffusive flux of Cdc42-GDP towards the active cell tip that is balanced by diffusive flux and dissociation of Cdc42-GTP away from the tip. In the MPS states, the lagging tip maintained active Cdc42 at lower concentrations compared to the dominant tip.

The profiles of Figure 2 also show the localization of GAP_I at the active cell tip and GAP_{II} in a collar/corset manner away from the cell tips. The concentration of GAP_{II} peaks around the active tip region, which is in qualitative agreement with profiles of Rga4 in microscopy images [36,38] (though perhaps more exaggerated compared to experiments).

Another common solution of the system was states exhibiting symmetric anticorrelated oscillations, as shown in Figure 3A and Video S1. These BPO (bipolar oscillatory) states had similar active tip concentration profiles to the stationary states of Figure 2; however, the dominant active tip kept switching from one tip to the other, as observed during Cdc42 fluctuations and oscillations after NETO [22].

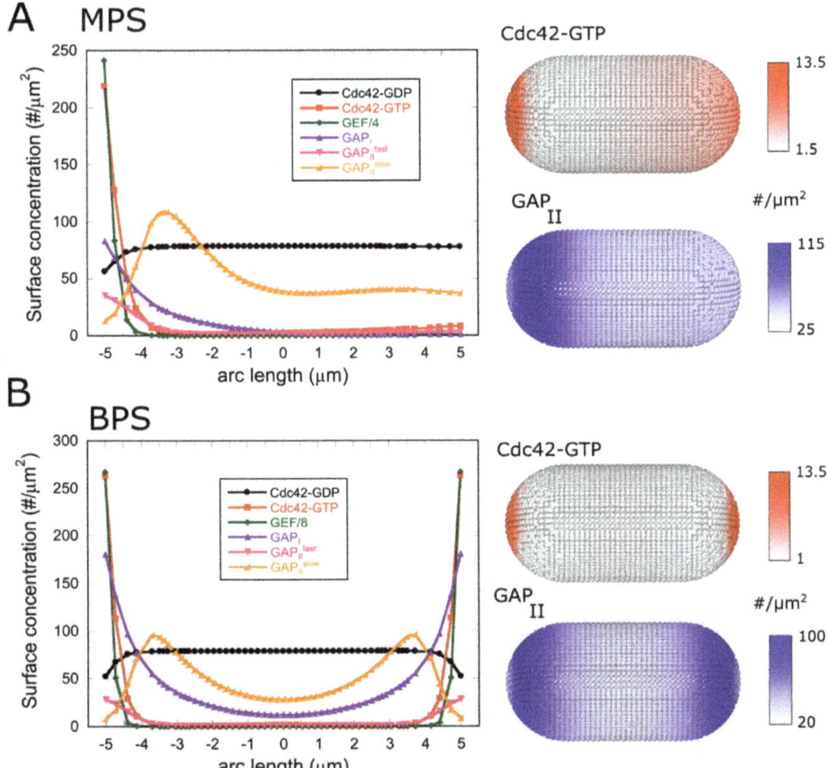

Figure 2. Examples of asymmetric stationary (monopolar stable, MPS) and symmetric stationary (bipolar stable, BPS) solutions. The system evolved to these stationary states over a time course in the order of hundreds of seconds from an initially unpolarized state. (**A**) Concentrations as a function of arc length distance from cell tip and snapshots showing the concentration of Cdc42-GTP and $\mathrm{GAP_{II}}$ (sum of $C^{\mathrm{slow}}_{\mathrm{GAP_{II}}}$ and $C^{\mathrm{fast}}_{\mathrm{GAP_{II}}}$). Parameter values as in Table 1, except for $k_{\mathrm{sat}} = 900/\mu\mathrm{m}^2$ and $E^{\mathrm{tot}}_{\mathrm{C}} = 700$. (**B**) Same as panel A but for a BPS state. Parameter values same as in Table 1, except for $E^{\mathrm{tot}}_{\mathrm{C}} = 1800$.

A characteristic feature of Cdc42 dynamics in fission yeast is the anticorrelated fluctuations and oscillations of Cdc42-GTP before NETO, a period of the cell cycle where cells grow in a monopolar manner and Cdc42 activity is larger at the old end [22]. We found that such dynamical states are also observed by our proposed dynamics, though a finer tuning of parameters is required as compared to MPS, BPS and BPO states. We found that the system of Equations (1)–(7) can generate asymmetric oscillations that persist undamped; however, within our desired set of criteria listed at the beginning of the Results section, we only found asymmetric damped oscillatory states (MPDO), as shown in the example of Figure 3B. Additional sources of noise in cells (not included in our model) may convert those states to long lived anti-correlated fluctuations and oscillations [43,53]; thus we interpret the existence of MPDO dynamics as consistent with the asymmetric oscillations and fluctuations of Cdc42-GTP before NETO.

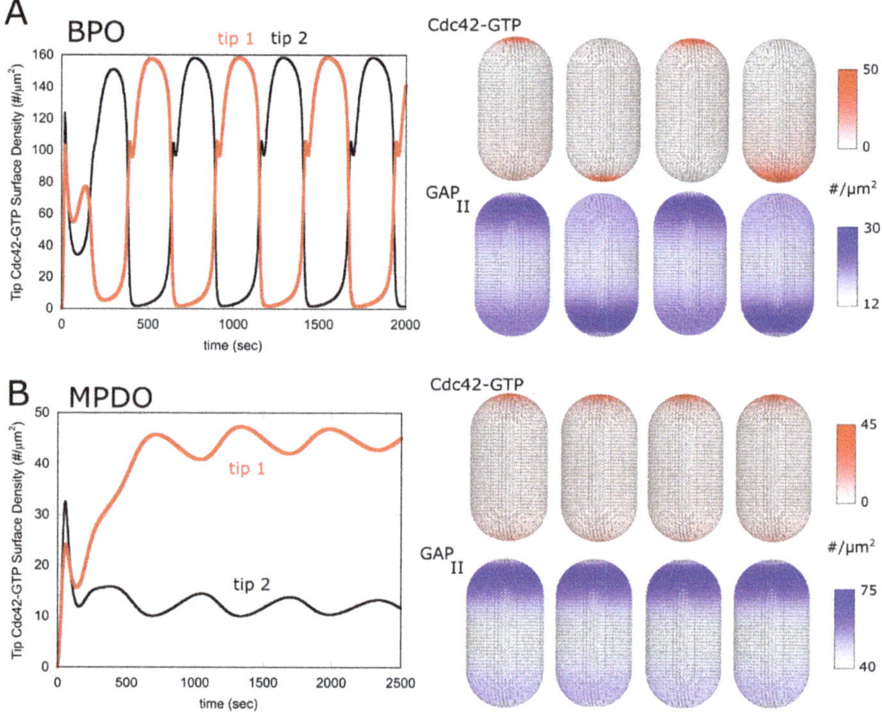

Figure 3. Examples of symmetric oscillations (bipolar oscillatory, BPO) and asymmetric damped oscillations (monopolar with damped oscillations, MPDO) solutions. (**A**) Graph shows concentration of Cdc42-GTP at each tip, as a function of time starting from an unpolarized state with both tips inactive. The concentration profiles of the other components in the system that are not shown follow the same trends as in the stationary states of Figure 2. Snapshots show Cdc42-GTP and total GAP_{II} profiles every 200 s once the system evolved to a periodic pattern. Parameter values same as shown in Table 1 except $k_5^n = 0.01$/s and $E_C^{tot} = 500$. (**B**) Same as panel A, but for an MPDO case where the system eventually evolves to a stationary asymmetric state through damped, anticorrelated oscillations. Snapshots show Cdc42-GTP and total GAP_{II} profiles every 200 s, starting at 1000 s. Parameter values same as shown in Table 1, except $E_C^{tot} = 210$.

3.3. *Structure of Solutions in Parameter Space*

Since fission yeast cells transition from asymmetric Cdc42 oscillations and fluctuations prior to NETO to symmetric oscillations and fluctuations after NETO [22], the MPDO and BPO states are biologically relevant. To better understand the requirements to observe such states in the simulations, we varied two parameters, k_0^p and k_2^n, describing the strength of positive and negative feedbacks in the system, focusing on a region around the rarer MPDO solutions, as shown in Figure 4. We found that all of the dynamical states of Figures 2 and 3 arose in the neighborhood of MPDO, with higher values of both k_0^p and k_2^n resulting in BPO. MPDO states were observed in between MPS and BPO states when the examined positive feedback parameter was above a threshold; below the threshold, the system results in weakly active BPS states. We note that along the boundaries of dynamical state regions of Figure 4, the system might settle to a different pattern from run to run, which indicates regions with multiple solutions may "co-exist" [22]. We did not explore the coexistence behavior in detail in this work.

We next asked if the model is consistent with transition from MPDO to BPO around NETO. We anticipate the MPDO to BPO transition to occur as a result of the increase of a limiting component with cell growth. We thus, varied the parameter that describes the total amount of GEF in the system, E_C^{tot}, together with a parameter that influences the negative feedback strength, k_{sat}, to help unfold the pattern of dynamical transitions in Figure 5A. We observed that, indeed, for k_{sat} above 600/μm^2 there is a large region in parameter space corresponding to MPDO, shifting to BPO as the total amount of GEF in the system is doubled. Interestingly, at even larger E_C^{tot} (above 1400) the system reverts to BPS. This is consistent with the uncorrelated small fluctuations of active Cdc42 in long *cdc25-22* cells [22].

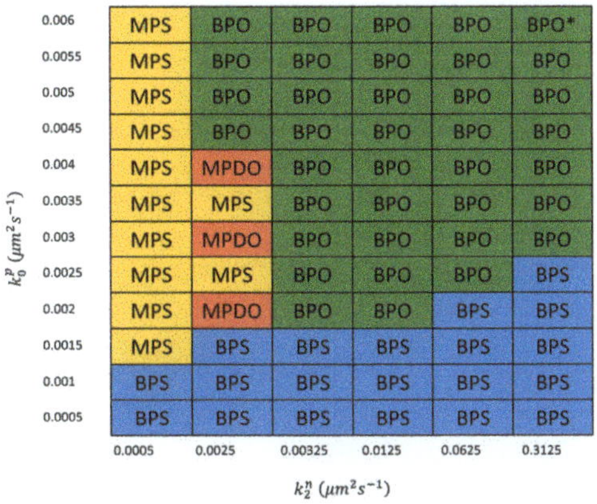

0.006 MPS BPO BPO BPO BPO BPO*
0.0055 MPS BPO BPO BPO BPO BPO
0.005 MPS BPO BPO BPO BPO BPO
0.0045 MPS BPO BPO BPO BPO BPO
0.004 MPS MPDO BPO BPO BPO BPO
0.0035 MPS MPS BPO BPO BPO BPO
0.003 MPS MPDO BPO BPO BPO BPO
0.0025 MPS MPS BPO BPO BPO BPS
0.002 MPS MPDO BPO BPO BPS BPS
0.0015 MPS BPS BPS BPS BPS BPS
0.001 BPS BPS BPS BPS BPS BPS
0.0005 BPS BPS BPS BPS BPS BPS
0.0005 0.0025 0.00325 0.0125 0.0625 0.3125
k_0^p ($\mu m^2 s^{-1}$)
k_2^n ($\mu m^2 s^{-1}$)

Figure 4. Structure of dynamical states reached, as function of a positive (k_0^p) and negative (k_2^n) feedback rate constants. Results show the final state reached by a single simulation for each set of parameters over time course of 800–1200 s, starting from an unpolarized state. In state marked with *, the Cdc42-GTP patch regions becomes as small as a single Voronoi region of the discretized cell surface. Other parameters same as Table 1.

A $k_5^n = 0.03$ /s
900 BPS BPS MPS MPS MPS MPS MPS MPS MPS MPDO BPO BPO BPO BPS BPS BPS BPS
750 BPS BPS MPS MPS MPDO MPDO MPDO BPO BPO BPO BPO BPO BPS BPS BPO BPS BPO
600 BPS BPS MPS MPDO BPO BPO BPO BPO BPO BPO BPO BPO BPO BPS BPS BPS BPS
450 BPS BPS BPO BPO BPO BPO BPO BPO BPO BPO BPO BPO BPS BPS BPO BPS BPS
300 BPS BPS BPO BPO BPO BPO BPO BPO BPO BPO BPO BPO BPO BPS BPS BPS BPO
100 150 200 250 300 400 500 600 700 800 900 1000 1400 1800 2200 2600 3000 E_C^{tot}
k_{sat} (#/μm^2)
B
$k_5^n = 0.015$ /s
900 BPS BPS BPS MPS BPO BPO BPO BPO BPO BPO BPO BPO BPO BPO BPS BPS BPS
750 BPS BPS BPS BPO BPO BPO BPO BPO BPO BPO BPO BPO BPS BPS BPS BPS BPS
600 BPS BPS BPS BPO BPO BPO BPO BPO BPO BPO BPO BPO BPO BPO BPS BPS BPS
450 BPS BPS BPS BPO BPO BPO BPO BPO BPO BPO BPO BPO BPO BPO BPS BPO BPS
300 BPS BPS BPS BPO BPO BPO BPO BPO BPO BPO BPO BPO BPO BPS BPO BPS BPS
100 150 200 250 300 400 500 600 700 800 900 1000 1400 1800 2200 2600 3000 E_C^{tot}
k_{sat} (#/μm^2)

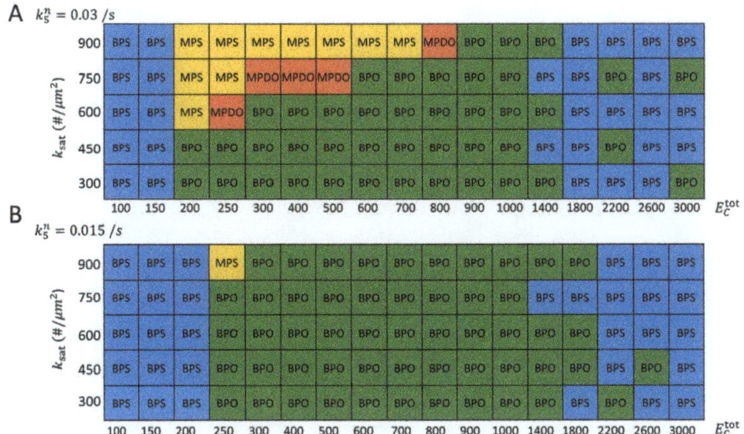

Figure 5. Structure of dynamical states reached, as function of total GEF amount (E_C^{tot}) and negative feedback parameter k_{sat}. Results show the final state reached by a single simulation for each set of parameters over time course of 800–1200 s, starting from an unpolarized state. Other parameters same as Table 1. (**A**) Reference simulations. For k_{sat} = 600–900/μm^2, the system can transition from MPDO to BPO with increasing E_C^{tot}, a behavior consistent with the dynamical change observed under cell growth around NETO. (**B**) Same as panel A but reducing the rate of GAP_{II} recruitment to the cell tips, as might occur in *rga4*Δ or *rga6*Δ cells. The BPO region is observed to expand compared to panel A, while the region corresponding to MPS and MPDO states shrinks.

To further demonstrate how the model captures the transition around NETO, we used the parameter values suggested from Figure 5A, to perform simulations at different fixed cell lengths with $k_{sat} = 650/\mu m^2$ and increasing the total GEF in proportion to cell volume such that $E_C^{tot} = 300$ for cell length $L = 7$ μm. Indeed, the simulations of Figure 6A show a transition from MPDO to BPO at around the cell length where NETO occurs (9.5 μm [20]). Very long cells with $L = 35$ μm in Figure 6A revert to BPS, consistent with the behavior of long *cdc25-22* cells [22], as mentioned in the preceding paragraph. The distribution of Cdc42, GEF and GAPs maintains the same features with increasing length, with the Cdc42 patch becoming more intense and slightly narrower with increasing length (Figure 6B).

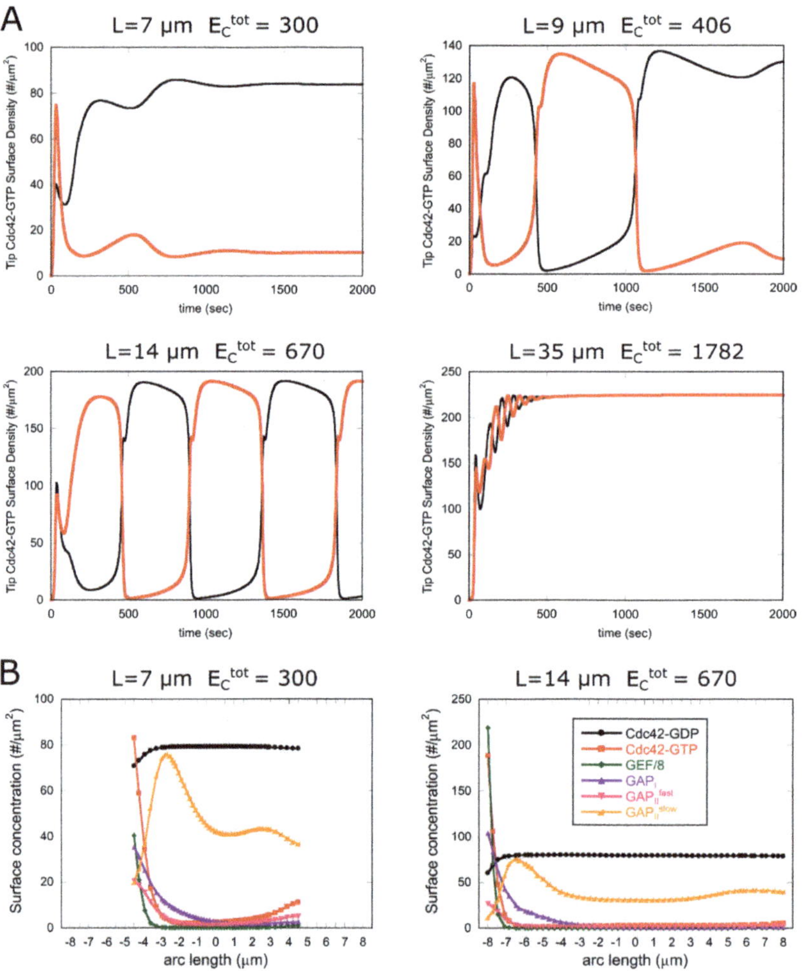

Figure 6. Results of simulations at different fixed cell lengths, L, and increasing the total GEF in proportion to cell volume, as indicated in each graph. Other parameters same as Table 1, except for $k_{sat} = 650/\mu m^2$. (**A**) Plots of Cdc42-GTP concentration at each tip (red and black curve) as a function of time starting from an unpolarized state with both tips inactive. System transitions from MPDO to BPO and BPS with increasing L. (**B**) Concentrations as a function of arc length distance from cell tip for a cell of length 7 μm (left, at steady asymmetric state) and length 14 μm (right, after 1060 s of simulation of cell undergoing symmetric oscillations, at an instant with a dominant left tip).

3.4. Simulations with Reduced GAP_{II} Recruitment

Motivated by prior experimental studies of Rga4 and Rga6 deletion mutants [22,36,39], in our model we reduced the rate of GAP_{II} recruitment to the cell tips by a factor of two (described by parameter k_5^n). We repeated the scan of the values of E_C^{tot} and k_{sat} and plot the resulting states in Figure 5B. Interestingly, this change lead to an expansion of the BPO region, a significant shrinkage of the MPS region and disappearance of MPDO states (within the resolution of the scan).

The behavior observed in Figure 5, when reducing parameter k_5^n, does not appear to closely relate to prior experimental observations: *rga4*Δ cells demonstrate more pronounced symmetric oscillations compared to wild-type; however, they also exhibit asymmetric states [22,36]. Meanwhile *rga6*Δ cells have been found to be slightly more monopolar, with smaller relative Cdc42 fluctuations compared to wild-type [39]. This comparison to experiments suggests that there is more to polarity regulation of these mutant cells than simply changing a single rate constant of our model (see recent results in [54]).

The model, nevertheless, captures some of the properties of *rga4*Δ and *rga6*Δ cells when plotting the Cdc42-GTP and GEF profiles around an active tip (Figure 7). We find a wider profile of Cdc42 activation, as observed in *rga4*Δ cells that have larger cell diameter compared to wild-type cells [32,52]. Similarly, *rga6*Δ have been found to be wider than wild-type cells and *rga4*Δ*rga6*Δ double mutants are wider than either single mutants [39].

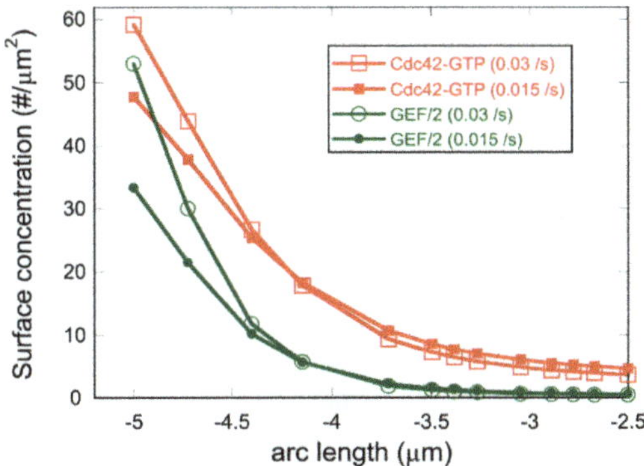

Figure 7. Plot of Cdc42-GTP and GEF concentration gradient when the rate of GAP_{II} recruitment to the cell tips is $k_5^n = 0.03\ /s$ (empty symbols) and for the same parameters with reduced rate of $k_5^n = 0.015\ /s$ (solid symbols), as might occur in *rga4*Δ cells. Other parameters are the same as Table 1 except $k_{sat} = 450\ /\mu m^2$, such cases result in BPO oscillations (see Figure 5B). This plot shows the concentration profiles at the dominant tip ~ 20 s after Cdc42-GTP tip concentration peaks, and after at least 1000 s of simulation time. The half-widths are larger with reduced GAP_{II} recruitment rate: An exponential fit to the Cdc42-GTP profile gives a decay length 0.64 ± 0.01 μm and 0.72 ± 0.01 μm, respectively. The left-most point on the x axis corresponds to the left tip, as in Figure 2.

4. Discussion

The model presented in this work generalizes the delayed differential equations (DDE) model of Das et al. [22] for fission yeast. To reproduce observed Cdc42 dynamics, the previous model included competition between the two tips for a limiting component, assumed the existence of a positive activation feedback, assumed a maximum (or saturation) of tip activity, and included negative feedback through an explicit time delay. By assuming that the limiting component increases in amount with cell growth, this prior DDE model generated the transition from asymmetric to symmetric oscillations (identified here as MPDO and BPO states, respectively) that occurs with cell growth. The properties of such a DDE system together with diffusion along one dimension have been studied in detail by

Xu and Bressloff [42]. Cerone et al. [23] also performed a detailed analysis of tip competition without oscillations at a level of ordinary differential equations.

In the current system of partial differential equations formulated on a surface in the shape of fission yeast, we associate the limiting component of Das et al. [22] with the GEF system, assuming a functional form for the positive feedback borrowed from studies of budding yeast polarization [11]. This limiting component enables transition from asymmetric to symmetric states with increasing amount of GEF. Unlike Das et al. we do not have an explicit parameter to saturate tip concentration and we do not have explicit time delay. These properties are assumed to be provided by the GAP system, for which we assume nonlinear dynamics similar to a prior model of transient Cdc42 patch competition in budding yeast [51]. Motivated by Khalili et al. [41], where we studied how Ras1 patch scans the cell membrane through appearance and disappearance, we assumed that GAPs of Cdc42 are recruited to the cell tip through Cdc42-GTP, similar to the recruitment of Gap1 to the mating patch, by Ras1-GTP [41,46].

Here we associated the nonlinear negative feedback necessary for emergence of oscillations with the function of tip-localized GAPs ("GAP_I"). However we note that deletion of the tip-localized Rga3 did not significantly change Cdc42 oscillations compared to wild-type cells [40]. While it is also conceivable that negative regulation occurs by diffusion-limited supply of Cdc42-GDP at activated cell tips, this depletion is relatively small in Figure 2. In this figure, Cdc42-GTP accumulates at cell tips at concentrations that exceed Cdc42-GDP concentration at cell sides by a factor of 2–3, as observed experimentally [33] when using diffusion coefficients close to experimental estimates [33]. One important component that we did not include explicitly is Pak1 kinase, which has been proposed to mediate negative regulation through GEF phosphorylation in both fission and budding yeast [22,55,56], and, more recently, in positive feedback regulation in fission yeast [35].

We modeled accumulation of Rga4 and Rga6 along the cell sides by recruitment of a fast-diffusing GAP_{II} (representing both of these proteins) at cell tips, spontaneous conversion of fast-diffusing GAP_{II} to a slow-diffusing form, and conversion of slow- to fast-diffusing GAP_{II} in the activated tip region. While such a mechanism can lead to the desired effect of a collar of enhanced concentration of GAP_{II} around the growing tip, additional experimental support of such differential mobility is still needed. The postulated GAP_{II} mechanism may relate to how proteins Pom1 and Tea4, which peak at active cell tips, bind Rga4 and negatively regulate it away from active cell tips [48,50], and the actin-dependence of Rga6 recruitment [39], though we did not explicitly include them in the model. Direct binding of Rga4 and Rga6 to the cell sides from the cytoplasm could be an additional mechanism that should be incorporated into the model to better capture the polarity process.

An important assumption in the model was that Cdc42 activation is biased towards cell tips. We can relax this assumption by taking the limit of parameter λ being very large. With the reference parameter values of Table 1, this results in uniform Cdc42 activation and loss of polarization. A further increase in positive feedback rate constant k_0^p enables symmetry breaking: the model can then readily generate localized stable or oscillating patches of Cdc42-GTP, GEF, and GAP_I, surrounded by a ring of high concentration of GAP_{II}, forming at random locations on the simulated membrane (Figure 8A). This behavior of the model may relate to how loss of Orb6 kinase results in round cell morphologies by directing cell growth to the cell sides: Orb6 has been implicated in excluding Gef1 from accumulation to cell sides [31] and enhancing positive feedback at cell tips through Ras1 [57], in addition to other functions [58]. Stable side patch formation in the absence of cell tip bias may also relate to how T-shaped mutants initiate side projections [21].

The plots of Figures 2–7 show estimated concentrations; however, we note that these numbers can be rescaled by adjustment of the unknown rate constants. In the model we accounted for whole-cell mass conservation of the limiting component, the GEF, but assumed a constant cytoplasmic concentration for everything else. Future improvements of the model would include accounting for mass conservation of all the system's components, as well as for the free energy flow associated with GTP hydrolysis and non-equilibrium transport required to maintain concentration gradients. The model proposed here, together with previous modeling efforts, could also serve as a starting point for further quantitative investigations that include additional biological components that influence polarization, including the cytoskeleton [33], Gef1 phosphorylation [52], vesicle trafficking, and ER [49], as well as independent consideration of Ras1 from Cdc42 [35].

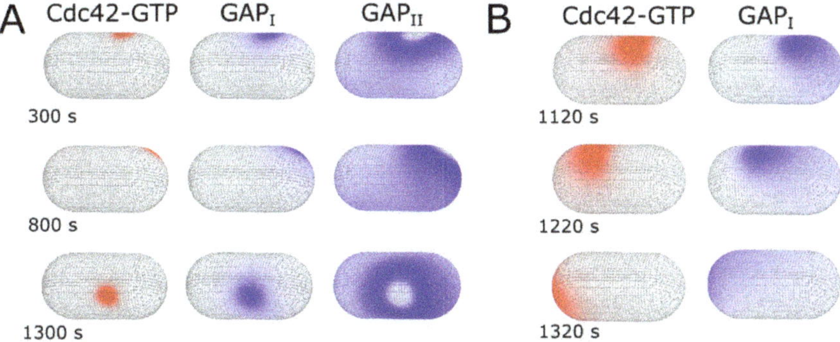

Figure 8. Snapshots of model results when Cdc42 activation is not biased at cell tips (limit of $\lambda \to \infty$). With other parameters same as Table 1, the system towards uniform activation. (**A**) Patch appearance and disappearance for k_0^p = 0.01 μm²/s increased four times compared to Table 1. (**B**) Traveling wave behavior with parameters from Table 1 except k_0^p = 0.01 μm²/s and $k_5^n = 0$, which eliminates GAP_{II}.

Our model implements similar mechanisms proposed for budding yeast polarization [11,51,59], including formation of active Cdc42-GTP patch through GEF-mediated nonlinear positive feedback, competition of different patches for a limiting component, and GAP-mediated negative feedback. This combination of feedbacks leads to patches of Cdc42-GTP that oscillate out of phase, as occurs transiently in budding yeast before the establishment of a dominant patch [51]. However, the overall dynamics we obtain are somewhat different and include the NETO polarity transition as well as stable symmetric and asymmetric oscillations. Another difference is the more prominent role of membrane diffusion of Cdc42-GDP and GAPs across the whole cell in our fission yeast model, as opposed to the diffusion in the cytoplasm and the patch region in budding yeast. How non-equilibrium fluxes through the cytoplasm versus the membrane impact pattern formation dynamics could be a topic for further investigation. Another similarity to prior budding yeast models, is the control of patch width through side GAP accumulation (implemented through different transport mechanisms): Budding yeast controls Cdc42 patch size in part through an inhibitory ring of septin-bound GAPs around the zone of Cdc42 activation [59].

A question of relevance in the broader context of cellular morphogenesis and its regulation by negative feedbacks, is why fission yeast uses GAPs with such different membrane localizations for its polarization process. In principle, one negative regulator could be sufficient for oscillations, for example models of the bacterial Min system, using similar reaction–diffusion mechanisms to this work, reproduce tip to tip MinD oscillations with MinE as a single negative inhibitor [27]. One negative inhibitor was also sufficient to contain and localize an activation zone in our prior model of Ras1 mating patch exploratory dynamics through fast diffusion of Gap1 around the active Ras1 zone [41]. Perhaps the combination of cell tip and cell side inhibitors allows for better control of a localized activation region with precise size over μm scales, as needed for stable tubular projections. Prior theoretical analysis further suggests that stability of cell diameter over successive divisions requires that the growth zone width do not vary strongly with cell diameter [60]. For example, Figure 7 shows how reduction of GAP_{II} recruitment rate can promote a wider Cdc42 patch, features seen in cells with deleted Rga4, which have a wider Cdc42 patch and cell diameter. To further illustrate the implications of GAP regulation in the model, consider the case where relaxing the assumption of tip-biased activation leads to a localized stable or oscillating patches of Cdc42-GTP surrounded by a ring of high concentration of GAP_{II} (Figure 8A). Further eliminating GAP_{II} by setting $k_5^n = 0$ results in a more diffuse Cdc42-GTP zone that moves as a traveling wave around the cell surface, chased by the only remaining inhibitor GAP_I (Figure 8B). Thus, the system changes qualitative behavior, similar to the traveling Rho waves in larger cells [61] and to the reconstituted traveling Min waves [27], which can also be described by reaction diffusion equations [27,61]. We thus, speculate that use of multiple GAPs allows for a robust dynamical landscape that suits fission yeast's tubular growth pattern, under conditions when this might not be possible by positive feedback and geometry alone [62,63].

Supplementary Materials: The following are available online at http://www.mdpi.com/2073-4409/9/8/1769/s1, Video S1: Simulation of Cdc42-GTP and total GAP_{II} corresponding to Figure 3A.

Author Contributions: Conceptualization, B.K. and D.V.; investigation, B.K, H.D.L., D.M.R., D.H. and D.V.; software, B.K. and D.M.R.; writing, D.V.; editing, B.K, H.D.L., D.M.R., D.H. and D.V. All authors have read and agreed to the published version of the manuscript.

Funding: This work was supported by National Institutes of Health Grants R01GM098430, R01GM114201, and R35 R35GM136372, and by the National Science Foundation Lehigh Physics REU grant PHY-1359195 that supported H.L.

Conflicts of Interest: The authors declare no conflict of interest. The funders had no role in the design of the study; in the collection, analyses, or interpretation of data; in the writing of the manuscript, or in the decision to publish the results.

References

1. Devreotes, P.N.; Bhattacharya, S.; Edwards, M.; Iglesias, P.A.; Lampert, T.; Miao, Y. Excitable Signal Transduction Networks in Directed Cell Migration. *Annu. Rev. Cell Dev. Biol.* **2017**, *33*, 103–125. [CrossRef]
2. Pichaud, F.; Walther, R.F.; Nunes de Almeida, F. Regulation of Cdc42 and its effectors in epithelial morphogenesis. *J. Cell Sci.* **2019**, *132*. [CrossRef]
3. Holcomb, P.S.; Deerinck, T.J.; Ellisman, M.H.; Spirou, G.A. Construction of a polarized neuron. *J. Physiol.* **2013**, *591*, 3145–3150. [CrossRef]
4. Martin, S.G.; Arkowitz, R.A. Cell polarization in budding and fission yeasts. *FEMS Microbiol. Rev.* **2014**, *38*, 228–253. [CrossRef]
5. Goryachev, A.B.; Leda, M. Many roads to symmetry breaking: molecular mechanisms and theoretical models of yeast cell polarity. *Mol. Biol. Cell* **2017**, *28*, 370–380. [CrossRef]
6. Halatek, J.; Brauns, F.; Frey, E. Self-organization principles of intracellular pattern formation. *Philos. Trans. R. Soc. Lond B. Biol. Sci.* **2018**, *373*. [CrossRef]
7. Pruyne, D.; Bretscher, A. Polarization of cell growth in yeast. *J. Cell Sci.* **2000**, *113*, 571–585.
8. Bi, E.; Park, H.O. Cell polarization and cytokinesis in budding yeast. *Genetics* **2012**, *191*, 347–387. [CrossRef]
9. Martin, S.G. Spontaneous cell polarization: Feedback control of Cdc42 GTPase breaks cellular symmetry. *Bioessays* **2015**, *37*, 1193–1201. [CrossRef]
10. Chiou, J.G.; Balasubramanian, M.K.; Lew, D.J. Cell Polarity in Yeast. *Annu. Rev. Cell Dev. Biol.* **2017**, *33*, 77–101. [CrossRef]
11. Goryachev, A.B.; Pokhilko, A.V. Dynamics of Cdc42 network embodies a Turing-type mechanism of yeast cell polarity. *FEBS Lett.* **2008**, *582*, 1437–1443. [CrossRef] [PubMed]
12. Jilkine, A.; Edelstein-Keshet, L. A Comparison of Mathematical Models for Polarization of Single Eukaryotic Cells in Response to Guided Cues. *PLoS Comput. Biol.* **2010**, *7*, e1001121. [CrossRef] [PubMed]
13. Edelstein-Keshet, L.; Holmes, W.R.; Zajac, M.; Dutot, M. From simple to detailed models for cell polarization. *Philos. Trans. R. Soc. Lond B. Biol. Sci.* **2013**, *368*, 20130003. [CrossRef] [PubMed]
14. Savage, N.S.; Layton, A.T.; Lew, D.J. Mechanistic mathematical model of polarity in yeast. *Mol. Biol. Cell* **2012**, *23*, 1998–2013. [CrossRef] [PubMed]
15. Dyer, J.M.; Savage, N.S.; Jin, M.; Zyla, T.R.; Elston, T.C.; Lew, D.J. Tracking shallow chemical gradients by actin-driven wandering of the polarization site. *Curr. Biol.* **2013**, *23*, 32–41. [CrossRef]
16. Freisinger, T.; Klunder, B.; Johnson, J.; Muller, N.; Pichler, G.; Beck, G.; Costanzo, M.; Boone, C.; Cerione, R.A.; Frey, E.; et al. Establishment of a robust single axis of cell polarity by coupling multiple positive feedback loops. *Nat. Commun.* **2013**, *4*, 1807. [CrossRef]

17. Klunder, B.; Freisinger, T.; Wedlich-Soldner, R.; Frey, E. GDI-mediated cell polarization in yeast provides precise spatial and temporal control of Cdc42 signaling. *PLoS Comput. Biol.* **2013**, *9*, e1003396. [CrossRef]
18. Woods, B.; Lai, H.; Wu, C.F.; Zyla, T.R.; Savage, N.S.; Lew, D.J. Parallel Actin-Independent Recycling Pathways Polarize Cdc42 in Budding Yeast. *Curr. Biol.* **2016**, *26*, 2114–2126. [CrossRef]
19. Verde, F.; Mata, J.; Nurse, P. Fission yeast cell morphogenesis: identification of new genes and analysis of their role during the cell cycle. *J. Cell Biol.* **1995**, *131*, 1529–1538. [CrossRef]
20. Mitchison, J.M.; Nurse, P. Growth in cell length in the fission yeast Schizosaccharomyces pombe. *J. Cell Sci.* **1985**, *75*, 357–376.
21. Csikász-Nagy, A.; Gyorffy, B.; Alt, W.; Tyson, J.J.; Novák, B. Spatial controls for growth zone formation during the fission yeast cell cycle. *Yeast* **2008**, *25*, 59–69. [CrossRef] [PubMed]
22. Das, M.; Drake, T.; Wiley, D.J.; Buchwald, P.; Vavylonis, D.; Verde, F. Oscillatory dynamics of Cdc42 GTPase in the control of polarized growth. *Science* **2012**, *337*, 239–243. [CrossRef] [PubMed]
23. Cerone, L.; Novak, B.; Neufeld, Z. Mathematical model for growth regulation of fission yeast Schizosaccharomyces pombe. *PLoS ONE* **2012**, *7*, e49675. [CrossRef]
24. Xu, B.; Jilkine, A. Modeling the Dynamics of Cdc42 Oscillation in Fission Yeast. *Biophys. J.* **2018**, *114*, 2025. [CrossRef] [PubMed]
25. Das, M.; Verde, F. Role of Cdc42 dynamics in the control of fission yeast cell polarization. *Biochem. Soc. Trans.* **2013**, *41*, 1745–1749. [CrossRef]
26. Raskin, D.M.; de Boer, P.A. Rapid pole-to-pole oscillation of a protein required for directing division to the middle of Escherichia coli. *Proc. Natl. Acad. Sci. USA* **1999**, *96*, 4971–4976. [CrossRef]
27. Loose, M.; Kruse, K.; Schwille, P. Protein self-organization: lessons from the min system. *Annu. Rev. Biophys.* **2011**, *40*, 315–336. [CrossRef]
28. Bendezu, F.O.; Martin, S.G. Cdc42 explores the cell periphery for mate selection in fission yeast. *Curr. Biol.* **2013**, *23*, 42–47. [CrossRef]
29. Haupt, A.; Ershov, D.; Minc, N. A Positive Feedback between Growth and Polarity Provides Directional Persistency and Flexibility to the Process of Tip Growth. *Curr. Biol.* **2018**, *28*, 3342–3351. [CrossRef]
30. Rincon, S.; Coll, P.M.; Perez, P. Spatial regulation of Cdc42 during cytokinesis. *Cell Cycle* **2007**, *6*, 1687–1691. [CrossRef]
31. Das, M.; Wiley, D.J.; Chen, X.; Shah, K.; Verde, F. The conserved NDR kinase Orb6 controls polarized cell growth by spatial regulation of the small GTPase Cdc42. *Curr. Biol.* **2009**, *19*, 1314–1319. [CrossRef]
32. Kelly, F.D.; Nurse, P. Spatial control of Cdc42 activation determines cell width in fission yeast. *Mol. Biol. Cell* **2011**, *22*, 3801–3811. [CrossRef]
33. Bendezu, F.O.; Vincenzetti, V.; Vavylonis, D.; Wyss, R.; Vogel, H.; Martin, S.G. Spontaneous Cdc42 polarization independent of GDI-mediated extraction and actin-based trafficking. *PLoS Biol.* **2015**, *13*, e1002097. [CrossRef]
34. Hercyk, B.S.; Rich-Robinson, J.; Mitoubsi, A.S.; Harrell, M.A.; Das, M.E. A novel interplay between GEFs orchestrates Cdc42 activity during cell polarity and cytokinesis in fission yeast. *J. Cell Sci.* **2019**, *132*. [CrossRef]
35. Lamas, I.; Merlini, L.; Vjestica, A.; Vincenzetti, V.; Martin, S.G. Optogenetics reveals Cdc42 local activation by scaffold-mediated positive feedback and Ras GTPase. *PLoS Biol.* **2020**, *18*, e3000600. [CrossRef]
36. Das, M.; Wiley, D.J.; Medina, S.; Vincent, H.A.; Larrea, M.; Oriolo, A.; Verde, F. Regulation of cell diameter, For3p localization, and cell symmetry by fission yeast Rho-GAP Rga4p. *Mol. Biol. Cell* **2007**, *18*, 2090–2101. [CrossRef] [PubMed]
37. Tang, H.; Bidone, T.C.; Vavylonis, D. Computational model of polarized actin cables and cytokinetic actin ring formation in budding yeast. *Cytoskeleton (Hoboken)* **2015**, *72*, 517–533. [CrossRef]
38. Tatebe, H.; Nakano, K.; Maximo, R.; Shiozaki, K. Pom1 DYRK regulates localization of the Rga4 GAP to ensure bipolar activation of Cdc42 in fission yeast. *Curr. Biol.* **2008**, *18*, 322–330. [CrossRef]
39. Revilla-Guarinos, M.T.; Martin-Garcia, R.; Villar-Tajadura, M.A.; Estravis, M.; Coll, P.M.; Perez, P. Rga6 is a Fission Yeast Rho GAP Involved in Cdc42 Regulation of Polarized Growth. *Mol. Biol. Cell* **2016**, *27*, 1524–1535. [CrossRef]
40. Gallo Castro, D.; Martin, S.G. Differential GAP requirement for Cdc42-GTP polarization during proliferation and sexual reproduction. *J. Cell Biol.* **2018**, *217*, 4215–4229. [CrossRef] [PubMed]

41. Khalili, B.; Merlini, L.; Vincenzetti, V.; Martin, S.G.; Vavylonis, D. Exploration and stabilization of Ras1 mating zone: A mechanism with positive and negative feedbacks. *PLoS Comput. Biol.* **2018**, *14*, e1006317. [CrossRef] [PubMed]
42. Xu, B.; Bressloff, P.C. A PDE-DDE Model for Cell Polarization in Fission Yeast. *SIAM J. Appl. Math.* **2016**, *76*, 1844–1870. [CrossRef]
43. Xu, B.; Kang, H.W.; Jilkine, A. Comparison of Deterministic and Stochastic Regime in a Model for Cdc42 Oscillations in Fission Yeast. *Bull. Math. Biol.* **2019**, *81*, 1268–1302. [CrossRef] [PubMed]
44. Paquin-Lefebvre, F.; Xu, B.; DiPietro, K.L.; Lindsay, A.E.; Jilkine, A. Pattern formation in a coupled membrane-bulk reaction-diffusion model for intracellular polarization and oscillations. *J. Theor. Biol.* **2020**, *497*, 110242. [CrossRef]
45. Geymonat, M.; Chessel, A.; Dodgson, J.; Punter, H.; Horns, F.; Nagy, A.C.; Carazo Salas, R.E. Activation of polarized cell growth by inhibition of cell polarity. *bioRxiv* **2018**. [CrossRef]
46. Merlini, L.; Khalili, B.; Dudin, O.; Michon, L.; Vincenzetti, V.; Martin, S.G. Inhibition of Ras activity coordinates cell fusion with cell-cell contact during yeast mating. *J. Cell Biol.* **2018**, *217*, 1467–1483. [CrossRef]
47. Novak, I.L.; Gao, F.; Choi, Y.S.; Resasco, D.; Schaff, J.C.; Slepchenko, B.M. Diffusion on a Curved Surface Coupled to Diffusion in the Volume: Application to Cell Biology. *J. Comput. Phys.* **2007**, *226*, 1271–1290. [CrossRef]
48. Kokkoris, K.; Gallo Castro, D.; Martin, S.G. The Tea4-PP1 landmark promotes local growth by dual Cdc42 GEF recruitment and GAP exclusion. *J. Cell Sci.* **2014**, *127*, 2005–2016. [CrossRef]
49. Ng, A.Y.E.; Ng, A.Q.E.; Zhang, D. ER-PM Contacts Restrict Exocytic Sites for Polarized Morphogenesis. *Curr. Biol.* **2018**, *28*, 146–153. [CrossRef]
50. Tay, Y.D.; Leda, M.; Goryachev, A.B.; Sawin, K.E. Local and global Cdc42 guanine nucleotide exchange factors for fission yeast cell polarity are coordinated by microtubules and the Tea1-Tea4-Pom1 axis. *J. Cell Sci.* **2018**, *131*. [CrossRef]
51. Howell, A.S.; Jin, M.; Wu, C.F.; Zyla, T.R.; Elston, T.C.; Lew, D.J. Negative feedback enhances robustness in the yeast polarity establishment circuit. *Cell* **2012**, *149*, 322–333. [CrossRef]
52. Das, M.; Nunez, I.; Rodriguez, M.; Wiley, D.J.; Rodriguez, J.; Sarkeshik, A.; Yates, J.R., 3rd; Buchwald, P.; Verde, F. Phosphorylation-dependent inhibition of Cdc42 GEF Gef1 by 14-3-3 protein Rad24 spatially regulates Cdc42 GTPase activity and oscillatory dynamics during cell morphogenesis. *Mol. Biol. Cell* **2015**, *26*, 3520–3534. [CrossRef] [PubMed]
53. Pablo, M.; Ramirez, S.A.; Elston, T.C. Particle-based simulations of polarity establishment reveal stochastic promotion of Turing pattern formation. *PLoS Comput. Biol.* **2018**, *14*, e1006016. [CrossRef] [PubMed]
54. Pino, M.R.; Nuñez, I.; Chen, C.; Das, M.E.; Wiley, D.J.; D'Urso, G.; Buchwald, P.; Vavylonis, D.; Verde, F. Cdc42 GTPase Activating Proteins (GAPs) Maintain Generational Inheritance of Cell Polarity and Cell Shape in Fission Yeast. *bioRxiv* **2020**. [CrossRef]
55. Kuo, C.C.; Savage, N.S.; Chen, H.; Wu, C.F.; Zyla, T.R.; Lew, D.J. Inhibitory GEF phosphorylation provides negative feedback in the yeast polarity circuit. *Curr. Biol.* **2014**, *24*, 753–759. [CrossRef]
56. Rapali, P.; Mitteau, R.; Braun, C.; Massoni-Laporte, A.; Unlu, C.; Bataille, L.; Arramon, F.S.; Gygi, S.P.; McCusker, D. Scaffold-mediated gating of Cdc42 signalling flux. *Elife* **2017**, *6*. [CrossRef]
57. Chen, C.; Rodriguez Pino, M.; Haller, P.R.; Verde, F. Conserved NDR/LATS kinase controls RAS GTPase activity to regulate cell growth and chronological lifespan. *Mol. Biol. Cell* **2019**, *30*, 2598–2616. [CrossRef]
58. Tay, Y.D.; Leda, M.; Spanos, C.; Rappsilber, J.; Goryachev, A.B.; Sawin, K.E. Fission Yeast NDR/LATS Kinase Orb6 Regulates Exocytosis via Phosphorylation of the Exocyst Complex. *Cell Rep.* **2019**, *26*, 1654–1667. [CrossRef]
59. Okada, S.; Leda, M.; Hanna, J.; Savage, N.S.; Bi, E.; Goryachev, A.B. Daughter cell identity emerges from the interplay of Cdc42, septins, and exocytosis. *Dev. Cell* **2013**, *26*, 148–161. [CrossRef]
60. Drake, T.; Vavylonis, D. Model of fission yeast cell shape driven by membrane-bound growth factors and the cytoskeleton. *PLoS Comput. Biol.* **2013**, *9*, e1003287. [CrossRef]
61. Bement, W.M.; Leda, M.; Moe, A.M.; Kita, A.M.; Larson, M.E.; Golding, A.E.; Pfeuti, C.; Su, K.C.; Miller, A.L.; Goryachev, A.B.; et al. Activator-inhibitor coupling between Rho signalling and actin assembly makes the cell cortex an excitable medium. *Nat. Cell Biol.* **2015**, *17*, 1471–1483. [CrossRef] [PubMed]

62. Vandin, G.; Marenduzzo, D.; Goryachev, A.B.; Orlandini, E. Curvature-driven positioning of Turing patterns in phase-separating curved membranes. *Soft Matter* **2016**, *12*, 3888–3896. [CrossRef] [PubMed]
63. Brauns, F.; Pawlik, G.; Halatek, J.; Kerssemakers, J.; Frey, E.; Dekker, C. Bulk-surface coupling reconciles Min-protein pattern formation in vitro and in vivo. *bioRxiv* **2020**. [CrossRef]

Essay

Symmetry Breaking during Cell Movement in the Context of Excitability, Kinetic Fine-Tuning and Memory of Pseudopod Formation

Peter J.M. van Haastert

Department of Cell Biochemistry, University of Groningen, Nijenborgh 7, 9747 AG Groningen, The Netherlands; p.j.m.van.haastert@rug.nl; Tel.: +31-50-3634209

Received: 5 July 2020; Accepted: 29 July 2020; Published: 30 July 2020

Abstract: The path of moving eukaryotic cells depends on the kinetics and direction of extending pseudopods. Amoeboid cells constantly change their shape with pseudopods extending in different directions. Detailed analysis has revealed that time, place and direction of pseudopod extension are not random, but highly ordered with strong prevalence for only one extending pseudopod, with defined life-times, and with reoccurring events in time and space indicative of memory. Important components are Ras activation and the formation of branched F-actin in the extending pseudopod and inhibition of pseudopod formation in the contractile cortex of parallel F-actin/myosin. In biology, order very often comes with symmetry. In this essay, I discuss cell movement and the dynamics of pseudopod extension from the perspective of symmetry and symmetry changes of Ras activation and the formation of branched F-actin in the extending pseudopod. Combining symmetry of Ras activation with kinetics and memory of pseudopod extension results in a refined model of amoeboid movement that appears to be largely conserved in the fast moving *Dictyostelium* and neutrophils, the slow moving mesenchymal stem cells and the fungus *B.d. chytrid*.

Keywords: pseudopod; Ras activation; cytoskeleton; *Dictyostelium*; chemotaxis; neutrophils

---

1. Introduction

Amoeboid cells move by extending protrusions [1]. The shape of amoeboid cells is very flexible with frequent changes of extensions and directions of movement. These cells have an asymmetric appearance with seemingly infinite ways to construct the amoeboid body. However, careful analysis of how cells extend protrusions and the resulting trajectories have shown that movement is not random [2,3]. Many amoeboid cells have the tendency to persist in the direction of movement. Cells have a front and a rear that, although flexible, are relatively stable on a minute time scale. Such cells have a longitudinal axis with (imperfect) symmetry. Here symmetry is used to describe order; the infinite ways to construct an asymmetrical amoeboid body are restricted by the extending pseudopod. The extending pseudopod brings order and breaks the symmetry of the cell. This symmetry breaking is not static, such as the creation of the stable bilateral body plan during animal development, but is very dynamic, since cells extend a new pseudopod every ~15 s.

After working for many years on the biochemistry of cell movement and chemotaxis, and after a period serving the university as dean of education, I returned to science performing experiments on what seemed to me the fundament of cell movement: how do cells extend pseudopods and what is the underlying mechanism? In this essay, I combine the results of these recent studies [4–7] to present my view on the mechanism of pseudopod extension, discussed from the perspective of symmetry breaking. Pseudopod extension is regulated by many signaling molecules, especially during directed cell movement guided by chemoattractant. In *Dictyostelium*, members of the Ras family of GTPases that are detected with RBD-Raf-GFP appear to play an important role, being an upstream signal with strong local activation, both during cell movement in chemotactic gradients and in buffer [5,8]. Here, I use active Ras-GTP as a molecular anchor for symmetry. In the next paragraphs, I first give some background on symmetry forms, and their changes in time and space, on the cytoskeleton, and on the coupling of excitable Ras-GTP activation and the cytoskeleton. The core of the essay is the discussions on symmetry during cell movement, that starts with uniform Ras activation in a round cell, to which different components of the cytoskeleton are added based on experiments, thereby obtaining the complex symmetry form of polarized cells. The resulting model of amoeboid movement reveals a very rich pallet of regulatory mechanisms that provide amoeboid cells with complex symmetry forms, memory and refined kinetics of cell movement. Interestingly, the fundaments of symmetry for cell movement uncovered in *Dictyostelium* are also detectable in very different cells, such as the fast moving neutrophils, the slow moving mesenchymal stem cells or the fungus *B.d. chytrid*.

2. Fundaments of Symmetry and Symmetry Breaking in Cell Movement

Symmetry means the existence of different viewpoints from which the system appears the same. Symmetry breaking is the process by which the number of these viewpoints (the order of symmetry) is reduced, to generate a more ordered, structured and improbable state [9]. The term order is easily confusing, because it has opposite meaning in thermodynamics and symmetry. For instance, as the symmetry breaks from rotation symmetry to reflection symmetry, the system becomes more ordered thermodynamically, while the order of symmetry (the number of viewpoints) decreases. To avoid this confusion, I will use the term complexity to indicate that both the structure and the symmetry become more complex. In biology, symmetry is not precise as in physics. Furthermore, symmetry is sometimes permanent, such as the shape of a sea star, but in cell movement and many other processes in biology, symmetry is dynamic. When symmetry is disturbed, the system may stay disturbed, it may return to the original symmetry state, or it may adopt a more complex symmetry, depending on the underlying mechanisms. To understand the role of dynamic symmetry in biology, it is essential to understand these underlying mechanisms. Some of these aspects of symmetry are illustrated in Figure 1a,b using objects that are constructed with elements from my greenhouse. Firstly, symmetry is not exact and may depend on the point of interest. The object of Figure 1a clearly appears to have 5-fold rotational symmetry: the five elements are approximately equal in size and arranged at approximately equal angles. However, in its details the elements have different width, color and curvature, and the five elements are oriented differently with curvature to the right in three elements and to the left in two elements. Therefore, the object is symmetric in some studies e.g., how local activators and inhibitors lead to rotational symmetry in a sea star, but the object is non-symmetric in other studies e.g., how elements in the object can get a left- or right-handed curvature such as in the rotation of runner beans and French beans. Secondly, symmetry is often dynamic. When one element disappears (Figure 1b) the object gets imperfect 5-fold rotational symmetry (one element has zero size) or imperfect 4-fold rotational symmetry (all four angles are different from 90 degrees). Interestingly the object gets reflection symmetry that is nearly perfect. If the underlying mechanism is rotational symmetry, the object may recover 5-fold symmetry requiring the growth of a new element, or 4-fold symmetry requiring that the position of the four elements adopt 90 degrees angles. However, if the system has the ability to form a polarity axis, the reflection symmetry may be enforced by specific growth of some elements and shrinkage of other elements. Thus, the disturbed structure dynamically changes to a regular structure with a symmetry that depends on the underlying mechanisms.

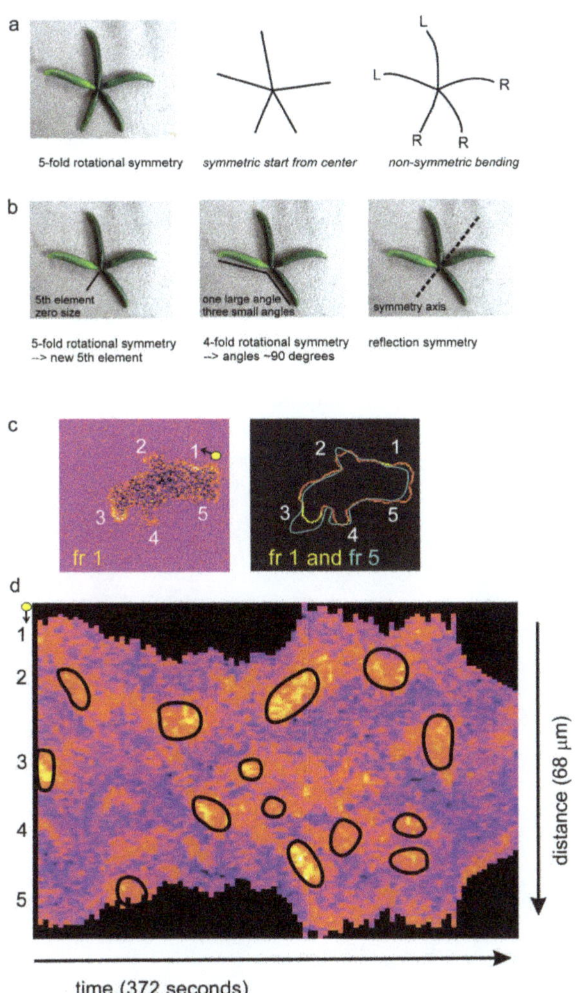

Figure 1. Fundaments of symmetry and dynamic symmetry breaking. (**a**) Symmetry in biology may depend on the viewpoint. The image is an object with 5 elements. The drawing in the middle connects the center of the object with the ends of the elements, showing approximately 5-fold rotational symmetry. The drawing on the right follows the curvature of the elements, showing that the object has poor 5-fold symmetry. (**b**) Distortion of symmetry. Removing one element leads to poor 4-fold and poor 5-fold rotational symmetry, but to very good reflection symmetry. Symmetry can recover in different ways, depending on the underlying molecular mechanism. (**c**,**d**) Unpolarized early *Dictyostelium* cells expressing RBD-Raf-GFP and cytosolic-RFP were followed in time at 4 s per frame, providing a very sensitive sensor for active Ras-GTP. The intensity at the boundary of the cell was measured and is presented in the kymograph. (**c**) Image of frame 1 reveals multiple Ras patches with approximately 5-fold rotational symmetry; the outline of the cell in frame 5 reveals that a pseudopod was extended at Ras-GTP patch 3. In frame 6 a pseudopod will start in patch 2. (**d**) The kymograph reveals about 53 Ras-GTP patches and 14 extending pseudopods (indicated by the back circles); (**c**,**d**) are redrawn from [4].

These aspects of symmetry are illustrated in Figure 1c,d for a *Dictyostelium* cell expressing a sensor for active Ras-GTP and extending pseudopods [4]. Depending on the viewpoint (pseudopod extension or Ras-GTP patches; Figure 1c), the same object may have different symmetry forms. The cell has multiple patches of Ras-GTP that are distributed nearly evenly around the cell: Ras-GTP patches have rotational symmetry. Often the cell is somewhat elongated with only one extending pseudopod: movement has reflection symmetry. Shape and movement has fewer viewpoints of symmetry than the Ras-GTP patches, therefore shape is a more complex symmetry state than Ras-GTP patches. Furthermore, the more complex symmetry state may depend on a symmetry state with less complexity. If this cell is followed in time using the kymograph of Figure 1d, it appears that the 53 Ras-GTP patches are dynamic with a life time of about 24 s; on average the cell has 3 to 4 patches, and when a Ras-GTP patch disappears a new Ras-GTP patch is initiated and the cell keeps rotational symmetry of Ras-GTP patches. This cell extends only 14 pseudopods with a life time of about 15 s; usually a cell extends only one pseudopod at the same time, far less than the 3 to 4 Ras-GTP patches. Importantly, when a new pseudopod is made, it always starts at a place of a Ras-GTP patch, and nearly always at the Ras-GTP patch with the highest intensity. Thus, although the shape of the cell has reflection symmetry, the underlying mechanism is rotational symmetry of Ras-GTP patches. Consequently, the pseudopods start at different sides of the cell, and these briefly starved cell moves in nearly random direction. When *Dictyostelium* cells are starved for prolonged periods, they become polarized in shape and Ras activation: Cells still have multiple Ras-GTP patches, but the intensity is much higher in the patch at the current front of the cell: Ras-GTP patches have reflection symmetry (see also below in Figure 2). New pseudopods are still formed at the strongest Ras-GTP patch and therefore all pseudopods start in the front near the existing pseudopod. Consequently, the cell moves with persistence. The difference in persistent starved cells and random movement of non-starved cells is not the shape of the cell, but the underlying symmetry form of the activating Ras-GTP patches.

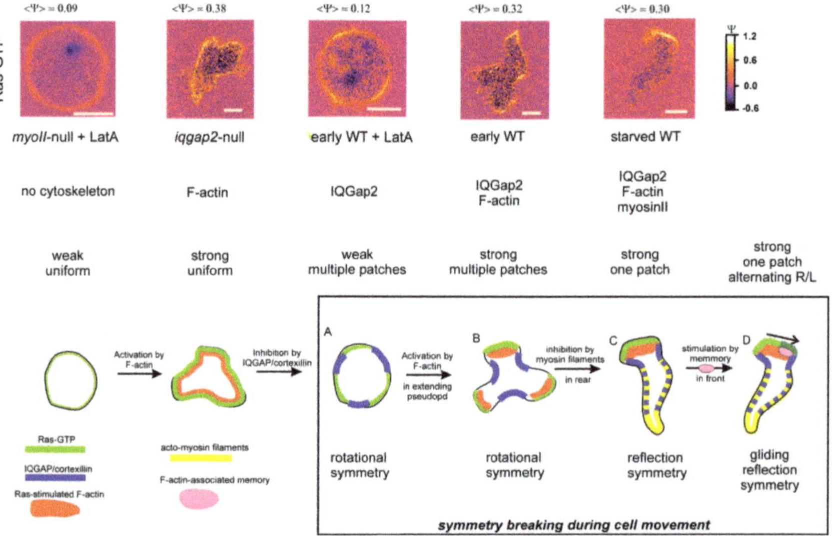

Figure 2. Symmetry and symmetry breaking of Ras-GTP localization in a series of mutants in the absence and presence of the F-actin inhibitor LatA. The mutants are ordered so they form a sequence of symmetry breaking. Top shows images of representative cells. <Ψ> is the average fluorescence intensity at the boundary of the cell (see [4] for definition). Bottom shows schematics with the localization of key components to establish the different forms of symmetry. The box represents the symmetry forms and transitions in wild type *Dictyostelium* cells. The figure is redrawn from [5].

3. The Cytoskeleton of Moving Cells

The two major parts of the cytoskeleton of moving cells are branched F-actin filaments in the extending protrusion (bF-actin) and parallel F-actin filaments (pF-actin) in the contractile cortex [1,10–12]. Cryogenic Electron Microscopy has revealed that an emerging pseudopod contains in the order of 4000 branched actin filaments directing towards the membrane of the extending pseudopod [6,13]. The bF-actin is regulated by nucleation of Arp2/3 branch points that generate new filament ends from which polymerization can occur. Extension of the pseudopod will continue as long as new branches and filaments are formed, and as long as sufficient place is available for sufficient time between the end points of these filaments and the plasma membrane. At some moment, the balance between elongation and counter forces reverses, further polymerization comes to an arrest, and the pseudopod stops. Nucleation of the Arp2/3 is induced by Scar, a complex of five proteins that is regulated by Rac-GTP and its upstream activator Ras-GTP [14].

The contractile cortex is a thin 100 nm thick layer under the plasma membrane. It consists of long parallel F-actin filaments, cross linkers such as α-actinin, membrane anchor proteins, and myosin II filaments that can provide contractile force [12,15–17]. In regions with a strong contractile cortex, it is difficult to generate a new protrusion of branched F-actin [18–20]. In this respect, bF-actin in the extending pseudopod and pF-actin in the contractile cortex have opposite function, but both are detected with many F-actin sensors such as LimE-GFP and lifeact-GFP [21]. Myosin II filaments disassemble by phosphorylation of the heavy chain by myosin heavy chain kinase (MHCK) [22,23]. Disassembly of myosin filaments weakens the contractile cortex thereby increasing the probability that branched F-actin can be formed inducing a new pseudopod. Polymerization of pF-actin in the cortex is regulated by formins, ring-like structures that are activated by the small GTPase RacE. Mutants lacking three formins (*forAEH*-null) or RacE have a very weak contractile cortex and extend multiple pseudopods [16].

4. Coupled Excitable Ras/bF-Actin

Many observations suggest that F-actin forms an excitable medium [4,24–28]. Small fluctuations are damped, but fluctuations above a threshold lead to strong amplification with the characteristics of standing waves. The small GTPase Ras can exist in the inactive Ras-GDP and active Ras-GTP form. Sensitive sensors for the active Ras-GTP form suggest that, in the absence of F-actin, activation of Ras is also excitable leading to multiple patches of Ras-GTP with a diameter of about 3 μm and a life time of about 16 s [4]. Interestingly, excitable Ras and excitable F-actin are coupled: Ras induces F-actin, and F-actin further increases Ras activation. Cells usually have only one extending pseudopod enriched with F-actin, but multiple Ras-GTP patches. Therefore, when a new pseudopod is formed it is induced at the position of the strongest Ras-GTP patch, and upon induction of F-actin in this emerging pseudopod Ras is further activated to a very strong Ras-GTP patch with increased size (from 3 to 7 μm) and increased life-time (from 16 to 42 s). In a coupled excitable system, fluctuations above a threshold of either component can lead to the excitation of both. Thus a new pseudopod can start with an increase of F-actin in a patch of Ras-GTP (that further activates Ras-GTP/F-actin) or with an increase of Ras-GTP (that induces F-actin and further activation of Ras-GTP). Indeed, detailed analysis revealed that new pseudopods in the front of the cell (that contains bF-actin and lacks a contractile cortex) start with an increase of F-actin (presumably bF-actin), while a new pseudopod that occasionally is formed in the contractile cortex can only start with a strong patch of Ras-GTP that must first locally weaken the contractile cortex by disassembly of myosin filaments, followed by induction of branched F-actin [4,6,7].

5. Symmetry of Moving Cells

The symmetry of Ras patches, F-actin and cell shape has been analyzed in wild-type cells and a series of mutants with defects in components of the cytoskeleton or its regulation [5]. The role of

F-actin (both bF-actin in the front and pF-actin in the contractile cortex) was studied with the use of the inhibitor Latrunculin A (LatA) and the role of myosin II in the contractile cortex in the rear of the cell with a null mutant. The border between bF-actin in the front and pF-actin in the rear is enriched with IQGap2/cortexillin/myosin II [29]. From this information a hierarchy of symmetry forms can be deduced (Figure 2).

Cells lacking functional F-actin, myosin and IQGAP (*myoII*-null + LatA) are round and have uniform low levels or Ras-GTP; i.e., they are in a low basal state. Incorporating F-actin (*iqgap2*-null without latA) induces some deformation in an otherwise rather round cell; importantly Ras-GTP levels are strongly elevated but still nearly uniform. In both mutant situations the cell has indefinite rotational symmetry. Cells that do have IQGap2 but no functional F-actin and myosin filaments (early wild-type + LatA) are also round, but now Ras-GTP is located in patches. These patches are rather uniform in size (3 μm) and life-time (16 s); most cells have 3 to 5 Ras-GTP patches that are approximately equally distributed over the circumference of the cell [4]. Thus, these cells have 3–5 fold rotational symmetry, and symmetry is due to inhibition of Ras-GTP formation in between the patches by IQGap2. Cells lacking only functional myosin II filaments (early wild-type or *myoII*-null cells) also have multiple Ras-GTP patches approximately symmetric around the cell. However, these cells do extend a protrusion by which one of the Ras-GTP patches in the protrusion is stronger than the others (see also the kymograph of early wild-type in Figure 1c,d). With myosin II filaments, and thereby an active contractile cortex, the starved wild-type cell gets a relatively stable front to rear axis: protrusions appear predominantly in the front. Therefore, these cells adopt reflection symmetry along the front-to-rear axis. Cells do make Ras-GTP patches at the side and in the rear, but usually these patches are not strong enough to overcome the inhibition of the contractile cortex to induce a branched F-actin filled protrusion. In polarized wild-type cells, the strongest Ras-GTP patches and associated protrusions appear in the front alternatingly to the right and left. Thus, in time, gliding reflection symmetry is detectable, comparable to the footsteps of a walker in the snow; two objects with gliding reflection symmetry are identical after one object is glided back opposite the other object and then reflected. A mutant expressing a phospho-mimic form of Scar (ScarS55D [30]) lacks this form of symmetry: Ras patches and pseudopods still appear in the front, but not alternatingly to the right and left, suggesting that Scar or a downstream component mediates the breaking of reflection symmetry into gliding reflection symmetry [4,7].

The experimenter can further intervene with these symmetry forms to induce specific responses; for instance the expression of dominant active Rap1G12V in polarized cells leads to massive destabilization of the contractile cortex (thereby losing longitudinal symmetry) and uniform activation of Ras-GTP (thereby losing rotational symmetry). These cells extend multiple pseudopods in all directions, and since Rap1-GTP induces strong adhesion of the cell to the substratum, these cells get a pancake-like appearance [6,28].

In summary, the Ras/cytoskeleton system can adopt a series of symmetry forms of activated Ras-GTP depending on the presence of different elements of the cytoskeleton. In wild-type cells (Figure 2A–D) the basis is rotational symmetry of Ras-GTP in unpolarized cells (A,B), converting to reflection symmetry of polarized cells (C), and finally to gliding reflection symmetry in polar wild-type cells with positional memory (D). And the coupled excitable Ras/F-actin system then induces a pseudopod at the position of the strongest Ras-GTP patch.

6. Kinetic Fine-Tuning of Pseudopod Formation and Symmetry Breaking

Each time the cell extends a pseudopod the symmetry gets broken in a new way. Therefore the principles underlying symmetry breaking are relevant for the kinetics of pseudopod formation and *vice versa*. Detailed kinetics of pseudopod formation revealed that a cell without pseudopod has a stochastic probability of 15%/s to extend a pseudopod. In wild-type cells the probability to extend a second pseudopod is strongly inhibited 3.5 fold in the entire cell, and the probability to extend a third pseudopod is inhibited even 13-fold [6]. Similar complex kinetics of pseudopod extension was also observed in neutrophils, mesenchymal stem cells and the fungus *B.d. chytrid* [6]. In *Dictyostelium* mutants *forAEH*-null and *racE*-null the start of a new pseudopod is inhibited only 1.2 fold by an extending pseudopod, demonstrating that inhibition requires the contractile cortex [6,16]. The contractile cortex generates the longitudinal axis of polarity. This suggests that the extending pseudopod strongly enhances this longitudinal axis of reflection symmetry. As a consequence, cells generally extend only one pseudopod at a time, not zero (then the probability to start a pseudopod is high) and not more than one (then the probability is strongly inhibited).

7. Memory and Symmetry Breaking

The extension of pseudopods is also non-random in space. Cells have persistent movement, meaning that new pseudopods are extended in a similar direction as previous pseudopods. Since pseudopods are extended perpendicular to the cell surface, the tendency to move in the same direction implies that new pseudopods start nearby previous pseudopods [2,31–34]. Somehow cells have a memory of the place in the cell where they extended previous pseudopod(s). Detailed analysis uncovered two types of memory: a long term memory of a polarity axis related to longitudinal front to rear reflection symmetry, and short term memory of position related to the alternating right/left gliding reflection symmetry. The long term memory stores the global position of the last ~11 pseudopods forming a polarity axis from front to rear with a contractile cortex in the rear half of the cell [3,7,35]. New pseudopods are preferentially made in the front 30% of the cell, which provides both persistence of movement and strengthens the polarity axis. Myosin II filaments in the contractile cortex are essential to establish a polarity axis. It is evident that longitudinal symmetry breaking generating reflection symmetry forms the basis for the memory of the polarity axis. Therefore, it appears that longitudinal symmetry breaking leads to a rather stable front-rear polarized cell.

The short term memory of position remembers only the position of the last previous pseudopod, which increases the probability to start the next-next pseudopod at that position [2,5,7,31]. This memory generates two series of pseudopods: odd pseudopods (1,3,5 etc.) starting at the same position and even pseudopods (2, 4, 6 etc.) all starting from another position. In polarized cells with their front-rear polarity axis, these two positions are both in the front 30% of the cell and cells move by alternating right/left pseudopods leading to persistent zig-zag trajectories [2]. A mutant expressing a phospho-mimick form of Scar (ScarS55D) has a polarity axis but no memory of position and therefore extend pseudopods somewhere in the front 30% of the cell but not in alternating right/left order [5,7]. It is evident that gliding reflection symmetry breaking generating excitable hotspots and is the basis for the memory of the position. Therefore, it appears that gliding reflection symmetry breaking is very dynamic with a life-time of only one pseudopod.

8. Conclusions

Symmetry in biology is approximate and not exact as in physics. Two aspects of symmetry and symmetry breaking are important during cell movement. Firstly, the symmetry forms are metastable: in rotational symmetry unpolarized cells may have a number of Ras-GTP patches that are approximately equally distributed around the cell. The number of patches may increase or decrease, but the cell retains rotational symmetry. Secondly, transitions to a more complex symmetry are transient and dynamic: an unpolarized cell with rotational symmetry of Ras-GTP patches may start a pseudopod at one of these patches thereby getting an elongated shape with reflection symmetry; when the pseudopod stops, the cell returns to a round shape, but still has rotational symmetry of Ras-GTP patches, and later a pseudopod may start from another Ras-GTP patch. Understanding the molecular basis how rotational symmetry of Ras-GTP patches is modified by cytoskeleton and signaling pathways is instrumental for our understanding how cells can use these symmetry forms to generate specific locomotion behavior (Figure 3). The coupled excitable Ras-GTP/F-actin system triggers at the strongest Ras-GTP patch the nucleation and further polymerization of branched F-actin, leading to the extension of a pseudopod. The extending pseudopod activates two systems: first, the local modification of presumably Scar providing a short term memory that later primes the Ras-GTP/F-actin excitatory to start a new pseudopod at that position. The second system activated by the extending pseudopod is a combination of the global induction of myosin II filaments in the entire cell and its inhibition at the place of the extending pseudopod. In *Dictyostelium* global activation myosin filament formation is mediated by the rapidly diffusing cGMP, produced by a guanylyl cyclase that is activated in the extending pseudopod [7,22], while local inhibition is mediated by Rap1-GTP that is activated in the pseudopod by Ras-GTP [36]. As a consequence of global activation and local inhibition, myosin II filaments are formed in the rear, thereby inhibiting pseudopod formation in the rear and providing a longitudinal symmetry axis. Since this longitudinal symmetry form is relatively stable, it leads to a long-term memory of pseudopods formation in the front of the cell, which is the fundament of intrinsic persistent movement for efficient food searching [2,37] and chemotaxis [38]. The scheme depicted in Figure 3 for *Dictyostelium* probably also hold for other organisms, suggested by comparing kinetics and memory of cell movement in *Dictyostelium*, the fast moving neutrophils, the slow moving mesenchymal stem cells or the fungus *B.d. chytrid* [6,7,39]. Although the molecules and regulatory mechanisms may be different in these organisms, such as the role of Ras versus CDC42 for actin polymerization [40] or cGMP versus Rho-Kinase for myosin polymerization [41], the fundaments of symmetry and symmetry breaking for cell movement may be conserved.

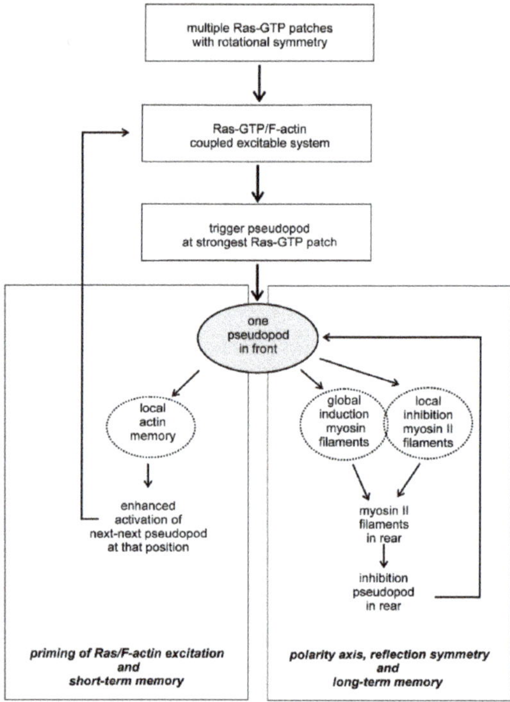

Figure 3. Flow diagram of symmetry, excitability and memory for cell movement.

Funding: This research received no external funding.

Conflicts of Interest: The author declares no conflict of interest.

References

1. Ridley, A.J. Life at the leading edge. *Cell* **2011**, *145*, 1012–1022. [CrossRef] [PubMed]
2. Li, L.; Nørrelykke, S.F.; Cox, E.C. Persistent cell motion in the absence of external signals: A search strategy for eukaryotic cells. *PLoS ONE* **2008**, *3*, e2093. [CrossRef] [PubMed]
3. Bosgraaf, L.; Van Haastert, P.J.M. The ordered extension of pseudopodia by amoeboid cells in the absence of external cues. *PLoS ONE* **2009**, *4*, e5253. [CrossRef] [PubMed]
4. Van Haastert, P.J.M.; Keizer-Gunnink, I.; Kortholt, A. Coupled excitable Ras and F-actin activation mediates spontaneous pseudopod formation and directed cell movement. *Mol. Biol. Cell* **2017**, *28*, 922–934. [CrossRef] [PubMed]
5. van Haastert, P.J.M.; Keizer-Gunnink, I.; Kortholt, A. The cytoskeleton regulates symmetry transitions in moving amoeboid cells. *J. Cell Sci.* **2018**, *131*, jcs208892. [CrossRef]
6. van Haastert, P.J.M. Unified control of amoeboid pseudopod extension in multiple organisms by branched F-actin in the front and parallel F-actin/myosin in the cortex. **2020**, submitted.
7. van Haastert, P.J.M. Short- and long-term memory of moving amoeboid cells. **2020**, submitted.
8. Kortholt, A.; Keizer-Gunnink, I.; Kataria, R.; Van Haastert, P.J.M. Ras activation and symmetry breaking during Dictyostelium chemotaxis. *J. Cell Sci.* **2013**, *126*, 4502–4513. [CrossRef]
9. Gross, D.J. The role of symmetry in fundamental physics. *Proc. Natl. Acad. Sci. USA* **1996**, *93*, 14256–14259. [CrossRef]
10. Pollard, T.D.; Borisy, G.G. Cellular motility driven by assembly and disassembly of actin filaments. *Cell* **2003**, *112*, 453–465. [CrossRef]
11. Insall, R.H.; Machesky, L.M. Actin dynamics at the leading edge: From simple machinery to complex networks. *Dev. Cell* **2009**, *17*, 310–322. [CrossRef]
12. Chugh, P.; Paluch, E.K. The actin cortex at a glance. *J. Cell Sci.* **2018**, *131*, jcs186254. [CrossRef]
13. Urban, E.; Jacob, S.; Nemethova, M.; Resch, G.P.; Small, J.V. Electron tomography reveals unbranched networks of actin filaments in lamellipodia. *Nat. Cell Biol.* **2010**, *12*, 429–435. [CrossRef] [PubMed]
14. Davidson, A.J.; Insall, R.H. SCAR/WAVE: A complex issue. *Commun. Integr. Biol.* **2013**, *6*, e27033. [CrossRef] [PubMed]
15. Nishikawa, M.; Naganathan, S.R.; Julicher, F.; Grill, S.W. Controlling contractile instabilities in the actomyosin cortex. *Elife* **2017**, *6*, e19595. [CrossRef] [PubMed]
16. Litschko, C.; Bruhmann, S.; Csiszar, A.; Stephan, T.; Dimchev, V.; Damiano-Guercio, J.; Junemann, A.; Korber, S.; Winterhoff, M.; Nordholz, B.; et al. Functional integrity of the contractile actin cortex is safeguarded by multiple Diaphanous-related formins. *Proc. Natl. Acad. Sci. USA* **2019**, *116*, 3594–3603. [CrossRef] [PubMed]
17. Rottner, K.; Faix, J.; Bogdan, S.; Linder, S.; Kerkhoff, E. Actin assembly mechanisms at a glance. *J. Cell Sci.* **2017**, *130*, 3427–3435. [CrossRef] [PubMed]
18. Lomakin, A.J.; Lee, K.-C.; Han, S.J.; Bui, D.A.; Davidson, M.; Mogilner, A.; Danuser, G. Competition for actin between two distinct F-actin networks defines a bistable switch for cell polarization. *Nat. Cell Biol.* **2015**, *17*, 1435–1445. [CrossRef]
19. Ngo, K.X.; Umeki, N.; Kijima, S.T.; Kodera, N.; Ueno, H.; Furutani-Umezu, N.; Nakajima, J.; Noguchi, T.Q.P.; Nagasaki, A.; Tokuraku, K.; et al. Allosteric regulation by cooperative conformational changes of actin filaments drives mutually exclusive binding with cofilin and myosin. *Sci. Rep.* **2016**, *6*, 35449. [CrossRef]
20. Davidson, A.J.; Wood, W. Unravelling the Actin Cytoskeleton: A New Competitive Edge? *Trends Cell Biol.* **2016**, *26*, 569–576. [CrossRef]
21. Riedl, J.; Crevenna, A.H.; Kessenbrock, K.; Yu, J.H.; Neukirchen, D.; Bista, M.; Bradke, F.; Jenne, D.; Holak, T.A.; Werb, Z.; et al. Lifeact: A versatile marker to visualize F-actin. *Nat. Methods* **2008**, *5*, 605–607. [CrossRef]
22. Bosgraaf, L.; van Haastert, P.J.M. The regulation of myosin II in Dictyostelium. *Eur. J. Cell Biol.* **2006**, *85*, 969–979. [CrossRef]
23. Steimle, P.A.; Yumura, S.; Côté, G.P.; Medley, Q.G.; Polyakov, M.V.; Leppert, B.; Egelhoff, T.T. Recruitment of a myosin heavy chain kinase to actin-rich protrusions in Dictyostelium. *Curr. Biol.* **2001**, *11*, 708–713. [CrossRef]
24. Gerisch, G.; Ecke, M. Wave Patterns in Cell Membrane and Actin Cortex Uncoupled from Chemotactic Signals. *Methods Mol. Biol.* **2016**, *1407*, 79–96. [PubMed]

25. Devreotes, P.; Horwitz, A.R. Signaling Networks that Regulate Cell Migration. *Cold Spring Harb. Perspect. Biol.* **2015**, *7*, a005959. [CrossRef] [PubMed]
26. Gerhardt, M.; Ecke, M.; Walz, M.; Stengl, A.; Beta, C.; Gerisch, G. Actin and PIP3 waves in giant cells reveal the inherent length scale of an excited state. *J. Cell Sci.* **2014**, *127*, 4507–4517. [CrossRef]
27. Vicker, M.G. F-actin assembly in Dictyostelium cell locomotion and shape oscillations propagates as a self-organized reaction-diffusion wave. *FEBS Lett.* **2002**, *510*, 5–9. [CrossRef]
28. Miao, Y.; Bhattacharya, S.; Edwards, M.; Cai, H.; Inoue, T.; Iglesias, P.A.; Devreotes, P.N. Altering the threshold of an excitable signal transduction network changes cell migratory modes. *Nat. Cell Biol.* **2017**, *19*, 329–340. [CrossRef]
29. Faix, J.; Weber, I. A dual role model for active Rac1 in cell migration. *Small GTPases* **2013**, *4*, 110–115. [CrossRef]
30. Ura, S.; Pollitt, A.Y.; Veltman, D.M.; Morrice, N.A.; Machesky, L.M.; Insall, R.H. Pseudopod growth and evolution during cell movement is controlled through SCAR/WAVE dephosphorylation. *Curr. Biol.* **2012**, *22*, 553–561. [CrossRef]
31. Cooper, R.M.; Wingreen, N.S.; Cox, E.C. An excitable cortex and memory model successfully predicts new pseudopod dynamics. *PLoS ONE* **2012**, *7*, e33528. [CrossRef]
32. Van Haastert, P.J.M. A model for a correlated random walk based on the ordered extension of pseudopodia. *PLoS Comput. Biol.* **2010**, *6*, e1000874. [CrossRef]
33. Makarava, N.; Menz, S.; Theves, M.; Huisinga, W.; Beta, C.; Holschneider, M. Quantifying the degree of persistence in random amoeboid motion based on the Hurst exponent of fractional Brownian motion. *Phys. Rev. E. Stat. Nonlin. Soft Matter Phys.* **2014**, *90*, 42703. [CrossRef] [PubMed]
34. Andrew, N.; Insall, R.H. Chemotaxis in shallow gradients is mediated independently of PtdIns 3-kinase by biased choices between random protrusions. *Nat. Cell Biol.* **2007**, *9*, 193–200. [CrossRef] [PubMed]
35. Skoge, M.; Yue, H.; Erickstad, M.; Bae, A.; Levine, H.; Groisman, A.; Loomis, W.F.; Rappel, W.-J. Cellular memory in eukaryotic chemotaxis. *Proc. Natl. Acad. Sci. USA* **2014**, *111*, 14448–14453. [CrossRef] [PubMed]
36. Jeon, T.J.; Lee, D.J.; Merlot, S.; Weeks, G.; Firtel, R.A. Rap1 controls cell adhesion and cell motility through the regulation of myosin II. *J. Cell Biol.* **2007**, *176*, 1021–1033. [CrossRef] [PubMed]
37. Van Haastert, P.J.M.; Bosgraaf, L. Food searching strategy of amoeboid cells by starvation induced run length extension. *PLoS ONE* **2009**, *4*, e6814. [CrossRef]
38. Bosgraaf, L.; Van Haastert, P.J.M. Navigation of chemotactic cells by parallel signaling to pseudopod persistence and orientation. *PLoS ONE* **2009**, *4*, e6842. [CrossRef]
39. Fritz-Laylin, L.K.; Lord, S.J.; Mullins, R.D. WASP and SCAR are evolutionarily conserved in actin-filled pseudopod-based motility. *J. Cell Biol.* **2017**, *216*, 1673–1688. [CrossRef]
40. Yang, H.W.; Collins, S.R.; Meyer, T. Locally excitable Cdc42 signals steer cells during chemotaxis. *Nat. Cell Biol.* **2016**, *18*, 191–201. [CrossRef]
41. Kimura, K.; Ito, M.; Amano, M.; Chihara, K.; Fukata, Y.; Nakafuku, M.; Yamamori, B.; Feng, J.; Nakano, T.; Okawa, K.; et al. Regulation of myosin phosphatase by Rho and Rho-associated kinase (Rho-kinase). *Science* **1996**, *273*, 245–248. [CrossRef]

Review

A "Numerical Evo-Devo" Synthesis for the Identification of Pattern-Forming Factors

Richard Bailleul [1,2], Marie Manceau [2,*] and Jonathan Touboul [3,*]

1. Developmental Biology & Cell Biology and Biophysics Units, European Molecular Biology Laboratory, Meyerhofstraße 1, 69117 Heidelberg, Germany; richard.bailleul@embl.de
2. Centre for Interdisciplinary Research in Biology, CNRS UMR 7241, INSERM U1050, Collège de France, 75005 Paris, France
3. Department of Mathematics and Volen National Center for Complex Systems, Brandeis University, 415 South Street, Waltham, MA 02453, USA

* Correspondence: marie.manceau@college-de-france.fr (M.M.); jtouboul@brandeis.edu (J.T.); Tel.: +33-1-44-27-15-22 (M.M.); +1-781-736-3080 (J.T.)

Received: 22 June 2020; Accepted: 23 July 2020; Published: 5 August 2020

Abstract: Animals display extensive diversity in motifs adorning their coat, yet these patterns have reproducible orientation and periodicity within species or groups. Morphological variation has been traditionally used to dissect the genetic basis of evolutionary change, while pattern conservation and stability in both mathematical and organismal models has served to identify core developmental events. Two patterning theories, namely instruction and self-organisation, emerged from this work. Combined, they provide an appealing explanation for how natural patterns form and evolve, but in vivo factors underlying these mechanisms remain elusive. By bridging developmental biology and mathematics, novel frameworks recently allowed breakthroughs in our understanding of pattern establishment, unveiling how patterning strategies combine in space and time, or the importance of tissue morphogenesis in generating positional information. Adding results from surveys of natural variation to these empirical-modelling dialogues improves model inference, analysis, and in vivo testing. In this evo-devo-numerical synthesis, mathematical models have to reproduce not only given stable patterns but also the dynamics of their emergence, and the extent of inter-species variation in these dynamics through minimal parameter change. This integrative approach can help in disentangling molecular, cellular and mechanical interaction during pattern establishment.

Keywords: pattern formation; natural variation; modelling

---

1. Introduction

Natural patterns have been classified in symmetries, trees, fractals, spirals, meanders, waves, foams, tessellations, cracks, stripes, and dotted arrays [1], and strikingly, examples for most can be found in animals. A long-standing scientific challenge has been to unravel the chain of developmental steps through which these intricate motifs emerge, from first symmetry-breaking events in initially homogeneous structures to the timely propagation of pattern-forming competence and the differentiation and spatial arrangement of patterned characters in typical geometries. Uncovering these processes implies bridging phenomena occurring at the molecular, cellular, tissue, organismal, and population levels, and together, developmental genetics, evolutionary biology and mathematics paved the way to identifying pattern-forming factors in biological systems, respectively taking advantage of model organisms, natural variation, and modelling. We discuss here how empirical and theoretical approaches led to the formulation of two major patterning theories, and review recent studies in model and non-model vertebrates, showing that combining these approaches sets a foundation for powerful work in pattern-forming studies.

2. Diversity vs. Stability of Natural Patterns: A Paradox Guiding Methods and Model Choices

The striking diversity of motifs created by the spatial arrangement of appendages, pigments, segments, etc., partly results from the complex combination and superimposition of many patterns across the body. It has been associated to the multiplicity of physiological and adaptive functions at the basis of evolutionary change [2–9]. However, within species or taxa, most patterns display similar periodicity (i.e., number of repetitions within a period of space or time) or orientation along body axes, a stability thought to guarantee survival and reproductive success in a given niche. This apparent paradox raised enormous interest from developmental biologists, evolutionary biologists and mathematicians, with pattern conservation being rooted in the spatio-temporal hierarchy of pattern-forming events during embryonic development [10], and the extent of pattern variation reflecting developmental constraints to evolution. Researchers from these different fields tackled pattern formation using methodologies based on an opportunistic choice of study systems. Developmental studies have taken advantage of easily tractable designs present in model systems and sharply defined in space and time. Emblematic examples include the antero-posterior distribution of segments and the spatial arrangement of bristles, wing veins, and ommatidia in drosophila (e.g., [11–13]), or the production of skeletal structures, teeth circumvolutions, appendage arrays, or colour patterns in fish, birds or rodents (e.g., [14–17]). The availability of genetic tools and the possibility to perform functional experiments in model organisms enabled functionally dissecting core developmental mechanisms and linked those to the production of conserved attributes of natural patterns. In particular, morphogens, molecules that diffuse from developing sources and form local concentration gradients, have been repeatedly implicated in pattern formation ([18] and see below). However, their expression profiles do not necessarily correlate spatially with patterns, complicating the selection of candidates. In addition, analysing pattern phenotypes through functional tests or genetic screens rarely allows disentangling mechanisms necessary for character production from those involved in their spatial distribution. Using groundwork from developmental genetics, many "evo-devo" studies recently focused on pinpointing the molecular or tissue basis of visible differences observed between homologous characters of non-model organisms. Quantitative genetics and comprehensive comparative expression analyses have been performed in a wide range of invertebrates, from ladybugs to butterflies or nematods [19–21], and vertebrates such as stickleback, cichlid, or cave fish, and African striped or deer mice, etc. [22–25]. This work provided insights into the genetic basis and late acting cellular/tissue events governing evolutionary changes in the orientation, repetition or geometry of patterned characters [22,23,26]. Because of technical challenges in natural populations, evo-devo studies, however, face the general difficulty of functional validation, and are often limited to large-scale correlations or to a few varying species. Finally, mathematical modelling has been central to pattern formation studies. Natural designs are spatial combinations of basic units, arranged so as to optimize steric space [27] or accommodate physical constraints [28], and corresponding to mathematically describable geometries: circles, straight/curved lines, squares, or polygons. A few mathematical concepts may thus describe numerous biological observations. Partial differential equations (PDEs) provide a useful framework, since by nature they create spatio-temporal dynamics. Their simulations in silico reproduce changes in tissue morphology, molecular concentrations or cell motility, naturally generating stable patterns of basic units as observed in vivo. Modelling PDEs thus enabled predicting biological conditions necessary to pattern formation such as the action and interaction of molecular and cellular factors ([29,30] and see below). In theoretical fields, choosing models also imposes limitations: several mathematical models with different behaviours, potentially representing different biological processes, can accurately reproduce a single pattern. Conversely, a single model can often generate a variety of patterns depending on parameters. In addition, the logic of using in silico simulations is often used to reproduce final pattern states but rarely the developmental paths to their formation. In sum, biological work identified putative patterning factors, but rarely allowed in depth mechanistic understanding of their mode of action. Conversely, patterning mechanisms were theoretically predicted, but deriving prediction for precise biological insights has been difficult to extract from modelling. Research fields and their inherent choices of models inevitably restricted and biased technical and conceptual approaches, which led to different views on the nature and mode of action of patterning factors. We present below the two main patterning theories formulated to date.

3. Two Long-Opposed Theories Explain Pattern Formation In Vivo

Alternative patterning dynamics have been proposed to explain the emergence of patterns. In instructional patterning, cells adopt fates according to the amounts of positional information acquired from an external source. Lewis Wolpert conceptualized in 1969 the "French flag model", in which morphogen gradients and differentiation thresholds create distinct compartments in a developing tissue. Biological evidence ensued: arguably the most emblematic is the periodic expression of *Gap* genes providing antero-posterior identity to segments in the drosophila embryo, governed by gradients of *Bicoid* maternal mRNA [31]. Morphogens have since been involved in invertebrate and vertebrate patterning and shown to possess properties compatible with instructional signalling [32–34]). A major strength of this theory is that it intuitively explains the orientation of many periodic patterns along body axes: pattern directionality would be given by candidate positional signals emanating from early axial structures (e.g., neural tube, somite). However, instructive signalling hardly reconciles with the diversity and complexity of patterns, nor does it provide mechanistic understanding of some attributes such as periodicity. In self-organisation, intrinsic instabilities within the initially homogeneous tissue spontaneously cause its arrangement in a pattern. Alan Turing first formulated this theory in 1952 with a "reaction-diffusion" model describing the interaction of an activator and long-range diffusing inhibitor. Most work has since assumed a molecular basis for self-organisation. Morphogens, in particular, diffuse at long range and have expression levels or degradation rates compatible with self-organising dynamics [34]. Self-organising models have been repeatedly used to recreate animal designs in numerical simulations. Their instabilities have inherently stochastic parameters (e.g., diffusion, attraction) with excitable or oscillatory behaviours able to generate repetition, and universal properties such that a few simple conditions, related to the model's equations evaluated at the homogeneous state, can lead to symmetry breaking and pattern emergence. In addition to being efficient at producing periodic patterns, self-organisation also provides an appealing explanation for how natural variation arises, sometimes rapidly at an evolutionary scale, as minimal parameter variation often results in important pattern differences in simulations (e.g., [35]). The malleability of self-organising dynamics, however, does not account for the pattern reproducibility and directionality seen in nature.

Long opposed, instructive signalling and self-organisation are now both viewed as major patterning processes, likely combining in space and time to form many patterns [36]. This synthesis was made possible through the combination of numerical and empirical work, which integrated molecular, cellular, and mechanical components of tissue development. For example, Turing models can recover the longitudinal orientation of fish coloured stripes in silico when simulated in frames seeded with non-homogeneous axial initial conditions [37], or when modulated by production/degradation gradients or tissue anisotropy [38], all providing instructive information. Similarly, combining PDEs describing competition for gene expression, auto-catalytic activity and small diffusion of molecular factors with positional information provided by transcription factors recreates in silico the regular boundaries observed between brain areas in the developing nervous system [39,40]. Biological work also partly reconciled the two patterning strategies. Teeth spatially arrange through Shh-dependent self-organising events arising from so-called "signalling centers" acting as instructive sources [41], and the patterning waves that characterise the timely segmentation of the vertebrate mesoderm in somites are triggered by early instruction and once propagating, generate periodic patterns de novo (e.g., [37,42]). In juvenile poultry birds, longitudinal bands of agouti expression foreshadow periodic coloured stripes adorning the dorsum: their width relies on a late control of agouti's dose by pigment cell self-organisation, while their absolute position is controlled by early instructive signals from the somite [43]. Taken together, results show that pattern formation combines molecular induction, cell/tissue autonomous processes such as self-organisation, and a morphogenetic response to positional cues (i.e., the integration of positional information by cells creating tissue shape changes). For given organs displaying motifs, these processes repeat over time and at different spatial resolutions, in a progressive refinement of space. A current challenge is thus to understand how patterning strategies emerge and combine in space and time to create robust designs in vivo.

4. Novel Integrative Frameworks Bridge Molecular and Cellular Patterning Strategies

In this endeavour, recent inter-disciplinary dialogues combined developmental biology with comprehensive numerical approaches, increasing predictive power and allowing better design of empirical tests, thereby providing novel insights into how patterning strategies combine. We mention here four study systems recently used to implement such approach in mice and birds. A bulk of work first took advantage of digit patterning during limb growth in mice. This process involves dynamic interaction between molecular factors, including Bmp2, Sox9, and Wnt, whose expression form striped patterns, and Hox and Fgf, which form gradients from a distal source. A mathematical model comprising Turing-like reaction-diffusion, where the molecular species are diffusive Bmp2 and Wnt and non-diffusive Sox9, external cues describing Hox and Fgf gradients [44] and realistic tissue growth [45], reproduced proper dynamics of Sox9 expression. Here, and consistent with the recent synthesis of patterning theories presented above, self-organisation is combined with instructive gradients to control the wavelength of the striped pattern [46]. Sadier and colleagues studied the formation of rodent teeth, which emerge through complex timely dynamics prior to stabilization in a final pattern. Sequential *Edar* expression in signalling centres was used to build a mathematical model composed of Turing-type dynamics (short range activation and long-range inhibition triggered by asymmetric bi-stable regime) and cell density and movement towards the activator through chemotaxis [47,48]. Simulations were run on growing domains corresponding to the antero-posterior axis of the dental epithelium. This reproduced the sequence of tooth centre emergence, predicting that newly formed structures can "erase" previously formed ones, a hypothesis that was then functionally validated in vivo. In both studies, modelling was used to recreate observed dynamics of molecular expression and character production rather than only a final pattern. Predictions thereby suggested that simple rules govern apparently complex spatio-temporal regulation during tissue development, allowing authors to explain the outcomes of experimental inhibition of Bmp or Wnt [49] or functional effects of *Edar* loss-of-function [41].

Similar approaches were used to study the timely establishment of feather follicle arrays in avian skin. In poultry birds, feather follicles arrange in a hexagonal array (each follicle is surrounded by six neighbours), and differentiate through dynamic molecular-cellular interaction: FGF signals from the epidermis cause local aggregation of mesenchymal cells, which activates BMP signals, and in turn, further FGF production in the epidermis. By integrating reaction-diffusion akin to Turing's model and chemotaxis, mathematical modelling predicted this interplay, providing self-organising conditions sufficient to break symmetry and create a periodic dotted pattern visually resembling the nascent follicle array [50]. Ho and colleagues then applied to this model a mathematical wave fitting simulation to the dynamic emergence of the follicle array in the domestic chicken, in which previous morphological description showed that follicles individualise in longitudinal rows through a medial-to-lateral wave of differentiation occurring within "tracts", which are areas of the skin which are able to form feathers [27]. Using this prediction as head-start, they searched for in vivo markers of a wave, and showed that both cell density and *Eda* expression increase progressively, travelling laterally prior to follicle formation. This suggested that follicle-forming competence is attained when cell density reaches a threshold whose value depends on the presence of *Eda* signalling. In this study, empirical work was built on modelling predictions to identify molecular and cellular dynamics characterising a morphogenetic wave [51]. Taking advantage of the same bulk of theoretical work, our laboratory studied how the follicle formation wave is initiated and progresses in a timely fashion. We first worked out to recreate the follicle array and dynamics of its emergence without resorting to an extrinsic mathematical wave. We combined the two preeminent processes previously used, namely reaction-diffusion and chemotaxis, with a third mathematical term explicitly describing cell proliferation as an autonomous logistic function. This term, inspired from modelling studies of virus spreading and population behaviour, implements growth in systems with resource limitations [52,53], and classical theories of diffusion equations with logistic growth are known to develop traveling waves, spreading initial perturbations through directional nearest-neighbour contagion [29,30]. The unified model of three partial differential equations applied to simulations frames seeded with axial initial conditions intrinsically reproduces sequential dynamics and the directionality of follicle array formation. This observation produced the first predictions: coupling self-organisation with proliferation triggers timely emergence of follicles; adding initial conditions provides directionality, together creating a patterning wave. Analysing and testing parameters of the model provided a second prediction: while self-organisation controls follicle individualisation, as previously shown, the proliferation rate of the logistic source controls the duration of the patterning process. We investigated the relevance of this model-based prediction by analysing dynamics of cell proliferation throughout tract differentiation in the Japanese quail, a species closely related to the domestic chicken, and confirmed the existence of a travelling front of increased cell density that progressively provides follicle-forming competence to longitudinal domains as it propagates laterally. Strikingly, the ratio of proliferative cells linearly decreased with respect to cell density, ceasing when the skin tissue attains a certain carrying capacity (i.e., a cell density threshold), showing that cell proliferation occurs according to predictions of a logistic growth term—a unique in vivo validation of this widely used mathematical equation. We used drug treatments on skin explants, inhibiting proliferation without altering the carrying capacity of the skin, to demonstrate that proliferation rate governs pattern duration, and also supporting the predictions of the model. In our study, accuracy of modelling was necessary to help design empirical experiments. Together, these three studies uncover similarities between patterning mechanisms in different systems (e.g., the important roles of morphogen and Eda/EdaR signalling, cell density/proliferation, and self-organisation; Figure 1). They illustrate the strength of establishing numerical-empirical crosstalk in pattern-forming studies. We argue below that adding elements of natural variation to in silico work is key to improving this crosstalk.

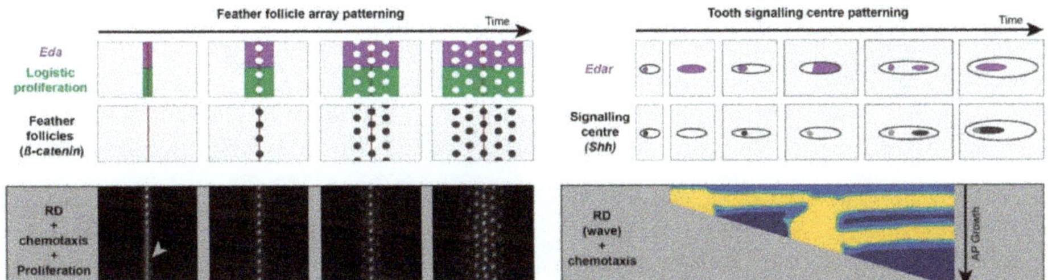

Figure 1. Similarities in molecular and cellular bases of patterning in different systems. During the patterning of both feather follicle arrays (marked by *β-catenin* expression) or tooth signalling centres (marked by *Shh* expression), *Eda/EdaR* allow visualizing morphogenetic waves, which are also characterised by a local increase in cell density and proliferation. In the first case, the wave travels along the medio-lateral axis, while in the second, timely regulation occurs during antero-posterior growth. Simulations of mathematical models composed of Turing-type reaction-diffusion, chemotaxis (and logistic proliferation in the case of follicles) can reproduce spatial dynamics of *EdaR*, *Eda*, or *β-catenin* expression and proliferation during tissue differentiation (high levels of *β-catenin* or *EdaR* expression respectively appear in white and yellow in simulations). Adapted from [37,41].

5. A New Synthesis: Integrating Natural Variation to Numerical-Empirical Crosstalk

Ho and colleagues extended observations to *Paleognathae*, an ancestral group comprising flightless birds. They found that both Eda signalling and the follicle patterning wave are absent in the ostrich, which forms follicles simultaneously within tracts, and in the emu, which forms follicles with a delay, with earlier lower cell density skin impeding competence for pattern formation. The *Paleognathae* follicle pattern is irregular, and authors thus proposed that the progressive acquisition of competence ensures hexagonal fidelity to the follicle pattern in domestic chickens. This work exemplifies the value of taking into account natural variation in a given pattern, a concept we extended and systematized ahead of modelling: in addition to final patterns and their dynamics of formation, the objective of our unified model was to accurately anticipate inter-species differences in this process. We performed a comparative survey by marking nascent feather follicles with *β-catenin* in the emu, representing *Paleognathae*, the domestic chicken, the Japanese quail, and their relative in the *Galleoanserae* group, the common pheasant, and two members of the third, species-rich group *Neoaves*, namely the zebra finch, passerine songbird, and a species of emblematic penguins, devoid of flight abilities. We observed variation in their formed follicle pattern, as well as in the location and shape of early *β-catenin* expression in the un-patterned tract: *β-catenin* marked segments of variable shapes in all poultry birds and the zebra finch, and larger bilateral areas in emus and penguins. In addition, the follicle array emerged in a row-by-row sequence in the first group, and apparently simultaneously in emus and penguins. This observation was consistent with the proposed link between a loss of flight abilities and of the patterning wave, but as we found the penguin pattern to be extremely regular, pattern regularity may be controlled by other means. Surveying a large number of species allowed for correlating differences in initial *β-catenin* expression to pattern sequentiality: we thus ran simulations of our model on simulation frames seeded with species-specific initial conditions. This recreated the species-specific dynamics of pattern emergence, producing a third prediction: initial conditions control wave sequentiality; their variation between birds is responsible for differences in the timing of the follicle patterning process (Figure 2). We thus varied parameters of initial conditions and showed that row-by-row dynamics occur only when initial conditions are spatially restricted and sharp enough. Further work is now necessary to identify in vivo factors displaying such expression profiles and test their role in the sequential control of wave initiation and/or progression. Appealing possibilities include molecules emanating from neighbouring neural tube or somites, structures that differentiate longitudinally and instruct the formation of epidermal and dermal derivatives, including the striped colour pattern adorning the dorsum of poultry birds [44,54,55]. In this study, tracking various developmental paths to pattern establishment and using these as basis for modelling increased the precision of simulations and better guided empirical work. In addition to generating hypotheses with key implication in evolutionary biology and ecology (e.g., here, on the loss of flight abilities), natural variation is thus a powerful tool, raising the bar in the extent to which biological data can be used for model building, and conversely, in the accuracy of numerical simulations.

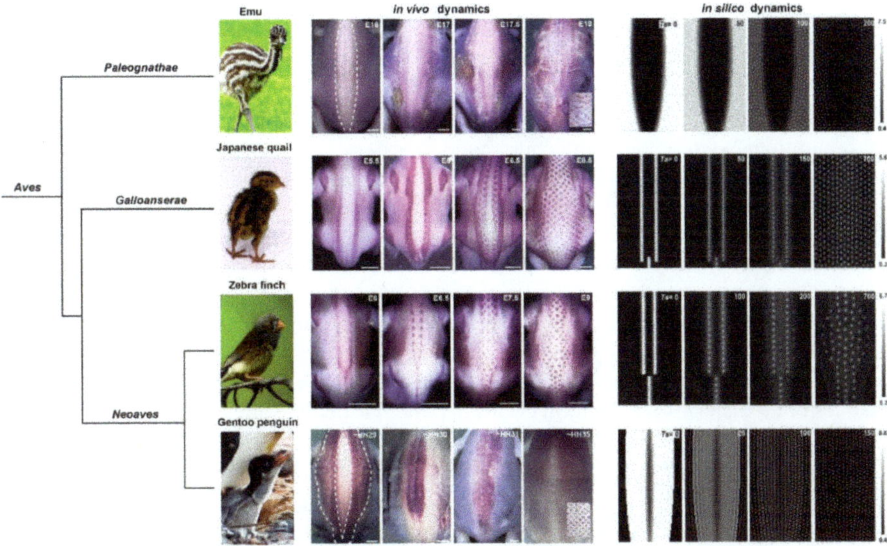

Figure 2. In vivo/in silico dynamics of dorsal feather follicle patterning in avian species. In the emu embryo, feather follicles appear simultaneously within large *β-catenin*-expressing areas initially covering the dorsal skin region expect for its medial-most part. In the Japanese quail and the zebra finch, dorsal follicles appear in a sequential manner from *β-catenin*-expressing longitudinal lines. In the Gentoo penguin, *β–catenin* is expressed in two bilateral expressing areas followed by a central area; follicles rapidly appear in the herein defined zones. Computer simulations of a unified model combining reaction-diffusion, chemotaxis and logistic proliferation recapitulate all dynamics when applied to species-specific axial initial conditions. This predicted that patterning sequentiality is due to spatially confined symmetry breaking in the dorsal skin of bird embryos. Adapted from [37]. Photo credits: Manceau laboratory, Paris, France (Japanese quail), Wikimedia (finch), Jooin (emu) and ©Raphaël Sané (www.raphaelsane.com; Gentoo penguin).

6. Designing Numerical Evo-Devo Approaches to Study Tissue Mechanics

The unified models described above are not exhaustive, and do not account for a number of other events experimentally shown to contribute to pattern formation. Tissue growth, for instance, was modelled in studies of mouse digit and teeth patterns [41,48,56,57] or periodic stripes in zebra fish tailfins [58,59], but reciprocal interactions between molecular, cellular, and mechanical mechanisms remain absent from most models. For example, all known cellular and molecular markers of follicle differentiation, from cell proliferation to the fluctuation of molecular expression, may be generated or affected by intrinsic mechanical properties of the cutaneous tissue. The onset of follicle emergence is indeed characterised by the nuclear translocation of β-catenin in epidermal cells, an event suggested to be triggered by contractile properties of aggregating dermal cells [60]. Similarly, epidermis compression has been shown to intensify FGF expression [51]. Tissue properties consequentially shape the resulting patterns. For example, modifying dermal cell contractility or substrate stiffness ex vivo alters the size and spacing of follicles [60]. More generally, in several other systems, the skin physical forces exerted on/by cells by/on the extra-cellular environment (e.g., shear stress, compression, tension, traction, adhesion) have been linked to changes in extra-cellular matrix architecture, cell cycle, cell motility and signalling [61–66], likely playing a defining role in patterning processes. With the advent of biophysical tools to measure physical parameters in vivo [67] and the ever growing amount of theoretical frameworks integrating biomechanics [68–71], it now becomes possible to explore the role of tissue mechanics with comprehensive experimental modelling approaches. One may infer previous unified models by adding explicit dependence of some parameters of reaction-diffusion and chemotaxis terms on mechanical parameters (e.g., molecular diffusion could be a function of substrate stiffness [72]). A more direct approach would be to use purely mechanical models, previously shown to produce periodic designs such as dotted arrays in one or two dimensional spaces [73,74]. Alternatively, mechanical and chemical couplings proved powerful: necessary Turing model's conditions were obtained by coupling biomechanical parameters promoting short-range activation and long-range inhibition to only one morphogen, suggesting a crucial role for tissue mechanics in pattern formation [69,75], and providing a new paradigm relaxing the necessity of an inhibitor diffusing at a long range—an important constraint of Turing's model which is yet to be supported in many in vivo systems [76]. In light of the studies described above, numerical methodologies integrating molecular, cellular and mechanical elements may greatly benefit from taking into account the developmental dynamics of pattern formation, and adding elements of variation in the architecture and bio-physical properties of the developing tissue as observed between species displaying relevant pattern differences. Such approaches will pave the way to identifying the relative implications of mechanical and chemical events in pattern formation and evolution.

7. Conclusions

The recent literature shows that biological and computational research fields can establish mutually beneficial crosstalk to uncover mechanisms at play in pattern formation. The extensive knowledge acquired in developmental genetics and increasing feasibility of in vivo tests, thanks to technical advances in molecular, imaging and bio-physical tools, better defines biological parameters in mathematical models, whose simulations increase in resemblance to patterns and patterning dynamics observed in living organisms. Conversely, the unification of partial differential equations and progresses in model analyses provide predictions which are less prone to ambiguity to guide empirical tests. When theory is guided by alterations shaped by evolution, including results from now-feasible functional experimentation in non-model organisms [77,78], or data from surveys of natural variation [37], it allows for shedding many pattern-forming dynamics to their greatest simplicity. Such "numerical evo-devo" synthesis is the promise of novel insights in pattern-forming studies in the very near future.

Funding: This work research was funded by Labex Memolife, ERC (Starting Grant #639060), and the Schlumberger Foundation for Education and Research.

Acknowledgments: We thank S. Pantalacci for useful comments.

Conflicts of Interest: The authors declare no conflicts of interest.

References

1. Stevens, P. *Patterns in Nature*; Little Brown: New York, NY, USA, 1974.
2. Pettingill, O.S. *Ornithology in Laboratory and Field*; Elsevier: Amsterdam, The Netherlands, 1985; ISBN 978-0-12-552455-1.
3. Allen, W.L.; Cuthill, I.C.; Scott-Samuel, N.E.; Baddeley, R. Why the leopard got its spots: Relating pattern development to ecology in felids. *Proc. R. Soc. B Biol. Sci.* **2011**, *278*, 1373–1380. [CrossRef]
4. Ortolani, A. Spots, stripes, tail tips and dark eyes: Predicting the function of carnivore colour patterns using the comparative method. *Biol. J. Linn. Soc.* **1999**, *67*, 433–476. [CrossRef]
5. Stevens, M.; Cuthill, I.C.; Alejandro Párraga, C.; Troscianko, T. The effectiveness of disruptive coloration as a concealment strategy. *Prog. Brain Res.* **2006**, *155*, 49–64. [PubMed]
6. Barbosa, A.; Mäthger, L.M.; Buresch, K.C.; Kelly, J.; Chubb, C.; Chiao, C.-C.; Hanlon, R.T. Cuttlefish camouflage: The effects of substrate contrast and size in evoking uniform, mottle or disruptive body patterns. *Vis. Res.* **2008**, *48*, 1242–1253. [CrossRef] [PubMed]
7. Stre, G.; Moum, T.; Bures, S.; Kràl, M.; Adamjan, M.; Moreno, J. A sexually selected character displacement in flycatchers reinforces premating isolation. *Nature* **1997**, *387*, 589–592. [CrossRef]
8. Caro, T. The Adaptive Significance of Coloration in Mammals. *BioScience* **2005**, *55*, 125. [CrossRef]

9. Caro, T.; Izzo, A.; Reiner, R.C.; Walker, H.; Stankowich, T. The function of zebra stripes. *Nat. Commun.* **2014**, *5*, 3535. [CrossRef]
10. Wolpert, L. Positional information and the spatial pattern of cellular differentiation. *J. Theor. Biol.* **1969**, *25*, 1–47. [CrossRef]
11. Cagan, R. Chapter 5 Principles of Drosophila Eye Differentiation. In *Current Topics in Developmental Biology*; Elsevier: Amsterdam, The Netherlands, 2009; Volume 89, pp. 115–135. ISBN 978-0-12-374902-4.
12. Crozatier, M.; Glise, B.; Vincent, A. Patterns in evolution: Veins of the Drosophila wing. *Trends Genet.* **2004**, *20*, 498–505. [CrossRef]
13. Smith, J.M.; Sondhi, K.C. The arrangement of bristles in Drosophila. *J. Embryol. Exp. Morphol.* **1961**, *9*, 661–672.
14. Dubrulle, J. Coupling segmentation to axis formation. *Development* **2004**, *131*, 5783–5793. [CrossRef]
15. Neguer, J.; Manceau, M. Embryonic Patterning of the Vertebrate Skin. *Rev. Cell Biol. Mol. Med.* **2017**, 3. [CrossRef]
16. Nüsslein-Volhard, C.; Singh, A.P. How fish color their skin: A paradigm for development and evolution of adult patterns: Multipotency, plasticity, and cell competition regulate proliferation and spreading of pigment cells in Zebrafish coloration. *BioEssays* **2017**, *39*, 1600231. [CrossRef] [PubMed]
17. Tucker, A.S.; Sharpe, P.T. Molecular Genetics of Tooth Morphogenesis and Patterning: The Right Shape in the Right Place. *J. Dent. Res.* **1999**, *78*, 826–834. [CrossRef] [PubMed]
18. Wartlick, O.; Kicheva, A.; Gonzalez-Gaitan, M. Morphogen Gradient Formation. *Cold Spring Harb. Perspect. Biol.* **2009**, *1*, a001255. [CrossRef] [PubMed]
19. Gautier, M.; Yamaguchi, J.; Foucaud, J.; Loiseau, A.; Ausset, A.; Facon, B.; Gschloessl, B.; Lagnel, J.; Loire, E.; Parrinello, H.; et al. The Genomic Basis of Color Pattern Polymorphism in the Harlequin Ladybird. *Curr. Biol.* **2018**, *28*, 3296.e7–3302.e7. [CrossRef]
20. Gaertner, B.E.; Phillips, P.C. Caenorhabditis elegans as a platform for molecular quantitative genetics and the systems biology of natural variation. *Genet. Res.* **2010**, *92*, 331–348. [CrossRef] [PubMed]
21. Paulsen, S.M. Quantitative genetics of butterfly wing color patterns. *Dev. Genet.* **1994**, *15*, 79–91. [CrossRef]
22. Kratochwil, C.F.; Liang, Y.; Gerwin, J.; Woltering, J.M.; Urban, S.; Henning, F.; Machado-Schiaffino, G.; Hulsey, C.D.; Meyer, A. Agouti-Related peptide 2 facilitates convergent evolution of stripe patterns across cichlid fish radiations. *Science* **2018**, *362*, 457–460. [CrossRef]
23. Mallarino, R.; Henegar, C.; Mirasierra, M.; Manceau, M.; Schradin, C.; Vallejo, M.; Beronja, S.; Barsh, G.S.; Hoekstra, H.E. Developmental mechanisms of stripe patterns in rodents. *Nature* **2016**, *539*, 518–523. [CrossRef]
24. Peichel, C.L.; Marques, D.A. The genetic and molecular architecture of phenotypic diversity in sticklebacks. *Philos. Trans. R. Soc. B Biol. Sci.* **2017**, *372*, 20150486. [CrossRef] [PubMed]
25. McGaugh, S.E.; Gross, J.B.; Aken, B.; Blin, M.; Borowsky, R.; Chalopin, D.; Hinaux, H.; Jeffery, W.R.; Keene, A.; Ma, L.; et al. The cavefish genome reveals candidate genes for eye loss. *Nat. Commun.* **2014**, *5*, 5307. [CrossRef]
26. Barsh, G.S. The genetics of pigmentation: From fancy genes to complex traits. *Trends Genet.* **1996**, *12*, 299–305. [CrossRef]
27. Sengel, P. *Morphogenesis of Skin*; Cambridge University Press: Cambridge, UK, 1976.
28. Thompson, D.W. *On Growth and Form*, 1st ed.; Cambridge University Press: Cambridge, UK, 1917.
29. *Mathematical Biology: II: Spatial Models and Biomedical Applications*; Murray, J.D., Ed.; Interdisciplinary Applied Mathematics; Springer: New York, NY, USA, 2003; Volume 18, ISBN 978-0-387-95228-4.
30. *Mathematical Biology: I. An Introduction*; Murray, J.D., Ed.; Interdisciplinary Applied Mathematics; Springer: New York, NY, USA, 2002; Volume 17, ISBN 978-0-387-95223-9.
31. Nüsslein-Volhard, C.; Wieschaus, E. Mutations affecting segment number and polarity in Drosophila. *Nature* **1980**, *287*, 795–801. [CrossRef]
32. Driever, W.; Nüsslein-Volhard, C. The bicoid protein determines position in the Drosophila embryo in a concentration-dependent manner. *Cell* **1988**, *54*, 95–104. [CrossRef]
33. He, S.; del Viso, F.; Chen, C.-Y.; Ikmi, A.; Kroesen, A.E.; Gibson, M.C. An axial Hox code controls tissue segmentation and body patterning in *Nematostella vectensis*. *Science* **2018**, *361*, 1377–1380. [CrossRef]
34. Muller, P.; Rogers, K.W.; Jordan, B.M.; Lee, J.S.; Robson, D.; Ramanathan, S.; Schier, A.F. Differential Diffusivity of Nodal and Lefty Underlies a Reaction-Diffusion Patterning System. *Science* **2012**, *336*, 721–724. [CrossRef] [PubMed]

35. Kondo, S.; Miura, T. Reaction-Diffusion Model as a Framework for Understanding Biological Pattern Formation. *Science* **2010**, *329*, 1616–1620. [CrossRef]
36. Green, J.B.A.; Sharpe, J. Positional information and reaction-diffusion: Two big ideas in developmental biology combine. *Development* **2015**, *142*, 1203–1211. [CrossRef]
37. Bailleul, R.; Curantz, C.; Dinh, C.D.-T.; Hidalgo, M.; Touboul, J.; Manceau, M. Symmetry breaking in the embryonic skin triggers directional and sequential plumage patterning. *PLoS Biol.* **2019**, *17*, e3000448. [CrossRef] [PubMed]
38. Hiscock, T.W.; Megason, S.G. Orientation of Turing-like Patterns by Morphogen Gradients and Tissue Anisotropies. *Cell Syst.* **2015**, *1*, 408–416. [CrossRef] [PubMed]
39. Perthame, B.; Quiñinao, C.; Touboul, J. Competition and boundary formation in heterogeneous media: Application to neuronal differentiation. *Math. Models Methods Appl. Sci.* **2015**, *25*, 2477–2502. [CrossRef]
40. Quininao, C.; Prochiantz, A.; Touboul, J. Local homeoprotein diffusion can stabilize boundaries generated by graded positional cues. *Development* **2015**, *142*, 1860–1868. [CrossRef] [PubMed]
41. Sadier, A.; Twarogowska, M.; Steklikova, K.; Hayden, L.; Lambert, A.; Schneider, P.; Laudet, V.; Hovorakova, M.; Calvez, V.; Pantalacci, S. Modeling Edar expression reveals the hidden dynamics of tooth signaling center patterning. *PLoS Biol.* **2019**, *17*, e3000064. [CrossRef] [PubMed]
42. Tsiairis, C.D.; Aulehla, A. Self-Organization of Embryonic Genetic Oscillators into Spatiotemporal Wave Patterns. *Cell* **2016**, *164*, 656–667. [CrossRef]
43. Haupaix, N.; Curantz, C.; Bailleul, R.; Beck, S.; Robic, A.; Manceau, M. The periodic coloration in birds forms through a prepattern of somite origin. *Science* **2018**, *361*, eaar4777. [CrossRef]
44. Sheth, R.; Marcon, L.; Bastida, M.F.; Junco, M.; Quintana, L.; Dahn, R.; Kmita, M.; Sharpe, J.; Ros, M.A. Hox Genes Regulate Digit Patterning by Controlling the Wavelength of a Turing-Type Mechanism. *Science* **2012**, *338*, 1476–1480. [CrossRef]
45. Boehm, B.; Westerberg, H.; Lesnicar-Pucko, G.; Raja, S.; Rautschka, M.; Cotterell, J.; Swoger, J.; Sharpe, J. The Role of Spatially Controlled Cell Proliferation in Limb Bud Morphogenesis. *PLoS Biol.* **2010**, *8*, e1000420. [CrossRef]
46. Miura, T. Turing and Wolpert Work Together During Limb Development. *Sci. Signal.* **2013**, *6*, pe14. [CrossRef]
47. Hillen, T.; Painter, K.J. A user's guide to PDE models for chemotaxis. *J. Math. Biol.* **2008**, *58*, 183. [CrossRef]
48. Painter, K.J. Mathematical models for chemotaxis and their applications in self-organisation phenomena. *J. Theor. Biol.* **2019**, *481*, 162–182. [CrossRef] [PubMed]
49. Raspopovic, J.; Marcon, L.; Russo, L.; Sharpe, J. Digit patterning is controlled by a Bmp-Sox9-Wnt Turing network modulated by morphogen gradients. *Science* **2014**, *345*, 566–570. [CrossRef] [PubMed]
50. Painter, K.J.; Ho, W.; Headon, D.J. A chemotaxis model of feather primordia pattern formation during avian development. *J. Theor. Biol.* **2018**, *437*, 225–238. [CrossRef] [PubMed]
51. Ho, W.K.W.; Freem, L.; Zhao, D.; Painter, K.J.; Woolley, T.E.; Gaffney, E.A.; McGrew, M.J.; Tzika, A.; Milinkovitch, M.C.; Schneider, P.; et al. Feather arrays are patterned by interacting signalling and cell density waves. *PLoS Biol.* **2019**, *17*, e3000132. [CrossRef]
52. Verhulst, P.F. Notice sur la loi que la population poursuit dans son accroissement. *Corresp. Math. Phys.* **1838**, *10*, 113–121.
53. Bologna, M.; Aquino, G. Deforestation and world population sustainability: A quantitative analysis. *Sci. Rep.* **2020**, *10*, 7631. [CrossRef] [PubMed]
54. Rios, A.C.; Serralbo, O.; Salgado, D.; Marcelle, C. Neural crest regulates myogenesis through the transient activation of NOTCH. *Nature* **2011**, *473*, 532–535. [CrossRef] [PubMed]
55. Gong, H.; Wang, H.; Wang, Y.; Bai, X.; Liu, B.; He, J.; Wu, J.; Qi, W.; Zhang, W. Skin transcriptome reveals the dynamic changes in the Wnt pathway during integument morphogenesis of chick embryos. *PloS ONE* **2018**, *13*, e0190933. [CrossRef]
56. Marin-Riera, M.; Moustakas-Verho, J.; Savriama, Y.; Jernvall, J.; Salazar-Ciudad, I. Differential tissue growth and cell adhesion alone drive early tooth morphogenesis: An ex vivo and in silico study. *PLoS Comput. Biol.* **2018**, *14*, e1005981. [CrossRef]
57. Häkkinen, T.J.; Sova, S.S.; Corfe, I.J.; Tjäderhane, L.; Hannukainen, A.; Jernvall, J. Modeling enamel matrix secretion in mammalian teeth. *PLoS Comput. Biol.* **2019**, *15*, e1007058. [CrossRef]
58. Volkening, A.; Abbott, M.R.; Chandra, N.; Dubois, B.; Lim, F.; Sexton, D.; Sandstede, B. Modeling Stripe Formation on Growing Zebrafish Tailfins. *Bull. Math. Biol.* **2020**, *82*, 56. [CrossRef]

59. Volkening, A. Linking genotype, cell behavior, and phenotype: Multidisciplinary perspectives with a basis in zebrafish patterns. *Curr. Opin. Genet. Dev.* **2020**, *63*, 78–85. [CrossRef]
60. Shyer, A.E.; Rodrigues, A.R.; Schroeder, G.G.; Kassianidou, E.; Kumar, S.; Harland, R.M. Emergent cellular self-organization and mechanosensation initiate follicle pattern in the avian skin. *Science* **2017**, *357*, 811–815. [CrossRef] [PubMed]
61. Sun, M.; Chi, G.; Li, P.; Lv, S.; Xu, J.; Xu, Z.; Xia, Y.; Tan, Y.; Xu, J.; Li, L.; et al. Effects of Matrix Stiffness on the Morphology, Adhesion, Proliferation and Osteogenic Differentiation of Mesenchymal Stem Cells. *Int. J. Med. Sci.* **2018**, *15*, 257–268. [CrossRef] [PubMed]
62. Razinia, Z.; Castagnino, P.; Xu, T.; Vázquez-Salgado, A.; Puré, E.; Assoian, R.K. Stiffness-dependent motility and proliferation uncoupled by deletion of CD44. *Sci. Rep.* **2017**, *7*, 16499. [CrossRef] [PubMed]
63. Sitarska, E.; Diz-Muñoz, A. Pay attention to membrane tension: Mechanobiology of the cell surface. *Curr. Opin. Cell Biol.* **2020**, *66*, 11–18. [CrossRef] [PubMed]
64. De Pascalis, C.; Etienne-Manneville, S. Single and collective cell migration: The mechanics of adhesions. *Mol. Biol. Cell* **2017**, *28*, 1833–1846. [CrossRef] [PubMed]
65. Lange, J.R.; Fabry, B. Cell and tissue mechanics in cell migration. *Exp. Cell Res.* **2013**, *319*, 2418–2423. [CrossRef] [PubMed]
66. Pandya, P.; Orgaz, J.L.; Sanz-Moreno, V. Actomyosin contractility and collective migration: May the force be with you. *Curr. Opin. Cell Biol.* **2017**, *48*, 87–96. [CrossRef]
67. Diz-Muñoz, A.; Weiner, O.D.; Fletcher, D.A. In pursuit of the mechanics that shape cell surfaces. *Nat. Phys.* **2018**, *14*, 648–652. [CrossRef]
68. Germann, P.; Marin-Riera, M.; Sharpe, J. ya||a: GPU-Powered Spheroid Models for Mesenchyme and Epithelium. *Cell Syst.* **2019**, *8*, 261–266.e3. [CrossRef] [PubMed]
69. Brinkmann, F.; Mercker, M.; Richter, T.; Marciniak-Czochra, A. Post-Turing tissue pattern formation: Advent of mechanochemistry. *PLoS Comput. Biol.* **2018**, *14*, e1006259. [CrossRef] [PubMed]
70. Okuda, S.; Inoue, Y.; Watanabe, T.; Adachi, T. Coupling intercellular molecular signalling with multicellular deformation for simulating three-dimensional tissue morphogenesis. *Interface Focus* **2015**, *5*, 20140095. [CrossRef]
71. Okuda, S.; Miura, T.; Inoue, Y.; Adachi, T.; Eiraku, M. Combining Turing and 3D vertex models reproduces autonomous multicellular morphogenesis with undulation, tubulation, and branching. *Sci. Rep.* **2018**, *8*, 1–15.
72. Hormuth, D.A.; Weis, J.A.; Barnes, S.L.; Miga, M.I.; Rericha, E.C.; Quaranta, V.; Yankeelov, T.E. A mechanically coupled reaction–Diffusion model that incorporates intra-Tumoural heterogeneity to predict in vivo glioma growth. *J. R. Soc. Interface* **2017**, *14*, 128. [CrossRef]
73. Oster, G.F.; Murray, J.D.; Harris, A.K. Mechanical aspects of mesenchymal morphogenesis. *J. Embryol. Exp. Morphol.* **1983**, *78*, 83–125.
74. Vaughan, B.L.; Baker, R.E.; Kay, D.; Maini, P.K. A Modified Oster–Murray–Harris Mechanical Model of Morphogenesis. *SIAM J. Appl. Math.* **2013**, *73*, 2124–2142. [CrossRef]
75. Mercker, M.; Brinkmann, F.; Marciniak-Czochra, A.; Richter, T. Beyond Turing: Mechanochemical pattern formation in biological tissues. *Biol. Direct* **2016**, *11*, 22. [CrossRef]
76. Bode, H.R. The head organizer in Hydra. *Int. J. Dev. Biol.* **2012**, *56*, 473–478. [CrossRef]
77. Salis, P.; Lorin, T.; Lewis, V.; Rey, C.; Marcionetti, A.; Escande, M.; Roux, N.; Besseau, L.; Salamin, N.; Sémon, M.; et al. Developmental and comparative transcriptomic identification of iridophore contribution to white barring in clownfish. *Pigment Cell Melanoma Res.* **2019**, *32*, 391–402. [CrossRef]
78. Ikmi, A.; McKinney, S.A.; Delventhal, K.M.; Gibson, M.C. TALEN and CRISPR/Cas9-Mediated genome editing in the early-branching metazoan Nematostella vectensis. *Nat. Commun.* **2014**, *5*, 5486. [CrossRef] [PubMed]

Perspective

Compete or Coexist? Why the Same Mechanisms of Symmetry Breaking Can Yield Distinct Outcomes

Andrew B. Goryachev * and Marcin Leda

SynthSys, Centre for Synthetic and Systems Biology, Institute for Cell Biology, University of Edinburgh, Edinburg EH9 3BD, UK

* Correspondence: Andrew.Goryachev@ed.ac.uk

Received: 12 August 2020; Accepted: 28 August 2020; Published: 1 September 2020

Abstract: Cellular morphogenesis is governed by the prepattern based on the symmetry-breaking emergence of dense protein clusters. Thus, a cluster of active GTPase Cdc42 marks the site of nascent bud in the baker's yeast. An important biological question is which mechanisms control the number of pattern maxima (spots) and, thus, the number of nascent cellular structures. Distinct flavors of theoretical models seem to suggest different predictions. While the classical Turing scenario leads to an array of stably coexisting multiple structures, mass-conserved models predict formation of a single spot that emerges via the greedy competition between the pattern maxima for the common molecular resources. Both the outcome and the kinetics of this competition are of significant biological importance but remained poorly explored. Recent theoretical analyses largely addressed these questions, but their results have not yet been fully appreciated by the broad biological community. Keeping mathematical apparatus and jargon to the minimum, we review the main conclusions of these analyses with their biological implications in mind. Focusing on the specific example of pattern formation by small GTPases, we speculate on the features of the patterning mechanisms that bypass competition and favor formation of multiple coexisting structures and contrast them with those of the mechanisms that harness competition to form unique cellular structures.

Keywords: symmetry breaking; activator-substrate mechanism; mass-conserved models; cell polarity; pattern formation; small GTPases

---

1. Introduction

Cellular morphogenesis proceeds through a sequence of symmetry-breaking events resulting in the progressive increase in the complexity of cellular organization and function. For example, a mature neuron with its complex asymmetric arbor made of one long axon and multiple branched dendrites develops from a round precursor cell via such a sequence. Like the construction of a large and complex building begins with laying out of its foundation, cellular morphogenesis starts with a prepattern. This prepattern is typically made of proteins that were first distributed spatially homogeneously in the cytoplasm or nucleus and are then laid out in the shape of a desired pattern on a suitably "solid" support, such as the cell membrane. To explain morphological prepatterning, a powerful theory of activator-inhibitor and activator-substrate systems had been developed by the efforts of several generations of researchers [1,2]. For details, the reader is referred to several excellent reviews and books written on this sprawling topic [3–11]. The most popular patterns generated by these systems emerge via the diffusion-driven (aka Turing) instability of the spatially homogeneous state. They typically consist of spots and stripes and are known simply as "Turing patterns". These patterns have two common properties that had been thought of as limiting their applicability to cellular morphogenesis. Firstly, once the conditions for the Turing instability are met, the spots or stripes emerge simultaneously everywhere there they can exist, for example, on the entire cell surface. Secondly, they form with an intrinsically defined characteristic size and, consequently, their number is determined by how many of them can fill out the whole cell surface. Contrary to these predictions, biological cells are known to develop fewer structures than what would fill the entire available space. At times, a cell produces only a single structure, such as the axon of a neuron [12]. Furthermore, those structures that are generated in multiple copies, like microvilli and filopodia, often appear to emerge sequentially, being added one by one to the preexisting number [13,14].

One avenue to resolving this discrepancy between the theory and the reality has emerged with the coming of age of fluorescence microscopy and the technology of fluorescent proteins. These experimental developments permitted researchers to closely follow cellular morphogenesis as it happens in live cells. This research showed that the morphogenetic prepattern, such as the polarized state of a chemotactic cell, can develop surprisingly fast, within just a few minutes. On such a short time scale, production and degradation of proteins comprising the prepattern is negligible and, thus, their cellular copy number can be considered constant throughout the establishment of the prepattern. This conservation of the total number (or the total mass) of the patterning proteins dramatically changes the character of the ensuing protein pattern. In brief, as the spots of the Turing pattern begin to develop by accumulating the patterning protein, they rapidly deplete the cytoplasmic store of the protein, making continued unrestrained growth of all spots impossible [15]. Larger spots manage to grow further but only at the expense of the smaller clusters that release their proteins back to the cytoplasm and eventually completely dissolve. This process of spot competition may proceed all the way to a single large spot that accrues all available protein. This ultimate super spot has no resemblance to the initial Turing pattern and its size and concentration profile cannot be predicted from the linear stability analysis of the homogeneous steady state that was disrupted by the Turing instability. This difference in the behavior of open and mass-conserved systems had long been generally known to mathematicians and physicists [16] but the recognition of its biological significance has not begun until recently [17,18]. Biological systems appear to exhibit the full spectrum of behavior from a vigorous competition invariably ending in the establishment of a unique cellular structure, such as the axon, to the formation of large arrays of similar structures, like microvilli, where no apparent competition between the individual structures is observed. Which properties of the symmetry-breaking mechanisms that underlie the formation of these structures are responsible for this difference? An appealingly simple answer to this question would be to propose that mathematically distinct mechanisms are responsible for different outcomes. Indeed, some early theoretical efforts apparently promoted this explanation. Recent comprehensive analyses of minimal two-variable models have greatly improved our quantitative understanding of the patterns formed by the mass-conserved systems and the phenomenon of competition between them. The picture that emerged from these analyses is more complex and nuanced. In a nutshell, quantitative differences, such as reaction constants, protein quantities and their ratios, rather than qualitative differences between the reaction mechanisms determine whether the system chooses competition or coexistence [19]. Unfortunately, a steep learning curve associated with the technical complexity of this research has inevitably reduced the ability of the broad research community to appreciate these results. Here we leave out rigorous derivations and, instead, review the main conclusions of the recent theoretical analyses focusing on their biological implications.

2. Yeast Bud and the Emergence of the Competition Concept in the Yeast Cell Polarity

To put abstract theoretical results into a concrete biological context, it is useful to introduce a specific cellular system in which the concept of competition plays an important role. Formation of the bud by the baker's yeast, *Saccharomyces cerevisiae*, is arguably the best studied example of cellular morphogenesis where the mechanism of prepattern formation is known in sufficient detail. Since several excellent reviews are available on this topic [20,21], only the aspects essential to this contribution will be briefly outlined below.

Already early genetic screens identified two proteins that appeared to be right at the apex of the molecular network responsible for the bud formation—a small Rho GTPase Cdc42 and its activator, a guanine nucleotide exchange factor (GEF), Cdc24 [22,23]. Currently, informed by the years of cell biology research, we are confident that the prepattern for the insipient bud is a round spot formed on the budding yeast plasma membrane by the active, GTP-bound, form of Cdc42. Importantly, the peak concentration of total Cdc42 within the cluster is substantially higher than its background level elsewhere on the plasma membrane [24–26]. Thus, within this spot, Cdc42 is not only activated but is also highly concentrated (enriched).

From the theoretical point of view, Cdc42 and small GTPases in general are perfect molecular tools for the membrane-localized prepatterning of cellular morphogenesis [27–29]. Firstly, their ability to hydrolyze GTP and convert the released energy into a change in the protein conformation satisfies the essential prerequisite postulated already at the dawn of nonequilibrium thermodynamics [30,31]. Indeed, the reduction in entropy, which is associated with the formation of a pattern (i.e., symmetry breaking), needs to be coupled to the consumption of energy. Secondly, molecular properties of small GTPases allow them to shuttle between the cytoplasm and the membrane surfaces. Importantly, this shuttling is coupled to the activation-inactivation cycle (nucleotide cycling) so that, in general, the inactive form is cytoplasmic, while the active GTP-bound form is membranous [32]. This makes inactive GTPase an ideal embodiment of the "substrate" in the activator-substrate mechanism. All that remains to be done to crown active GTPase as the prototypical "activator" is to enable it with the property of autocatalytic activation. Many small GTPases, including Cdc42, indeed possess this crucial ability: they recruit onto the membrane their own activators, GEFs, thus enabling a positive feedback loop of autoactivation (for a recent review see [33]).

Specific to the field of bud morphogenesis, a dramatic breakthrough in the understanding of Cdc42 pattern formation was made by the discovery that Cdc42-GTP can recruit into the Cdc42 cluster its GEF Cdc24 with the help of Cdc42 effectors, molecules that specifically recognize and bind to the active GTPase [34–37]. The first mechanistic model of self-organized Cdc42 pattern formation [18] demonstrated that the experimentally characterized molecular interactions are both necessary and sufficient for the formation of the Cdc42 cluster via the Turing instability. It also demonstrated that the coupling of the activation-inactivation cycle of Cdc42 to its membrane-cytoplasmic shuttling leads to the substantial concentration of Cdc42 and its activator Cdc24 within the spot. As the total cellular number (mass) of these molecules is conserved, this accumulation leads to the depletion of their cytoplasmic pool—an important prerequisite for the competition. Finally, a biochemically detailed eight-variable model was systematically reduced to the two-variable model of the activator-substrate type. This model was distinct from the recently analyzed two-variable models in one important aspect. It conserved not only the total quantity of Cdc42 but also that of its GEF Cdc24. This cannot be achieved in a model that, from the beginning, has only two variables, the activator (Cdc42-GTP) and the substrate (Cdc42-GDP). However, since the two-variable model of [18] was derived from a mechanistically complete eight-variable model, conservation of GEF Cdc24 appeared in it in the form of a global negative feedback represented by an integral term that, incidentally, complicated the analysis of the model. As we will see in Section 4, this property largely determined the type of patterns produced by the model and their competition.

Analysis of the formation of a single Cdc42 spot in the model [18] showed that it emerges only as a final outcome of the temporary evolution of a fast succession of states with multiple small spots whose number rapidly decreased as their sizes increased. This coarsening behavior was explained by the vigorous competition of spots for the common cytoplasmic molecular pool (Cdc42 itself, but more importantly, its GEF Cdc24, and its effector Bem1) [18]. Furthermore, it was proposed that this competition, in fact, explains the normal uniqueness of the yeast bud. The fact that such a uniqueness (frequently named "singularity" in the literature) is necessary was well known, since the genetic material of the cell can be robustly divided between only two cells. However, the molecular mechanism of the uniqueness remained unknown. Interestingly, a few known genetic mutants breaking the rule of uniqueness pointed to the role of Cdc42 [38]. The modeling prediction of competition as the cause of the bud uniqueness was unexpected as no intermediate states with multiple Cdc42 spots had been observed before. Nonetheless, a dedicated effort to detect them by fast live-cell fluorescence imaging was successful, albeit it showed only a fleeting coexistence of two spots under normal (wild-type) conditions [39]. However, two different types of genetic perturbation aimed at weakening the spot competition produced the desired result—the simultaneous formation of two or more buds. Since these pioneering analyses, our understanding of competition between the yeast Cdc42 clusters has been substantially advanced both experimentally and theoretically, nevertheless, the problem continues to attract the interest of researchers [19,40–43].

3. Lessons from the Mass-Conserved Two-Variable Models

In this section we will survey the results of the theoretical analyses of the simplified two-variable activator-substrate models focusing on the most recent developments in this field. Keeping mathematical apparatus and jargon to the minimum, we will generally follow the Brauns et al. [44,45] approach that arguably provides the most comprehensive and self-consistent treatment of pattern formation in the mass-conserved activator-substrate (MCAS) models as seen from the physicist's point of view.

A typical two-variable MCAS model consists of two partial differential equations:

$$\begin{cases} \dot{u} = D_u \Delta u + F(u,v) \\ \dot{v} = D_v \Delta v - F(u,v) \end{cases}, \tag{1}$$

where u and v are the spatially and time dependent concentrations of the activator and the substrate, $\dot{u}$ and $\dot{v}$ their rates of change, $D_u \Delta u$ and $D_v \Delta v$ are the terms describing their diffusion, and $F(u,v)$ is the reaction function. The fact that $F(u,v)$ enters both equations but with opposite signs represents the mass-conserved nature of the model: any reaction that increases u, reduces v by the same amount, and vice versa. Typically, the reaction function can be written as:

$$F(u,v) = A(u)v - I(u), \tag{2}$$

where $A(u)$ and $I(u)$ are two positive functions of the activator concentration that, keeping in mind our specific example of the small GTPase patterning system, can be called activation and inactivation functions, respectively [19,46]. Note that, because the MCAS model (1) is defined in a single spatial compartment (usually a 1D or 2D spatial domain that represents both the membrane surface and the mathematically "projected" on it cytoplasm), it cannot explicitly distinguish the membrane-cytoplasmic shuttling of GTPases from their activation-inactivation cycle. Instead, model (1) tacitly assumes that activation of a GTPase molecule attaches it to the membrane while inactivation instantaneously puts it back into the fast diffusing cytoplasmic pool.

In a spatially homogeneous steady state (SHSS) of model (1) the concentrations of u and v are constant in space and time and, thus, $F(u,v)$ must be identically 0. Each SHSS is represented by just a single point on the phase plane (u,v) and we can find the positions of all SHSSs by solving equation $F(u,v) = 0$. From (2) we readily get that all such states lie on the curve

$$v = f(u) = I(u)/A(u). \tag{3}$$

The properties of this function that is but an explicit form of the reaction function $F(u,v)$ largely determine the behavior of the model and, thus, $f(u)$ deserves a closer look.

Figure 1A shows several characteristic examples of $f(u)$ with different shapes. Surprisingly, a simple mathematical analysis [45] provides an important result:

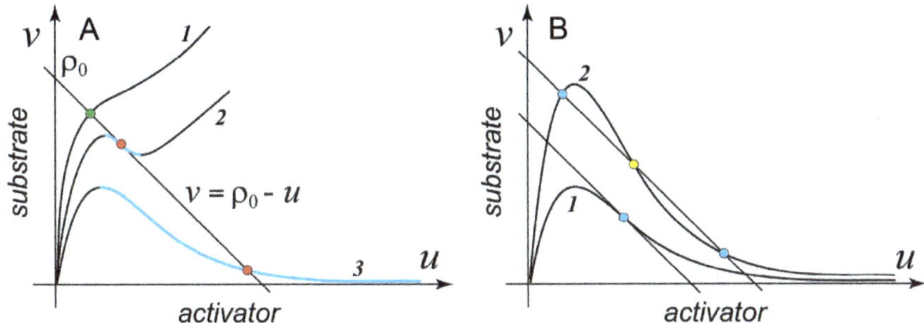

Figure 1. Properties of the reaction function $F(u,v)$ on the phase plane (u,v). (**A**) Shape of the reaction function determines the behavior of the model (see text). Intersections of the reaction function with the mass conservation line $v = \rho_0 - u$ define phase space locations of the model's spatially homogeneous steady states. Green circle, SHSS (spatially homogeneous steady state) stable to the Turing instability. Red circles mark SHSS that lie on the descending fragment of the reaction function (cyan) and are unstable to the Turing instability. (**B**) Any MCAS (mass-conserved activator-substrate) model (1) whose reaction function has a descending fragment has a domain of parameters in which it is monostable in the well-mixed regime (case 1) and also a domain of parameters in which it is bistable (case 2). Stable states are shown by blue circles, unstable is shown by yellow circle.

Result 1. Only those MCAS models whose reaction function $v = f(u)$ has a segment with negative slope on the (u,v) phase plane (descending segment, shown in color in Figure 1A) have the ability to form spatial patterns.

Thus, all models with a monotonously increasing $f(u)$, such as case 1 in Figure 1A, cannot form a spatial pattern, while cases 2 and 3 can. More rigorously, for the spontaneous symmetry breaking to occur via the Turing instability, the negative slope of the descending segment should exceed $-D_u/D_v$. However, in most practical cases, the difference between the diffusion coefficients of the activator (a membrane protein) and that of the substrate (a cytoplasmic protein) is so large that this requirement does not change Result 1 appreciably, for example, for a molecule of small GTPase, typical diffusion coefficients are $D_u \approx 0.1\mu m^2/s$ and $D_v \approx 10\ \mu m^2/s$, so that $D_u/D_v = 10^{-2} \approx 0$. We now introduce two specific examples of the two-variable MCAS models that are frequently discussed and contrasted in the literature. Model A is a highly simplified version of the reduced two-variable model of the Cdc42 cluster formation introduced in [18]:

$$Model\ A \left| F_A = (1 + au^2)v - bu;\ f_A(u) = \frac{bu}{1 + au^2}\right., \tag{4}$$

where a and b are the effective strengths of the positive feedback activation and inactivation, respectively. The activation part of its reaction function consists of two parts with distinct meaning. First term, a nonspecific constitutive background term (in this case, simply, 1), does not depend on the concentration of the activator and may represent a sum of spontaneous activation and activation by GEFs not involved in the positive feedback. Indeed, spontaneous, i.e., unassisted by enzymes, activation and inactivation are not negligible contributions to the nucleotide cycle of a small GTPase [47]. Second term, the classical autocatalytic function au^2v [48,49] formally represents a nominal trimolecular reaction between the activator and the substrate:

$$2U + V \rightarrow 3U\,, \tag{5}$$

but in reality it stands for the autocatalytic activation that involves additional molecules, for example, the positive-feedback GEF (see [18] for an example of rigorous reduction of a detailed biochemical model with many variables). It is easy to see that $f_A(u)$ is Λ-shaped and behaves as case 3 in Figure 1A.

Model B is a dimensionless version [45] of the model first introduced in [50]:

$$Model\ B \left| \; F_B(u,v) = \left(k + \frac{u^2}{1+u^2}\right)v - u; \;\; f_B(u) = \frac{u(1+u^2)}{k+(1+k)u^2} \right. . \tag{6}$$

Here, a single nondimensional parameter k formally represents a constitutive substrate activation, which is independent of the activator. Comparing models A and B, we see that the sole difference between them is that the activation function in the model B is explicitly saturated and thus, at high concentrations of the activator, the rate of substrate activation is no longer dependent on the activator. As a result, $f_B(u)$ is generally N-shaped (case 2 in Figure 1A). However, $f_B(u)$ has a descending segment only for $k < 1/8$. For larger k, this segment vanishes altogether and $f_B(u)$ acquires the shape shown as case 1, in other words, at $k > 1/8$ model B can no longer form the pattern.

Prior to analyzing spatial patterns formed by the MCAS models, it is useful to have a closer look at the behavior of model (1) in the so-called "well-mixed regime". Indeed, imagine that all biochemical reactions are happening in a shaken test tube. In this regime, spatial gradients are no longer possible, diffusion terms in the system (1) vanish, and the concentrations of the activator $u(t)$ and inhibitor $v(t)$ are only functions of time but not space. It is easy to see that the dynamics of the model in this regime is fully determined by the reaction function $F(u,v)$ and the mass conservation relation:

$$u(t) + v(t) = \rho_0 \, . \tag{7}$$

On the phase plane (u,v), the mass conservation relation (7) is represented by a straight line with the slope -1 and the intercept ρ_0, a constant determined entirely by the initial concentrations $u(0)$ and $v(0)$ (see Figure 1A). Because of the mass conservation, in a well-mixed system the concentrations of the activator and inhibitor must remain on line (7) at all times. Therefore, all possible steady states of the well-mixed system lie on the intersection of the function $f(u)$ and the mass conservation line (7). Importantly, the steady states of the model (1) in the well-mixed regime and SHSSs of the non-mixed system (1) are the same points on the phase plane (u,v). We now can extend Result 1 by stating that only those SHSSs of system (1) that lie on the descending segment of $f(u)$ (but see the caveat following Result 1) can be destabilized by the diffusion-driven (Turing) instability and will *spontaneously* give rise to a spatial pattern (Figure 1A).

How many steady states can the MCAS model (1) have in the well-mixed regime? Figure 1A shows that $f(u)$ always makes at least one intersection with the line of mass conservation. This, in fact, is required if model (1) is to describe a valid biochemical system. However, more intersections than one are possible. Consider model A at varying values of the inactivation rate b. Case 1 in Figure 1B shows that at sufficiently small b values there is only one intersection between the unstable (descending) segment of $f_A(u)$ and the mass conservation line. However, as b increases, the negative slope of $f_A(u)$ at its inflection point also increases [45] and at $b > 8$ becomes more negative than the slope of the mass conservation line. For all larger b, model A is bistable: $f_A(u)$ intersects with the mass conservation line three times giving rise to two stable states and one (intervening) unstable state (Figure 1B). Analogously, in model B the function $f_B(u)$ intersects the mass-conservation line thrice for all k values smaller than ~0.0607 (precisely $3/2\sqrt{2} - 1$). Thus, in the interval $0.0607 < k < 1/8$ model B can form patterns but is not bistable. Therefore, both models have parameter domains where they can form spatial patterns but are not bistable in the well-mixed regime. The following important result [45,48] generalizes these examples:

Result 2. Bistability of the MCAS model (1) in the well-mixed regime is not necessary for the formation of spatial patterns at the same parameter values in the model without mixing. However, any MCAS model that can form patterns (has a descending segment of $f(u)$, Result 1) is also bistable in the well-mixed regime within a parameter domain that overlaps but does not coincide with that of pattern formation.

This result finally settles the confusion that emerged in the early analyses of the two-variable MCAS model where it had been claimed that bistability of the reaction dynamics is required for the formation of patterns [50,51].

Now we discuss properties of the steady state patterns formed by the MCAS model (1) on extended spatial domains (of course, in the absence of mixing). If we add the two equations of model (1) together, we get that

$$\dot{u} + \dot{v} = D_u \Delta u + D_v \Delta v \,. \tag{8}$$

For any steady state pattern (including SHSSs), both $\dot{u}$ and $\dot{v}$ are 0 by definition and we find that any steady state pattern must satisfy the equation

$$D_u \Delta u + D_v \Delta v = 0 \,. \tag{9}$$

By a straightforward mathematical manipulation, (9) can be converted into

$$v = \eta_0 - \frac{D_u}{D_v} u \,, \tag{10}$$

an equation of another straight line on the phase plane (u, v) with a small negative slope $(-D_u/D_v \approx 0)$ and positive intercept η_0. Surprisingly, the fact that (9) holds true for any steady state pattern of MCAS (1) means that any such pattern, once projected onto the phase plane (u, v), must reside on line (10), which we can therefore call the "line of patterns". To appreciate the importance of this line, let us consider an example. Figure 2A shows model B in the parameter domain ($k = 0.04$) where it can form pattern and is also bistable in the well-mixed regime. Given that $f_B(u)$ is N-shaped, the line of patterns intersects $f_B(u)$ in three points. Since the reaction function $F(u, v)$ becomes 0 everywhere on $f_B(u)$ and we consider a steady state (pattern), both the reaction and diffusion terms of model (1) must be identically 0 at the intersections of $f(u)$ and the line of patterns (10). The two conditions can be satisfied together only if these points in the phase space (L and H on Figure 2A) correspond on the spatial domain to two flat plateaus with low and high concentrations of the activator and (nearly) the same concentration of the substrate (Figure 2B). The two plateaus are separated by an S-shaped interface and the third (intervening) intersection point of $f_B(u)$ and the line of patterns (point I on Figure 2A) corresponds precisely to the inflection point of this interface. Together with its mirror-symmetric counterpart, the interface shown in Figure 2B defines a pattern with a flat-topped profile, which is frequently referred to in the literature as mesa (Spanish for "table"). Note that the low and high plateaus of the mesa pattern are not identical to the two stable states of the well-mixed MCAS model (compare points L and H with SHSS1 and SHSS2 on Figure 2A). They coincide only if D_u is equal to D_v, in which case the line of mass conservation and the line of patterns are the same line.

We now arrive at next important milestone [45] in the analysis of model (1)

Result 3. Mesa patterns are generic for the two-variable MCAS model (1). The only characteristic spatial size of the mesa pattern is the width ℓ_{int} of the interface between the low and the high plateaus of the mesa pattern and it is determined by the model's intensive parameters, i.e., reaction constants and diffusion coefficients.

Indeed, the length (or the area in 2D) of the high plateau depends on the total quantity of the pattern forming protein in the system, in other words, an extensive parameter of the model. What does it mean, in practical terms, that these patterns are *generic* for the model? Firstly, we must assume that the linear size of the spatial domain L (e.g., the equatorial circumference of a spherical cell) is much larger than ℓ_{int} to permit formation of full-fledged mesas. Secondly, we stipulate that the total quantity of the patterning protein within the system should be sufficient to build at least one "minimal" mesa. Such a pattern would consist of two interfaces mirror-juxtaposed to each other with essentially no high plateau in between. Interestingly, several biologically important cases lie well outside of the parameter domain in which the requirements for the generic behavior are satisfied. Indeed, what if ℓ_{int} is on the same order as L and the total amount of the patterning protein in the cell is not sufficient to build even a minimal mesa pattern?

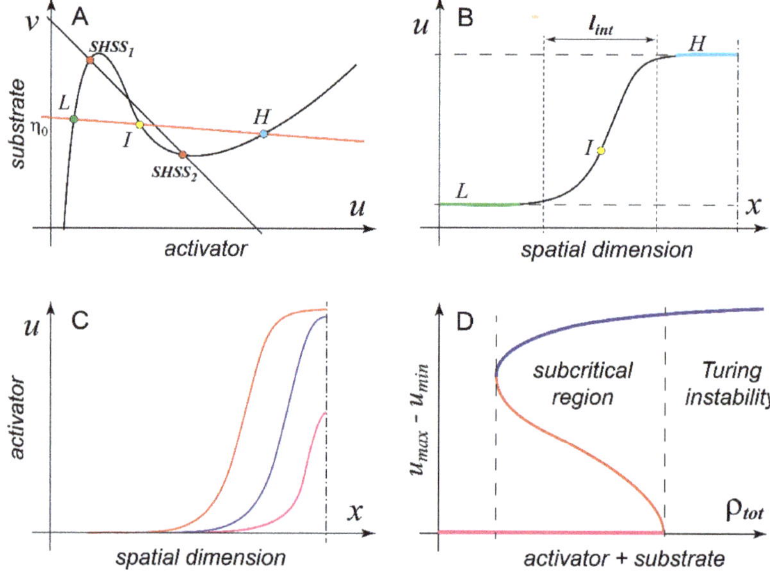

Figure 2. Spatial patterns formed by model (1). (**A**) Pattenrs formed by model B on a large spatial domain are determined in the phase plane (u, v) by the intersections of the reaction function and the line of patterns $v = \eta_0 - D_u/D_v v$ shown in red (see text). Circles L and H represent low and high plateaus of the spatial pattern, circle I corresponds to the inflection point of the mesa pattern. Red circles represent steady states of the model in the well-mixed regime. (**B**) Mesa pattern formed by model B consist of low and high plateaus connected by interfaces with characteristic width ℓ_{int}. (**C**) Model (1) forms stable peaks rather than mesas then it cannot reach saturation. Red curve represents mesa; blue curve represents a peak very close to saturation; magenta curve represents a peak far from saturation. Dash-dotted line in (**B**,**C**) represents an axis of pattern mirror symmetry. (**D**) Model (1) typically possesses a domain of parameters where both the SHSS (magenta) and the pattern (blue) are stable. Red curve represents unstable pattern (see text).

At this point in the discussion it is customary to switch from model B to model A. In model A, $f_A(u)$ and the line of patterns also have three points of intersection (unless D_u is identically 0, or D_v is infinite), but since $f_A(u)$ is Λ-shaped, rather than N-shaped, the third point of intersection that corresponds to the high plateau may lie at the unrealistically large value of the activator concentration. If we also assume that ℓ_{int} is smaller than L but is of the same order, a peak pattern whose maximum never reaches saturation (i.e., the value of the high plateau) can stably form in the system with mass conservation. If the pattern forming protein is continuously added to the system, the peak will also continue to grow both in height and in width until it finally reaches the saturation and becomes a mesa (Figure 2C). Unfortunately, peaks are not mathematically universal in the same sense as the interfaces of the mesa pattern and their analytical analysis is limited [45].

Can MCAS model (1) have a stable spatial pattern outside of the parameter domain of the Turing instability? In the above discussion, we concluded (Result 1) that only those SHSSs of the model that lie on the descending branch of $f(u)$ can spontaneously break symmetry due to the amplification of the diffusion-induced spatial fluctuations. However, if we drop the requirement for the spontaneous symmetry breaking, stable patterns can be readily induced by finite perturbations well outside of the domain of Turing instability. This ability is associated with the so-called phenomenon of subcriticality, which effectively stands for the bistability between spatial patterns and should not be confused with the bistability of the well-mixed system. Figure 2D shows a schematic diagram illustrating this concept (see also Figure 1E and the accompanying discussion in [46]). Here a suitable order parameter, for example, the difference between the maximal and the minimal concentrations of the activator within the stable pattern, is plotted against one of the model parameters, for example, the total quantity of the pattern-forming protein in the cell. Figure 2D shows that subcriticality of the pattern corresponds to the interval of parameter in which both the SHSS and the pattern are stable. As is the case of all bistable systems, two stable states (a pattern and an SHSS, in this case) have to always be separated by an unstable state, which is also a pattern in this situation and can be numerically calculated with some computational tricks [45]. To induce a stable pattern in the bistable regime, the perturbation must exceed the unstable pattern and cannot be "small" in either its amplitude or spatial extent. In fact, it becomes larger the further the parameter value is from the boundary of the Turing instability. On the left (opposite to the Turing) end of the bistability interval, the required "perturbation" is essentially equal to the pattern itself. Subcritical patterns induced in model (1) by finite perturbations are also peaks and mesas and generally cannot be distinguished from their counterparts induced by the Turing instability. Rigorous analysis [45,48,52] shows that

Result 4. Bistability of the spatially homogeneous steady state and pattern is generic for MCAS model (1). Furthermore, any two-variable MCAS model (1) in which pattern can be induced by a finite perturbation also must have a domain of parameters in which it forms pattern via the Turing instability.

This result implies that, regardless of the exact mathematical formulation of the model, for example, the shape of $f(u)$, the parameter domain in which model (1) forms pattern via the Turing instability is surrounded by a broader area of parameters in which the pattern can be induced by finite perturbations. Due to the nature of subcriticality (see Figure 2D), it is always found "on the fringes" of the domain of Turing instability. Perhaps, even more important is the second part of Result 4: patterns induced by spontaneous symmetry breaking (Turing instability) and the patterns induced by finite perturbations must always coexist in any model that is capable of forming pattern (albeit at different parameter values). In the words of Frank Sinatra, "you can't have one without the other". This means, for example, that there is no such model (1) that could only form patterns in response to finite perturbations but has no Turing instability. This settles another confusion in the field that has emerged from the claims that the subcritical formation (i.e., by finite perturbation) of mesa patterns in model B, so-called "wave pinning", is a scenario fundamentally distinct from the Turing instability [50,51]. We now know that model B generates mesa patterns generically, via both Turing instability and by finite perturbations. Furthermore, like model A, it can also generate peaks, but unlike model A, it does so apparently only in the subcritical regime [45].

Finally, we are armed with sufficient apparatus to discuss competition between multiple maxima ("spots") of the pattern generated by model (1). The interest in understanding the final outcome of pattern formation in the two-variable MCAS models has been largely, if not solely, driven by biological applications. However, this mathematical problem turned out to be a hard nut to crack. From the analysis of a few specific models, it had been conjectured that the coexistence of multiple spots in model (1) is always unstable and only a single spot would eventually emerge as the final outcome of the pattern coarsening process [17,18,53]. However, whether this hypothesis holds true for all possible models of type (1), remained unknown. Furthermore, the speed with which multiple spots can resolve into one is also of great importance for the biological systems. The prepattern for cellular morphogenesis is typically short-lived and is rapidly "fixed" by the downstream processes. Thus, if there is a strong biological reason why only a single spot needs to be generated, the competition must be really fast. For example, the Cdc42 clusters of the budding yeast had been observed to coexist for just a couple of minutes [39,42], while the leading protrusions of chemotactic *Dictyostelium* ameba coexist for an even shorter time, less than one minute (P.J.M. van Haastert, personal communication). Thus, what determines the kinetics of pattern coarsening was another unanswered question of high biological importance.

A major advance in the understanding of competition in model (1) was provided by the pioneering effort of Chiou et al. [19]. Firstly, the conjecture of ultimate instability of multiple spots in model (1) was mathematically proven for any reaction function $F(u,v)$ in the practically important limit of infinitely fast diffusion of the substrate ($D_v \to \infty$). Secondly, and perhaps more importantly, it was demonstrated that the kinetics of competition is fundamentally dependent on the type of pattern. While peaks can compete essentially as fast as they form, mesas, once formed, are very static and can coexist for very long times. Ironically, mesa patterns were originally proposed as the model for the highly dynamic and rapidly competing protrusions of migrating cells [50,51]. Of course, peaks close to saturation exhibit intermediate competition kinetics. In fact, multiple peaks close to saturation will slowly compete to form several mesas that "compete" so slowly that, for all practical purposes, can be considered stably coexisting. Complementing results of [19], another recent work [44] built scaling arguments to approach competition from the physicist's point of view. Using the average separation Λ between the maxima of the pattern as an effective order parameter that increases as the number of spots decreases, they showed that, for peaks, it increases as a power of the time of competition t_{comp}

$$\Lambda \sim t_{comp}^{\alpha}\,, \quad \alpha < 1\,, \tag{11}$$

but for mesas it grows only as a logarithm of this time. Unfortunately, the mathematically non-universal nature of peaks, as opposed to mesas, showed itself also in this analysis. It transpired that the exponent α of the power law (11) is dependent on the exact definition of the reaction function $F(u,v)$. This suggests a tantalizing hypothesis that some reaction mechanisms can provide faster peak competition than others. These results of [19] and [44] can be consolidated as the following

Result 5. All patterns with multiple maxima (spots) that are formed by two-variable MCAS model (1) are unstable, i.e., after passing of infinite amount of time they are guaranteed to resolve into a single spot (peak or mesa). However, the kinetics of this resolution process is qualitatively different for peaks and mesas.

This theoretical result comes with an important caveat. The validity of the assumption of mass-conservation demands that the characteristic time of pattern formation from its inception (either spontaneous or induced by a finite perturbation) to the resolution of competition should be much shorter than the characteristic times of production and degradation of proteins that are used to form the pattern. Ironically, it is not uncommon for the mathematical models to consider competition on time scales by far exceeding the entire lifetime of a biological organism. This raises an important question: how does the validity of Result 5 change if the mass-conservation is not (strictly) observed? Several works began to address this question by introducing production and degradation of the pattern-forming protein as small perturbations to the reactive dynamics [44,54,55]. Brauns et al. [44] demonstrated that this dramatically affects the first part of Result 5. Indeed, as the contribution of production and degradation to the reaction dynamics gradually increases, a pattern with multiple mesas becomes stable (i.e., mesas can coexist indefinitely). At even higher rates of production and degradation, mesas begin to split, making more, not less, spots. Taken together, these results cannot help but proffer an epistemological conclusion that theoretical models give simple predictions only when they are themselves, well, simple.

4. Implications for Cellular Morphogenesis

What can a biologist, who studies cellular morphogenesis in one of model organisms, learn about the competition and coexistence of cellular structures from the above theoretical results? Firstly, we conclude that, as far as the two-variable mass-conserved activator-substrate models are concerned, we are presently confident that all of them will eventually end up in the state with a single maximum. However, the time it takes the model system to arrive at this state could be anywhere between seconds and millennia. In real-life biological systems, "competition" and "coexistence" are more binary outcomes. If the "spots" of the prepattern compete any slower than the downstream morphogenetic processes that fix them, they might as well be considered as coexisting. Thus, in the budding yeast, formation of the Cdc42-GTP cluster triggers a cascade of processes, such as secretion of membrane vesicles, formation of septin cytoskeleton, softening and deformation of the cell wall, that together sculpt the nascent bud. If the competition between multiple Cdc42 clusters is slower than these processes, the cell will inevitably form multiple buds [39,41]. Thus, only "very fast" competition is of biological significance when a unique structure, for example, axon, bud, or protruding pseudopod, must form to fulfill its biological function within the cell.

It has been rightfully noted that the coexistence of multiple maxima in a pattern could be a stable state and not just a state of the frozen in time competition, as is the case for all two-variable MCAS models [40]. Several avenues to achieving stable coexistence are possible. For example, the requirement for the strict mass conservation can be relaxed [44,55]. Alternatively, additional negative feedback variables can be introduced [40,55]. Negative feedback is known to endow biological systems with homeostatic properties required for the stabilization of multiple structures. Specific to the small GTPase cell-patterning systems, negative feedback has been shown to generate oscillations and waves as well as blinking, jumping and moving spots of GTPase activity [42,56–62]. However, these rich dynamics and numerous models describing it lie outside of the scope of the present contribution.

While the two-variable MCAS models might provide a somewhat oversimplified description of actual cell-patterning systems, they offer clear recipes for designing patterning mechanisms with specific purpose. If multiple coexisting structures are the target, the "activator" state of the pattern-forming protein P must have minimal, if any, mobility $D_u \approx 0$. This can be readily achieved by polymerization, phase separation or simply by binding to the actin cytoskeleton and will ensure a sharp interface of the prepattern $\ell_{int} \ll L$, where L is a characteristic linear dimension of the cell. Reaction rates of the autocatalytic assembly of the "activator" cluster should evolve so that its "saturated" concentration (the high plateau of a mesa) is only moderately higher than the ground state concentration (the low plateau). Together, the two requirements will make for low and narrow mesas, hence, resulting in only a small number of patterning proteins per spot. If, in addition, P is an abundant cellular protein, formation of new spots will barely affect the existing ones.

If, however, only a unique cellular structure is required, which is the case of the establishment of cellular polarity, all requirements must be reversed. The "activator" should freely diffuse on the membrane and have a capacity to accumulate to high concentrations before reaching saturation. If not P itself, some component of the activation complex (see below) must be a small copy number protein, so that its accumulation within the pattern is acutely sensed by its cytoplasmic pool. These requirements are, indeed, highly consistent with the known experimental facts. In retrospect, it is not surprising, that much is known about the prepattern for the unique cellular structures, because they are themselves large, conspicuous entities marked by the high concentration contrast of the localization of marker proteins, which is readily detectable by the fluorescence microscopy. On the contrary, little is known about the prepattern for small repetitive structures, such as podosomes or microvilli. It is possible that, at least in part, this is because the fine-scale variation of the spatial distribution of the respective pattern-forming proteins is hard to resolve under the microscope and distinguish from the molecular noise.

Additional conclusions can be made for the specific case of small GTPase patterning systems, for which our understanding of the reaction mechanisms is much better than for other patterning systems. While in model (1) the type of pattern depends only on the total conserved quantity of the patterning protein, in the real GTPase systems, this will also be controlled by the ratio of the quantities of GTPase and its positive-feedback GEF. In the analysis of the behavior of the multivariable model for the Cdc42 cluster formation, it has been observed that the component present in the least amount defines the competition behavior of the pattern [40]. This fully confirms the early results of [18] that, following the contemporary experimental data, assumed that Cdc42 is in large molar excess over its positive-feedback activator GEF Cdc24 (~300:1). Because of that choice of parameters, pattern formation in model [18] stopped upon the depletion of the cytoplasmic store of the GEF and well below the saturation limit for Cdc42 insuring that only highly competitive peaks can be formed by this model. Relaxing this condition by reversing the GTPase:GEF ratio produced multiple coexisting mesa patterns, despite the reaction kinetics remaining the same. These modeling results are in good agreement with the results of [63] that, for the first time in any known system, succeeded in the in vitro reconstitution of the GTPase pattern formation using only a minimal set of recombinant proteins and lipid-coated beads. This remarkable effort produced multiple mesa-type spots of Rab5-GTP that stably coexisted during the duration of experiment. Remarkably, to achieve these patterns, the authors used molar excesses of the positive-feedback GEF Rabex-5 over the GTPase ranging from 5:1 to 50:1. At the latter value, the high plateau of the pattern covered most of the bead surface in perfect agreement with the theoretical predictions.

As a final remark, we would like to point out that the results discussed above need not be restricted only to the GTPase patterning systems or to the surface of membranes. A remarkable symmetry-breaking phenomenon in cells is the formation of a daughter centriole on the side of the mother centriole [64–66]. In this case, the pattern-forming protein is the kinase PLK4 and the "activator" is its phosphorylated and kinase-active state. Recently a competition-based model has been proposed to explain why under normal conditions only a single centriole forms, while overexpression of PLK4 generates multiple centrioles [67]. Interestingly, although this model is not mass-conserved, and, in fact, the growing PLK4 mass is the bifurcation parameter, the principle of competition can still be applied as far as the characteristic time scales of protein production and competition remain well separated.

Author Contributions: Conceptualization, A.B.G. and M.L.; writing, revision and editing, A.B.G. and M.L. All authors have read and agreed to the published version of the manuscript.

Funding: This research was funded by the Biotechnology and Biological Sciences Research Council of UK grants number BB/P01190X and BB/P006507.

Conflicts of Interest: The authors declare no conflict of interest.

References

1. Turing, A.M. The chemical basis of morphogenesis. *Philos. Trans. R. Soc. Lond. Ser. B-Biol. Sci.* **1952**, *237*, 37–72. [CrossRef]
2. Gierer, A.; Meinhardt, H. A theory of biological pattern formation. *Kybernetik* **1972**, *12*, 30–39. [CrossRef] [PubMed]
3. Meinhardt, H.; Gierer, A. Pattern formation by local self-activation and lateral inhibition. *BioEssays* **2000**, *22*, 753–760. [CrossRef]
4. Murray, J.D. *Mathematical Biology*; Springer: Berlin, Germany, 1993; p. 770.
5. Landge, A.N.; Jordan, B.M.; Diego, X.; Müller, P. Pattern formation mechanisms of self-organizing reaction-diffusion systems. *Dev. Biol.* **2020**, *460*, 2–11. [CrossRef] [PubMed]
6. Schweisguth, F.; Corson, F. Self-organization in pattern formation. *Dev. Cell* **2019**, *49*, 659–677. [CrossRef] [PubMed]
7. Marcon, L.; Sharpe, J. Turing patterns in development: What about the horse part? *Curr. Opin. Genet. Dev.* **2012**, *22*, 578–584. [CrossRef] [PubMed]
8. Howard, J.; Grill, S.W.; Bois, J.S. Turing's next steps: The mechanochemical basis of morphogenesis. *Nat. Rev. Mol. Cell Biol.* **2011**, *12*, 392–398. [CrossRef]
9. Grecco, H.E.; Schmick, M.; Bastiaens, P.I. Signaling from the living plasma membrane. *Cell* **2011**, *144*, 897–909. [CrossRef]
10. Baker, R.E.; Schnell, S.; Maini, P.K. Waves and patterning in developmental biology: Vertebrate segmentation and feather bud formation as case studies. *Int. J. Dev. Biol.* **2009**, *53*, 783–794. [CrossRef]
11. Halatek, J.; Brauns, F.; Frey, E. Self-organization principles of intracellular pattern formation. *Philos. Trans. R. Soc. Lond. B Biol. Sci.* **2018**, *373*. [CrossRef] [PubMed]
12. Craig, A.M.; Banker, G. Neuronal polarity. *Annu. Rev. Neurosci.* **1994**, *17*, 267–310. [CrossRef] [PubMed]
13. Sauvanet, C.; Wayt, J.; Pelaseyed, T.; Bretscher, A. Structure, regulation, and functional diversity of microvilli on the apical domain of epithelial cells. *Annu. Rev. Cell Dev. Biol.* **2015**, *31*, 593–621. [CrossRef]

14. Mattila, P.K.; Lappalainen, P. Filopodia: Molecular architecture and cellular functions. *Nat. Rev. Mol. Cell Biol.* **2008**, *9*, 446–454. [CrossRef] [PubMed]
15. Goehring, N.W.; Hyman, A.A. Organelle growth control through limiting pools of cytoplasmic components. *Curr. Biol.* **2012**, *22*, R330–R339. [CrossRef]
16. Cross, M.C.; Hohenberg, P.C. Pattern formation outside of equilibrium. *Rev. Mod. Phys.* **1993**, *65*, 851–1112. [CrossRef]
17. Otsuji, M.; Ishihara, S.; Co, C.; Kaibuchi, K.; Mochizuki, A.; Kuroda, S. A mass conserved reaction-diffusion system captures properties of cell polarity. *PLoS Comput. Biol.* **2007**, *3*, e108. [CrossRef]
18. Goryachev, A.B.; Pokhilko, A.V. Dynamics of Cdc42 network embodies a Turing-type mechanism of yeast cell polarity. *FEBS Lett.* **2008**, *582*, 1437–1443. [CrossRef] [PubMed]
19. Chiou, J.G.; Ramirez, S.A.; Elston, T.C.; Witelski, T.P.; Schaeffer, D.G.; Lew, D.J. Principles that govern competition or co-existence in Rho-GTPase driven polarization. *PLoS Comput. Biol.* **2018**, *14*, e1006095. [CrossRef]
20. Chiou, J.G.; Balasubramanian, M.K.; Lew, D.J. Cell polarity in yeast. *Annu. Rev. Cell Dev. Biol.* **2017**, *33*, 77–101. [CrossRef]
21. Bi, E.; Park, H.O. Cell polarization and cytokinesis in budding yeast. *Genetics* **2012**, *191*, 347–387. [CrossRef]
22. Hartwell, L.H.; Culotti, J.; Pringle, J.R.; Reid, B.J. Genetic control of the cell division cycle in yeast. *Science* **1974**, *183*, 46–51. [CrossRef] [PubMed]
23. Adams, A.E.; Johnson, D.I.; Longnecker, R.M.; Sloat, B.F.; Pringle, J.R. CDC42 and CDC43, two additional genes involved in budding and the establishment of cell polarity in the yeast Saccharomyces cerevisiae. *J. Cell Biol.* **1990**, *111*, 131–142. [CrossRef] [PubMed]
24. Lu, M.S.; Drubin, D.G. Cdc42 GTPase regulates ESCRTs in nuclear envelope sealing and ER remodeling. *J. Cell Biol.* **2020**, *219*. [CrossRef]
25. Richman, T.J.; Sawyer, M.M.; Johnson, D.I. Saccharomyces cerevisiae Cdc42p localizes to cellular membranes and clusters at sites of polarized growth. *Eukaryot. Cell* **2002**, *1*, 458–468. [CrossRef] [PubMed]
26. Ziman, M.; Preuss, D.; Mulholland, J.; O'Brien, J.M.; Botstein, D.; Johnson, D.I. Subcellular localization of Cdc42p, a Saccharomyces cerevisiae GTP-binding protein involved in the control of cell polarity. *Mol. Biol. Cell* **1993**, *4*, 1307–1316. [CrossRef]
27. Heasman, S.J.; Ridley, A.J. Mammalian Rho GTPases: New insights into their functions from in vivo studies. *Nat. Rev. Mol. Cell Biol.* **2008**, *9*, 690–701. [CrossRef]
28. Cherfils, J.; Zeghouf, M. Regulation of small GTPases by GEFs, GAPs, and GDIs. *Physiol. Rev.* **2013**, *93*, 269–309. [CrossRef]
29. Park, H.O.; Bi, E. Central roles of small GTPases in the development of cell polarity in yeast and beyond. *Microbiol. Mol. Biol. Rev.* **2007**, *71*, 48–96. [CrossRef]
30. Prigogine, I.; Nicolis, G. On symmetry-breaking instabilites in dissipative systems. *J. Chem. Phys.* **1967**, *46*, 3542–3550. [CrossRef]
31. Prigogine, I.; Lefever, R. Symmetry breaking instabilities in dissipative systems. II. *J. Chem. Phys.* **1968**, *48*, 1695–1700. [CrossRef]
32. Johnson, J.L.; Erickson, J.W.; Cerione, R.A. New insights into how the Rho guanine nucleotide dissociation inhibitor regulates the interaction of Cdc42 with membranes. *J. Biol. Chem.* **2009**, *284*, 23860–23871. [CrossRef] [PubMed]
33. Goryachev, A.B.; Leda, M. Autoactivation of small GTPases by the GEF-effector positive feedback modules. *F1000Res.* **2019**, *8*. [CrossRef] [PubMed]
34. Butty, A.C.; Perrinjaquet, N.; Petit, A.; Jaquenoud, M.; Segall, J.E.; Hofmann, K.; Zwahlen, C.; Peter, M. A positive feedback loop stabilizes the guanine-nucleotide exchange factor Cdc24 at sites of polarization. *Embo J.* **2002**, *21*, 1565–1576. [CrossRef]
35. Irazoqui, J.E.; Gladfelter, A.S.; Lew, D.J. Scaffold-mediated symmetry breaking by Cdc42p. *Nat. Cell Biol.* **2003**, *5*, 1062–1070. [CrossRef] [PubMed]
36. Bose, I.; Irazoqui, J.E.; Moskow, J.J.; Bardes, E.S.; Zyla, T.R.; Lew, D.J. Assembly of scaffold-mediated complexes containing Cdc42p, the exchange factor Cdc24p, and the effector Cla4p required for cell cycle-regulated phosphorylation of Cdc24p. *J. Biol. Chem.* **2001**, *276*, 7176–7186. [CrossRef]
37. Kozubowski, L.; Saito, K.; Johnson, J.M.; Howell, A.S.; Zyla, T.R.; Lew, D.J. Symmetry-breaking polarization driven by a Cdc42p GEF-PAK complex. *Curr. Biol.* **2008**, *18*, 1719–1726. [CrossRef]

38. Caviston, J.P.; Tcheperegine, S.E.; Bi, E. Singularity in budding: A role for the evolutionarily conserved small GTPase Cdc42p. *Proc. Natl. Acad. Sci. USA* **2002**, *99*, 12185–12190. [CrossRef]
39. Howell, A.S.; Savage, N.S.; Johnson, S.A.; Bose, I.; Wagner, A.W.; Zyla, T.R.; Nijhout, H.F.; Reed, M.C.; Goryachev, A.B.; Lew, D.J. Singularity in polarization: Rewiring yeast cells to make two buds. *Cell* **2009**, *139*, 731–743. [CrossRef]
40. Chiou, J.-g.; Moran, K.D.; Lew, D.J. How cells determine the number of polarity sites. *BioRxiv* **2020**. [CrossRef]
41. Wu, C.F.; Chiou, J.G.; Minakova, M.; Woods, B.; Tsygankov, D.; Zyla, T.R.; Savage, N.S.; Elston, T.C.; Lew, D.J. Role of competition between polarity sites in establishing a unique front. *Elife* **2015**, *4*. [CrossRef]
42. Howell, A.S.; Jin, M.; Wu, C.F.; Zyla, T.R.; Elston, T.C.; Lew, D.J. Negative feedback enhances robustness in the yeast polarity establishment circuit. *Cell* **2012**, *149*, 322–333. [CrossRef] [PubMed]
43. Pablo, M.; Ramirez, S.A.; Elston, T.C. Particle-based simulations of polarity establishment reveal stochastic promotion of Turing pattern formation. *PLoS Comput. Biol.* **2018**, *14*, e1006016. [CrossRef] [PubMed]
44. Brauns, F.; Weyer, H.; Halatek, J.; Yoon, J.; Frey, E. Wavelength selection by interrupted coarsening in reaction-diffusion systems. *arXiv* **2020**, arXiv:2005.01495v2.
45. Brauns, F.; Halatek, J.; Frey, E. Phase-space geometry of reaction-diffusion dynamics. *arXiv* **2018**, arXiv:1812.08684v1.
46. Goryachev, A.B.; Leda, M. Many roads to symmetry breaking: Molecular mechanisms and theoretical models of yeast cell polarity. *Mol. Biol. Cell.* **2017**, *28*, 370–380. [CrossRef]
47. Goryachev, A.B.; Pokhilko, A.V. Computational model explains high activity and rapid cycling of Rho GTPases within protein complexes. *PLoS Comput. Biol.* **2006**, *2*, e172. [CrossRef] [PubMed]
48. Trong, P.K.; Nicola, E.M.; Goehring, N.W.; Kumar, K.V.; Grill, S.W. Parameter-space topology of models for cell polarity. *New J. Phys.* **2014**, *16*, 065009. [CrossRef]
49. Tyson, J.J.; Light, J.C. Properties of 2-component bimolecular and trimolecular chemical-reaction systems. *J. Chem. Phys.* **1973**, *59*, 4164–4173. [CrossRef]
50. Mori, Y.; Jilkine, A.; Edelstein-Keshet, L. Wave-pinning and cell polarity from a bistable reaction-diffusion system. *Biophys. J.* **2008**, *94*, 3684–3697. [CrossRef]
51. Mori, Y.; Jilkine, A.; Edelstein-Keshet, L. Asymptotic and bifurcation analysis of wave-pinning in a reaction-diffusion model for cell polarization. *SIAM J. Appl. Math.* **2011**, *71*, 1401–1427. [CrossRef]
52. Gross, P.; Kumar, K.V.; Goehring, N.W.; Bois, J.S.; Hoege, C.; Jülicher, F.; Grill, S.W. Guiding self-organized pattern formation in cell polarity establishment. *Nat. Phys.* **2019**, *15*, 293–300. [CrossRef] [PubMed]
53. Ishihara, S.; Otsuji, M.; Mochizuki, A. Transient and steady state of mass-conserved reaction-diffusion systems. *Phys. Rev. E Stat. Nonlin. Soft Matter. Phys.* **2007**, *75*, 015203. [CrossRef] [PubMed]
54. Verschueren, N.; Champneys, A. A Model for Cell Polarization Without Mass Conservation. *SIAM J. Appl. Dyn. Syst.* **2017**, *16*, 1797–1830. [CrossRef]
55. Jacobs, B.; Molenaar, J.; Deinum, E.E. Small GTPase patterning: How to stabilise cluster coexistence. *PLoS ONE* **2019**, *14*, e0213188. [CrossRef]
56. Sakumura, Y.; Tsukada, Y.; Yamamoto, N.; Ishii, S. A molecular model for axon guidance based on cross talk between rho GTPases. *Biophys. J.* **2005**, *89*, 812–822. [CrossRef]
57. Okada, S.; Leda, M.; Hanna, J.; Savage, N.S.; Bi, E.; Goryachev, A.B. Daughter cell identity emerges from the interplay of Cdc42, septins, and exocytosis. *Dev. Cell* **2013**, *26*, 148–161. [CrossRef]
58. Bement, W.M.; Leda, M.; Moe, A.M.; Kita, A.M.; Larson, M.E.; Golding, A.E.; Pfeuti, C.; Su, K.C.; Miller, A.L.; Goryachev, A.B.; et al. Activator-inhibitor coupling between Rho signalling and actin assembly makes the cell cortex an excitable medium. *Nat. Cell Biol.* **2015**, *17*, 1471–1483. [CrossRef]
59. Khalili, B.; Lovelace, H.D.; Rutkowski, D.M.; Holz, D.; Vavylonis, D. Fission yeast polarization: Modeling Cdc42 oscillations, symmetry breaking, and zones of activation and inhibition. *Cells* **2020**, *9*, 1769. [CrossRef]
60. Khalili, B.; Merlini, L.; Vincenzetti, V.; Martin, S.G.; Vavylonis, D. Exploration and stabilization of Ras1 mating zone: A mechanism with positive and negative feedbacks. *PLoS Comput. Biol.* **2018**, *14*, e1006317. [CrossRef]
61. Das, M.; Drake, T.; Wiley, D.J.; Buchwald, P.; Vavylonis, D.; Verde, F. Oscillatory dynamics of Cdc42 GTPase in the control of polarized growth. *Science* **2012**, *337*, 239–243. [CrossRef]
62. Moran, K.D.; Kang, H.; Araujo, A.V.; Zyla, T.R.; Saito, K.; Tsygankov, D.; Lew, D.J. Cell-cycle control of cell polarity in yeast. *J. Cell Biol.* **2019**, *218*, 171–189. [CrossRef] [PubMed]

63. Cezanne, A.; Lauer, J.; Solomatina, A.; Sbalzarini, I.F.; Zerial, M. A non-linear system patterns Rab5 GTPase on the membrane. *Elife* **2020**, *9*. [CrossRef] [PubMed]
64. Nigg, E.A.; Holland, A.J. Once and only once: Mechanisms of centriole duplication and their deregulation in disease. *Nat. Rev. Mol. Cell Biol.* **2018**, *19*, 297–312. [CrossRef] [PubMed]
65. Banterle, N.; Gönczy, P. Centriole biogenesis: From identifying the characters to understanding the plot. *Annu. Rev. Cell Dev. Biol.* **2017**, *33*, 23–49. [CrossRef] [PubMed]
66. Conduit, P.T.; Wainman, A.; Raff, J.W. Centrosome function and assembly in animal cells. *Nat. Rev. Mol. Cell Biol.* **2015**, *16*, 611–624. [CrossRef]
67. Leda, M.; Holland, A.J.; Goryachev, A.B. Autoamplification and competition drive symmetry breaking: Initiation of centriole duplication by the PLK4-STIL network. *iScience* **2018**, *8*, 222–235. [CrossRef] [PubMed]

Article

CDC-42 Interactions with Par Proteins Are Critical for Proper Patterning in Polarization

Sungrim Seirin-Lee [1,*], Eamonn A. Gaffney [2] and Adriana T. Dawes [3,4,*]

1 Department of Mathematics and Department of Mathematical and Life Sciences, Graduate School of Integrated Sciences for Life, Hiroshima University, Higashi-Hiroshima 739-8530, Japan
2 Mathematical Institute, University of Oxford, Oxford OX2 6GG, UK; gaffney@maths.ox.ac.uk
3 Department of Mathematics, The Ohio State University, Columbus, OH 43210, USA
4 Department of Molecular Genetics, The Ohio State University, Columbus, OH 43210, USA
* Correspondence: seirin@hiroshima-u.ac.jp (S.S.-L.); dawes.33@osu.edu (A.T.D.)

Received: 10 August 2020; Accepted: 2 September 2020; Published: 5 September 2020

Abstract: Many cells rearrange proteins and other components into spatially distinct domains in a process called polarization. This asymmetric patterning is required for a number of biological processes including asymmetric division, cell migration, and embryonic development. Proteins involved in polarization are highly conserved and include members of the Par and Rho protein families. Despite the importance of these proteins in polarization, it is not yet known how they interact and regulate each other to produce the protein localization patterns associated with polarization. In this study, we develop and analyse a biologically based mathematical model of polarization that incorporates interactions between Par and Rho proteins that are consistent with experimental observations of CDC-42. Using minimal network and eFAST sensitivity analyses, we demonstrate that CDC-42 is predicted to reinforce maintenance of anterior PAR protein polarity which in turn feedbacks to maintain CDC-42 polarization, as well as supporting posterior PAR protein polarization maintenance. The mechanisms for polarity maintenance identified by these methods are not sufficient for the generation of polarization in the absence of cortical flow. Additional inhibitory interactions mediated by the posterior Par proteins are predicted to play a role in the generation of Par protein polarity. More generally, these results provide new insights into the role of CDC-42 in polarization and the mutual regulation of key polarity determinants, in addition to providing a foundation for further investigations.

Keywords: intracellular polarization; partial differential equations; sensitivity analysis

---

1. Introduction

Intracellular polarization, whereby a cell establishes a pattern and specifies a spatial axis by segregating proteins and other factors to distinct domains, is a fundamental and ubiquitous process. Polarization is implicated in a wide variety of biological phenomena including asymmetric cell division, cell migration, wound healing, and embryonic development [1–3]. Aberrant polarization is also thought to play a role in disease progression: a hallmark of the epithelial to mesenchymal transition (EMT) in malignant cells is the acquisition of a polarized migratory phenotype [4,5]. The same key polarity determinants, including the Par and Rho protein families, are required for forming the pattern associated with polarization in virtually all cell types and organisms [3]. Despite the importance of polarization and the highly conserved nature of the proteins involved, the mechanisms and signalling networks regulating this patterning process are not completely understood.

Two main protein families have been intensively studied for their role in polarization: Rho proteins and Par proteins (Figure 1). Rho proteins, also referred to as Rho GTPases, are monomeric G proteins that interconvert between an active GTP-bound state, and an inactive GDP-bound state. The active form of the protein can associate with the membrane at the periphery of the cell, while the inactive form is found diffusing in the cytoplasm [6]. Three members of this family, Rac, Rho and Cdc-42, have been extensively studied for their role in cell migration [6–9]. Polarized migrating cells establish front and back domains, with Cdc-42 and Rac segregating to the front, and Rho associating with the rear of the cell. Par proteins were first identified in the early embryos of the nematode worm *C. elegans* for their role in properly patterning the embryo prior to first division [3,10,11]. The asymmetric localization of the Par proteins specifies the anterior/posterior axis of the developing embryo, and is required for the asymmetric first division. PAR-3, PAR-6 and the atypical protein kinase aPKC bind to the membrane on the periphery of the cell and specify the anterior half of the cell, while PAR-1 and PAR-2 bind to the membrane on the posterior half. Loss of polarity in the first cell cycle is lethal to the embryo [12]. Recent experimental observations suggest that Par and Rho proteins rely on mutual interaction and feedback to produce the pattern of proteins associated with polarization: T cells require both Cdc-42 and Par proteins for polarization [13], and *C. elegans* embryos require CDC-42 for proper patterning of PAR-2 and PAR-6 [14,15]. Despite their involvement in polarization, the dynamics of Par and Rho proteins have been studied largely independently from each other.

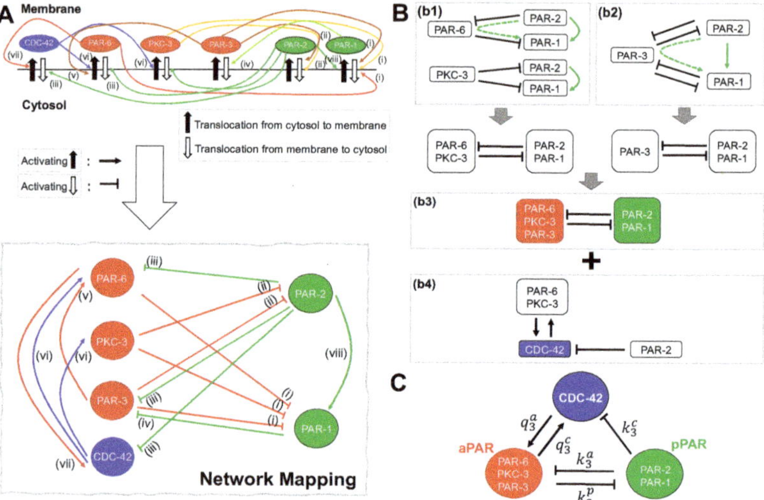

Figure 1. Interaction network of PAR proteins and the Rho GTPase CDC-42. (**A**) Top: Interaction network between PAR proteins and CDC-42 consistent with experimental results. References for interactions (i)–(viii) are given in Section 2.1 of the main text. Bottom: Interaction network used in this study, with cytosol to membrane translocation represented by activation, and membrane to cytosol translocation represented by inhibition. (**B**) Reduction of interaction network in (**A**) into subnetworks. The interaction network is separated into three modules: (b1) shows the interactions between the anterior Par proteins PAR-6 and PKC-3 and the posterior Par proteins PAR-1 and PAR-2, (b2) shows the interactions between the anterior Par protein PAR-3 and the posterior Par proteins PAR-1 and PAR-2, and (b4) shows the interactions between the anterior and posterior Par proteins and the Rho GTPase CDC-42. Note that the upregulation of PAR-6 by PAR-3 is implicitly captured by the inhibitory action of PAR-3 on PAR-2, since the latter in turn downregulates PAR-6. As described in the text, (b1) and (b2) can be reduced to (b3). The green dotted arrows indicates two sequential inhibition interactions ($\top\perp$) that are equivalent to an activating interaction ($\uparrow$, solid green line). (**C**) Network (b3) merged with (b4) is the fundamental network investigated here. Rate constants $q_3^a, q_3^c, k_3^p, k_3^a$, and k_3^c appear in the corresponding model Equations (4).

While polarization is observed in many different systems, we focus on protein patterning in the early *C. elegans* embryo [16] (Figure 2). The establishment phase, which is initiated by fertilization, moves symmetrically distributed Par proteins, including PAR-3, PAR-6 and aPKC, to the anterior half of the cell. The cleared area in the posterior half is then available for binding by the posterior Par proteins, including PAR-1 and PAR-2. Once the polarized domains are established, the cell transitions into the maintenance phase, where it maintains the asymmetric protein pattern as the cell prepares for first division. While the protein transport associated with polarization is important for establishment of polarization [17], we focus on the role of biochemical interactions in the generation and maintenance of polarized domains, independent of advective flow. Members of the Rho protein family, notably CDC-42, are thought to be important for polarization in the early embryo, but how they interact with and regulate Par proteins is not clear.

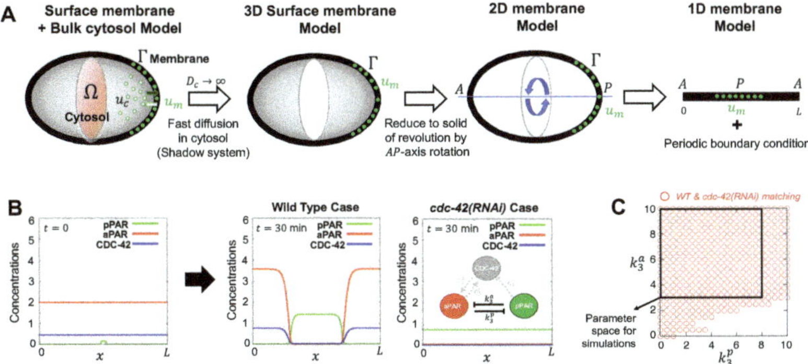

Figure 2. Schematic of model components and geometry. (**A**) Model reduction from three dimensions to one dimension in space. A cross section of the bulk cytosol region is coloured by red, the surface is coloured by gray, with u_c, u_m denoting generic cytosolic and membrane densities, respectively. Fast cytosolic diffusion timescales compared to the timescales of interest lead to effectively homogeneous cytosolic concentrations, that can be captured by membrane concentrations via conservation principles, as discussed in Section 2.2. This allows the model to be reduced to membrane equations, with axisymmetry about the major axis of the cell for a simple geometry yielding the final simplification, as detailed in Appendix A.1. (**B**) Representative simulations for wild type case and *cdc-42(RNAi)* case. Simulations start with initial conditions given by initial conditions, Equation (5). After 30 min, the solutions shown have reached steady state. Here, and throughout the paper, concentrations are in units of $\mathcal{M}/L$, where $\mathcal{M}$ is the characteristic protein number of Table A1, and L is the perimeter of the projected membrane in (**A**). (**C**) The range of values for the parameters k_3^p, and k_3^a that satisfy both wild type and *cdc-42(RNAi)* behaviours which serves as a validation of the model against observations. The red circles indicate the parameter region matching both wild type and *cdc-42(RNAi)* behaviours. The black square indicates the range of (k_3^p, k_3^a) used for sensitivity analysis.

In this paper, we develop a continuum model of Par and Rho protein dynamics in the generation and maintenance of polarization in the early *C. elegans* embryo from the rational simplification for the underlying interactions among CDC-42 and the PAR proteins that have been uncovered by numerous experimental studies. This model also utilizes the large ratio of diffusive transport scales between the cell cytosol and membrane and a simple representation of the cell geometry. We demonstrate that the resulting partial differential equation model is consistent with observations of CDC-42, highlighting the requirement for CDC-42 during the maintenance of polarization in the early *C. elegans* [18,19]. To elucidate the detailed mechanism of interactions capable of reproducing experimental observations, we perform sensitivity analysis and minimal network analysis to identify and characterize predictions for key cross talk and mutual regulation interactions among CDC-42 and the PAR proteins in controlling the generation and maintenance of cellular polarization. Due to the conserved nature of both the process and proteins involved in polarization, the insights gained in this study apply broadly to other biological systems beyond *C. elegans*.

2. The Mathematical Model

2.1. Network of Par Protein and Cdc-42 Interactions

To investigate the role potential role of interactions between Par and Rho protein family members in the generation and maintenance of distinct spatial domains, we construct a network of interactions that is consistent with experimental data. Justification for each arrow in Figure 1A is as follows:

(i) PAR-6/PKC-3/PAR-3 promotes dissociation of PAR-1 from the membrane to the cytosol [10,16,20].
(ii) PKC-3/PAR-3 promotes dissociation of PAR-2 from the membrane to the cytosol [20–22].
(iii) PAR-2 promotes dissociation of PAR-6, PAR-3, and CDC-42 from the membrane to the cytosol [22].
(iv) PAR-1 promotes dissociation of PAR-3 from the membrane to the cytosol [23].
(v) PAR-3 associates with PAR-6, allowing PAR-6 to be maintained in the membrane [22].
(vi) CDC-42 associates with PAR-6/PKC-3, allowing PAR-6/PKC-3 to be maintained in the membrane [18,20,22–24].
(vii) PAR-6 associates with CDC-42, allowing CDC-42 to be maintained in the membrane [18,22,24,25].
(viii) PAR-2 associates with PAR-1, preventing PAR-1 dissociation from the membrane by PAR-6/PKC-3/PAR-3 [10,26].

Figure 1A shows a full schematic diagram and the network of interactions corresponding to the descriptions (i)–(viii). Some of these interactions may not be direct, but are mediated by other proteins. For instance, enhanced cortical dissociation of the anterior Par proteins by PAR-2 is likely mediated by PAR-1 [27]. In addition, the role of CDC-42 and its interactions with the Par proteins are not entirely clear, and some observations are not accounted for in our model, such as the appearance of CDC-42 on the cortex even in the absence of PAR-6 [14]. Nevertheless, experimental observations of all interactions listed above have been reported in the literature.

To determine the fundamental network of interactions based on the above experimental evidence, we separated the interaction network in Figure 1A into three subnetworks, Figure 1B(b1,b2,b4). In Figure 1B(b1), we consider the interactions between [PAR-6, PAR-2 and PAR-1] and [PKC-3, PAR-2 and PAR-1]. In this subnetwork, the inhibition of PAR-1 by PAR-6 and inhibition of PAR-6 by PAR-2 (dotted green line) is equivalent to PAR-2 activation of PAR-1 (solid green line). Since PAR-6 and PKC-3 only dissociate PAR-1 and PAR-2 when PAR-6 and PKC-3 are bound together in a complex, we can reduce this subnetwork to a mutual inhibition between [PAR-6 and PKC-3] and [PAR-1 and PAR-2] (Figure 1B(b1)). In a similar manner, the subnetwork containing [PAR-3, PAR-2 and PAR-1] can be reduced to a network of mutual inhibition between [PAR-3] and [PAR-1 and PAR-2] (Figure 1B(b2)). We further simplify these subnetworks by grouping the Par proteins according to their localization: anterior Par proteins (aPAR) and posterior Par proteins (pPAR) (Figure 1B(b3)). Finally, we combine the mutual Par protein inhibition subnetwork with the [CDC-42, PAR-6/PKC-3 and PAR-2] subnetwork (Figure 1B(b4)) to arrive at the fundamental interaction network of [aPAR, pPAR, and CDC-42] (Figure 1C).

2.2. Model Equations for Apar, Ppar and Cdc-42 Network

We define the cytosol by $\Omega \subset \mathbb{R}^3$ and the membrane by $\partial\Omega(\equiv \Gamma)$ so that $\Omega \cup \partial\Omega$ represents *C. elegans* single cell embryo (Figure 2A). We represent the concentrations of anterior proteins (aPARs, i.e., PAR-6, PAR-3, and PKC-3) in the membrane and the cytosol by $[A_m(\mathbf{x},t)]$ and $[A_c(\mathbf{x},t)]$, respectively, and the concentrations of posterior proteins (pPARs, i.e., PAR-1 and PAR-2) in the membrane and the cytosol by $[P_m(\mathbf{x},t)]$ and $[P_c(\mathbf{x},t)]$, respectively, and the concentrations of CDC-42 in the membrane and the cytosol by $[C_m(\mathbf{x},t)]$ and $[C_c(\mathbf{x},t)]$, respectively, where $\mathbf{x} \in \mathbb{R}^3$ and $t \in [0,\infty)$. The dynamics of the network shown in Figure 1C can then be described as follows:

$$\begin{aligned}
\frac{\partial[A_m]}{\partial t} &= D_m^A \nabla_\Gamma^2 [A_m] + \{\gamma_a + F_{\text{on}}^A([C_m])\}[A_c] - \{\alpha_a + F_{\text{off}}^A([P_m])\}[A_m] && \text{on } \mathbf{x} \in \partial\Omega, \\
\frac{\partial[A_c]}{\partial t} &= D_c^A \nabla^2 [A_c] && \text{on } \mathbf{x} \in \Omega, \\
D_c^A \frac{\partial[A_c]}{\partial \mathbf{n}} &= -\{\gamma_a + F_{\text{on}}^A([C_m])\}[A_c] + \{\alpha_a + F_{\text{off}}^A([P_m])\}[A_m] && \text{on } \mathbf{x} \in \partial\Omega, \\
\\
\frac{\partial[P_m]}{\partial t} &= D_m^P \nabla_\Gamma^2 [P_m] + \gamma_p [P_c] - \{\alpha_p + F_{\text{off}}^P([A_m])\}[P_m] && \text{on } \mathbf{x} \in \partial\Omega, \\
\frac{\partial[P_c]}{\partial t} &= D_c^P \nabla^2 [P_m] && \text{on } \mathbf{x} \in \Omega, \\
D_c^P \frac{\partial[P_c]}{\partial \mathbf{n}} &= -\gamma_p [P_c] + \{\alpha_a + F_{\text{off}}^P([A_m])\}[P_m] && \text{on } \mathbf{x} \in \partial\Omega, \\
\\
\frac{\partial[C_m]}{\partial t} &= D_m^C \nabla_\Gamma^2 [C_m] + \gamma_c [C_c] - \{\alpha_c + F_{\text{off}}^C([P_m],[A_m])\}[C_m] && \text{on } \mathbf{x} \in \partial\Omega, \\
\frac{\partial[C_c]}{\partial t} &= D_c^C \nabla^2 [C_c] && \text{on } \mathbf{x} \in \Omega, \\
D_c^C \frac{\partial[C_c]}{\partial \mathbf{n}} &= -\gamma_c [C_c] + \{\alpha_c + F_{\text{off}}^C([P_m],[A_m])\}[C_m] && \text{on } \mathbf{x} \in \partial\Omega,
\end{aligned} \tag{1}$$

where D_m^A, D_m^P and D_m^C are the diffusion coefficients of aPARs, pPARs and CDC-42 in the membrane, respectively, and D_c^A, D_c^P and D_c^C are the diffusion coefficients of aPARs, pPAR and CDC-42 in the cytosol, respectively. $\mathbf{n}$ is the inner normal vector on $\partial\Omega$. We assume that F_{on}^A, F_{off}^A and F_{off}^P are increasing functions of CDC-42 concentration, pPAR concentration, and aPAR concentration, respectively. We also assume that F_{off}^C is an increasing function of pPAR concentration but a decreasing function of aPAR concentration. For the purposes of this study, we choose the following functional forms [22,28];

$$\begin{aligned}
F_{\text{on}}^A([C_m]) &= \frac{q_3^a [C_m]^2}{q_1^a + q_2^a [C_m]^2}, \quad F_{\text{off}}^A([P_m]) = \frac{k_3^a [P_m]^2}{k_1^a + k_2^a [P_m]^2}, \\
F_{\text{off}}^P([A_m]) &= \frac{k_3^p [A_m]^2}{k_1^p + k_2^p [A_m]^2}, \quad F_{\text{off}}^C([P_m],[A_m]) = \frac{k_3^c [P_m]^2}{k_1^c + k_2^c [P_m]^2} + \frac{q_3^c}{q_1^c + q_2^c [A_m]^2}.
\end{aligned} \tag{2}$$

Rate constants of the form k_i^j denote interactions that include pPAR while rate constants of the form q_i^j denote interactions strictly between aPAR and CDC-42.

Parameter values corresponding to the final simplified model are summarized in Table A1. Note that some parameters and variables are redefined during this simplification process. The first step in this simplification exploits differences in diffusion dynamics. We require $D_c^A \gg D_m^A$, $D_c^P \gg D_m^P$ and $D_c^C \gg D_m^C$ since diffusion in the cytosol is much faster than diffusion in the membrane [29,30]. Furthermore, this fast cytosolic diffusion, if sufficiently large, results in homogeneous spatial distribution of cytoplasmic proteins over the time scale of interest (Figure A1) and allows us to reduce the model (1) to a shadow system. By taking $D_c^A, D_c^P, D_c^C \to \infty$, the leading order approximations for cytosolic protein concentrations quickly approach a homogeneous steady states such that

$$A_c(\mathbf{x},t) \to \frac{1}{|\Omega|}\int_\Omega [A_c](\mathbf{x},t)d\mathbf{x}, \quad P_c(\mathbf{x},t) \to \frac{1}{|\Omega|}\int_\Omega [P_c](\mathbf{x},t)d\mathbf{x}, \quad C_c(\mathbf{x},t) \to \frac{1}{|\Omega|}\int_\Omega [C_c](\mathbf{x},t)d\mathbf{x}.$$

The following conservation relations hold:

$$\frac{d}{dt}\left(\int_{\Omega}[A_c](\mathbf{x},t)d\mathbf{x}+\int_{\Gamma}[A_m](\mathbf{x},t)d\mathbf{x}\right)=0,$$
$$\frac{d}{dt}\left(\int_{\Omega}[P_c](\mathbf{x},t)d\mathbf{x}+\int_{\Gamma}[P_m](\mathbf{x},t)d\mathbf{x}\right)=0,$$
$$\frac{d}{dt}\left(\int_{\Omega}[C_c](\mathbf{x},t)d\mathbf{x}+\int_{\Gamma}[C_m](\mathbf{x},t)d\mathbf{x}\right)=0,$$

and the total mass of the model (1) is conserved. We define the conserved total concentrations as follows:

$$A_{tot}\equiv\int_{\Omega}[A_c](\mathbf{x},t)d\mathbf{x}+\int_{\Gamma}[A_m](\mathbf{x},t)d\mathbf{x},\qquad P_{tot}\equiv\int_{\Omega}[P_c](\mathbf{x},t)d\mathbf{x}+\int_{\Gamma}[P_m](\mathbf{x},t)d\mathbf{x},$$
$$C_{tot}\equiv\int_{\Omega}[C_c](\mathbf{x},t)d\mathbf{x}+\int_{\Gamma}[C_m](\mathbf{x},t)d\mathbf{x},$$

giving us

$$A_c(\mathbf{x},t)\approx\frac{1}{|\Omega|}\left(A_{tot}-\int_{\Gamma}[A_m](\mathbf{x},t)d\mathbf{x}\right),\quad P_c(\mathbf{x},t)\approx\frac{1}{|\Omega|}\left(P_{tot}-\int_{\Gamma}[P_m](\mathbf{x},t)d\mathbf{x}\right),$$
$$C_c(\mathbf{x},t)\approx\frac{1}{|\Omega|}\left(C_{tot}-\int_{\Gamma}[C_m](\mathbf{x},t)d\mathbf{x}\right),$$

at leading order in an asymptotic approximation based on the cytosolic diffusion dominating all other possible diffusive scales in the model. Thus, working to this level of approximation, model (1) is reduced to the surface model:

$$\begin{aligned}
\frac{\partial[A_m]}{\partial t}&=D_m^A\nabla_\Gamma^2[A_m]+\{\gamma_a+F_{\text{on}}^A([C_m])\}\left(\frac{A_{tot}}{|\Omega|}-\frac{1}{|\Omega|}\int_\Gamma[A_m]d\mathbf{x}\right)-\{\alpha_a+F_{\text{off}}^A([P_m])\}[A_m] && \text{on } \mathbf{x}\in\Gamma,\\
\frac{\partial[P_m]}{\partial t}&=D_m^P\nabla_\Gamma^2[P_m]+\gamma_p\left(\frac{P_{tot}}{|\Omega|}-\frac{1}{|\Omega|}\int_\Gamma[P_m]d\mathbf{x}\right)-\{\alpha_p+F_{\text{off}}^P([A_m])\}[P_m] && \text{on } \mathbf{x}\in\Gamma,\\
\frac{\partial[C_m]}{\partial t}&=D_m^C\nabla_\Gamma^2[C_m]+\gamma_c\left(\frac{C_{tot}}{|\Omega|}-\frac{1}{|\Omega|}\int_\Gamma[C_m]d\mathbf{x}\right)-\{\alpha_c+F_{\text{off}}^C([P_m],[A_m])\}[C_m] && \text{on } \mathbf{x}\in\Gamma.
\end{aligned}\tag{3}$$

We define the model geometry as the surface of a solid of revolution found by rotating the arc of the membrane about the AP-axis (Figure 2A). Since we are only interested in the dynamics on the membrane, we seek to reduce model (3) to a one dimensional model on $x\in[0,L]$ where x is the arclength along the perimeter of the projected membrane, depicted in Figure 2A, $x=0$ corresponds to the leftmost point on the membrane and L is the cell perimeter. This reduction to a 1D domain also ensures the minimal network and eFAST analysis can be accomplished in a reasonable amount of time. Periodic boundary conditions are imposed at $x=0,L$ (Figure 2A). We further approximate the geometry effectively by a cylinder of radius $H\ll L$ with asymptotically short caps and we define the tilde variables and parameters via

$$\tilde{A}_m=\pi HA_m,\ \tilde{P}_m=\pi HP_m,\ \tilde{C}_m=\pi HC_m,\ \tilde{\gamma}_a/\gamma_a=\tilde{\gamma}_p/\gamma_p=\tilde{\gamma}_c/\gamma_c=\tilde{q}_3^a/q_3^a=L\pi H/|\Omega|.$$

As detailed in Appendix A.1, we rewrite Equation (3) in terms of these new parameters, together with the assumption of axisymmetry and geometrical approximations that are stated and justified in Appendix A.1. On dropping tildes, we have

$$\begin{aligned}
\frac{\partial[A_m]}{\partial t} &= D_m^A \frac{\partial^2}{\partial x^2}[A_m] + \{\gamma_a + F_{\text{on}}^A([C_m])\}\left(\frac{A_{tot}}{L} - \frac{1}{L}\int_0^L [A_m]dx\right) - \{\alpha_a + F_{\text{off}}^A([P_m])\}[A_m], \\
\frac{\partial[P_m]}{\partial t} &= D_m^P \frac{\partial^2}{\partial x^2}[P_m] + \gamma_p\left(\frac{P_{tot}}{L} - \frac{1}{L}\int_0^L [P_m]dx\right) - \{\alpha_p + F_{\text{off}}^P([A_m])\}[P_m], \\
\frac{\partial[C_m]}{\partial t} &= D_m^C \frac{\partial^2}{\partial x^2}[C_m] + \gamma_c\left(\frac{C_{tot}}{L} - \frac{1}{L}\int_0^L [C_m]dx\right) - \{\alpha_c + F_{\text{off}}^C([P_m],[A_m])\}[C_m].
\end{aligned} \quad (4)$$

The parameter values in Table A1 are with respect to this final model (Equation (4)). The model was simulated by custom software written in C. The PDEs were solved via an implicit numerical scheme using standard finite-difference methods. Source code used to generate the results in this paper are available upon request.

2.3. Parameter Values

Before symmetry breaking in the *C. elegans* embryo, aPAR is spatially homogeneously distributed on the membrane and pPAR is spatially homogeneously distributed in the cytosol. Gotta et al [19] experimentally demonstrated that loss of CDC-42 by RNAi results in a loss of polarity, with low PAR-6 and high PAR-2 levels on the membrane. Thus, we choose representative kinetic parameters such that aPar, pPar and CDC-42 establish distinct spatial domains under wild type conditions, but fail to polarize when CDC-42 is absent (Figure 2B). For sensitivity analysis, we restrict our parameters as shown in Figure 2C, corresponding to parameter values that are consistent with both wild type and *cdc-42(RNAi)* experimental observations. The fact that such a large range of parameter values produces the appropriate model behaviours provides further validation of the model's predicted importance for CDC-42 in polarization. Parameter names and definitions for the final model, Equation (4), are summarized in Table A1, together with the non-dimensionalization used in the numerical investigation of the model.

2.4. Initial Conditions

We use two sets of initial conditions to evaluate either maintenance (Section 3.1.1) or generation (Section 3.2.1) of polarization. For simulating maintenance of polarization, we specify polarized initial conditions with aPAR and CDC-42 high in the anterior (left and right) region of the domain and low in the posterior (middle) region, and the reverse profile for pPAR (high in the posterior, low in the anterior). See Figure 2 for a schematic of the model geometry and the anterior/posterior regions. The high and low values are derived from the stationary long time asymptotic solution of Equation (4) (Figure 2B).

For simulating generation of polarization, we specify initial conditions with a small spatial perturbation:

$$[P_m](x,0) = \epsilon_p \delta(x - L/2),\ [A_m](x,0) = \frac{A_{tot}}{L}(1 + \epsilon\phi_a(x)),\ [C_m](x,0) = C_0(1 + \epsilon\phi_c(x)), \quad (5)$$

where $x \in [0,\ L]$, δ is the delta function, and C_0 is the equilibrium initial concentration of CDC-42. $\phi_a(x)$ and $\phi_c(x)$ are perturbation functions, ϵ_p is the strength of the initial external perturbation signal and ϵ is the magnitude of the perturbation.

2.5. Minimal Network Analysis

To determine the minimal set of interactions required for maintenance (Section 3.1.1) or generation (Section 3.2.1) of a polarization pattern, we devised a method of minimal network analysis.

Starting from the fundamental network containing all interactions (Figure 1C), we first remove individual interactions from the network. This reduced network is simulated and evaluated for the presence of a pattern using initial conditions described above. The total simulation time for pattern evaluation is 16 min for maintenance of polarization and 30 min for generation of polarization. While the numerical solution may not have converged to a stationary solution by the time of evaluation, it is clear whether a pattern has been initiated. We determine which interactions are common to all reduced networks that are capable of maintaining/generating polarization. A network consisting of these common interactions is tested for its capacity to maintain/generate a polarized pattern. If the common interaction network is not capable of maintaining/generating a pattern consistent with polarization, interactions are reintroduced into the model, first individually and then in pairs, and the model is simulated and evaluated for the presence of a polarized pattern. The reduced network(s) with the least number of interactions that is capable of maintaining/generating polarization is retained for further analysis.

2.6. eFAST Sensitivity Analysis

Sensitivity analysis is used to determine the influence of model parameters on the dynamics of the network via the use of Extended Fourier Amplitude Sensitivity Test (eFAST) [31,32]. In brief, we define sensitivity functions whose dynamics reveals self-organization and track their changes in response to changes in parameter values. In particular, the sensitivity functions for aPAR, pPAR, and CDC-42, denoted by F_S^A, F_S^P and F_S^C, respectively, are defined as

$$\begin{aligned} F_S^A &= \sqrt{\int_0^L \left| \frac{[A_m](x,t^*) - [A_m](x,0)}{t^*} \right|^2}, \\ F_S^P &= \sqrt{\int_0^L \left| \frac{[P_m](x,t^*) - [P_m](x,0)}{t^*} \right|^2}, \\ F_S^C &= \sqrt{\int_0^L \left| \frac{[C_m](x,t^*) - [C_m](x,0)}{t^*} \right|^2}, \end{aligned} \tag{6}$$

where t^* is the time scale of interest. This time scale is chosen as 16 min for maintenance of polarization, and 30 min for generation of polarization, consistent with experimental observations [33,34]. These sensitivity functions give a quantitative measure of how much the polarity pattern (the model output) has been altered in response to changes in parameter values (the model input). We calculate two sensitivity measures, the first order index (S_i) and total-effect index (S_{T_i}). These measures are defined as follows:

$$\begin{aligned} S_i &\equiv \frac{\text{Variance of the expected model output } y \text{ with respect to parameter } p_i}{\text{Total variance}} = \frac{V_i}{V_{total}}, \\ S_{T_i} &\equiv \text{Total effect (contribution) of parameter } p_i \text{ to the output variance} = 1 - \frac{V_{-i}}{V_{total}}, \end{aligned}$$

where V_{-i} is the effect of any order that does not include the factor i. The first order index indicates the influence of parameter p_i on the variance of the polarization measure (the model output), independent of interactions with the other parameters. The total-effect index indicates the effect of parameter p_i when interactions with the other parameters are included. These two measures give a full quantification of the importance of parameter p_i and whether the extent whether this is a direct influence or through interactions with other parameters or both. In this study, we focus on the parameters q_3^a, q_3^c, k_3^p, k_3^a and k_3^c which determine the magnitude of the interaction functions, Equation (2).

A more in depth discussion of the eFAST sensitivity method and its detailed accompanying calculations can be found in Appendix A.3. The values used for sensitivity analysis are given in Table A2.

3. Results

3.1. Critical Network Interactions and Parameters for Maintenance of Par Protein Polarization

It has been observed that CDC-42 is required during maintenance but not establishment of polarization in the early *C. elegans* embryo [18,19]. However, it is not clear how CDC-42 is interacting with or regulating the anterior and posterior Par proteins to ensure maintenance of polarization in the early embryo.

3.1.1. Minimal Network for Maintenance of Par Protein Polarization

Using our method of minimal network analysis, we aim to determine the minimal interaction network between anterior Par proteins (aPAR), CDC-42 and posterior Par proteins (pPAR) that may maintain spatial polarization. As shown in Figure 3A, the activation of aPAR by CDC-42 ((a1), q_3^a interaction) and inhibition of pPAR by aPAR ((a3), k_3^p interaction) are always present in reduced networks that are able to maintain polarization, suggesting these interactions are critical. The ability of the model to polarize in the absence of the other interactions ((a2), (a4), (a5)) show these interactions are less critical for polarity maintenance.

A minimal network consisting of only these two interactions (q_3^a and k_3^p, Figure 3C) is not sufficient to maintain polarization. As shown in Figure 3B, adding CDC-42 inhibition by pPAR ((b1), k_3^c interaction) or aPAR inhibition by pPAR ((b2), k_3^a interaction) back to the network also cannot maintain polarization. Adding CDC-42 activation by aPAR ((b3), q_3^c interaction) allows polarity maintenance, although pPAR does not vary much over the domain. Further reinforcing the importance of CDC-42 for maintenance of polarity, we find that polarization of aPAR can be maintained with only two interactions (Figure 3D, q_3^a and k_3^p networks) and without mutual inhibition by pPAR. This is consistent with previous results, showing that aPARs intially polarize during the establishment phase, transiently establishing an anterior Par protein domain. However, this polarization can not be maintained, and the aPAR domain gradually creeps back toward the posterior pole, resulting in an eventual loss of polarity [17,18]. Our model reproduces the initial aPAR polarization in the absence of pPAR.

Taken together, the results suggest that the minimal network that is capable of maintaining polarization of aPAR and pPAR involves mutual activation of aPAR and CDC-42 and aPAR-mediated inhibition of pPAR (Figure 3B(b3)).

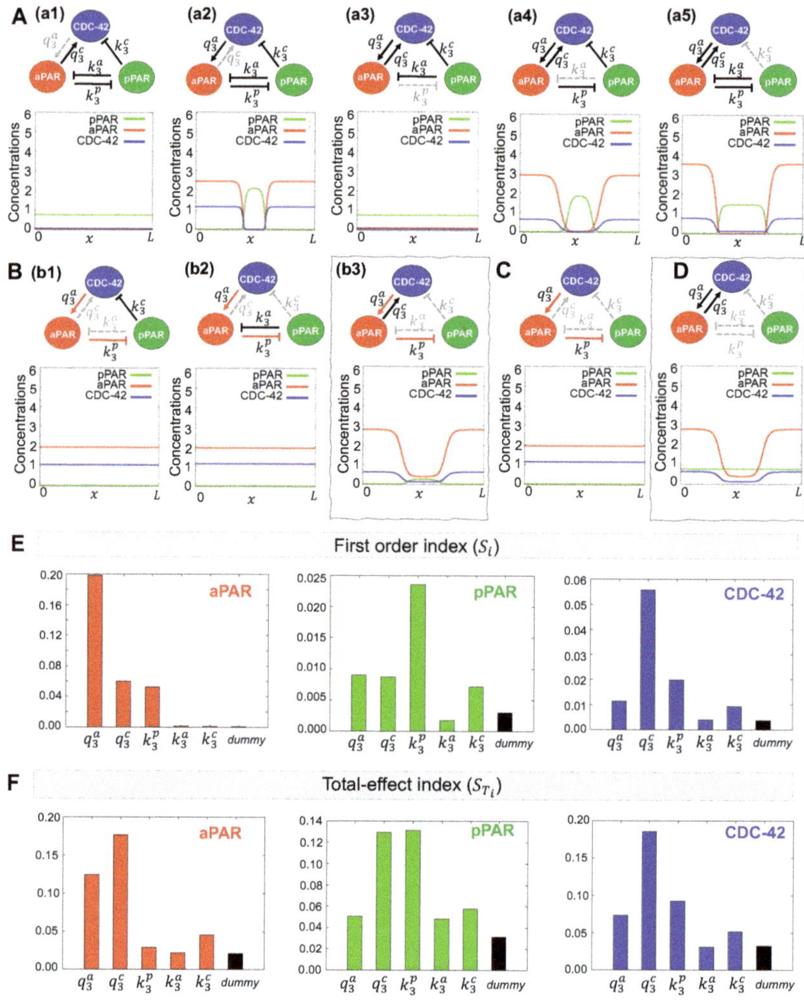

Figure 3. Minimal network needed for maintenance of Par protein polarization. (**A–D**) Representative simulations for the effects of network interactions. Dotted gray lines indicate interactions excluded from the network. Simulations are shown at $t = 16$ min, the approximate amount of time the early *C. elegans* embryo spends in the maintenance phase [33]. Initial conditions are given by the corresponding stationary solution (Figure 2B). (**A**) (a1)–(a5) Simulation results of networks omitting a single interaction. Minimal networks that produce the correct spatial pattern and are retained for further analysis are boxed in gray. (**B**) Red arrows indicate interactions identified in (**A**) that are retained in these additional networks. (b1)–(b3) Simulation results of networks omitting two interactions. (**E,F**) Variance-based sensitivity analysis results using eFAST. Parameters with higher values have a more significant effect on the network dynamics. (**E**) The first order index (the effect of a parameter on the model dynamics independent of the other parameters) and (**F**) The total effect index (the effect of a parameter including the interactions with other all model parameters) both indicate that CDC-42 interactions with aPAR are critical for producing the correct spatial pattern.

3.1.2. Critical Parameters for Maintenance of Par Protein Polarization

To determine the effect of the network interactions on maintenance of Par protein polarity with the fundamental network, we performed a sensitivity analysis using the eFAST method [32,35]. See Appendix A.3 for more details about this approach. We wish to compare the critical interactions found by minimal network analysis with the important parameters indicated by the sensitivity analysis. We evaluated the first order index, S_i, for the five parameters governing the magnitude of each possible interaction in the network (Figure 3E). We find that the most important parameters as determined by the sensitivity analysis depends on the choice of sensitivity function. When the sensitivity function for aPAR is used as the model output, we find that aPAR activation by CDC-42 (q_3^a) is the most important parameter, followed by CDC-42 activation by aPAR (q_3^c) and pPAR inhibition by aPAR (k_3^p). When the sensitivity function for pPAR is used as the model output, we find that pPAR inhibition by aPAR (k_3^p) is the most important parameter, followed by mutual activation of aPAR and CDC-42 (q_3^a and q_3^c) as well as CDC-42 inhibition by pPAR (k_3^c). Using the sensitivity function for CDC-42 as the model output, we find that CDC-42 activation by aPAR (q_3^c) is the most important parameter, followed by pPAR inhibition by aPAR (k_3^p), aPAR activation by CDC-42 (q_3^a), and CDC-42 inhibition by pPAR (k_3^c). Combining the most important parameters for each sensitivity function, we find that they correspond to the network interactions in our minimal network (q_3^a, q_3^c, and k_3^p interactions, Figure 3B(b3)). Since the minimal network was found independently of the sensitivity analysis, this suggests our minimal network contains only the most critical interactions.

We then evaluated the total-effect index, S_{T_i}. We find that CDC-42 activation by aPAR (q_3^c interaction) appears as a high index value with respect to both the aPAR and pPAR sensitivity functions, which we did not find with the first order index (see Appendix A.4 for further details of index value significance and meaning). When considering the aPAR sensitivity function, the mutual activation of aPAR and CDC-42 (q_3^a and q_3^c interactions) have a high total index, indicating that these two mutual activation interactions are very important for the maintenance of aPAR polarity (Figure 3D). Similarly, when considering the pPAR sensitivity function, the two high index values correspond to q_3^c and k_3^p interactions, suggesting that the activation of CDC-42 by aPAR plays an important role in the maintenance of pPAR polarity along with the inhibition of pPAR by aPAR. When considering the sensitivity function for CDC-42, the total-effect index followed the same trend as the first order index, with CDC-42 activation by aPAR (k_3^c interaction) being the most important parameter. When considering the total effect of a parameter, this suggests that CDC-42 plays a critical role for maintenance of both aPAR and pPAR polarity. We also found that the magnitude of the sensitivity indices for S_i and S_{T_i} are substantially different for the pPAR and CDC-42 cases, but not for the aPAR case, suggesting that maintenance of pPAR and CDC-42 polarity are more strongly affected by interactions with the other proteins in the network, but maintenance of aPAR polarity operates largely independently of the other proteins. Together, this indicates that aPAR may play a central role for the entire network during maintenance of polarity, and that maintenance of pPAR polarity may depend directly on maintenance of aPAR polarity.

3.2. Critical Network Interactions and Parameters for Generation of Par Protein Polarization

In the previous section, we determined the critical interactions and parameters in a minimal network for the maintenance of Par protein polarity in the early *C. elegans* embryo. Under wild type conditions, establishment of polarization in the early *C. elegans* embryo relies on actin and myosin based advective flow [17,36]. However, polarization can still occur even in the absence of cortical flow [30], although on a longer time scale. Thus, using a similar approach as in the previous section, we aim to determine if the interactions and parameters required for maintenance of polarity are different from those required to generate polarization in the absence of cortical flow.

3.2.1. Minimal Network for Generation of Par Protein Polarization

To determine the minimal network required to generate a polarity pattern, we simulated networks with individual interactions missing to determine which interactions are predicted to be necessary for self-organization. Simulations were given a local stimulus as an initial conditions, as shown in Figure 2B, first panel, and were assessed at 30 min for the presence of polarization. As shown in Figure 4A, the three interactions represented by the parameters q_3^a, q_3^c, and k_3^p play critical roles in generation of polarization as their absence leads to the absence of pattern generation and these are the same three, interactions required for maintenance of polarization (Figure 3B(b3)). However, the minimal network for maintenance of polarization (Figure 3B(b3)) has not been sufficient to generate a polarized pattern (Figure 4B), and the addition of either the k_3^a or k_3^c interaction was also important (Figure 4A(a4,a5)). We have also confirmed that the minimal network for aPAR polarity maintenance (q_3^a and q_3^c interactions, Figure 3D) is not sufficient to generate a polarity pattern even in the presence of CDC-42 inhibition by pPAR (Figure 4C). These results suggest that in addition to the minimal network for maintenance phase, aPAR inhibition by pPAR is critical for generating Par protein polarity, and the mutual inhibition between the anterior and posterior Par proteins directly and via CDC-42 is important in generating the pattern.

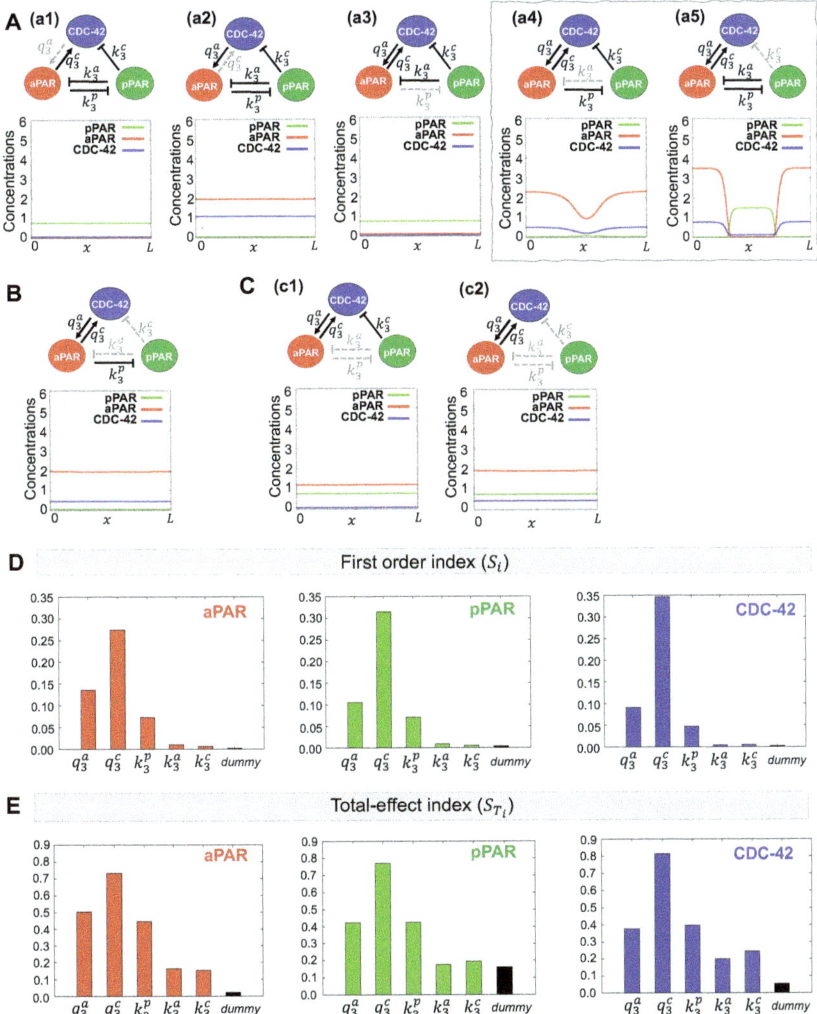

Figure 4. Minimal network needed for generation of Par protein polarization. (**A**–**C**) Representative simulations for the indicated network. Dotted gray lines indicate interactions excluded from the network. The initial condition provides a small local perturbation, as discussed in Section 2.4. Simulation results are shown at $t = 30$, longer than the time scale for polarity emergence reported experimentally [30]. (**A**) (a1)–(a5) Simulation results of networks omitting a single interaction. Networks boxed in gray are minimal networks capable of generating a polarization pattern. (**B**) and (**C**) (c1)–(c2) Simulation results of networks omitting two or more interactions. (**E**–**F**) Variance-based sensitivity analysis results using the eFAST method. Both the first order and total effect index indicate the importance of CDC-42 in producing the correct spatial pattern.

3.2.2. Critical Parameters for Generation of Par Protein Polarization

We then used sensitivity analysis to determine which parameters are critical for generation of polarization. We used the same sensitivity functions as above (Equation (6)), and assessed the model simulation at $t^* = 30$ min. In contrast to our sensitivity analysis results for the maintenance of polarization, we find that CDC-42 activation by aPAR (q_3^c interaction) is the most important parameter for aPAR, pPAR and CDC-42 with respect to both the first order and total-effect index. Although the interactions represented by the parameters q_3^a, q_3^c, and k_3^p tended to show higher importance than the other two interactions, k_3^a and k_3^c, in both first order and total-effect indices, the effect of the k_3^a and k_3^c interactions are not negligible in the total-effect index as compared to the first order index for aPAR and CDC-42. This indicates that the k_3^a and k_3^c interactions may influence the generation of the polarity pattern, consistent with our minimal network analysis (Figure 4A(a4,a5)).

4. Discussion

Through minimal network analysis and eFAST sensitivity analysis, we identified different roles for aPAR, pPAR and CDC-42 in maintaining or generating the spatial pattern associated with polarization. To focus on the biochemical interactions, we focus on the dynamics of polarization in the absence of cortical flow, and neglect interactions between CDC-42 and the Par proteins with mechanical proteins including actin and myosin. Results are summarized in Figure 5 and Table 1. CDC-42 is required in both pattern maintenance and generation: CDC-42 reinforces maintenance of aPAR polarity which in turn directs pPAR polarization as indicated by the high sensitivities of the latter to the up-regulation of CDC-42 and down-regulation of pPAR by aPAR. Thus, the entire system of interactions relies on aPAR polarity to maintain polarization in all variables. However, the minimal network capable of maintaining Par protein polarization is insufficient to generate the polarization pattern, and additional inhibition of aPAR or CDC-42 by pPAR is required. This additional interaction acts to balance the mutual inhibition between aPAR and pPAR, allowing a pattern to be generated. In other words, the posterior Par proteins play a more significant role when generating the polarity pattern as compared to maintaining an established pattern. Interactions between CDC-42 and the anterior Par proteins are critical in both situations (Figure 5 and Table 1). This is consistent with observations suggesting the importance of anterior Par proteins and CDC-42 in generating polarization [37]. This suggests that pPAR, with the support of CDC-42, plays a critical role in generating polarization, though, unlike maintenance, we found all interactions between CDC-42, pPAR and aPAR suggested by previous observation, and thus under consideration, can have some influence on polarization generation.

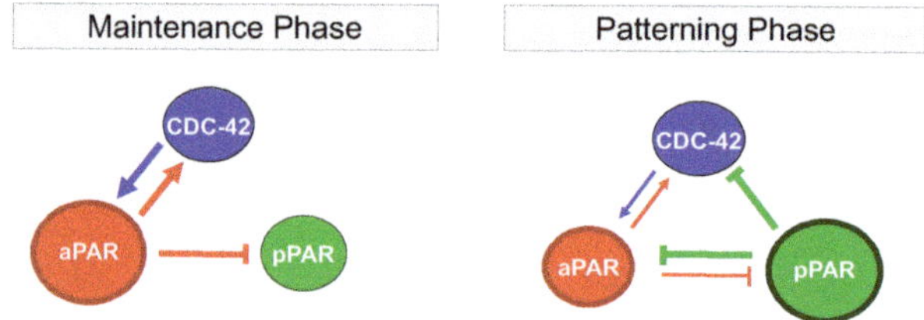

Figure 5. Critical interactions during maintenance and generation of Par protein polarization. A bigger circle indicates a key role for that protein in the indicated phase. In the maintenance phase, aPAR plays a key role in maintaining spatial polarity via interactions with CDC-42. aPAR enforces pPAR polarity through mutual inhibition. In the patterning phase, pPAR is playing the key role by inhibiting aPAR through either direct or CDC-42-mediated inhibition. CDC-42 is indispensable for both the maintenance and emergence of polarization, but does not play a critical role in the dynamics of either phase.

Remarkably, the minimal network analysis and eFAST both selected the same critical interactions and parameters in polarization maintenance and generation. This independent determination suggests the minimal networks found here represents the key interactions involved in patterning associated with polarization.

Table 1. Summary of minimal network and sensitivity analyses results. MNA: Minimal Network Analysis, S_i: First order index, S_{T_i}: Total effect index. For MNA, a small checkmark indicates additional interactions required for pattern generation. For sensitivity measures, a/p/c denote the aPAR/pPAR/CDC-42 sensitivity functions, respectively, and a capital letter indicates the parameter with the highest index for that function.

	Maintenance			Generation		
Parameter/Interaction	**MNA**	S_i	S_{T_i}	**MNA**	S_i	S_{T_i}
q_3^a : CDC-42→aPAR	✓	Apc	a	✓	apc	apc
q_3^c : aPAR→CDC-42	✓	apC	ApC	✓	APC	APC
k_3^p : aPAR⊣pPAR	✓	aPc	Pc	✓	apc	apc
k_3^a : pPAR⊣aPAR			p	✓		ac
k_3^c : pPAR⊣CDC-42		pc	apc	✓		ac

In this study, we have made a number of simplifying assumptions to investigate the core interactions between the Rho protein CDC-42 and members of the Par protein family. In future investigations, we will explicitly consider separate members of the Par protein family, each of which have distinct dynamics. For instance, the anterior Par protein PAR-3 appears to compete with CDC-42 to bind to a complex of PAR-6 and aPKC [18], and in the posterior, PAR-2 must be recruited to the membrane before PAR-1 can bind [10]. We will add other members of the Rho protein family, including Rac and Rho, which are known to act in a network with CDC-42 [38]. In this investigation, we have explored protein interactions within the pattern maintenance phase of early embryo development, when advective flow has largely ceased [17,22], and in addition considered the differences in protein interactions required for pattern generation in the absence of cortical flow, as observed in select experimental systems. Future extensions of the model will include the spatial dynamics of advective flow, which may act with the biochemical interaction network to establish polarity. In this study, we have reduced the geometry of the embryo to a 1D domain for computational efficiency to facilitate the exploration of a five dimensional parameter space. However, other studies have studied polarization dynamics of simplified two variable models on fully 3D domains [39] or explicitly included the geometry of the embryo [40] to investigate the spatial orientation of the polarity axis. Noting the computational demands, we leave minimal network and eFAST analysis of higher dimensional parameter spaces in more complex geometries and 3D domains for future work.

In this investigation, we have developed and analysed a mathematical model of the biochemical network that integrates components of two protein families, Par proteins and Rho GTPases, that are known polarity determinants. The minimal network and sensitivity analyses have identified critical interactions in this network, providing testable hypotheses for future experimental work. By resolving the critical role of CDC-42 and highlighting the most important PAR interactions in this network, we have extended our understanding of the signalling network responsible for polarization, and laid the foundation for further investigations into patterning associated with polarization.

Author Contributions: Conceptualization, S.S.-L., E.A.G., A.T.D.; methodology, S.S.-L., E.A.G., A.T.D.; investigation, S.S.-L., E.A.G., A.T.D.; writing—original draft preparation, S.S.-L., E.A.G., A.T.D.; writing—review and editing, S.S.-L., E.A.G., A.T.D. All authors have read and agreed to the published version of the manuscript.

Funding: This work was supported by a Grant-in-Aid for Scientific Research from the Ministry of Education, Culture, Sports, Science and Technology, Japan, to S.S.L. (JP19H01805 and JP17KK0094) and funding from the National Science Foundation (USA) to A.T.D. (DMS-1554896).

Conflicts of Interest: The authors declare no conflict of interest. The funders had no role in the design of the study; in the collection, analyses, or interpretation of data; in the writing of the manuscript, or in the decision to publish the results.

Appendix A

Appendix A.1. Model Reduction

The model reduction is summarized schematically in Figure 2A; here we provide the associated simplifications of the governing equations, starting from Equation (3), that is

$$
\begin{aligned}
\frac{\partial [A_m]}{\partial t} &= D_m^A \nabla_\Gamma^2 [A_m] + \{\gamma_a + F_{\text{on}}^A([C_m])\} \left(\frac{A_{tot}}{|\Omega|} - \frac{1}{|\Omega|} \int_\Gamma [A_m] d\mathbf{x} \right) - \{\alpha_a + F_{\text{off}}^A([P_m])\}[A_m] \quad \text{on } \mathbf{x} \in \Gamma, \\
\frac{\partial [P_m]}{\partial t} &= D_m^P \nabla_\Gamma^2 [P_m] + \gamma_p \left(\frac{P_{tot}}{|\Omega|} - \frac{1}{|\Omega|} \int_\Gamma [P_m] d\mathbf{x} \right) - \{\alpha_p + F_{\text{off}}^P([A_m])\}[P_m] \quad \text{on } \mathbf{x} \in \Gamma, \\
\frac{\partial [C_m]}{\partial t} &= D_m^C \nabla_\Gamma^2 [C_m] + \gamma_c \left(\frac{C_{tot}}{|\Omega|} - \frac{1}{|\Omega|} \int_\Gamma [C_m] d\mathbf{x} \right) - \{\alpha_c + F_{\text{off}}^C([P_m],[A_m])\}[C_m] \quad \text{on } \mathbf{x} \in \Gamma.
\end{aligned}
\tag{A1}
$$

Axisymmetry of the cell geometry is assumed about the major axis of symmetry of the cell in Figure 2A, and the objective is to further simplify model (3) to the one dimensional model of (4). To proceed, let $R(s)$ denote the radial distance of the membrane from the major axis of symmetry, with $s \in [0, L]$ the arclength along the perimeter of the depicted, projected membrane in Figure 2A and L the total perimeter length. In addition, we take $s = 0$ to be the leftmost point on the membrane, and impose periodic boundary conditions at $s = 0, L$. Note that the metric tensor components for this surface with respect to coordinates $X_1 = s$ and $X_2 = \phi$, the angle of rotation about the major axis of symmetry, are given by $g_{ij} = \text{diag}(1, R^2(s))_{ij}$. Hence the Laplacian operation for an axisymmetric membrane concentration, denoted $u_m(s)$ in general, is given by

$$
\nabla_\Gamma^2 u_m = \frac{1}{g^{1/2}} \frac{\partial}{\partial X_i} \left(g^{1/2} g^{ij} \frac{\partial u_m}{\partial X_j} \right) = \frac{1}{R(s)} \frac{\partial}{\partial s} \left(R(s) \frac{\partial u_m}{\partial s} \right) = \frac{\partial^2 u_m}{\partial s^2} + \frac{R_s}{R} \frac{\partial u_m}{\partial s}, \tag{A2}
$$

where the subscript s denotes the ordinary derivative with respect to arclength, summation convention is used, $g = \det(g_{ij})$, $g^{ij} = (g_{ij})^{-1}$ and noting that ϕ-derivatives generate no contribution by axisymmetry.

We also note that the mechanism of patterning for the above equations is widely recognized to be wave pinning, due to the multiple steady states associated with the kinetics. The pattern organization that emerges corresponds to the emergence of a wave transitioning between different steady states, with homogeneous concentrations elsewhere [41]. In particular, the location where the transition wave halts is approximated by an estimate of where the speed for wave-like solutions of the governing equations drops to zero, so that the final pattern is a standing wave, rather than, for instance, a Turing pattern [41].

Furthermore, for the presented model, these transitions between steady states are sharp for the parameter values of Table A1, as also seen *a posteriori* in Figure 2B for example. This is a consequence of the small size of the *non-dimensional* diffusion coefficients relative to the *non-dimensional* inverse timescales in Table A1, with $\epsilon^2 \sim D_m^A/\gamma_a \sim 1\times10^{-3}$, so that the lengthscale of these transitions corresponds to an order unity coefficient multiplying $\epsilon \ll 1$.

Thus either side of a transition is a homogeneous solution, whereby u_m is at a constant steady state for each concentration in the model and the advective term $[R_s/R]\partial u_m/\partial s$ is not important. Within a transition, there are steep gradients generating a dominant balance between the Laplacian and the kinetics. In particular for any given concentration, after non-dimensionalization and with the non-dimensional kinetics generically denoted by K_*, we have equations of the form

$$
\eta \frac{\partial u_{*m}}{\partial t_*} = \epsilon^2 \nabla_{*\Gamma}^2 u_{*m} + K_* = \epsilon^2 \left(\frac{\partial^2 u_{*m}}{\partial s_*^2} + \frac{R_{*s_*}}{R_*} \frac{\partial u_{*m}}{\partial s_*} \right) + K_*, \tag{A3}
$$

where asterisks indicate non-dimensionalized variables and η is a non-dimensional constant that emerges from the rescaling. With the further change of variable $S = (s_* - s_0)/\epsilon$, where s_0 is the location of the transition, to make the dominant balance explicit by matching the transport and kinetic terms and to generate order unity derivatives, we have that the inner region equations for the transition layer in a matched asymptotic expansion approximation are given by

$$\eta \frac{\partial u_{*m}}{\partial t_*} = \frac{\partial^2 u_{*m}}{\partial S^2} + \epsilon \left(\left. \frac{R_{*s_*}}{R_*} \right|_{s_0} + O(\epsilon S) \right) \frac{\partial u_{*m}}{\partial S} + K_* = \frac{\partial^2 u_{*m}}{\partial S^2} + K_* + O(\epsilon), \tag{A4}$$

with the last step valid provided $R_{*s_*}/R_* \not\sim O(1/\epsilon)$. Under such circumstances the advection term does not contribute to the structure of the equations either in the outer regions where the solution is at steady state, or the inner region. In turn, this justifies approximating the the Laplacian with the second derivative $\nabla_\Gamma^2 u_m \approx \partial^2 u_m/\partial s^2$. Furthermore, with this scaling, where all terms are order unity, it can be seen that the advective term will only generate a subleading contribution to the estimates of where the transitions become fixed as standing waves [41], so that the temporal dynamics of interest, that fix the location of the pattern, are also well approximated in the absence of the advective term away from regions where $R_{*s_*}/R_* \not\sim O(1/\epsilon)$.

However, sufficiently near the poles at $s = 0, L/2$, $1/R$ becomes arbitrarily large. With X denoting distance along the horizontal axis of symmetry in Figure 2A, we have $s_R^2 = 1 + 1/R_X^2$ and, noting R_X blows upon approaching the poles, we have $R(s) \approx s$ and, in turn $R_* \approx s_*$, for $s_* \ll \epsilon$. Thus $R_{*s_*}/R_* \approx 1/s_* \sim O(1/\epsilon)$ once $s_* \sim O(\epsilon)$ near the left hand pole and analogously near the right hand pole. Thus, approximating the Laplacian by second derivatives only breaks down within a distance of ϵ of the poles in the non-dimensional model, and only if there is a transition in this region. In particular, with the appropriate scaling of $S = s_*/\epsilon$ for a potential transition within ϵ of the left hand pole, we have

$$\eta \frac{\partial u_{*m}}{\partial t_*} = \frac{\partial^2 u_{*m}}{\partial S^2} + \frac{1}{S}\frac{\partial u_{*m}}{\partial S} + K_*, \tag{A5}$$

and the equations take the form of the well-studied, radially symmetric, system near the origin of the 2D plane. The impact of the advective term here is to slow the transition wave, though its wavespeed asymptotes to that associated with Equation (A4) once the transition has propagated into a region with $s_* \gg \epsilon$ (e.g., Eqn. 11.20 *et. seq.* in Murray's textbook (1993)), with analogous conclusions for the right hand pole at $s = Ls_* = L/2$.

In the presented model we only consider the second derivatives. Then with patterns emerging from one of the poles (for example as in Figure 2B, with an emerging pattern at $x \equiv s = Ls_* = L/2$), the presented model will underestimate the timescale for the emergence of the wave, but otherwise will capture its dynamics once the transition is beyond a distance of ϵ from $x = L/2$. Furthermore, unless a steady state transition occurs ϵ-close to the left or right poles of the cell, its location and spatial variation will be well approximated even in the absence of the advective term in the model equations, as motivated above. Thus, and in summary, in the simplified model of (4) we take $\nabla_\Gamma^2 u_m = \partial^2 u_m/\partial s^2$ for all concentrations, assuming that the membrane changes sufficiently slowly in shape to ensure $R_{*s_*}/R_* = LR_s/R \not\sim O(1/\epsilon)$ holds away from the poles. Then apart from the temporal dynamics as a transition passes through, or emerges from, a pole the simplified model yields an accurate picture, especially for the final steady-state pattern, providing a transition does not occur at the poles (and a posteriori the results presented do not indicate steady state transitions at the poles).

The approximation in particular greatly simplifies the model in that it sidesteps dealing with coordinate singularities at the poles. While these are possible to accommodate, this entails additional numerical complexity, such as Taylor expanding and asymptotically approximating near the poles (Morton and Mayers [42], Woolley et al. [43]) or, when such approximations are not feasible, borrowing from basic differential geometry and working with an atlas of charts and the associated multiple coordinate systems [44], which in this model would result in the loss of the simplifications from axisymmetry. Such complexities are not warranted for the current study given the need for extensive computation with the global sensitivity analyses and the minor impact of the approximation $\nabla_\Gamma^2 u_m \approx \partial^2 u_m/\partial s^2$, where $x \equiv s$ in the main text.

We further approximate the geometry by a slender cylinder radius H, length $\mathcal{L} \approx L/2 - 2H$, with essentially neglected asymptotically short caps of length δ with $\delta \ll H \ll L$. Then, on use of axisymmetry, the surface integrals within the governing Equation (3) simplify via

$$\begin{aligned}\int_\Gamma u_m \mathrm{d}\mathbf{x} &= 2\pi \int_0^{L/2} u_m R \mathrm{d}x = 2\pi H \int_0^{L/2} u_m \mathrm{d}x + 2\pi \int_0^P u_m (R-H)\mathrm{d}x + 2\pi \int_{L/2-P}^{L/2} u_m (R-H)\mathrm{d}x \\ &= 2\pi H \int_0^{L/2} u_m \mathrm{d}x \left(1 + O\left(\frac{H}{L}\right)\right) = \int_0^L \pi H u_m \mathrm{d}x \left(1 + O\left(\frac{H}{L}\right)\right),\end{aligned} \quad (A6)$$

where x is dimensional arclength, $P \approx H$ is the contribution to the cell perimeter arclength from the left hand pole to the first point where $R = H$. Noting $R \approx x$ for the left hand cap and similarly for the right hand cap the integrals can be approximated as above to accuracy H/L given the separation of the geometrical scales. These corrections are also dropped in the final dimensional model given $H \ll L$. Finally it is easier to work with $[\tilde{A}_m] = \pi H [A_m]$, which is the density per unit length away from the caps, and we analogously define $[\tilde{P}_m]$, $[\tilde{C}_m]$. Then

$$\begin{aligned}\frac{\partial[\tilde{A}_m]}{\partial t} &= D_m^A \frac{\partial^2}{\partial x^2}[\tilde{A}_m] + \{\tilde{\gamma}_a + F_{\mathrm{on}}^A([\tilde{C}_m])\}\left(\frac{A_{tot}}{L} - \frac{1}{L}\int_0^L [\tilde{A}_m]dx\right) - \{\alpha_a + F_{\mathrm{off}}^A([P_m])\}[\tilde{A}_m], \\ \frac{\partial[\tilde{P}_m]}{\partial t} &= D_m^P \frac{\partial^2}{\partial x^2}[\tilde{P}_m] + \tilde{\gamma}_p \left(\frac{P_{tot}}{L} - \frac{1}{L}\int_0^L [\tilde{P}_m]dx\right) - \{\alpha_p + F_{\mathrm{off}}^P([\tilde{A}_m])\}[\tilde{P}_m], \\ \frac{\partial[\tilde{C}_m]}{\partial t} &= D_m^C \frac{\partial^2}{\partial x^2}[\tilde{C}_m] + \tilde{\gamma}_c \left(\frac{C_{tot}}{L} - \frac{1}{L}\int_0^L [\tilde{C}_m]dx\right) - \{\alpha_c + F_{\mathrm{off}}^C([\tilde{P}_m],[\tilde{A}_m])\}[\tilde{C}_m],\end{aligned} \quad (A7)$$

where $\tilde{\gamma}_a = \gamma_a L\pi H/|\Omega|$, with analogous definitions for $\tilde{\gamma}_p, \tilde{\gamma}_c$ and also for $\tilde{q}_3^a = q_3^a \pi HL/|\Omega|$. Dropping tildes, and thus redefining $[A_m], [P_m], [C_m], \gamma_a, \gamma_p, \gamma_c, q_3^a$, gives the final dimensional model, Equation (4) of Section 2.2.

Appendix A.2. Additional Simulations and Representative Parameter Set

The simulation results for Model (1) with fast cytosol diffusions are shown in Figure A1. The representative parameter sets which we used in simulations are given in Table A1.

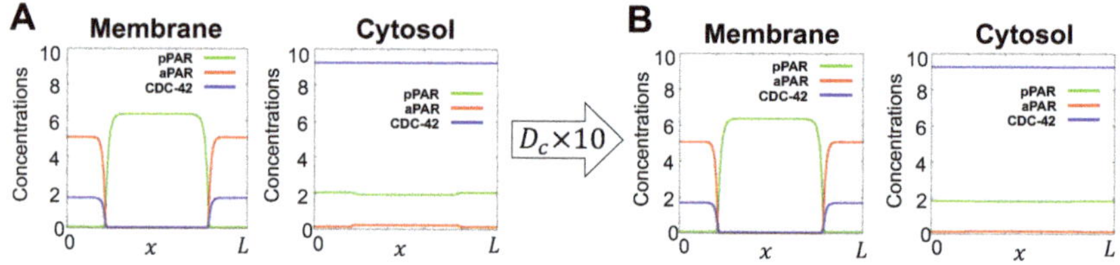

Figure A1. Simulations results for Model (1) on 1D. (**A**) The simulation results of Model (1) with biologically feasible cytosol diffusion coefficient. (**B**) The simulation results for the case where the cytosol diffusions are ten times larger than the case of (**A**). Both cases (**A**) and (**B**) show almost homogeneous concentrations for cytosol proteins and with a sufficiently large diffusion coefficient, we can assume that the asymptotic dynamics of cytosol proteins is homogeneous steady state. The simulation results are reported at 30 min and the detailed parameters are given in Table A1.

Table A1. Representative parameter set. The parameters L, τ, $\mathcal{M}$ are respectively a lengthscale, timescale and a scale of protein number used to non-dimensionalize the model. The dimensional parameters are defined in terms of the protein number scale, so that the non-dimensional model is independent of $\mathcal{M}$. In practice, this means the parameters associated with the non-linear kinetics are such that the non-linear reactions are significant and also in balance with other reactions at typical cellular numbers of proteins. It should be noted that the parameter set below is inherently non-identifiable as, for example, the model depends on the three parameters q_1^a, q_2^a, q_3^a only via the two degrees of freedom $q_1^a/q_3^a, q_2^a/q_3^a$. For presentational simplicity we have not eliminated such superfluous degrees of freedom from the parameter set, though this would be necessary if parameter inference from experimental data was pursued in future work.

Parameter	Dimensional Value	Non-Dim. Value	Parameter	Dimensional Value	Non-Dim. Value
†1 L	142.75 (μm)	1.0	τ	2.00 (s)	1.0
†2 D_m^A	0.28 (μm^2/s)	2.748×10^{-5}	†2 D_m^P	0.15 (μm^2/s)	1.472×10^{-5}
†3 D_m^C	0.10 (μm^2/s)	0.981×10^{-5}	D_c^A	14.0 (μm^2/s)	1.374×10^{-3}
D_c^P	7.50 (μm^2/s)	0.736×10^{-3}	D_c^C	10.0 (μm^2/s)	0.981×10^{-3}
γ_a	0.015 (s^{-1})	0.03	γ_p	0.100 (s^{-1})	0.20
γ_c	0.015 (s^{-1})	0.03	α_a	0.05 (s^{-1})	0.10
α_p	0.03 (s^{-1})	0.06	α_c	0.05 (s^{-1})	0.10
q_1^a	0.20 ($\mathcal{M}/L^2$)	0.20	q_2^a	1.00 ($\mathcal{M}^{-1}$)	1.00
q_3^a	2.60 ($\mathcal{M}^{-1}s^{-1}$)	5.20	k_1^a	1.25 ($\mathcal{M}/L^2$)	1.25
k_2^a	1.00 ($\mathcal{M}^{-1}$)	1.00	k_3^a	2.60 ($\mathcal{M}^{-1}s^{-1}$)	5.20
k_1^p	1.25 ($\mathcal{M}/L^2$)	1.25	k_2^p	1.00 ($\mathcal{M}^{-1}$)	1.00
k_3^p	2.60 ($\mathcal{M}^{-1}s^{-1}$)	5.20	k_1^c	1.25 ($\mathcal{M}/L^2$)	1.25
k_2^c	1.00 ($\mathcal{M}^{-1}$)	1.00	k_3^c	2.60 ($\mathcal{M}^{-1}s^{-1}$)	5.20
q_1^c	1.00 ($\mathcal{M}/L^2$)	1.00	q_2^c	1.00 ($\mathcal{M}^{-1}$)	1.00
q_3^c	0.50 ($\mathcal{M}s^{-1}L^{-2}$)	1.00	A_{tot}	$\mathcal{M}$	1.00
P_{tot}	2$\mathcal{M}$	2.00	C_{tot}	5$\mathcal{M}$	5.00

Parameter range for sensitivity analysis

Parameter	Dimensional Range	Non-Dim. Range	Parameter	Dimensional Range	Non-Dim. Range
k_3^a	[1.5, 5.0] ($\mathcal{M}^{-1}s^{-1}$)	[3, 10]	k_3^p	[0.0, 4.0] ($\mathcal{M}^{-1}s^{-1}$)	[0, 8]
k_3^c	[0.0, 5.0] ($\mathcal{M}^{-1}s^{-1}$)	[0, 10]	q_3^a	[0.0, 5.0] ($\mathcal{M}^{-1}s^{-1}$)	[0, 10]
q_3^c	[0.0, 5.0] ($\mathcal{M}s^{-1}L^{-2}$)	[0, 10]			

†1 [30], †2 [34], †3 [45].

Appendix A.3. Variance-Based Sensitivity Analysis by Using Extended Fourier Amplitude Sensitivity Test (eFAST) Method

For a large enough sample size, the variance-based sensitivity analysis provides a quantitative measure for how much factor A is more important than factor B. It generally requires extensive computation and the FAST method has been demonstrated as a method to reduce the cost effectively by exploring the multidimensional space of the input factors via a suitably defined search-curve [31,32]. eFAST is a generalization the FAST method that provides the total effect index, defined below.

Let us define y as the expected model output and $p_i (i = 1 \cdots \bar{N})$ are the input factors (model parameters). The effect (contribution) of each parameter p_i is considered as the variance of the expected model output, namely, $V_i = V(E(y|p_i))$. The effect of the interaction between two orthogonal inputs p_i and p_j on the output y is defined in terms of conditional variances as $V_{ij} = V(E(y|p_i, p_j)) - V_i - V_j$ (Sections 5.9 and 5.10, [31]). Similarly, the effect of the interaction among three distinct inputs p_i, p_j and p_k on the output y is given as $V_{ijk} = V(E(y|p_i, p_j, p_k)) - V_{ij} - V_{ik} - V_{jk} - V_i - V_j - V_k$ and so on. Then the total output variance V_{total} for a model with $\bar{N}$ input factors is given by

$$V_{total} = \sum_i V_i + \sum_i \sum_{j>i} V_{ij} + \sum_i \sum_{j>i} \sum_{k>j} V_{ijk} + \cdots + V_{1 \ldots \bar{N}}.$$

The first order sensitivity index is defined by

$$S_i = \frac{V_i}{V_{total}},$$

and the total-effect index is defined by

$$S_{T_i} = \frac{V_{total} - V(E(y|p_{-i}))}{V_{total}} = 1 - \frac{V_{-i}}{V_{total}},$$

where $-i$ stands for *all but i* and V_{-i} is given to

$$V_{-i} = V_{total} - V_i - \sum_j V_{ij} - \cdots - V_{1\cdots\tilde{N}}.$$

The sensitivity analysis using eFAST has been carried out according to the following steps [32,35].

STEP 1: Sampling for search-curve

Each parameter value was sampled according to

$$p_i(s) = p_i^{\min} + (p_i^{\max} - p_i^{\min})\left[\frac{1}{2} + \frac{1}{\pi}\sin^{-1}(\sin(\omega_i s + \psi_i))\right],$$

where $s \in (-\pi, \pi)$ and ψ_i is a random phase-shift chosen in $[0, 2\pi)$ which is used for making different curves (i.e., resampling) and we carried out two different curves by resampling. $\{\omega_i\}$ is a set of different angular frequencies associated with each input factor. The sample size (N_s) is given by $N_s = (2M\omega_{\max} + 1)N_r$ where M is the interference factor (usually 4 or higher), $\omega_{\max}$ is highest frequency and N_r is the number of resamplings. The detailed values which we chosen are shown in Table A2.

Table A2. The values used for sensitivity analysis.

Index	ω_i	M	N_r	N_s
S_i	$\{59, 113, 143, 149, 161\}$	4	2	2578
S_{T_i}	$\{3, 7, 11, 15, 128\}$	4	2	2050

STEP 2: Calculating output of sensitivity function

With respect to the parameter samples, we numerically solved the main model (4) and calculated each sensitivity function, $\mathbf{y}(s) = F_S(\mathbf{p}(s))$, given by the Equation (6), where $\mathbf{p}(s) = (p_1(s), \cdots, p_{\tilde{N}}(s))$ and $\mathbf{y} = (y_1(s), \cdots, y_{\tilde{N}}(s))$, with s partitioned via the nodes $s_k = \pi(2k - N_s - 1)/N_s, k \in \{1, \cdots, N_s\}$.

STEP 3: Calculating the power spectrum in Fourier series

We expand $F_S(s)$ in a Fourier series such that

$$F_S(s) = \sum_{j=-\infty}^{+\infty} \{A_j \cos(js) + B_j \sin(js)\},$$

where the Fourier coefficients A_j and B_j are defined as

$$A_j = \frac{1}{2\pi}\int_{-\pi}^{\pi} F_S(s)\cos(js)ds = \frac{1}{2\pi}\sum_{k=1}^{Ns} F_S(s_k)\cos(js_k)\triangle s = \frac{1}{N_s}\sum_{k=1}^{Ns} F_S(s_k)\cos(js_k),$$

$$B_j = \frac{1}{2\pi}\int_{-\pi}^{\pi} F_S(s)\sin(js)ds = \frac{1}{2\pi}\sum_{k=1}^{Ns} F_S(s_k)\sin(js_k)\triangle s = \frac{1}{N_s}\sum_{k=1}^{Ns} F_S(s_k)\sin(js_k),$$

and $\triangle s = 2\pi / N_s$. Then the power spectrum is calculated by

$$\Lambda^2(j) = A_j^2 + B_j^2.$$

STEP 4: Calculating S_i and S_{T_i}

The total variance (V_{total}), the variance of i factor (V_i), and the variance of *all but i* (V_{-i}) are calculated by

$$V_{total} \approx \hat{V}_{total} = 2 \sum_{\omega=1}^{M\omega_{\max}} \Lambda^2(\omega),$$

$$V_i \approx \hat{V}_i = 2 \sum_{p=1}^{M} \Lambda^2(p\omega_i),$$

$$V_{-i} \approx \hat{V}_{-i} = 2 \sum_{p=1}^{\frac{\omega_i}{2}} \Lambda^2(p\omega),$$

as in [35]. We average the variances above over the sets of resampling and calculate the indices

$$S_i = \frac{V_i}{V_{total}} \qquad \text{and} \qquad S_{T_i} = 1 - \frac{V_{-i}}{V_{total}}.$$

Note that we need to give the maximal frequency value for ω_i of ith factor in calculating V_{-i}, so that total-effect variance is calculated separately for each factor i.

Appendix A.4. First Order Index and Total Index for the Dummy Parameter

Since the eFAST method artifactually produces small but non-zero sensitivity indices, we have calculated the first order index of the dummy parameter (S_{dummy}) and the total index of the dummy parameter ($S_{T_{\text{dummy}}}$) in order to confirm that the index values which we obtained in Figures 3 and 4 give reliable data to see the significance of model parameters [46]. If the model parameters with a total index less than or equal to that of the dummy parameter, the model parameters should be considered not significantly different from zero.

For the calculations of the first order index for the dummy parameter, we put the frequency of the dummy parameter to $\omega_{\text{dummy}} = 91$ with the same frequency set given in Table A2. In the calculation of the total index for the dummy parameter, we used additional frequency $\omega = 13$ together with the same frequency set given in Table A2. The sensitivity results for the dummy parameter are shown in Table A3.

Table A3. First order index and total index for the dummy parameter.

Maintenance			
	aPAR	**pPAR**	**CDC-42**
S_{dummy}	0.000869	0.002956	0.003730
$S_{T_{\text{dummy}}}$	0.021139	0.031567	0.032265
Generation			
	aPAR	**pPAR**	**CDC-42**
S_{dummy}	0.001711	0.004020	0.002282
$S_{T_{\text{dummy}}}$	0.023502	0.160468	0.055702

References

1. Hodge, R.G.; Ridley, A.J. Regulating Rho GTPases and their regulators. *Nat. Rev. Mol. Cell Biol.* **2016**, *17*, 496–510. [CrossRef] [PubMed]
2. Campanale, J.P.; Sun, T.Y.; Montell, D.J. Development and dynamics of cell polarity at a glance. *J. Cell Sci.* **2017**, *130*, 1201–1207. [CrossRef] [PubMed]
3. Suzuki, A.; Ohno, S. The PAR-aPKC system: Lessons in polarity. *J. Cell Sci.* **2006**, *119*, 979–987. [CrossRef] [PubMed]
4. Yilmaz, M.; Christofori, G. EMT, the cytoskeleton, and cancer cell invasion. *Cancer Metastasis Rev.* **2009**, *28*, 15–33. [CrossRef]
5. Muthuswamy, S.K.; Xue, B. Cell polarity as a regulator of cancer cell behavior plasticity. *Annu. Rev. Cell Dev. Biol.* **2012**, *28*, 599–625. [CrossRef]
6. Ridley, A.J. Rho GTPases and cell migration. *J. Cell Sci.* **2001**, *114*, 2713–2722.
7. Johnson, D.I. Cdc42: An essential Rho-type GTPase controlling eukaryotic cell polarity. *Microbiol. Mol. Biol. Rev. MMBR* **1999**, *63*, 54–105. [CrossRef]
8. Amano, M.; Nakayama, M.; Kaibuchi, K. Rho-kinase/ROCK: A key regulator of the cytoskeleton and cell polarity. *Cytoskeleton* **2010**, *67*, 545–554. [CrossRef]
9. Nobes, C.D.; Hall, A. Rho GTPases Control Polarity, Protrusion, and Adhesion during Cell Movement. *J. Cell Biol.* **1999**, *144*, 1235–1244. [CrossRef]
10. Goldstein, B.; Macara, I.G. The PAR proteins: Fundamental players in animal cell polarization. *Dev. Cell* **2007**, *13*, 609–622. [CrossRef]
11. Macara, I.G. Parsing the Polarity Code. *Nat. Rev. Mol. Cell Biol.* **2004**, *5*, 220–231. [CrossRef] [PubMed]
12. Kemphues, K.J.; Priess, J.R.; Morton, D.G.; Cheng, N.S. Identification of genes required for cytoplasmic localization in early *C. elegans* embryos. *Cell* **1988**, *52*, 311–320. [CrossRef]
13. Gérard, A.; Mertens, A.; van der Kammen, R.A.; Collard, J.G. The Par polarity complex regulates Rap1- and chemokine-induced T cell polarization. *J. Cell Biol.* **2007**, *176*, 863–875. [CrossRef] [PubMed]
14. Schonegg, S.; Hyman, A. CDC-42 and RHO-1 coordinate acto-myosin contractility and PAR protein localization during polarity establishment in *C. elegans* embryos. *Dev. Camb. Engl.* **2006**, *133*, 3507–3516. [CrossRef] [PubMed]
15. Kumfer, K.T.; Cook, S.J.; Squirrell, J.M.; Eliceiri, K.W.; Peel, N.; O'Connell, K.F.; White, J.G. CGEF-1 and CHIN-1 Regulate CDC-42 Activity during Asymmetric Division in the Caenorhabditis elegans Embryo. *Mol. Biol. Cell* **2010**, *21*, 266–277. [CrossRef]
16. Cuenca, A.A.; Schetter, A.; Aceto, D.; Kemphues, K.; Seydoux, G. Polarization of the *C. elegans* zygote proceeds via distinct establishment and maintenance phases. *Development* **2002**, *130*, 1255–1265. [CrossRef]
17. Munro, E.; Nance, J. Cortical Flows Powered by Asymmetrical Contraction Transport PAR Proteins to Establish and Maintain Anterior-Posterior Polarity in the Early *C. elegans* Embryo. *Dev. Cell* **2004**, *7*, 413–424. [CrossRef]
18. Aceto, D.; Beers, M.; Kemphues, K.J. Interaction of PAR-6 with CDC-42 is required for maintenance but not establishment of PAR asymmetry in *C. elegans*. *Dev. Biol.* **2006**, *299*, 386–397. [CrossRef]
19. Gotta, M.; Abraham, M.C.; Ahringer, J. CDC-42 controls early cell polarity and spindle orientation in *C. elegans*. *Curr. Biol.* **2001**, *11*, 482–488. [CrossRef]
20. Lang, C.F.; Munro, E. The PAR proteins: From molecular circuits to dynamic self-stabilizing cell polarity. *Co. Biol.* **2017**, *144*, 3405–3416. [CrossRef]
21. Rodriguez, J.; Peglion, F.; Martin, J.; Hubatsch, L.; Reich, J.; Hirani, N.; Gubieda, A.G.; Roffey, J.; Fernandes, A.R.; Johnston, D.S.; et al. aPKC Cycles between Functionally Distinct PAR Protein Assemblies to Drive Cell Polarity. *Dev. Cell* **2017**, *42*, 400–415. [CrossRef]
22. Small, L.E.; Dawes, A.T. PAR proteins regulate maintenance-phase myosin dynamics during Caenorhabditis elegans zygote polarization. *Mol. Biol. Cell* **2017**, *28*, 2220–2231. [CrossRef] [PubMed]
23. Arata, Y.; Hiroshima, M.; Pack, C.G.; Ramanujam, R.; Motegi, F.; Nakazato, K.; Shindo, Y.; Wiseman, P.W.; Sawa, H.; Kobayashi, T.J.; et al. Cortical Polarity of the RING Protein PAR-2 Is Maintained by Exchange Rate Kinetics at the Cortical-Cytoplasmic Boundary. *Cell Rep.* **2016**, *16*, 2156–2168. [CrossRef] [PubMed]
24. Kay, A.J.; Hunter, C.P. CDC-42 regulates PAR protein localization and function to control cellular and embryonic polarity in *C. elegans*. *Curr. Biol. CB* **2001**, *11*, 474–481. [CrossRef]

25. Beatty, A.; Morton, D.G.; Kemphues, K. PAR-2, LGL-1 and the CDC-42 GAP CHIN-1 act in distinct pathways to maintain polarity in the *C. elegans* embryo. *Development* **2013**, *140*, 2005–2014. [CrossRef]
26. Boyd, L.; Guo, S.; Levitan, D.; Stinchcomb, D.T.; Kemphues, K.J. PAR-2 is asymmetrically distributed and promotes association of P granules and PAR-1 with the cortex in *C. elegans* embryos. *Development* **1996**, *122*, 3075–3084.
27. Hao, Y.; Boyd, L.; Seydoux, G. Stabilization of cell polarity by the *C. elegans* RING protein PAR-2. *Dev. Cell* **2006**, *10*, 199–208. [CrossRef]
28. Seirin-Lee, S.; Shibata, T. Self-organization and advective transport in the cell polarity formation for asymmetric cell division. *J. Theor. Biol.* **2015**, *382*, 1–14. [CrossRef]
29. Kuhn, T.; Ihalainen, T.O.; Hyvaluoma, J.; Dross, N.; Willman, S.F.; Langowski, J.; Vihinen-Ranta, M.; Timonen, J. Protein Diffusion in Mammalian Cell Cytoplasm. *PLoS ONE* **2011**, *6*, e22962. [CrossRef]
30. Goehring, N.W.; Trong, P.K.; Bois, J.S.; Chowdhury, D.; Nicola, E.M.; Hyman, A.A.; Grill, S.W. Polarization of PAR Proteins by Advective Triggering of a Pattern-Forming System. *Science* **2011**, *334*, 1137–1141. [CrossRef]
31. Saltelli, A.; Tarantola, S.; Campolongo, F.; Ratto, M. *Sensitivity Analysis in Practice: A Guide to Assessing Scientific Models*; Wiley: Hoboken, NJ, USA, 2004.
32. Saltelli, A.; Tarantola, S.; Chan, K.P.S. A Quantitative Model-Independent Method for Global Sensitivity Analysis of Model Output. *Technometrics* **1999**, *41*, 39–56. [CrossRef]
33. Gönczy, P. Asymmetric cell division and axis formation in the embryo. *WormBook* **2005**. [CrossRef]
34. Goehring, N.W.; Hoege, C.; Grill, S.W.; Hyman, A.A. PAR proteins diffuse freely across the anterior-posterior boundary in polarized *C. elegans* embryos. *J. Cell Biol.* **2011**, *193*, 583–594. [CrossRef] [PubMed]
35. Saltelli, A.; Tarantola, S.; Campolongo, F. Sensitivity Analysis as an Ingredient of Modeling. *Stat. Sci.* **2000**, *15*, 377–395.
36. Kachur, T.M.; Audhya, A.; Pilgrim, D. UNC-45 is required for NMY-2 contractile function in early embryonic polarity establishment and germline cellularization in *C. elegans*. *Dev. Biol.* **2008**, *314*, 287–299. [CrossRef] [PubMed]
37. Dickinson, D.J.; Schwager, F.; Pintard, L.; Gotta, M.; Goldstein, B. A Single-Cell Biochemistry Approach Reveals PAR Complex Dynamics during Cell Polarization. *Dev. Cell* **2017**, *42*, 416–434.e11. [CrossRef]
38. Mack, N.; Georgiou, M. The Interdependence of the Rho GTPases and Apicobasal Cell Polarity. *Small GTPases* **2014**, *5*, 10. [CrossRef]
39. Cusseddu, D.; Edelstein-Keshet, L.; Mackenzie, J.A.; Portet, S.; Madzvamusea, A. A coupled bulk-surface model for cell polarisation. *J. Theor. Biol.* **2019**, *481*, 119–135. [CrossRef]
40. Geßele, R.; Halatek, J.; Würthner, L.; Frey, E. Geometric cues stabilise long-axis polarisation of PAR protein patterns in *C. elegans*. *Nat. Commun.* **2020**, *11*, 539. [CrossRef]
41. Mori, Y.; Jilkine, A.; Edelstein-Keshet, L. Wave-Pinning and Cell Polarity from a Bistable Reaction-Diffusion System. *Biophys. J.* **2008**, *94*, 3684–3697. [CrossRef]
42. Morton, K.; Mayers, D. *Numerical Solution of Partial Differential Equations*; Cambridge University Press: Cambridge, UK, 1993.
43. Woolley, T.; Gaffney, E.; Goriely, A. Membrane shrinkage and cortex remodelling are predicted to work in harmony to retract blebs. *R. Soc. Open Sci.* **2016**, *2*, 150184. [CrossRef] [PubMed]
44. Walker, B.; Ishimoto, K.; Gaffney, E. A new basis for filament simulation in three dimensions. *arXiv* **2019**, arXiv:1907.04823.
45. Marée, A.F.M.; Jilkine, A.; Dawes, A.; Grieneisen, V.A.; Edelstein-Keshet, L. Polarization and Movement of Keratocytes: A Multiscale Modelling Approach. *Bull. Math. Biol.* **2006**, *68*, 1169–1211. [CrossRef] [PubMed]
46. Marino, S.; Hogue, I.B.; Ray, C.J.; Kirschner, D.E. A methodology for performing global uncertainty and sensitivity analysis in systems biology. *J. Theor. Biol.* **2008**, *254*, 178–196. [CrossRef] [PubMed]

Article

Activation of Cdc42 GTPase upon CRY2-Induced Cortical Recruitment Is Antagonized by GAPs in Fission Yeast

Iker Lamas, Nathalie Weber and Sophie G. Martin *

Department of Fundamental Microbiology, Faculty of Biology and Medicine, University of Lausanne, Biophore building, 1015 Lausanne, Switzerland; iker.lamasherrera@unil.ch (I.L.); nadliweber@hotmail.com (N.W.)

* Correspondence: Sophie.Martin@unil.ch

Received: 2 June 2020; Accepted: 7 September 2020; Published: 12 September 2020

Abstract: The small GTPase Cdc42 is critical for cell polarization in eukaryotic cells. In rod-shaped fission yeast *Schizosaccharomyces pombe* cells, active GTP-bound Cdc42 promotes polarized growth at cell poles, while inactive Cdc42-GDP localizes ubiquitously also along cell sides. Zones of Cdc42 activity are maintained by positive feedback amplification involving the formation of a complex between Cdc42-GTP, the scaffold Scd2, and the guanine nucleotide exchange factor (GEF) Scd1, which promotes the activation of more Cdc42. Here, we use the CRY2-CIB1 optogenetic system to recruit and cluster a cytosolic Cdc42 variant at the plasma membrane and show that this leads to its moderate activation also on cell sides. Surprisingly, Scd2, which binds Cdc42-GTP, is still recruited to CRY2-Cdc42 clusters at cell sides in individual deletion of the GEFs Scd1 or Gef1. We show that activated Cdc42 clusters at cell sides are able to recruit Scd1, dependent on the scaffold Scd2. However, Cdc42 activity is not amplified by positive feedback and does not lead to morphogenetic changes, due to antagonistic activity of the GTPase activating protein Rga4. Thus, the cell architecture is robust to moderate activation of Cdc42 at cell sides.

Keywords: Cdc42; GTPase activating protein (GAP); cell polarity; fission yeast *Schizosaccharomyces pombe*; CRY2-CIBN; optogenetics; clustering; positive feedback; pattern formation

---

1. Introduction

In eukaryotes, the small Rho-family GTPase Cdc42 is a highly conserved regulator of cell morphogenesis, proliferation, and differentiation. Prenylation of Cdc42's C-terminal CAAX motif underlies its association with the plasma membrane, where it functions as a molecular switch that alternates between GTP-bound, active and GDP-bound, inactive states. Activation of Rho GTPases relies on the activity of guanine nucleotide exchange factors (GEFs), while their intrinsic GTPase activity is enhanced by GTPase activating proteins (GAPs) to return them to the inactive state. GDP-bound Cdc42 also binds guanine-nucleotide dissociation inhibitors (GDI), which both block the exchange of GDP by GTP and solubilize Cdc42-GDP in the cytosol [1–3].

In the fission yeast *Schizosaccharomyces pombe*, Cdc42 is active at sites of polarized growth during vegetative and sexual life cycles. GTP-loading is promoted by two GEFs, Scd1 and Gef1. Scd1, which localizes to cell poles, receives information from the upstream Ras1 GTPase signal and mediates feedback control through the scaffolding activity of Scd2 [4–6]. For this, Scd1 forms a quaternary complex with Cdc42-GTP, the Pak1 kinase effector and Scd2 [7,8], which leads in vivo to the positive feedback activation of other Cdc42 molecules, as shown in our recent work using optogenetic strategies [6]. The second GEF, Gef1, which localizes to cell poles only in some conditions, promotes Cdc42 activation in response to stress and becomes essential only in absence of Scd1 [9–12]. Three GAPs, namely Rga4, Rga6, and Rga3, enhance the intrinsic GTP hydrolytic activity of Cdc42 [13–15]. Rga4 and Rga6 GAPs localize at cell sides, where growth does not occur in non-stressed cells, whereas Rga3 localizes at sites of active growth (cell poles). Fission yeast cells also express a GDI, called Rdi1, though Cdc42 localization and dynamics are not strongly perturbed in its absence [4,16].

Recently, optogenetic studies revealed a novel mechanism that triggers the activation of small GTPases in mammalian cells: Human Rac1 and RhoA, which belong to the same Rho GTPase family as Cdc42, were shown to become active at the cell cortex upon light-dependent cytosolic clustering [17]. In these experiments, the small GTPases were fused to CRY2PHR, the photolyase homology region of *A. thaliana* cryptochrome 2, which oligomerizes upon blue light exposure. Artificially clustered RhoA induced RhoA signalling-dependent cytoskeletal re-organization and membrane retraction in human cells, suggesting that oligomerization promotes RhoA activation [17]. Ras and Ras-like GTPases are well known to form nanoclusters and dimers at the membrane to activate signal transduction [18–20]. Several Rho-family GTPases, including RhoA, Rac1, Rac2 and Cdc42, were also shown to form dimers or oligomers through homophilic interactions of their polybasic region adjacent to the C-terminal CAAX motif [21,22]. While oligomerization of GTP-bound Cdc42 and Rac1 increases their GTPase activity in vitro, the physiological relevance of clustering of these small GTPases remains to be investigated [22]. In vivo, Rac1-GTP oligomers have been shown to contain several dozen Rac1 molecules together with charged phospholipids and appear to promote signal transduction [21–24]. Cdc42 also forms nanoclusters in *Saccharomyces cerevisiae* cells [25,26]. These nanoclusters show an anisotropic distribution: they accumulate and exhibit larger sizes at cortical sites of polarized growth, in a manner dependent on the scaffold protein Bem1 and anionic membrane lipids [25,27]. Because Bem1 also acts as scaffold that bridges Cdc42-GTP to its GEF and promotes positive feedback activation of Cdc42, Cdc42 nanoclusters may promote Cdc42 feedback activation, though this has not been tested.

In this work, we used an artificial optogenetic strategy to induce the recruitment and clustering of Cdc42 at the plasma membrane of fission yeast cells. We built on our recent work that used the CRY2-CIB1 optogenetic system to probe the positive feedback of Cdc42 [6]. The CRY2-CIB1 system relies on the blue light-induced binding of CRY2PHR (simply denoted CRY2 below) to the N-terminal part of CRY2-binding partner CIB1 (CIBN) [28]. Blue light also induces the formation of CRY2 oligomers [17]. We fused CRY2 to a cytosolic variant of Cdc42 (Cdc42$^{\Delta CaaX}$) and co-expressed CIBN linked to the membrane-associated RitC anchor. In our earlier study, we showed that cortical recruitment of a GTP-locked, constitutively active Cdc42 variant (CRY2-Cdc42$^{Q61L,\Delta CaaX}$) led to the Scd2-dependent co-recruitment of its GEF Scd1 and accumulation of endogenous Cdc42, demonstrating feedback amplification [6]. Surprisingly, we also found that cortical recruitment of CRY2-Cdc42$^{\Delta CaaX}$ (not GTP-locked) also induced the co-recruitment of Scd2, suggesting the activation of CRY2-Cdc42$^{\Delta CaaX}$. In this work, we confirm that CRY2-dependent recruitment of Cdc42$^{\Delta CaaX}$ at lateral sites, where Cdc42 is normally inactive, promotes its activation. We show that activated clustered Cdc42 is able to recruit its GEF Scd1 through the scaffold Scd2, suggesting that positive feedback is initiated. However, the activation is efficiently countered by Rga4 GAP-mediated Cdc42 inactivation, and does not lead to cell shape alteration, showing the robustness of the cell polarization system.

2. Results and Discussion

2.1. Weak Activation of CRY2-Cdc42 at the Cell Cortex

To better characterize CRY2-Cdc42$^{\Delta CaaX}$, we first measured its kinetics of recruitment to CIBN-RitC at the plasma membrane. Similar to rates measured for CRY2, CRY2-Cdc42$^{\Delta CaaX}$ showed a half-time of protein recruitment to the cortex < 1 s and independent of the length of the blue light (488 nm) pulses (30 GFP pulses of 50 ms = 0.92 s ± 0.24 s; 22 GFP pulses of 250 ms = 0.98 s ± 0.25 s; 17 GFP pulses of 500 ms = 0.99 s ± 0.31 s; Figure S1A,B). CRY2-Cdc42$^{\Delta CaaX}$ cells did not exhibit any morphological defects and grew in a bipolar fashion in the dark (Figure S1C). In blue-light, CRY2-Cdc42$^{\Delta CaaX}$ cells maintained their characteristic rod-shape and continued growing from the cell tips (Supplementary Figure S1D, green cells), while cells with GTP-locked CRY2-Cdc42$^{Q61L,\Delta CaaX}$ rounded up indicating isotropic growth (Supplementary Figure S1D, blue cells; [6]). These evidences initially suggested that the recruitment of CRY2-Cdc42$^{\Delta CaaX}$ to the cell cortex was innocuous and unable to bias the endogenous Cdc42 and its regulatory network.

We monitored the distribution of Cdc42-GTP using three GFP-tagged markers that specifically associate with Cdc42-GTP: the scaffold Scd2 [7,29], the CRIB bioreporter (Cdc42-Rac1-interactive-binding domain, [30]), and the Cdc42 effector Pak1, which also contains a CRIB domain. These proteins and probe are normally only detected at the poles and division sites of yeast cells, as well as weakly in the nucleus for the first two. As previously described [6], upon blue light-dependent recruitment of CRY2-Cdc42$^{\Delta CaaX}$ to the plasma membrane, Scd2-GFP formed stable foci at the cell sides, which increased progressively in intensity and coincided with CRY2-Cdc42$^{\Delta CaaX}$ clusters, while Scd2-GFP intensity decreased at cell poles (Figure 1A–D and Figure S2). We had previously shown that Scd2 was strongly recruited by GTP-locked CRY2- Cdc42$^{Q61L,\Delta CaaX}$ but not GDP-locked CRY2- Cdc42$^{T17N,\Delta CaaX}$ [6]. Indeed, we confirmed that CRY2-Cdc42$^{\Delta CaaX\text{-}T17N}$ does not lead to Scd2 foci at cell sides, indicating that CRY2- Cdc42$^{\Delta CaaX}$ must be in the GTP-bound form to recruit Scd2 (Figure 1A). CRIB-3GFP also formed dim foci at the cell sides, which became visible 40–60s after light stimulation (Figure 1A,E and Figure S3). The CRIB-3GFP side signal was however weaker and delayed relative to that observed upon light-induced recruitment of GTP-locked CRY2-Cdc42$^{Q61L,\Delta CaaX}$ (Figure 1E; [6]). Indeed, recruitment of CRIB-3GFP was only statistically significant in the second half of the 87 s-time-lapse (see materials and methods). This side recruitment was also mirrored by a reduction of the CRIB signal at the cell poles (Figure 1F). The Pak1-sfGFP traces also showed an upward trajectory on cell sides and a downward trajectory at cell poles but were not statistically different from negative control after 87 s (Figure 1A,G,H). We note, however, that the higher levels of both CRIB and Pak1 on cell sides were marginally statistically significant after 30 min illumination ($p = 0.04$; see Figure 5B,C). Based on the recruitment of Scd2 by CRY2-Cdc42$^{\Delta CaaX}$ on cell sides, these data suggest that the heterologous Cdc42 moiety within the CRY2-Cdc42$^{\Delta CaaX}$ system is transiently activated when recruited in clusters at the cell sides and sufficient to alter the endogenous sites of Cdc42 activity at cell poles. We hypothesize that the stronger Scd2 than CRIB and Pak1 signals reflect a more stable binding, likely stabilized by additional association, for instance, to anionic lipids [27].

As an alternative strategy to increase Cdc42 levels at the plasma membrane, we overexpressed Cdc42. In this experiment, we used the functional, internally tagged *cdc42-mCherry*^{SW} allele [4] expressed under the p^{act1} promoter in cells lacking the endogenous gene, which allowed us to quantify the global increase in expression levels at 3.3-fold (Figure S4A,B). The Cdc42 level increase was roughly uniform around the cell cortex (not shown). Cdc42 overexpression also led to a 1.2-fold increase in the expression of the CRIB-3GFP reporter (expressed under the p^{pak1} promoter, Figure S4B). Cdc42 overexpression led to a small increase in CRIB signal at cell poles (even after correction by the 1.2-fold increase in probe expression) and a small increase in cell length (p^{act1}-*cdc42* cell length = 14.2 ± 1 µm vs. WT cell length = 13.4 ± 1.1 µm, t-test *p*-value = 2.8×10^{-5}; p^{act1}-cdc42 cell width = 3.8 ± 0.3 µm vs. WT cell width = 3.8 ± 0.3 µm, *t*-test *p*-value = 0.96), suggesting increased Cdc42 activity at cell poles. However, Cdc42 overexpression had no effect on CRIB-3GFP levels at cell sides (Figure S4C,D). We conclude that activation of CRY2-Cdc42$^{\Delta CaaX}$ on cell sides is not simply a consequence of overexpression but may be due to other changes imposed by CRY2 activation. It is possible that CRY2-dependent clustering of Cdc42 directly causes the activation of the GTPase, as has been proposed for other GTPases [17], though unknown mechanism. Alternatively, clustering may have indirect effects, such as slowing down Cdc42 dynamics, which influence its activation cycle.

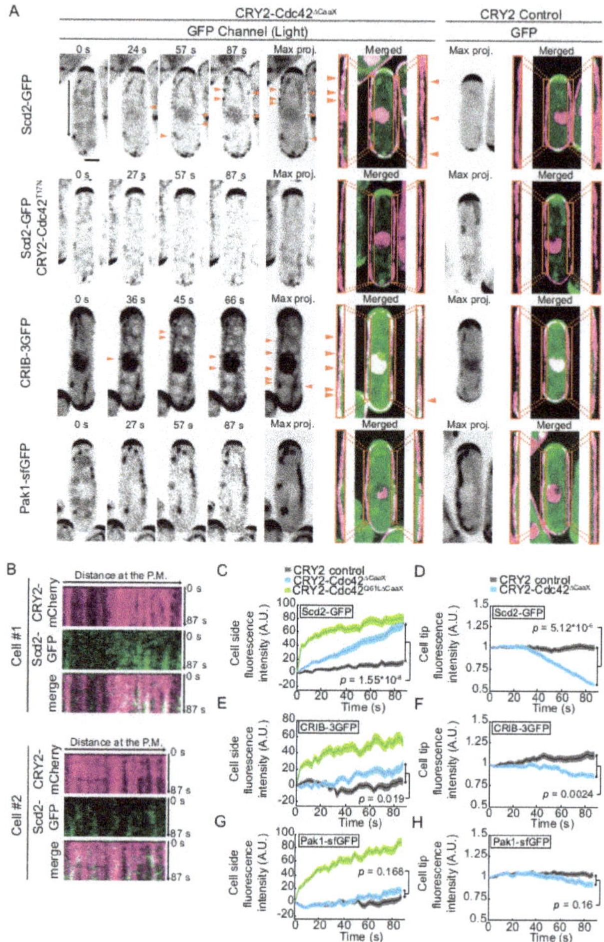

Figure 1. Ectopic sites of Cdc42 activation upon CRY2-Cdc42$^{\Delta CaaX}$ cell-sides recruitment. (**A**) Localization of Scd2-GFP, CRIB-3GFP and Pak1-sfGFP in CRY2-Cdc42$^{\Delta CaaX}$-expressing cells (B/W inverted images and green channel in merge). The GFP max projection ("max proj.") images show GFP maximum-intensity projections of 30 time points over 87 s. Merged images are composites of GFP and RFP max projections (t0 omitted from the RFP projection). Magnification of the lateral cortex is shown in the orange insets. Arrowheads point to lateral Scd2-GFP and CRIB-3GFP signal. The black arrow in Scd2-GFP panel indicates the cortical region in the kymograph shown in (B, cell #1). Note that autofluorescent organelles appear as linear and circular structures in some of the GFP channel images. (**B**) Kymograph at the lateral cell cortex for two cells over the 87 s (cell #1 corresponds to the cell shown in (**A**)). (**C**,**D**) Quantification of Scd2-GFP signal intensity at cell sides (**C**) and cell poles (**D**). (**E**,**F**) Quantification of CRIB-3GFP signal intensity at cell sides (**E**) and cell poles (**F**). (**G**,**H**) Quantification of Pak1-sfGFP signal intensity at cell sides (**G**) and cell poles (**H**). In (**C**–**H**), $n > 66$ cells. Exact numbers are listed in the methods. In (**C**,**E**,**G**), the CRY2-Cdc42$^{Q61L,\Delta CaaX}$ average trace is shown for cell-side recruitment comparison (data from (Lamas et al., 2020)). In all graphs, thick line = average; shaded area = standard error of the mean (SEM); WT, wild type; A.U., arbitrary units. Bars = 2 µm. Associated trace analysis is shown in Figure S2A for cell-side and Figure S2B for cell pole analyses.

2.2. CRY2-Cdc42 Activation in Absence of Cdc42 GEFs

To probe the mode of CRY2-Cdc42$^{\Delta CaaX}$ activation, we repeated the optogenetic experiments above in strains lacking the Cdc42 GEF Scd1. In *scd1Δ* cells, Pak1-sfGFP was not detected at cell sides, similar as in wildtype cells. We also observed only rare CRIB-3GFP dots, and no significant increase in CRIB levels at the sides nor decrease at cell poles of *scd1Δ* cells (Figure 2A–C), suggesting that Scd1 participates in CRY2-Cdc42$^{\Delta CaaX}$ activation. However, Scd2-GFP was still recruited to cell sides and decreased from cell poles (Figure 2A–C). Because Scd2 recruitment is strictly dependent on the GTP-bound form of Cdc42 (see Figure 1A; [6,7,29]), this suggests that CRY2-Cdc42$^{\Delta CaaX}$ may still be active in these cells. We thus probed the role of the second Cdc42 GEF Gef1. In *gef1Δ* cells, Scd2-GFP accumulation at cell sides and decrease at cell poles exhibited similar dynamics as in WT cells (Figure 2D–F). It is possible that the two GEFs work redundantly in this situation, a hypothesis difficult to test due to the lethality of *scd1Δ gef1Δ* double mutants [11,12]. An alternative hypothesis, which we do not favour, is that Cdc42 clustering through CRY2 binding may promote Scd2 recruitment independently of its activation.

2.3. CRY2-Cdc42 Promotes Recruitment of Its GEF Scd1 in Scd2 Scaffold-Dependent Manner

Because Cdc42-GTP promotes the recruitment of its GEF Scd1 for feedback amplification of Cdc42 activation [6], we probed whether CRY2-Cdc42$^{\Delta CaaX}$ induces Scd1 recruitment. Indeed, Scd1 formed weak foci at cell sides upon blue-light activation (Figure 3A,B), similar to the CRIB-3GFP foci observed in CRY2-Cdc42$^{\Delta CaaX}$ cells (see Figure 1A and Figure S3). The appearance of Scd1 foci at cell sides was also mirrored by a decrease of Scd1-3GFP at the cell tips (Figure 3C). Scd1 recruitment was dependent on the scaffold Scd2, as no cell side accumulation of Scd1-3GFP, nor decrease at cell tips, was detected in *scd2Δ* cells (Figure 3D–F). These data suggest that the activated CRY2-Cdc42$^{\Delta CaaX}$ is poised to trigger the positive feedback leading to recruitment of its GEF Scd1.

Although *scd2* deletion abolished Scd1 recruitment, it did not substantially affect the accumulation of the CRIB probe (Figure 3D,G), indicating that CRY2-Cdc42$^{\Delta CaaX}$ is still activated in these cells. This observation is in agreement with the finding that CRY2-Cdc42$^{\Delta CaaX}$ may be activated independently of Scd1. Because CRIB intensity on cell sides was not reduced in *scd2Δ* cells (Figure 3H), we conclude that the scaffold-dependent recruitment of the GEF by CRY2-Cdc42$^{\Delta CaaX}$ does not play a major role in amplifying Cdc42 activation at cell sides.

2.4. The Cdc42 GAP Rga4 Prevents Isotropic Growth of CRY2-Cdc42$^{\Delta CaaX}$ Cells

If CRY2-Cdc42$^{\Delta CaaX}$ is activated at cell sides and recruits its own GEF, why is Cdc42 activity not further amplified by the positive feedback mechanism and does not lead to cell shape changes? Indeed, even long-term growth of CRY2-Cdc42 cells in the light did not change their cell length and width, or aspect ratio, similar to control CRY2 cells. By contrast, constitutive cortical localization of CRY2-Cdc42$^{Q61L,\Delta CaaX}$ by growth in light conditions led to a significant increase in cell width and decrease in cell length, yielding a reduced aspect ratio (Figure 4A–C).

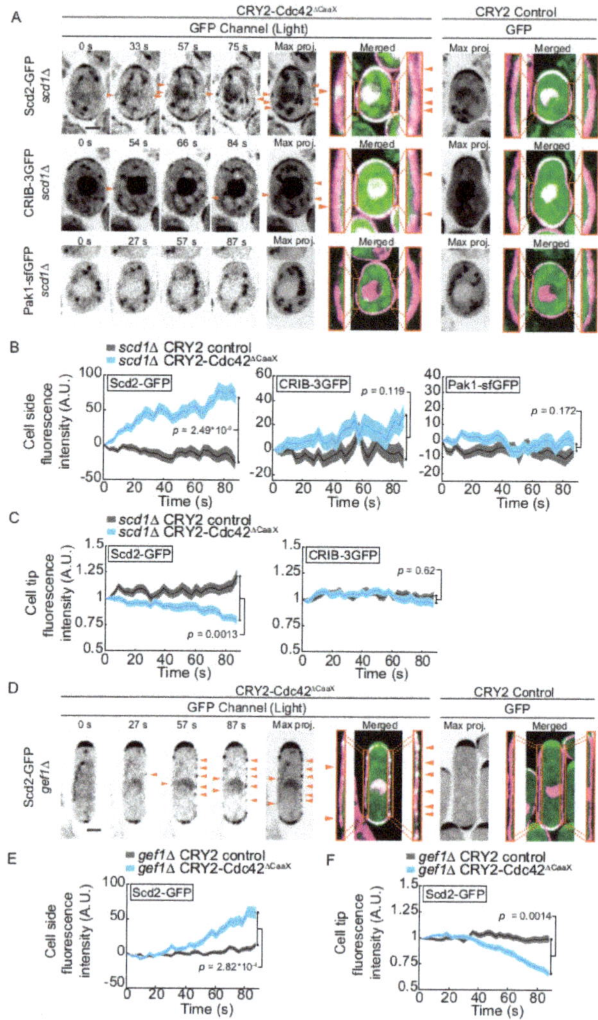

Figure 2. Role of Cdc42 GEFs in activation of CRY2-Cdc42$^{\Delta CaaX}$. (**A**) Localization of Scd2-GFP, CRIB-3GFP and Pak1-sfGFP in CRY2-Cdc42$^{\Delta CaaX}$-expressing *scd1*Δ cells (B/W inverted images and green channel in merge). The GFP max projection ("max proj.") images show GFP maximum-intensity projections of 30 time points over 87 s. Merged images are composites of GFP and RFP max projections (t0 omitted from the RFP projection). Magnification of the lateral cortex is shown in the orange insets. Arrowheads point to lateral Scd2-GFP and CRIB-3GFP signal. (**B**) Quantification of GFP signal intensities of Scd2-GFP, CRIB-3GFP and Pak1-sfGFP at cell sides of *scd1*Δ mutants. $n > 66$ cells. Exact numbers are listed in the methods. (**C**) Quantification of Scd2-GFP and CRIB-3GFP signal intensity at cell poles of *scd1*Δ mutants. $n > 62$ cells. Exact numbers are listed in the methods. (**D**) Localization of Scd2-GFP in CRY2-Cdc42$^{\Delta CaaX}$-expressing *gef1*Δ cells. Layout as in panel A. (**E**,**F**) Quantification of Scd2-GFP signal intensity at cell sides (**E**) and cell poles (**F**) of *gef1*Δ mutants. $n > 68$ cells. Exact numbers are listed in the methods. In all graphs, thick line = average; shaded area = standard error of the mean (SEM); WT, wild type; A.U., arbitrary units. Bars = 2 μm. Associated trace analysis is shown in Figure S5. Autofluorescent organelles appear as linear and circular structures in some of the GFP channel images.

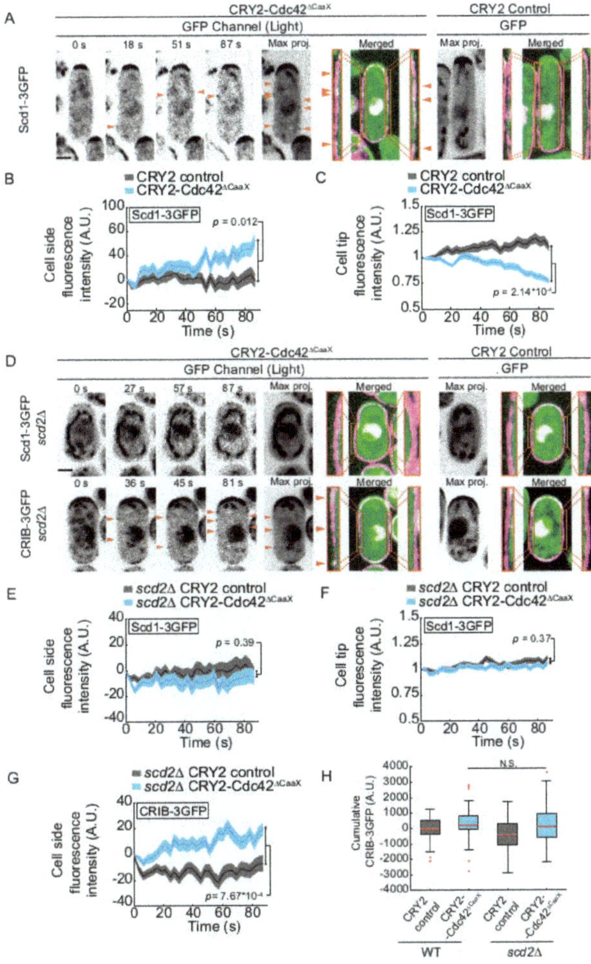

Figure 3. CRY2-Cdc42$^{\Delta CaaX}$ recruits the GEF Scd1 dependent on the scaffold Scd2. (**A**) Localization of Scd1-3GFP in CRY2-Cdc42$^{\Delta CaaX}$-expressing cells (B/W inverted images and green channel in merge). The GFP max projection ("max proj.") images show GFP maximum-intensity projections of 30 time points over 87 s. Merged images are composites of GFP and RFP max projections (t0 omitted from the RFP projection). Magnification of the lateral cortex is shown in the orange insets. Arrowheads point to lateral Scd1-3GFP signal. (**B**,**C**) Quantification of Scd1-3GFP signal intensity at cell sides (**B**) and cell poles (**C**). $n > 71$ cells. Exact numbers are listed in the methods. (**D**) Localization of Scd1-3GFP and CRIB-3GFP in CRY2-Cdc42$^{\Delta CaaX}$-expressing *scd2Δ* cells. Layout as in panel A. (**E**,**F**) Quantification of Scd1-3GFP at cell sides (**E**) and cell poles (**F**) of *scd2Δ* mutant. $n > 74$ cells. Exact numbers are listed in the methods. (**G**) Quantification of CRIB-3GFP at cell sides of *scd2Δ* mutants. $n > 84$ cells. Exact numbers are listed in the methods. (**H**) Comparison of cumulative CRIB-3GFP intensities in WT and *scd2Δ* CRY2 and CRY2-Cdc42$^{\Delta CaaX}$ cells; in *scd2*$^{+}$ p^{control} vs. $^{\text{CRY2-Cdc42}}$ = 0.019 (data from Figure 1E); in *scd2Δ* p^{control} vs. $^{\text{CRY2-Cdc42}}$ = 7.67×10^{-4} (data from panel 3G); $p^{\text{CRY2-Cdc42 in WT}}$ vs. $^{scd2\Delta}$ = 0.4. In all graphs, thick line = average; shaded area = standard error of the mean (SEM); WT, wild type; A.U., arbitrary units. Bars = 2 μm. Associated trace analysis is shown in Figure S6. Autofluorescent organelles appear as linear and circular structures in some of the GFP channel images.

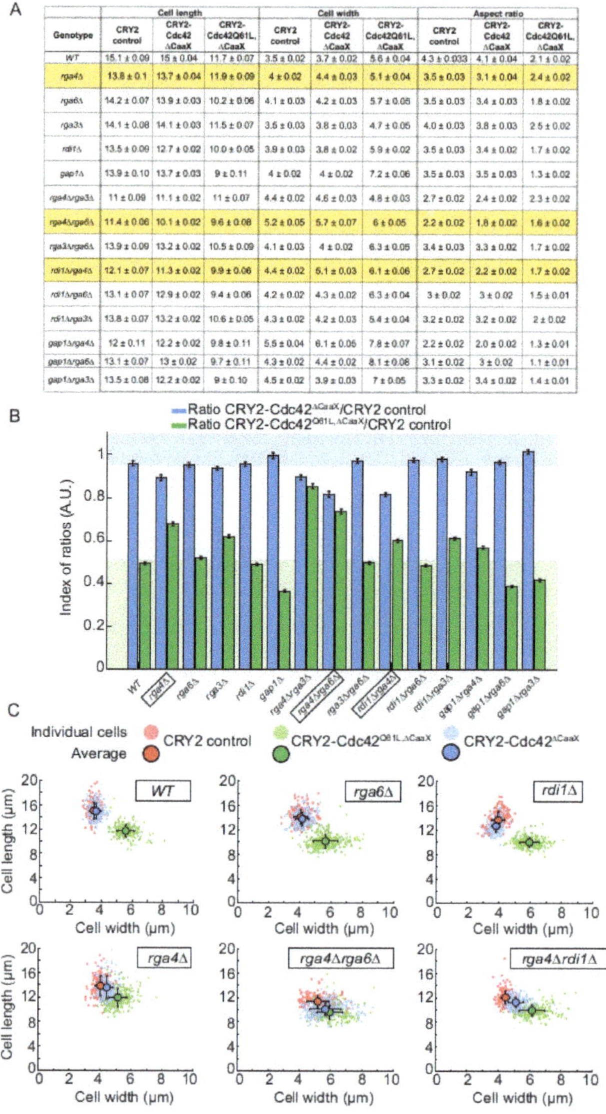

Genotype	Cell length			Cell width			Aspect ratio		
	CRY2 control	CRY2-Cdc42 ΔCaaX	CRY2-Cdc42Q61L, ΔCaaX	CRY2 control	CRY2-Cdc42 ΔCaaX	CRY2-Cdc42Q61L, ΔCaaX	CRY2 control	CRY2-Cdc42 ΔCaaX	CRY2-Cdc42Q61L, ΔCaaX
WT	15.1 ± 0.09	15 ± 0.04	11.7 ± 0.07	3.5 ± 0.02	3.7 ± 0.02	5.6 ± 0.04	4.3 ± 0.033	4.1 ± 0.04	2.1 ± 0.02
rga4Δ	13.8 ± 0.1	13.7 ± 0.04	11.9 ± 0.09	4 ± 0.02	4.4 ± 0.03	5.1 ± 0.04	3.5 ± 0.03	3.1 ± 0.04	2.4 ± 0.02
rga6Δ	14.2 ± 0.07	13.9 ± 0.03	10.2 ± 0.06	4.1 ± 0.03	4.2 ± 0.03	5.7 ± 0.05	3.5 ± 0.03	3.4 ± 0.03	1.8 ± 0.02
rga3Δ	14.1 ± 0.08	14.1 ± 0.03	11.5 ± 0.07	3.5 ± 0.03	3.8 ± 0.03	4.7 ± 0.05	4.0 ± 0.03	3.8 ± 0.03	2.5 ± 0.02
rdi1Δ	13.5 ± 0.09	12.7 ± 0.02	10.0 ± 0.05	3.9 ± 0.03	3.8 ± 0.02	5.9 ± 0.02	3.5 ± 0.03	3.4 ± 0.02	1.7 ± 0.02
gap1Δ	13.9 ± 0.10	13.7 ± 0.03	9 ± 0.11	4 ± 0.02	4 ± 0.02	7.2 ± 0.06	3.5 ± 0.03	3.5 ± 0.03	1.3 ± 0.02
rga4Δrga3Δ	11 ± 0.09	11.1 ± 0.02	11 ± 0.07	4.4 ± 0.02	4.6 ± 0.03	4.8 ± 0.03	2.7 ± 0.02	2.4 ± 0.02	2.3 ± 0.02
rga4Δrga6Δ	11.4 ± 0.06	10.1 ± 0.02	9.6 ± 0.08	5.2 ± 0.05	5.7 ± 0.07	6 ± 0.05	2.2 ± 0.02	1.8 ± 0.02	1.6 ± 0.02
rga3Δrga6Δ	13.9 ± 0.09	13.2 ± 0.02	10.5 ± 0.09	4.1 ± 0.03	4 ± 0.02	6.3 ± 0.05	3.4 ± 0.03	3.3 ± 0.02	1.7 ± 0.02
rdi1Δrga4Δ	12.1 ± 0.07	11.3 ± 0.02	9.9 ± 0.06	4.4 ± 0.02	5.1 ± 0.03	6.1 ± 0.06	2.7 ± 0.02	2.2 ± 0.02	1.7 ± 0.02
rdi1Δrga6Δ	13.1 ± 0.07	12.9 ± 0.02	9.4 ± 0.06	4.2 ± 0.02	4.3 ± 0.02	6.3 ± 0.04	3 ± 0.02	3 ± 0.02	1.5 ± 0.01
rdi1Δrga3Δ	13.8 ± 0.07	13.2 ± 0.02	10.6 ± 0.05	4.3 ± 0.02	4.2 ± 0.03	5.4 ± 0.04	3.2 ± 0.02	3.2 ± 0.02	2 ± 0.02
gap1Δrga4Δ	12 ± 0.11	12.2 ± 0.02	9.8 ± 0.11	5.5 ± 0.04	6.1 ± 0.05	7.8 ± 0.07	2.2 ± 0.02	2.0 ± 0.02	1.3 ± 0.01
gap1Δrga6Δ	13.1 ± 0.07	13 ± 0.02	9.7 ± 0.11	4.3 ± 0.02	4.4 ± 0.02	8.1 ± 0.08	3.1 ± 0.02	3 ± 0.02	1.1 ± 0.01
gap1Δrga3Δ	13.5 ± 0.08	12.2 ± 0.02	9 ± 0.10	4.5 ± 0.02	3.9 ± 0.03	7 ± 0.05	3.3 ± 0.02	3.4 ± 0.02	1.4 ± 0.01

Figure 4. Rga4 GAP prevents growth on cell sides in CRY2-Cdc42$^{\Delta CaaX}$ cells. (**A**) Average cell length (μm), cell width (μm) and aspect ratio of CRY2, CRY2-Cdc42$^{\Delta Caax}$ and CRY2-Cdc42$^{Q61L,\Delta CaaX}$ mutants. For all mutant $N = 2$, $n > 80$ cells per experiment; except for *gap1Δ* mutants, $N = 1$, $n > 80$ cells per experiment. (**B**) Aspect ratios of CRY2-Cdc42$^{\Delta Caax}$ and CRY2-Cdc42$^{Q61L,\Delta Caax}$ -expressing cells grown in the light normalized to the aspect ratios of CRY2-expressing cells for all the tested mutants. Bars = standard error. The green background indicates expected reduction in aspect ratio upon CRY2-Cdc42$^{Q61L,\Delta CaaX}$ recruitment in WT cells. The blue background indicates absence of change in aspect ratio upon CRY2-Cdc42^{Q61L} recruitment. Note that all *rga4Δ* mutants fall in the white intermediate space. (**C**) Cluster plot of length and width in single cells of WT, *rga4Δ*, *rga6Δ*, *rga4Δrga6Δ*, *rdi1Δ and rga4Δrdi1Δ* mutants expressing CRY2, CRY2-Cdc42$^{\Delta Caax}$ or CRY2-Cdc42$^{Q61L,\Delta CaaX}$. Small dots = single cells; Large, dark dots = average; bars = standard deviation; A.U., arbitrary units; WT, wild type.

We hypothesized that the activation of CRY2-Cdc42$^{\Delta CaaX}$ at cell sides is rapidly counteracted by negative regulators. We focused our attention on the three Cdc42 GAPs Rga3, Rga4 and Rga6, the GDI protein Rdi1 and the Ras1 GAP Gap1. Rga3, Rga4, and Rga6 directly promote Cdc42-GTP hydrolysis [13–15]. Rdi1 may promote Cdc42 extraction from the membrane, although previous work showed that it is largely dispensable for Cdc42 dynamics in *S. pombe* [4,16]. Gap1 directly promotes Ras1-GTP hydrolysis [31]. As Ras1 promotes Scd1 activation and is uniformly active at the plasma membrane in *gap1*Δ [6,31,32], we hypothesized Scd1 activation on cell sides may be amplified in this mutant. We constructed single and most double deletion mutants expressing either of CRY2-Cdc42$^{\Delta CaaX}$, CRY2-Cdc42$^{Q61L,\Delta CaaX,}$ as positive control or CRY2 alone as negative control. We then measured the cell length and cell width of calcofluor-stained dividing cells after at least 14h of exponential growth in light conditions and calculated aspect ratios (Figure 4A).

To estimate the change in aspect ratio upon Cdc42 lateral recruitment while taking into account the initial shape of the cell, we normalized the aspect ratios from cells recruiting Cdc42 to those expressing only CRY2 (Figure 4B). CRY2-Cdc42$^{Q61L,\Delta CaaX}$ led to a >2-fold reduction in aspect ratio in WT, *rga6*Δ, *rdi1*Δ and *gap1*Δ cells, but had less effect on cell shape change in single and double *rga4*Δ mutants, perhaps in part due to the already wider cell shape of *rga4*Δ cells [14]. Interestingly, CRY2-Cdc42$^{\Delta CaaX}$ had little effect on aspect ratio in WT or any single mutants, except in *rga4*Δ cells, which became significantly rounder. Similar, more marked effects were also observed in combinations of *rga4*Δ with *rga6*Δ or *rdi1*Δ. The effect of CRY2-Cdc42$^{\Delta CaaX}$ on the shape of these mutants can also readily be observed in plots of cell width to cell length, with the CRY2-Cdc42$^{\Delta CaaX}$ cell population placed at an intermediate position between the negative CRY2 and positive CRY2-Cdc42$^{Q61L,\Delta CaaX}$ controls (Figure 4C). These data indicate that the optogenetic-dependent Cdc42 activation is counteracted by Cdc42 GAPs placed at cell sides.

We hypothesized that Rga4 and Rga6 directly antagonize Cdc42 activity at cell sides. To probe this idea directly, we imaged CRIB-3GFP and Pak1-sfGFP for 87s immediately after CRY2-Cdc42$^{\Delta CaaX}$ recruitment. Both of these Cdc42-GTP interactors showed significantly increased levels at the sides of *rga4*Δ *rga6*Δ cells (Figure 5A). The data for Pak1 contrast with what we observed in wildtype cells, where no significant increase was detected at this timepoint (see Figure 1G). To strengthen these data further, we probed for CRIB-3GFP and Pak1-sfGFP localization at cell sides at a later timepoint, after 30 min of continuous white-light illumination, similar to the light conditions used to assess cell shape changes. Compared to the small increase of CRIB-3GFP and Pak1-sfGFP at the sides of wildtype cells in these conditions, *rga4*Δ *rga6*Δ mutants showed higher increase, which was highly statistically significant (Figure 5B,C). These data support the view that Cdc42 GAPs at cell sides antagonize the CRY2-Cdc42$^{\Delta CaaX}$ activation to prevent cell widening. A second, non-mutually exclusive possibility is that the cell shape change observed in *rga4*Δ cells reflects the weakening of polarity at the native sites at cell poles, as shown by the decrease of Scd2 and CRIB at cell poles (see Figure 1D,F). Thus, both local and competition effects with native polarity sites may combine to alter the shape of CRY2-Cdc42$^{\Delta CaaX}$-expressing *rga4*Δ cells.

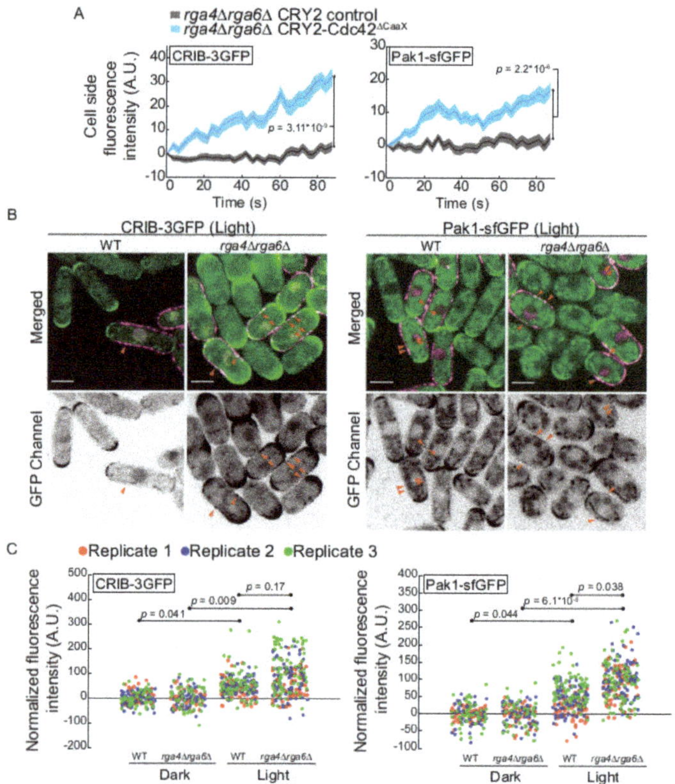

Figure 5. Stronger CRY2-Cdc42$^{\Delta CaaX}$ activation in cells lacking lateral Cdc42 GAPs. (**A**) Quantification of CRIB-3GFP and Pak1-sfGFP signal intensity at cell sides of *rga4Δrga6Δ* mutants, *n* > 80 cells. Exact numbers are listed in the methods. (**B**) Localization of CRIB-3GFP and Pak1-sfGFP in control and CRY2-Cdc42$^{\Delta CaaX}$-expressing cells upon 30 min-exposure to light (B/W inverted images and green channel in merge). Merged images are composites of GFP and RFP single middle-section images. Arrowheads point to lateral CRIB-3GFP and Pak1-sfGFP signal. Control cells lack the CRY2-Cdc42$^{\Delta CaaX}$ magenta signal in the merged images. (**C**) Cluster plots of CRIB-3GFP and Pak1-sfGFP signal intensity at cell sides of WT and *rga4Δrga6Δ* mutants, *n* > 138 cells. Exact numbers are listed in the methods. The three experimental replicates are plotted with different colours. Statistical *p*-values are from t-test across experimental replicates (*n* = 3). In all graphs, thick line = average; shaded area = standard error of the mean (SEM); WT, wild type; A.U., arbitrary units. Bars = 4 μm. Associated trace analysis is shown in Figure S8. Autofluorescent organelles appear as linear and circular structures in some of the GFP channel images.

In summary, the data presented in this work show that CRY2-clustered Cdc42 is ectopically activated at cell sides and leads to an alteration of the native polarity sites at cell poles. It is possible that CRY2-tagging interferes with Cdc42 GTPase activity, such that CRY2-Cdc42$^{\Delta CaaX}$ represents a slightly activated Cdc42 variant. Alternatively, the CRY2-dependent clustering may lead to Cdc42 activation, as proposed for other GTPases, such as Rac1 and RhoA in mammalian cells [17]. CRY2-clustered Cdc42 leads to very clear Scd2 recruitment, but much weaker CRIB and Pak1 recruitment, raising the question of why these Cdc42-GTP binding proteins behave differently. We hypothesize that this is explained by the stabilization of Scd2 localization through additional interactions for instance with membrane lipids. Interestingly, the activity level of CRY2-Cdc42$^{\Delta CaaX}$ at cell sides is sufficient to recruit the GEF Scd1, whose recruitment to Cdc42-GTP depends on the scaffold Scd2 [6]. However, positive feedback does not appear to become established as Cdc42 activity levels at cell sides are not altered in absence of Scd2. It is possible that the non-physiological linkage of Cdc42 to the plasma membrane (through RitC-CIBN-CRY2 binding instead of through the normal prenyl group) undermines the positive feedback, although we previously showed that a similarly engineered constitutively active Cdc42 construct did trigger feedback-dependent recruitment of endogenous Cdc42 [6]. We favour the view that cellular regions may not be equally permissive to Cdc42 activity feedback amplification. For instance, the absence of Ras1 activity on cell sides may reduce the effectiveness of the feedback [6]. Our data further indicate that the inefficient positive feedback at cell sides may be due to the action of the cell side-localized Cdc42 GAP Rga4, which we have shown antagonizes the effect of CRY2-Cdc42$^{\Delta CaaX}$ on effector recruitment and cell morphogenesis. Additional mechanisms preventing growth on cell sides are also being proposed [33]. Thus, one critical question for future research is to better understand the many layers that confer robustness to cell morphogenesis.

3. Materials and Methods

3.1. Strains, Media, and Growth Conditions

Strains used in this study are listed in Supplementary Table S1. Standard genetic manipulation methods for *S. pombe* transformation and tetrad dissection were used to generate the strains listed. For microscopy experiments, cells were first pre-cultured in 3 mL of Edinburgh minimal media (EMM) in dark conditions at 30 °C for 6–8 h. Once exponentially growing, pre-cultures were diluted (Optical Density (O.D.) $_{600nm}$ = 0.02) in 10 mL of EMM and incubated in dark conditions overnight at 30 °C. In order to allow proper aeration of the culture, 50 mL Erlenmeyer flasks were used. For cell size analyses cells were pre-cultured and diluted once in 3 mL of Edinburgh minimal media (EMM) in dark conditions at 30 °C for 6–8 h. Once exponentially growing, pre-cultures were diluted in 10 mL of EMM and incubated in light conditions for a minimum of 14 h. All live-cell imaging was performed on EMM-ALU agarose pads, except calcofluor-white experiments in which cells were placed directly on a slide [34]. Gene tagging was performed at endogenous genomic locus at the 3′ end, yielding C-terminally tagged proteins, as described [35]. Pak1 gene tagging was performed by transforming a WT strain with AfeI linearized (pBSII(KS$^+$))-based single integration vector (pAV72-3′UTR^{pak1}-Pak1-sfGFP-kanMX-5′UTR^{pak1}) targeting the endogenous locus [36]. The functional mCherry-tagged and sfGFP-tagged *cdc42* alleles *cdc42-mCherry^{sw}* and *cdc42-sfGFP^{sw}* were used as described in [4]. Gene tagging, deletion, and plasmid integration were confirmed by diagnostic PCR for both sides of the gene.

The construction of plasmids and strains expressing CIBN-mTagBFP2-Ritc, CRY2, CRY2-Cdc42$^{\Delta CaaX}$ and CRY2-Cdc42$^{Q61L,\Delta CaaX}$ was done as described in [6].

To generate the *Pact1-cdc42-mCh^{SW}* strain a pINT-ura4$^+$ integrative vector was generated. *Pact1* was amplified from gDNA using primers osm2378 (atgggcccgctagcatgcGATCTACGATAATGAGACGG) and osm2379 (ccggctcgagGGTCTTGTCTTTTGAGGGTT) and cloned using ApaI and XhoI. *Cdc42-mCh^{SW}-nmt$_{terminator}$* was amplified from pSM1224 using osm2343 (cccaagcttATGCCCACCATTAAGTGTGTCG) and osm2344 (gctctagaCTTCTAATTACACAAATTCCG) and cloned using HindIII and XbaI. As a results pSM1449 was generated. This plasmid was linearized with AvrII and integrated at *ura4* locus of YSM485 strain. The endogenous allele of *cdc42* was deleted using a hygromycin (*hph$^+$*) resistance cassette as described [35]. *Hph$^+$* cassette was amplified from pSM693 using osm2511 (TACTTAGGGGTTTGAACTTTCTAGGAATTCAATAAAGTGAAGCAAAGCTTTACGATTAATTAT TTTTTGTGAAATAGTcggatccccgggttaattaa) and osm2512 (AAGCTAAGACATTGTTTACTGTTGTAAA CTAGCTGTATTAAGGAAATTTCGGAAAAGGAAAGAAAACCAGGGGTTAAAgaattcgagctcgtttaaac). Finally, strain YSM3732 was generated by transforming the *Cdc42-mCh^{SW}-nmt$_{terminator}$* strain with the *hph$^+$* resistance cassette.

In primer sequences, restriction sites are underlined. Plasmid maps are available upon request.

3.2. Cell Length and Width Measurements

For cell length and width measurements shown in Figure 4 and Supplementary Figure S1, cells were grown at 30 °C in 10 mL EMM in light and dark conditions respectively. In this case, we used white light, which we found is sufficient to activate CRY2, shown by the cell rounding triggered by CRY2-Cdc42$^{Q61L,\Delta CaaX}$. Exponentially growing cells were stained with calcofluor to visualize the cell wall and imaged on a Leica epifluorescence microscope with 60× magnification platform described previously [34]. Measurements were made with ImageJ on septating cells. For each experiment, strains with identical auxotrophies were used.

Aspect ratios were calculated as:

$$Aspect\ ratio = \frac{Cell\ length}{Cell\ width} \quad (1)$$

The index of ratios shown in Figure 4B was calculated as:

$$Index_{Cdc42\Delta CaaX} = \frac{Aspect\ ratio(CRY2Cdc42\Delta CaaX)}{Aspect\ ratio\ (CRY2)}$$

$$Index_{Cdc42Q61L\Delta CaaX} = \frac{Aspect\ ratio(CRY2Cdc42\Delta Q61LCaaX)}{Aspect\ ratio\ (CRY2)} \quad (2)$$

3.3. Microscopy

Fluorescence microscopy experiments were done in a spinning disk confocal microscope, essentially as described [34,37]. Image acquisition was performed on a Leica DMI6000SD inverted microscope equipped with an HCX PL APO 100X/1.46 numerical aperture oil objective and a PerkinElmer Confocal system. This system uses a Yokagawa CSU22 real-time confocal scanning head, solid-state laser lines and a cooled 14-bit frame transfer EMCCD C9100-50 camera (Hamamatsu) and is run by Volocity (PerkinElmer). When imaging strains expressing the CRY2-Cdc42$^{\Delta CaaX}$ and/or CRY2 systems, an additional long-pass color filter (550 nm, Thorlabs Inc., Newton, NJ, USA) was used for bright-field (BF) image acquisition, in order to avoid photo-activation by white light.

Spinning disk confocal microscopy experiments shown in Figures 1–3, Supplementary Figures S1A,D, S3, and S7 were carried out using cell mixtures [6]. Cell mixtures were composed by one strain of interest (the sample optogenetic strain, expressing or not an additional GFP-tagged protein) and 2 control strains, namely:

(1) RFP control: An RFP bleaching correction strain, expressing cytosolic CRY2PHR-mCherry.
(2) GFP control: A wild type strain expressing the same GFP-tagged protein as the strain of interest but without the optogenetic system. This strain was used both as negative control for cell side re-localization experiments and as GFP bleaching correction strain (in Figures 1–3 and 5 and related Figures S2, S5, S6 and S8).

Strains were handled in dark conditions throughout. Red LED light was used in the room in order to manipulate strains and to prepare the agarose pads. Strains were cultured separately. Exponentially growing cells (O.D.$_{600nm}$ = 0.4–0.6) were mixed with 2:1:1 (strain of interest, RFP control and GFP control) ratio, and harvested by soft centrifugation (2 min at 1600 rpm). The cell mixture slurry (1 µl) was placed on a 2% EMM-ALU agarose pad, covered with a #1.5-thick coverslip, and sealed with VALAP (vaseline, lanolin, and paraffin). Samples were imaged after 5–10 min of rest in dark conditions.

The plasma membrane recruitment dynamics of CRY2-Cdc42$^{\Delta CaaX}$ and CRY2 systems were assessed using cell mixtures. Protein recruitment dynamics were assessed by applying the 3 different photo-activating cycles listed below. Lasers were set to 100%. Shutters were set to maximum speed and in all instances the RFP channel was imaged first, before the GFP channel. The duration of the experiment was equal regardless of the exposure time settings (≈ 15 s):

(1) 50 ms: RFP channel (200 ms), GFP channel (50 ms). This constitutes one cycle (≈ 0.5 s). 30 time points were acquired (≈ 0.5 s * 30 = 15.1 s).
(2) 250 ms: RFP channel (200 ms), GFP channel (250 ms). This constitutes one cycle (≈ 0.7 s). 22 time points were acquired (≈ 0.7 s * 22 = 15.1 s).
(3) 500 ms: RFP channel (200 ms), GFP channel (500 ms). This constitutes one cycle (≈ 0.9 s). 17 time points were acquired (0.9 s * 17 = 15.5 s).

Endogenous GFP-tagged protein re-localization experiments were carried out using cell mixtures. Lasers were set to 100%; shutters were set to sample protection and in all instances the RFP channel was imaged first and then the GFP channel. RFP exposure time was always set to 200 ms, whereas the GFP exposure time varied depending on the monitored protein. Cells were monitored in these conditions for 87 s.

Spinning disk confocal time (sum) projections of five consecutive single plane images are shown in Supplementary Figure S4. Z-stack images were acquired on a Spinning disk confocal microscope using an optimal z-spacing of 0.71 μm between successive stacks. 6 stacks were acquired (covered Z distance = 4 μm). In Figure 5, single plane GFP and RFP images were combined to generate the merged (Figure 5B) and used for the GFP signal analyses at the cell side (Figure 5C). Cells were cultured as stated above and placed on EMM-ALU pads. Pads were kept for 30 min in the dark or under white-light illumination for 30 min prior imaging.

3.4. *Image Analysis*

All image-processing analyses were performed with Image J software (http://rsb.info.nih.gov/ij/). Image and time-lapse recordings were imported to the software using the Bio-Formats plugin (http://loci.wisc.edu/software/bio-formats). Time-lapse recordings were aligned using the StackReg plugin (http://bigwww.epfl.ch/thevenaz/stackreg/) according to the rigid body method. All optogenetic data analyses were performed using MATLAB (R2019a), with scripts developed in-house. Figures were assembled with Adobe Photoshop CC2019 and Adobe Illustrator 2020.

3.4.1. CRY2-Cdc42$^{\Delta CaaX}$ and CRY2 Quantifications

The plasma membrane recruitment dynamics of CRY2-Cdc42$^{\Delta CaaX}$ and CRY2 systems was assessed by recording the fluorescence intensity over a 15-pixel long by 36-pixel wide ROI (roughly 1.25 μm by 3 μm), drawn perpendicular to the plasma membrane of sample cells, from outside of the cell towards the cytosol. The fluorescence intensity values across the length of the ROI were recorded over time in the RFP channel, in which each pixel represents the average of the width (36 pixels) of the ROI (3 replicates, 30 cells per replicate). Average background signal was measured from tag-free wild-type cells incorporated into the cell mixture. The total fluorescence of the control RFP strain was also measured over time in order to correct for mCherry fluorophore bleaching. In both cases, the ROI encompassed whole cells, where ROI boundaries coincide with the plasma membrane.

Photobleaching correction coefficient was calculated by the following formula:

$$RFP\ bleaching\ correction\ coefficient = \frac{(RFP\ Intensity_{tn} - NoGFPBckg_{tn})}{(RFP\ Intensity_{t0} - NoGFPBckg_{t0})} \quad (3)$$

where *RFP intensity* is the signal measured from single RFP control cells, *NoGFPBckg* is the average background signal measured from tag-free cells, t_n represents a given time point along the time course of the experiment and t_0 represents the initial time point (n = 30 time points). These coefficients were corrected by a moving average smoothing method (moving averaged values = 5). RFP bleaching correction coefficient values calculated for individual RFP control cells were averaged in order to correct for bleaching of the RFP signal.

The fluorescence intensity values of optogenetic cells were corrected at each time point with the following formula:

$$RFP\ intensity = ((Raw\ RFP\ signal_{tn} - NoGFPBckg_{tn})/RFP\ bleaching\ correction\ coefficient_{tn}) \quad (4)$$

where *Raw RFP signal* is for the RFP values measured from sample strains, *NoGFPBckg* is the average background signal measured from tag-free cells and t_n represents a given time point along the time course of the experiment (n = 30 time points). The profiles resulting from these analyses were used to get the net plasma membrane recruitment profiles (Figure S1A, [6]), the fluorescence intensities from the peak ± 1 pixel were averaged and plotted over time.

$$Peak\ RFP\ intensity_{tn} = \frac{\left(RFP\ intensity_{peak-1pixel\ tn} + RFP\ intensity_{peak\ tn} + RFP\ intensity_{peak+1pixel\ tn}\right)}{3} \quad (5)$$

$$Net\ P.M.\ recruitment\ Profile = (Peak\ RFP\ intensity_{tn} - Peak\ RFP\ intensity_{t0}) \quad (6)$$

Finally, the single-cell plasma membrane recruitment half-times were calculated by fitting the normalized recruitment profiles with the following formula:

$$RFP\ intensity\ (y) = a * \left(1 - e^{(-b*t)}\right) \quad (7)$$

$$Recruitment\ t_{1/2} = \frac{ln(0.5)}{b} \quad (8)$$

3.4.2. Quantifications of the Re-Localization of GFP-Tagged Proteins to Cell Sides

Endogenous GFP-tagged protein re-localization was assessed upon photo-activation of CRY2-Cdc42$^{\Delta CaaX}$ and CRY2 systems by recording the fluorescence intensity over a 3 pixel-wide by 36 pixel-long (≈ 0.25 µm by 3 µm) ROI drawn parallel to the cell side cortex of sample cells. The average fluorescence intensity values of both GFP and RFP channels were recorded over time from sample strains. In these particular experiments, a GFP control strain was included. These strains serve 2 purposes:

(1) Calculation of the GFP bleaching correction coefficient (see below).
(2) Negative control of the experiment: These strains carry the same endogenous GFP-tagged protein as the sample strain of the experiment, however lacking the optogenetic system. This controlled that GFP fluorescence changes were due to the optogenetic system and not caused by imaging per se. Control GFP strains were imaged in the same pad and analysed in the same way as optogenetic cells.

To derive photobleaching correction coefficients, the average camera background signals (*Bckg*) from 5 cell-free regions was measured as above, and fluorophore bleaching from RFP control and GFP control strains were measured at the cell side of control RFP and control GFP strains, for RFP and GFP channels, respectively.

$$RFP\ bleaching\ correction\ coefficient = \frac{(RFP\ Intensity_{tn} - Bckg_{tn})}{(RFP\ Intensity_{t0} - Bckg_{t0})} \quad (9)$$

$$GFP\ bleaching\ correction\ coefficient = \frac{(GFP\ Intensity_{tn} - Bckg_{tn})}{(GFP\ Intensity_{t0} - Bckg_{t0})} \quad (10)$$

where *RFP intensity* and *GFP intensity* stand for the signal measured from RFP control and GFP control cells, respectively, t_n represents a given time point along the time course of the experiment and t_0 represents the initial time point (n = 30 time points). These coefficients were corrected by a moving average smoothing method, as above.

The fluorescence intensity values of optogenetic cells in both GFP and RFP channels were independently analysed as follows. First, GFP and RFP signals were background and bleaching corrected, using Formulas (9) and (10) for the RFP and GFP channels, respectively:

$$P.M.\ GFP/RFP\ value_{tn} = ((Raw\ signal_{tn} - Bckg_{tn})/bleaching\ correction\ coefficient_{\ tn}) \quad (11)$$

where *Raw signal* intensity represents the GFP or RFP raw values at the cell side cortex, *Bckg* stands for the average fluorescence intensity of 5 independent cell-free regions and t_n represents a given time point along the time course of the experiment (n = 30 time points). The net fluorescence intensity at the cell side cortex was then calculated for both GFP and RFP signals.

$$Net\ P.M.\ GFP/RFP\ value_{tn} = (Fluorescence\ intensity_{tn} - Fluorescence\ intensity_{t0}) \quad (12)$$

From here on, RFP and GFP signals were treated differently. Single cell plasma membrane RFP profiles from Equation (10) were individually normalized and fitted to the Equation (5) in order to extrapolate the parameter b. Using the Equation (6), recruitment half times of CRY2 and CRY2-Cdc42$^{\Delta CaaX}$ were calculated. Because of lower signal-to-noise of the weak GFP fluorescence, plasma membrane GFP profiles from Equation (10) were averaged (n > 20 profiles per experiment). Measurements from cells obtained in 3 experimental replicates are plotted on Figure 1C (CRY2-Cdc42$^{\Delta CaaX}$ n = 84 cells; CRY2 control n = 84 cells, WT control n = 252 cells); Figure 1E (CRY2-Cdc42$^{\Delta CaaX}$ n = 89 cells, CRY2 control n = 78 cells, WT control n = 247 cells); Figure 1G (CRY2-Cdc42$^{\Delta CaaX}$ n = 90 cells, CRY2 control n = 85 cells, WT control n = 280 cells); Figure 2B (Scd2-GFP *scd1Δ*: CRY2-Cdc42$^{\Delta CaaX}$ n = 67 cells, CRY2 control n = 71 cells, WT control n = 226 cells; CRIB-3GFP *scd1Δ*: CRY2-Cdc42$^{\Delta CaaX}$ n = 75 cells, CRY2 control n = 69 cells, WT control n = 222 cells; Pak1-sfGFP *scd1Δ*: CRY2-Cdc42$^{\Delta CaaX}$ n = 78 cells, CRY2 control n = 67 cells, WT control n = 231 cells), Figure 2E (CRY2-Cdc42$^{\Delta CaaX}$ n = 74 cells, CRY2 control n = 86 cells, WT control n = 247 cells); Figure 3B (CRY2-Cdc42$^{\Delta CaaX}$ n = 79 cells, CRY2 control n = 72 cells, WT control n = 228 cells); Figure 3E (CRY2-Cdc42$^{\Delta CaaX}$ n = 81 cells, CRY2 control n = 75 cells, WT control n = 257 cells); Figure 3G (CRY2-Cdc42$^{\Delta CaaX}$ n = 86 cells, CRY2 control n = 85 cells, WT control n = 246 cells) and Figure 5A (CRIB-3GFP *rga4Δrga6Δ*: CRY2-Cdc42$^{\Delta CaaX}$ n = 83 cells, CRY2 control n = 80 cells, WT control n = 145 cells; Pak1-sfGFP *rga4Δrga6Δ*: CRY2-Cdc42$^{\Delta CaaX}$ n = 83 cells, CRY2 control n = 84 cells, WT control n = 163 cells).

3.4.3. Quantifications of the Re-Localization of GFP-Tagged Proteins from Cell Tips

Scd2-GFP, CRIB-3GFP, Pak1-sfGFP and Scd1-3GFP tip signal analyses were performed from the same time-lapse recordings as cell side re-localization experiments. GFP tip signals were recorded over a 3 pixel-wide by 6–12 pixel-long (≈ 0.25 µm by 0.5–1 µm) ROI drawn at the tip of the cells. To derive photobleaching correction coefficients, the average camera background signals (*Bckg*) from 5 cell-free regions was measured as before, and GFP bleaching from GFP control strain was measured at the cell tip.

$$Tip\ GFP\ bleaching\ correction\ coefficient = \frac{(GFP\ Intensity_{tn} - Bckg_{tn})}{(GFP\ Intensity_{t0} - Bckg_{t0})} \quad (13)$$

where *GFP intensity* stands for the signal measured from the tip of GFP control cells, t_n represents a given time point along the time course of the experiment and t_0 represents the initial time point (n = 30 time points). This coefficient was corrected by a moving average smoothing method, as before.

The tip GFP fluorescence intensity values of optogenetic cells was analysed as follows. First, GFP signals was background and bleaching corrected, using Formula (14):

$$Tip\ GFP\ value_{tn} = (Tip\ Raw\ signal_{tn} - Bckg_{tn})/Tip\ GFP\ bleaching\ correction\ coefficient_{\ tn} \quad (14)$$

where *Tip Raw signal* intensity represents the GFP raw values at the cell tip, *Bckg* stands for the average fluorescence intensity of 5 independent cell-free regions and t_n represents a given time point along the time course of the experiment ($n = 30$ time points). The tip fluorescence intensities of single optogenetic strains were then normalized relative to their GFP values at the initial time-point.

$$Normalized\ tip\ GFP\ value_{tn} = (Tip\ GFP\ value_{tn} / tip\ GFP_{t0}) \quad (15)$$

Measurements from cells obtained in 3 experimental replicates are plotted on Figure 1D (CRY2-Cdc42$^{\Delta CaaX}$ $n = 88$ cells; CRY2 control $n = 87$ cells, WT control $n = 146$ cells); Figure 1F (CRY2-Cdc42$^{\Delta CaaX}$ $n = 79$ cells; CRY2 control $n = 67$ cells, WT control $n = 147$ cells): Figure 1H (CRY2-Cdc42$^{\Delta CaaX}$ $n = 72$ cells; CRY2 control $n = 68$ cells, WT control $n = 203$ cells); Figure 2C (Scd2-GFP *scd1*Δ: CRY2-Cdc42$^{\Delta CaaX}$ $n = 68$ cells, CRY2 control $n = 63$ cells, WT control $n = 134$ cells; CRIB-3GFP *scd1*Δ: CRY2-Cdc42$^{\Delta CaaX}$ $n = 65$ cells, CRY2 control $n = 67$ cells, WT control $n = 131$ cells); Figure 2F (CRY2-Cdc42$^{\Delta CaaX}$ $n = 69$ cells; CRY2 control $n = 80$ cells, WT control $n = 160$ cells); Figure 3C (CRY2-Cdc42$^{\Delta CaaX}$ $n = 109$ cells; CRY2 control $n = 75$ cells, WT control $n = 259$ cells) and Figure 3F (CRY2-Cdc42$^{\Delta CaaX}$ $n = 85$ cells; CRY2 control $n = 78$ cells, WT control $n = 226$ cells).

3.4.4. Quantifications of CRIB-3GFP and Cdc42-mCh^{SW} Relative Expression and Distribution Profiles

CRIB-3GFP fluorescence intensity was measured from sum projection of 6 Z-stacks (Figure S4B). The background signal from cell-free regions were used to correct the data. The relative fluorescence intensities were calculated by dividing the single-cell CRIB-3GFP fluorescence intensity measurements of WT and *Pact1-cdc42-mCh*^{SW} cells by the average CRIB-3GFP fluorescence intensity of WT cells.

CRIB-3GFP distribution profiles were generated from sum projection images of 5 middle-sections (Figure S4C). 3-pixel wide ROIs were drawn from side to side following cell membrane contour. The background signal from cell-free regions were used to correct the data. Whole tip profiles were split in half based on the pixel position of their maximum CRIB-3GFP intensity (approximately in the middle of the profile), generating 60 half tips. To take into account the CRIB-3GFP expression level, the average CRIB-3GFP profiles from WT and *Pact1-cdc42-mCh*^{SW} cells were then normalized by dividing each value by the relative CRIB-3GFP fluorescence intensities values shown in Figure S4B.

3.4.5. Quantifications of CRIB-3GFP and Pak1-sfGFP at the Cell Sides of WT and *rga4*Δ*rga6*Δ Mutants

CRIB-3GFP and Pak1-sfGFP re-localization at the cell sides of WT and *rga4*Δ*rga6*Δ mutants were assessed upon 30 min of photo-activation by recording the average GFP fluorescence intensity over a 3 pixel-wide by 36 pixel-long (≈ 0.25 μm by 3 μm) ROI drawn parallel to the cell side cortex of CRY2-Cdc42$^{\Delta CaaX}$ and WT control cells. Control GFP strains were imaged in the same pad and analysed in the same way as optogenetic cells: CRIB-3GFP and Pak1-sfGFP fluorescence signals from CRY2-Cdc42$^{\Delta CaaX}$ and WT control cells were corrected for the average camera background signals, derived from 6 cell-free regions. The WT control strain, which carries CRIB-3GFP or Pak1-sfGFP like the sample strain of the experiment, but lack the optogenetic system, served as negative control. This controlled that GFP fluorescence changes were not caused by imaging per se. Measurements from these cells were also used as background signal and the average GFP value was subtracted from the the CRY2-Cdc42$^{\Delta CaaX}$ measurements for each experimental replicate. Measurements from cells obtained in 3 experimental replicates are plotted on Figure 5C (CRIB-3GFP WT control dark = 146 cell sides, light = 202 cell sides; CRIB-3GFP CRY2-Cdc42$^{\Delta CaaX}$ dark = 143 cell sides, light = 207 cell sides; CRIB-3GFP *rga4*Δ*rga6*Δ control dark = 145 cell sides, light = 209 cell sides; CRIB-3GFP *rga4*Δ*rga6*Δ CRY2-Cdc42$^{\Delta CaaX}$ dark = 145 cell sides, light = 211 cell sides; Pak1-sfGFP WT control dark = 138 cell sides, light = 230 cell sides; Pak1-sfGFP CRY2-Cdc42$^{\Delta CaaX}$ dark = 147 cell sides, light = 221 cell sides; Pak1-sfGFP *rga4*Δ*rga6*Δ control dark = 143 cell sides, light = 208 cell sides; Pak1-sfGFP *rga4*Δ*rga6*Δ CRY2-Cdc42$^{\Delta CaaX}$ dark = 144 cell sides, light = 222 cell sides).

3.4.6. Cell Size Measurements, Aspect Ratios and Index of Ratios

The aspect ratio of mutant cells was calculated by dividing the cell length by the cell width (Figure 4A). In Figure 4B, the index of ratios was calculated by dividing CRY2 control aspect ratios by CRY2-Cdc42$^{\Delta CaaX}$ and CRY2-Cdc42$^{Q61L,\Delta}$CaaX.

Figure were assembled with Adobe Photoshop CS5 and Adobe Illustrator CS5. All error bars on bar graphs are standard deviations. For statistical analysis, in Figures 1–3, cumulative GFP signal (addition of GFP signal along the 30 time points of the time-lapse) was calculated from single cell traces of CRY2, CRY2-Cdc42$^{\Delta CaaX}$ and GFP control cells.

3.4.7. Cell Images and Kymographs

Cell images shown in Figures 1A, 2A–D and 3A–D, and Supplementary Figures S3 and S7 were obtained from time lapses that were first aligned using the StackReg (http://bigwww.epfl.ch/thevenaz/stackreg/#Explanations) plugin on image J. Aligned time lapses were corrected for the average camera background by subtracting the average signal from 6 cell-free regions measured over the first frame (t0). Finally, time lapses were corrected for photo-bleaching using the Bleach correction plugin (https://www.embl.de/eamnet/html/bleach_correction.html) on image J. GFP maximum-intensity projections of entire time lapses (30 time points) were used to generate GFP max projection images. GFP max projection images were merged with RFP max maximum-intensity projections, in which the the first frame (t0) was omitted in order to segment the plasma membrane. Prior to blue light exposure CRY2-Cdc42$^{\Delta CaaX}$ remains cytosolic as shown in Figure S1D.

Kymographs shown in Figure 1B were generated with the MultipleKymograph (https://www.embl.de/eamnet/html/body_kymograph.html) ImageJ plugin. A single pixel-wide ROI was drawn along the cortex at the cell side (1 pixel = 0.083 μm).

3.4.8. Statistical Analysis

For statistical analysis, single cell cumulative GFP signals of the entire dataset (3 independent experiments combined) were considered, without averaging. For CRIB analysis, we also performed statistical tests on cumulative GFP signal of the second half of the time lapse relative to either the first half or the second half of the CRY2 control, both of which returned statistically significant values (p-values = 0.0008 and 0.0025, respectively). By contrast, a comparison of cumulative GFP signal over the first half (first 45s) with the corresponding time lapse in the control was non-significant (p-value = 0.45). We also did these comparisons on the Scd1-3GFP signal with similar results (2nd half of CRY2-Cdc42$^{\Delta CaaX}$ vs. 1st half p-value = 0.024; 2nd half of CRY2-Cdc42$^{\Delta CaaX}$ vs. 2nd half of CRY2 control p-value = 0.0035; 1st half of CRY2-Cdc42$^{\Delta CaaX}$ vs. 1st half of CRY2 control p-value = 0.12). We chose to report the tests on the cumulative signal for the entire time lapse, as these are more conservative, and thus less prone to over-interpretation. Data normality was assessed by the Lilliesfors test and significance by pairwise Kruskal–Wallis analysis. p values show significance of differences between CRY2-Cdc42$^{\Delta CaaX}$ and CRY2 cells, unless indicated otherwise. A T-test was used in Figures 4 and 5C. On each box, the central red mark indicates the median, while the bottom and the top edges indicate the 25th and 75th percentiles, respectively. The whiskers extend to the most extreme data points not considering outliers, which are plotted individually using the red '+' symbol. All experiments were done at least three independent times. For clarity of the average traces, the plots are shown with standard error of the mean (SEM) in the main figures, which shows the robustness of the average value. The same plots are available in the supplementary figures with standard deviation (SD) to show the biological variability.

Supplementary Materials: The following are available online at http://www.mdpi.com/2073-4409/9/9/2089/s1, Figure S1: Controls for CRY2-Cdc42$^{\Delta CaaX}$ optogenetic recruitment, Figure S2: Control and single-cell traces for optogenetic recruitment in WT cells, Figure S3: Additional examples for CRY2-Cdc42$^{\Delta CaaX}$ induced CRIB-3GFP cell side recruitment, Figure S4: Overexpression of Cdc42-mCh^{SW} in WT cells does not induce ectopic activation of Cdc42, Figure S5: Control and single-cell traces for optogenetic recruitment in *scd1*Δ and *gef1*Δ cells, Figure S6: Control and single-cell traces for optogenetic recruitment of Scd1-3GFP and CRIB-3GFP in WT *and sdc2*Δ cells, Figure S7: Additional examples for CRY2-Cdc42$^{\Delta CaaX}$ induced Scd1-3GFP cell side recruitment, Figure S8: Control and single-cell traces for optogenetic recruitment of CRIB-3GFP and Pak1-sfGFP in rga4Δ*rga6*Δ cells, Table S1: List of strains used in this study.

Author Contributions: I.L. and S.G.M. conceived the project. N.W. performed the experiments presented in Figure 4. I.L. performed all other experiments. I.L. and S.G.M. wrote the manuscript. S.G.M. acquired funding. All authors have read and agreed to the published version of the manuscript.

Funding: This work was funded by ERC Consolidator grant (CellFusion) to SGM.

Acknowledgments: We thank Aleksandar Vjestica (University of Lausanne) for gift of the CRY2-Cdc42$^{\Delta CaaX}$ strain and Serge Pelet (University of Lausanne) for help with MatLab.

Conflicts of Interest: The authors declare no conflict of interest.

References

1. Goryachev, A.B.; Leda, M. Autoactivation of small GTPases by the GEF-effector positive feedback modules. *F1000Research* **2019**, *8*, 1676. [CrossRef] [PubMed]
2. Woods, B.; Lew, D.J. Polarity establishment by Cdc42: Key roles for positive feedback and differential mobility. *Small GTPases* **2019**, *10*, 1301–1337. [CrossRef] [PubMed]
3. Martin, S.G. Spontaneous cell polarization: Feedback control of Cdc42 GTPase breaks cellular symmetry. *Bioessays* **2015**, *37*, 1193–1201. [CrossRef] [PubMed]
4. Bendezu, F.O.; Vincenzetti, V.; Vavylonis, D.; Wyss, R.; Vogel, H.; Martin, S.G. Spontaneous Cdc42 polarization independent of GDI-mediated extraction and actin-based trafficking. *PLoS Biol.* **2015**, *13*, e1002097. [CrossRef]
5. Kelly, F.D.; Nurse, P. Spatial control of Cdc42 activation determines cell width in fission yeast. *Mol. Biol. Cell* **2011**, *22*, 38013–38811. [CrossRef]
6. Lamas, I.; Merlini, L.; Vjestica, A.; Vincenzetti, V.; Martin, S.G. Optogenetics reveals Cdc42 local activation by scaffold-mediated positive feedback and Ras GTPase. *PLoS Biol.* **2020**, *18*, e3000600. [CrossRef]
7. Endo, M.; Shirouzu, M.; Yokoyama, S. The Cdc42 binding and scaffolding activities of the fission yeast adaptor protein Scd2. *J. Biol. Chem.* **2003**, *278*, 8438–8452. [CrossRef]
8. Chang, E.C.; Barr, M.; Wang, Y.; Jung, V.; Xu, H.P.; Wigler, M.H. Cooperative interaction of S. pombe proteins required for mating and morphogenesis. *Cell* **1994**, *79*, 1311–1341. [CrossRef]
9. Tay, Y.D.; Leda, M.; Goryachev, A.B.; Sawin, K.E. Local and global Cdc42 GEFs for fission yeast cell polarity are coordinated by microtubules and the Tea1/Tea4/Pom1 axis. *J. Cell Sci.* **2018**. [CrossRef]
10. Vjestica, A.; Zhang, D.; Liu, J.; Oliferenko, S. Hsp70-Hsp40 chaperone complex functions in controlling polarized growth by repressing Hsf1-driven heat stress-associated transcription. *PLoS Genet.* **2013**, *9*, e100. [CrossRef]
11. Coll, P.M.; Trillo, Y.; Ametzazurra, A.; Perez, P. Gef1p, a New Guanine Nucleotide Exchange Factor for Cdc42p, Regulates Polarity in Schizosaccharomyces pombe. *Mol. Biol. Cell* **2003**, *14*, 313–323. [CrossRef] [PubMed]
12. Hirota, K.; Tanaka, K.; Ohta, K.; Yamamoto, M. Gef1p and Scd1p, the Two GDP-GTP exchange factors for Cdc42p, form a ring structure that shrinks during cytokinesis in Schizosaccharomyces pombe. *Mol. Biol. Cell* **2003**, *14*, 36173–36627. [CrossRef] [PubMed]
13. Gallo Castro, D.; Martin, S.G. Differential GAP requirement for Cdc42-GTP polarization during proliferation and sexual reproduction. *J. Cell Biol.* **2018**, *217*, 42154–42229. [CrossRef] [PubMed]
14. Das, M.; Wiley, D.J.; Medina, S.; Vincent, H.A.; Larrea, M.; Oriolo, A.; Verde, F. Regulation of cell diameter, For3p localization, and cell symmetry by fission yeast Rho-GAP Rga4p. *Mol. Biol. Cell* **2007**, *18*, 2090–2101. [CrossRef] [PubMed]
15. Revilla-Guarinos, M.T.; Martin-Garcia, R.; Villar-Tajadura, M.A.; Estravis, M.; Coll, P.M.; Perez, P. Rga6 is a Fission Yeast Rho GAP Involved in Cdc42 Regulation of Polarized Growth. *Mol. Biol. Cell* **2016**, *27*, 1524–1535. [CrossRef] [PubMed]
16. Nakano, K.; Mutoh, T.; Arai, R.; Mabuchi, I. The small GTPase Rho4 is involved in controlling cell morphology and septation in fission yeast. *Genes Cells* **2003**, *8*, 357–370. [CrossRef] [PubMed]

17. Bugaj, L.J.; Choksi, A.T.; Mesuda, C.K.; Kane, R.S.; Schaffer, D.V. Optogenetic protein clustering and signaling activation in mammalian cells. *Nat. Methods* **2013**, *10*, 249–252. [CrossRef]
18. Nussinov, R.; Tsai, C.J.; Jang, H. Ras assemblies and signaling at the membrane. *Curr. Opin. Struct. Biol.* **2020**, *62*, 1401–1448. [CrossRef]
19. Inouye, K.; Mizutani, S.; Koide, H.; Kaziro, Y. Formation of the Ras dimer is essential for Raf-1 activation. *J. Biol. Chem.* **2000**, *275*, 37373–37740. [CrossRef]
20. Kang, P.J.; Beven, L.; Hariharan, S.; Park, H.O. The Rsr1/Bud1 GTPase interacts with itself and the Cdc42 GTPase during bud-site selection and polarity establishment in budding yeast. *Mol. Biol. Cell* **2010**, *21*, 3007–3016. [CrossRef]
21. Zhang, B.; Gao, Y.; Moon, S.Y.; Zhang, Y.; Zheng, Y. Oligomerization of Rac1 gtpase mediated by the carboxyl-terminal polybasic domain. *J. Biol. Chem.* **2001**, *276*, 89588–89967. [CrossRef] [PubMed]
22. Zhang, B.; Zheng, Y. Negative regulation of Rho family GTPases Cdc42 and Rac2 by homodimer formation. *J. Biol. Chem.* **1998**, *273*, 25728–25733. [CrossRef]
23. Remorino, A.; De Beco, S.; Cayrac, F.; Di Federico, F.; Cornilleau, G.; Gautreau, A.; Parrini, M.C.; Masson, J.B.; Dahan, M.; Coppey, M. Gradients of Rac1 Nanoclusters Support Spatial Patterns of Rac1 Signaling. *Cell Rep.* **2017**, *21*, 19221–19935. [CrossRef]
24. Maxwell, K.N.; Zhou, Y.; Hancock, J.F. Rac1 Nanoscale Organization on the Plasma Membrane Is Driven by Lipid Binding Specificity Encoded in the Membrane Anchor. *Mol. Cell Biol.* **2018**, *38*. [CrossRef] [PubMed]
25. Sartorel, E.; Unlu, C.; Jose, M.; Massoni-Laporte, A.; Meca, J.; Sibarita, J.B.; McCusker, D. Phosphatidylserine and GTPase activation control Cdc42 nanoclustering to counter dissipative diffusion. *Mol. Biol. Cell* **2018**, *29*, 1299–1310. [CrossRef]
26. Slaughter, B.D.; Unruh, J.R.; Das, A.; Smith, S.E.; Rubinstein, B.; Li, R. Non-uniform membrane diffusion enables steady-state cell polarization via vesicular trafficking. *Nat. Commun.* **2013**, *4*. [CrossRef] [PubMed]
27. Meca, J.; Massoni-Laporte, A.; Martinez, D.; Sartorel, E.; Loquet, A.; Habenstein, B.; McCusker, D. Avidity-driven polarity establishment via multivalent lipid-GTPase module interactions. *EMBO J.* **2019**, *38*. [CrossRef]
28. Kennedy, M.J.; Hughes, R.M.; Peteya, L.A.; Schwartz, J.W.; Ehlers, M.D.; Tucker, C.L. Rapid blue-light-mediated induction of protein interactions in living cells. *Nat. Methods* **2010**, *7*, 9739–9775. [CrossRef]
29. Wheatley, E.; Rittinger, K. Interactions between Cdc42 and the scaffold protein Scd2: Requirement of SH3 domains for GTPase binding. *Biochem. J.* **2005**, *388*, 1771–1784. [CrossRef]
30. Tatebe, H.; Nakano, K.; Maximo, R.; Shiozaki, K. Pom1 DYRK regulates localization of the Rga4 GAP to ensure bipolar activation of Cdc42 in fission yeast. *Curr. Biol.* **2008**, *18*, 3223–3230. [CrossRef]
31. Merlini, L.; Khalili, B.; Dudin, O.; Michon, L.; Vincenzetti, V.; Martin, S.G. Inhibition of Ras activity coordinates cell fusion with cell-cell contact during yeast mating. *J. Cell Biol.* **2018**, *217*, 1467–1483. [CrossRef]
32. Merlini, L.; Khalili, B.; Bendezu, F.O.; Hurwitz, D.; Vincenzetti, V.; Vavylonis, D.; Martin, S.G. Local Pheromone Release from Dynamic Polarity Sites Underlies Cell-Cell Pairing during Yeast Mating. *Curr. Biol.* **2016**, *26*, 1117–1125. [CrossRef] [PubMed]
33. Miller, K.E.; Magliozzi, J.O.; Picard, N.A.; Moseley, J. Sequestration of the exocytic SNARE Psy1 into multiprotein nodes reinforces polarized morphogenesis in fission yeast. *BioRxiv* **2020**. [CrossRef]
34. Dudin, O.; Bendezu, F.O.; Groux, R.; Laroche, T.; Seitz, A.; Martin, S.G. A formin-nucleated actin aster concentrates cell wall hydrolases for cell fusion in fission yeast. *J. Cell Biol.* **2015**, *208*, 897–911. [CrossRef] [PubMed]
35. Bähler, J.; Wu, J.Q.; Longtine, M.S.; Shah, N.G.; McKenzie, A., III; Steever, A.B.; Wach, A.; Philippsen, P.; Pringle, J.R. Heterologous modules for efficient and versatile PCR-based gene targeting in Schizosaccharomyces pombe. *Yeast* **1998**, *14*, 9439–9451. [CrossRef]
36. Vjestica, A.; Marek, M.; Nkosi, P.J.; Merlini, L.; Liu, G.; Berard, M.; Billault-Chaumartin, I.; Martin, S.G. A toolbox of Stable Integration Vectors (SIV) in the fission yeast Schizosaccharomyces pombe. *J. Cell Sci.* **2019**. [CrossRef]
37. Bendezu, F.O.; Martin, S.G. Actin cables and the exocyst form two independent morphogenesis pathways in the fission yeast. *Mol. Biol. Cell* **2011**, *22*, 44–53. [CrossRef]

MDPI

Review

The Path towards Predicting Evolution as Illustrated in Yeast Cell Polarity

Werner Karl-Gustav Daalman, Els Sweep and Liedewij Laan *

Department of Bionanoscience, TU Delft, 2629 HZ Delft, The Netherlands; W.K.Daalman@tudelft.nl (W.K.-G.D.); E.Sweep@tudelft.nl (E.S.)

* Correspondence: L.Laan@tudelft.nl; Tel.: +31-(0)15-27-82856

Received: 30 September 2020; Accepted: 21 November 2020; Published: 24 November 2020

Abstract: A bottom-up route towards predicting evolution relies on a deep understanding of the complex network that proteins form inside cells. In a rapidly expanding panorama of experimental possibilities, the most difficult question is how to conceptually approach the disentangling of such complex networks. These can exhibit varying degrees of hierarchy and modularity, which obfuscate certain protein functions that may prove pivotal for adaptation. Using the well-established polarity network in budding yeast as a case study, we first organize current literature to highlight protein entrenchments inside polarity. Following three examples, we see how alternating between experimental novelties and subsequent emerging design strategies can construct a layered understanding, potent enough to reveal evolutionary targets. We show that if you want to understand a cell's evolutionary capacity, such as possible future evolutionary paths, seemingly unimportant proteins need to be mapped and studied. Finally, we generalize this research structure to be applicable to other systems of interest.

Keywords: network evolution; *Saccharomyces cerevisiae*; polarity; modularity; neutrality; symmetry breaking

---

1. Introduction

How cells work and how they evolve is at the heart of cell biology. In this work we will review how cellular architecture ("how cells work") and its evolutionary properties ("how they evolve") are related to each other. Understanding evolution and possible mutational paths of protein networks, and especially the cell polarity network, is not only satisfying our curiosity but may also help us understand and possibly predict cancer progression [1].

Every cell consists of many different interconnected functional protein networks (for definitions, see Table 1), such as transcription, translation, or polarity establishment [2]. The network's architecture, (for example: which protein binds to/reacts with which other protein), impacts the evolutionary possibilities of a network in multiple ways. For example, hubs, proteins with many binding partners, tend to evolve slower [3]. Less connected proteins, that may be deleted in a cell without a detectable change in cell physiology, can permit duplication of other genes and thus promote evolution [4]: duplicates of a gene enable new options for diversification, which facilitate further evolution of a gene/protein and the surrounding network [5–7]. Interestingly, many mutations (from 3% of nonsilent mutations in bacteria to 30% in hominids [8]) in a cell show very weak, or no effect on the cell's function, a phenomenon called neutrality [9]. Thus, proteins that may seem unimportant for how the cell works now, in this environment, may become important when changes occur in the network architecture due to a mutation or a switch in environment [10,11].

Table 1. Definitions of important terms used throughout the introduction.

Term	Definition
Protein network	Group of proteins with physical interactions together performing a function
Connectivity	Degree to which parts of the network are embedded with other parts in the network. In this sense, it can be received as the reciprocal of modularity.
Modularity	Potential to group parts of a protein network given a certain representation of the protein network (e.g., in terms of mechanisms, genetic or physical interactions)
Hub protein	Highly connected protein in a network (often essential)
Neutrality	No consequence of a mutation to phenotype (in current environment)
Hierarchy	Clear layering of pathways inside a protein network
Redundancy	Multiple mechanisms that can to some extent interchangeably contribute to the same function

A well-studied model organism to concretize how these proteins, without a detectable phenotype, shape a network is *Saccharomyces cerevisiae*, or budding yeast. The organism generally exhibits many of the network properties defined in Table 1, such as hierarchy and the presence of hubs [2]. Here, neutrality is also pervasive, as only 40% of homozygous gene deletions for the entire organism initially had obvious phenotypes [12]. Moreover, the environment has been shown to have a notable influence on neutrality, as lethal heterozygous deletions can be compensated by poor medium [13]. As a general rule, both the network architecture and the environment can mask the function of many proteins.

Within this organism, a, to some extent, representative example of a protein network is the polarity network, which governs how the yeast chooses a direction in which to divide and involves directing dozens of proteins in a process of breaking its internal spherical symmetry (see, e.g., [14]). As required, we observe the presence of the common network properties demonstrated in Section 2, such as hierarchy and redundancy. Polarity is also one of many biological functions in yeast for which a subdivision of many proteins into a quasi-modular network proved possible [15]. Within polarity, even more detailed submodules can be distinguished [16]. Neutrality is also exhibited by several polarity proteins discussed in Section 3, in part responsible for the difficulty in determining the role of each protein. Lastly, polarity is a pattern formation process where, by definition, spatiotemporal dependencies are important, and understanding evolution generally relies on understanding this type of dependencies [17].

However, the polarity network is not a prime example of the sort of networks with abundant transcriptional regulation. Other templates are better suited for learning about the evolution of gene regulation, such as interaction networks centered around transcriptional regulators (e.g., Mcm1 [18]), ribosomal regulation (e.g., [19]) the stress response (e.g., [20]) or metabolic response (e.g., *GAL* pathway [21]). The existence of established regulatory templates thus conveniently complements our focus on symmetry breaking during polarity as a model for the protein interaction network, which is also a topic for evolutionary studies.

Concretely, in [22] a mutant strongly defective in polarity establishment was experimentally evolved and found to recover remarkably reproducibly, e.g., the first rescuing mutation to sweep the population was always the same. Because of this exhibited tractability of the adaptations, network structures within the polarity network that facilitated evolution could be concretely interpreted in terms of redundancies [23]. In another approach to determine the flexibility of the polarity network, historical evolution was studied for 40+ proteins in almost 300 fungal species in [11]. Again, the polarity network exhibited sufficient modularity so that studying its evolution separately from other functions still yielded interpretable results. For example, authors showed that polarity network size was shaped in part by fungal lifestyle (e.g., uni- or multicellular).

Nevertheless, clear justification for observed evolutionary trajectories in this network remain difficult to make. It is sometimes possible to generate more abstract predictions by linking network architecture to the evolvability of the polarity network using classical regulatory motifs. For example, the presence of positive and negative feedback in polarity establishment confers robustness to the network [24,25], which in turn may facilitate evolution. But in order to make more concrete and detailed predictions about evolutionary trajectories, an important insight is that we need mechanistic information of (parts of) the polarity network.

To illustrate this point, a bottom-up model was constructed in [26] where molecular details were coarse-grained following analysis on numerical simulations of multiple polarity mechanisms [23]. This approach was successful in quantitatively describing the fitness along the evolutionary trajectory in [22]. Furthermore, the predictions on epistasis, an important bottleneck for predicting evolution [27], can be extended to other modules, and although use of full mechanistic understanding is superior in quantitative assessments of epistasis, biofunctional information (viz. from GO-terms, in agreement with [28]) as input to the model suffices for epistasis sign predictions.

Instructive is the Nrp1 case where full information is absent, but some phenomenological information is available. Based on the latter, inclusion of Nrp1 into the bottom-up coarse-grained model of [26] is still worthwhile, but leaves room for improvement. This marks the importance of continuing to investigate protein networks until molecular mechanisms have been elucidated. In this review we advocate that obtaining this knowledge is (soon) feasible, motivating the use of the yeast polarity system for studies in network organization (with properties such as hierarchy, modularity/connectivity and redundancy that can "hide" proteins) as well as evolution.

We present three case studies in our review to reflect different stages in this knowledge quest; while literature on the first case, bud scar proteins, has been quite advanced for several years, only recently the GAP mechanism [23] (second case) has been revealed and for Nrp1 (third case) more work remains. These cases illustrate how to move from detecting the strong and more obvious phenotypes to unveiling evolutionary important hidden features (such as for Nrp1). The cases build on literature of the polarity network summarized in the form of a Venn diagram, ideal for depicting a hierarchical and semimodular protein grouping. The feasibility of deciphering a protein's role in this Venn diagram turns out to depend on how deep the protein is embedded inside the network, which can impose neutrality on mutations to hinder unique attribution of genotypes to phenotypes. In the sections thereafter, our three examples with varying depths of embedding serve to show how improved understanding of the nontrivial parts of the networks can elucidate evolutionary trajectories. Ultimately, we believe that in all the work done in yeast for many decades by many researchers there is a common recipe applicable and useful for many protein networks, as expanded upon in the outlook.

2. Polarity Overview

Within the yeast polarity network, four pathways to polarization exist which cannot easily be considered modular. Their interconnectivity can be conveniently visualized in the form of a Venn diagram (Figure 1, with references found in Appendix A Table A1). These pathways are hierarchically set up, in the following order: the mating pathway at the top, then the bud scar pathway, followed by the reaction–diffusion pathway and finally the actin pathway. In short, their function boils down to the act of condensing the GTPase Cdc42 bound to GTP molecules (i.e., active Cdc42) to one point on the plasma membrane, which can signal downstream effectors to proceed the cell cycle [29]. To prevent premature or overdue localization of active Cdc42 and allow some influence on the hierarchy of pathways, a fifth pathway exists to control the previous four, namely the timing pathway. The next section summarizes the most important interactions in and across all pathways, starting with the timing cue, before expanding upon the three examples. As a site note, we have done our best to include all relevant papers, but apologize for important papers we have missed.

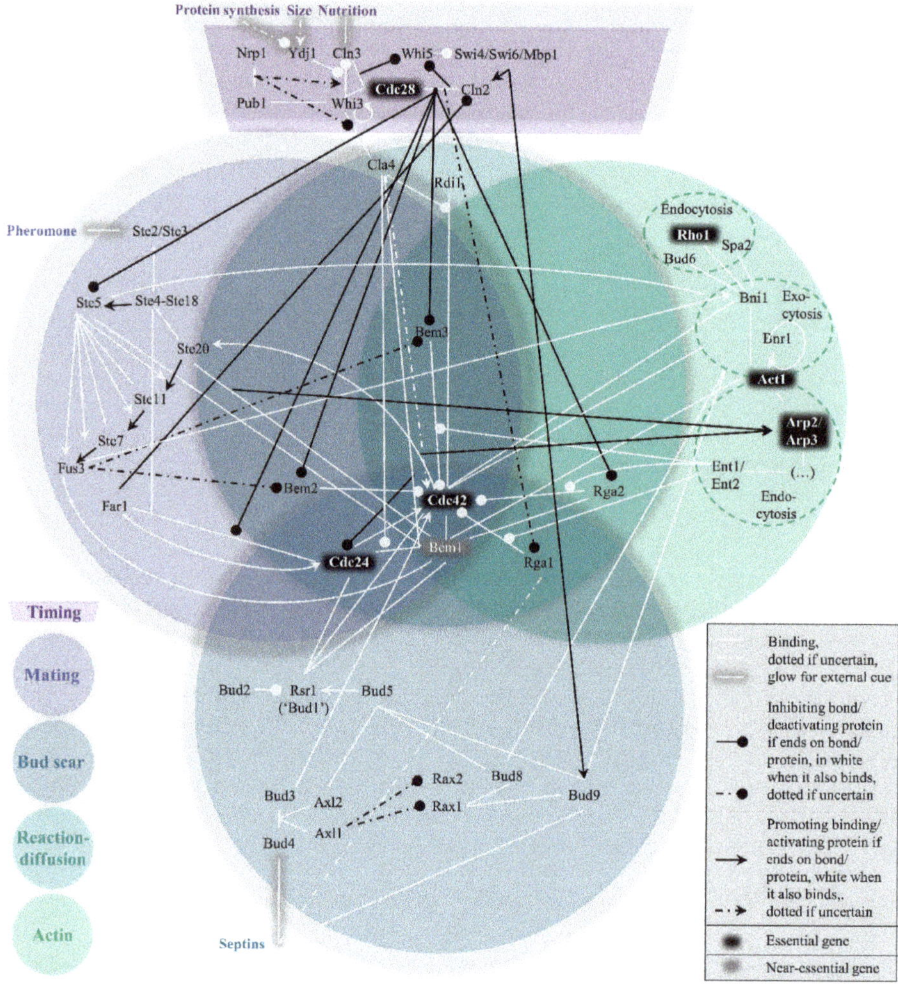

Figure 1. Venn diagram representation of the hierarchical, nonmodular pathway structure for protein–protein interactions in budding yeast polarity establishment (Table A1 for references).

2.1. Timing: The Control Knob

During isotropic growth in G1, active Cdc42 localization is suppressed by overactivity of its associated GTPase activating proteins (GAPs) and sequestration of its guanine nucleotide exchange factor (GEF). The consequence of both circumstances is the vast abundance of inactive Cdc42, which is bound to GDP instead of GTP [30], rendering it impossible to signal the polarity cue. The purpose of this pathway (see top dark purple region in Figure 1) is hence to timely reduce GAP activity and release the GEF, which must be in response to important physiological parameters that indicate the readiness of the cell: sufficient protein production, a sufficient size, and sufficient nutrition.

The physiological state of the cell enters the equation through nuclear levels of cyclin Cln3. Upon sufficient nutrition and size, Cln3 levels rise either more directly through higher Cln3 mRNA abundance [31], or more indirectly through Ydj1 disturbing Cln3 localization by Whi3 [32], the latter also being an inhibitor of Cln3 mRNA translation [33]. The arrival of nuclear Cln3 allows binding partner and cyclin-dependent kinase Cdc28 [34] to phosphorylate Whi5, which had inhibited expression of Cln2, another cyclin [35,36]. Cln2 can then reinforce its own expression, consolidating the original Cln3 signal [37].

Now, the Cdc28-Cln2 complex can distribute the physiological signal to the aforementioned targets, the GAPs and the GEF. The kinase Cdc28 phosphorylates all four GAPs Bem2, Bem3, Rga1 and Rga2 [38–41] and Far1, which was keeping the GEF Cdc24 in the nucleus [42]. Now cytoplasmic levels of active Cdc42 can rise, leading to polarity establishment through subsequent pathways.

Importantly, the completion of the timing pathway causes the hierarchy of the subsequent pathways to change. While the mating pathway is otherwise dominant, the kinase Cdc28 phosphorylates Ste5, a crucial hub in the mating pathway, to stop the mating in its tracks [43–45]. In the following discussion of the mating pathway, the situation is considered where the timing pathway did not overwrite its behavior.

2.2. Mating: Heavily Cross-Linked

The mating pathway is the dominant force across the four symmetry-breaking pathways. While polarization in a random orientation is possible after the timing cue (see the section on reaction–diffusion further on), the presence of pheromones of the opposite mating type (a or α) should redirect the Cdc42 localization to the side of the pheromone signal. This process revolves around Ste5, as also depicted in the left, blue-grey circle of Figure 1.

Briefly put, once pheromones bind membrane proteins Ste2 and Ste3 [46,47], Ste4 is released from the membrane [48,49] and binds Ste20 and scaffold Ste5. This scaffold binds Ste7, Ste11 and Fus3, which are activated by sequential phosphorylation [50–52]. Fus3 may inhibit the GAPs Bem2 and Bem3 [39], while Ste5 binds the GEF Cdc24 [53], replacing the absence of the timing pathway result to stimulate activity of Cdc42.

While this simplified view would suffice to redirect the Cdc42 localization, the mating pathway is much more intertwined with the other pathways than seemingly necessary, particularly with the actin pathway. The abundant mechanistic redundancies result in a more complex picture, obfuscating the role of the proteins involved. For example, active Cdc42 stimulates Ste11 phosphorylation/activation [54]. Another form of positive feedback, as well as a bridge to the actin pathway, is the Cdc42 recruitment of formin Bni1 [55]. The resulting nucleation of actin cables may transport Ste5-GEF Cdc24 complexes [56], possibly also through Bem1. This scaffold co-immunoprecipitates with Act1 [57] and Far1 [58], which is bound to the GEF Cdc24 [59], but is itself in turn also bound to Bem1 [60]. Another actin cross-link is the phosphorylation and localization of Bni1 through Fus3 [61]. Clearly, care must be taken in assigning roles to different proteins, as many are overloaded.

2.3. Bud Scar: Mostly Modular and Ordered

In the absence of a mating cue, the timing pathway reduces GAP activity and releases the GEF, while the mating pathway is repressed. Under the new hierarchy, the bud scar pathway is normally dominant. The scar refers to leftover proteins from the previous division, named septins [62,63]. This spatial cue can be exploited for polarity establishment; a new bud forms adjacent to the scar (axial budding, haploids) or also at the opposing side (bipolar budding, diploids) [64,65]. The bottom, dark blue circle of Figure 1 represents this path from septins to Cdc42 recruitment graphically. More background information about the core bud scar protein group Bud1 to Bud5 is discussed separately in one of the three case studies, and only a brief overview of the pathway as a whole is discussed in this section.

An important bud scar localization target is Bud5, which activates and recruits Bud1 [66,67], a protein also known as Rsr1 [65,68]. In haploids, where Axl1 is specifically expressed [69], the Bud5 localization occurs by relay of the septin signal to a protein complex of Bud3, Bud4, Axl1, Axl2 and Bud5 [70–73]. In diploids, functionality of Rax1 and Rax2 is not impaired, presumably by blocked expression of Axl1 [74], so these can localize Bud9 and Bud8 to the bud scar or the opposite end of the scar respectively [75], which in turn recruit Bud5 [76].

After Bud5, localization follows of Bud2 [77], the GAP for Bud1 [78], to complete the control of the GTPase cycle of Bud1. Finally, Bud1 binds GEF Cdc24 and Bem1 [79] (although Cdc24 has the strongest affinity with Bem1 [80]), to redirect the pattern formation made possible after the timing cue. As linkage of Cdc42 GAP Rga1 to septins prevents reuse of the previous location [81], the new bud forms adjacent to the bud scar.

As a whole, the bud scar pathway is not completely modular either. Aside from nudging the reaction–diffusion pathway (see next section), an example of a cross-link is that Bud8 and Bud9 are delivered by actin transport [82]. The highest position in the hierarchy in the absence of the mating cue is also not absolute; multiple ways to promote the subsequent reaction–diffusion pathway exist, such as deletion of Bud1 and Bud8 [83], Axl2 and Rax1 [84], or Bem1 [22]. Therefore, it has been possible to retrieve the information discussed in the following section.

2.4. Reaction–Diffusion: Ample Redundancy

Even in the absence of chemical or spatial cues, the shift in balance towards activation of Cdc42 induced by the timing pathway still provides the conditions for swift symmetry breaking. Theoretical models concerning this pathway have been subject to updates as more molecular details have been revealed. Central is the strong positive feedback generated by the Bem1-GEF Cdc24 complex, as modelled in, e.g., [85,86] with further refinement in [87]. This feedback is sufficient for polarity success, which becomes rather insensitive to GAP abundance. More details on the GAPs were uncovered in [23] and are placed in a broader context in the case study further on.

What makes this pathway special is the limited number of proteins that are unique to this pathway, as seen from the central, emerald circle in Figure 1, namely only Cla4 and Rdi1. The latter is the least cross-linked of the two, providing a possible justification for referring to the WT mechanism as the Rdi1 polarity mechanism, as in [24]. Cla4 is more context-dependent, possibly having two opposing roles, promoting and inhibiting polarity [23,88,89]. Yet both Rdi1 and Cla4 are dispensable for polarity [88,90].

An even stronger addition to the redundancy within this pathway is on the positive feedback side. Without Bem1, generic rescuing feedbacks suffice [23], among which Cla4 could account for 20% of their function [91]. More feedbacks may be found in the GAPs (see GAP case study) through actin transport as described in, e.g., [24]. This brings us to the actin pathway as the final layer to discuss.

2.5. Actin: The Mysterious Auxiliary Layer

The actin pathway (rightmost green circle in Figure 1) has featured several times already in the previously discussed pathways, but its individual role is still quite uncertain. Yeast formin Bni1 which nucleates actin cables, binds active Cdc42 [55], and is known to be involved in exo- and endocytosis [92,93]. This suggests transport of polarity proteins from and to the presumptive bud site. The resulting actin pathway has been confusingly implicated in two opposing roles; promoting Cdc42 polarization, see, e.g., [24,94,95], as well as negatively impacting Cdc42 polarization [96,97].

A way to reconcile these findings is that actin transport contributes to a process promoting Cdc42 polarization but without relying on significant transport of Cdc42 itself. As mentioned in the mating pathway, Bem1 and Act1 co-immunoprecipitate [57], suggesting that Bem1, and concordantly its multiple binding partners, might get transported through the actin pathway. However, in the absence of Bem1, 80% of the positive feedback is still unidentified [91], which may very well be actin-related. Instead, a prime candidate is the GAP group, which is known to bind the epsin-coating of actin cables involved in endocytosis [98]. This is further discussed in the case study on the GAPs. In any case, it is quite difficult to decipher the actin pathway, in large part due to its low positioning in the yeast polarity hierarchy.

3. Case Studies

In the introduction the need for determining the (potential) functions of a protein inside a complex network even when these are normally hidden, was explained as, evolution may exploit these later. In the following sections, the reconstitution of the mechanistic details involving three protein(s) (classes), that are currently at different stages in their discovery (see graphical mechanistic summary in Figure 2), is illustrated; the Bud1-5 bud scar protein subset, the GAP proteins for Cdc42, and finally Nrp1.

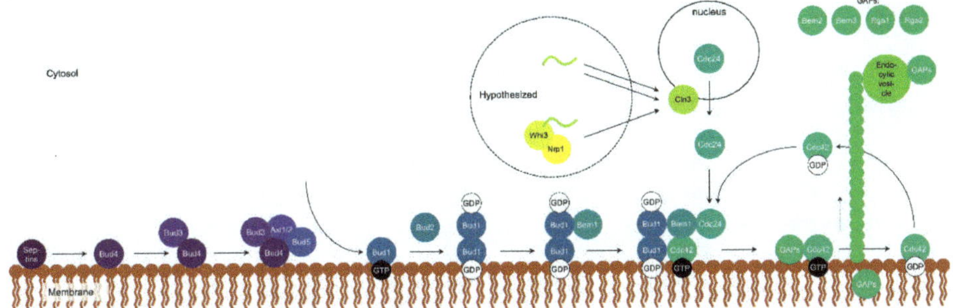

Figure 2. Mechanistic model for the haploid bud scar pathway, the reaction–diffusion pathway including the GAPs, and the hypothesized role of Nrp1 in the timing pathway.

In the first case, detailed knowledge has been established the longest, and has also been put in an evolutionary perspective as forming a "weak regulatory link" that facilitates phenotypic variation [99,100]. Therefore, relatively small gains are expected from the remaining open questions. For the GAP proteins, detailed knowledge is only very recently obtained. Together, the purpose of putting the literature of these cases into historical context is that these two examples form convenient templates to delineate the route towards complete understanding for proteins as in the final case, Nrp1, where knowledge is scarce. On the one hand, technical advances that improve detection of phenotypes were the most prominent method of progress for the bud scar proteins. On the other hand, the GAP protein and the Nrp1 case show that progress can rely more on very specific designs of experiments than on new technology. All three cases combined in turn illustrate the foundation under the general strategy to approach networks put forward in the outlook.

3.1. Bud Scar Proteins

One of the strongest phenotypes to observe for budding yeast is the location of the next bud. Already more than half a century ago, it was documented that *S. cerevisiae* exhibited two possible budding patterns [64]. Normally for haploids, the next bud grows next to the previous division site (axial budding), whereas diploids can also pick the location opposite the previous site (bipolar budding). Changes in this pattern are clear phenotypes, and are therefore a useful detection tool in bulk mutagenesis screens. In 1991, five genes were identified that affected (seemingly as their sole phenotype) the bud site selection, and were therefore named *BUD1* (formerly known as *RSR1* [68]), *BUD2*, *BUD3*, *BUD4* and *BUD5* [65,66]. These genes and their associated proteins are therefore straightforwardly localized in the bud scar circle in the Venn diagram (Figure 1).

The next step, retrieving information on physical interactions, is more elaborate. For example, authors in [66] reasoned the physical mechanism of Bud5 (amongst others a GEF for Bud1) based on sequence information, which was further substantiated in [78], together with the GAP action of Bud2. Bud2 and Bud5 hence underlie the relevant GTPase cycling of Bud1, as in vitro studies established that Bud1 when GTP-bound, mostly binds Cdc24, while binding and recruiting Bem1 when GDP-bound [79,101]. This provides the bridge to the previously described reaction–diffusion pathway. With the advent of GFP [102], it also became possible to get in vivo spatiotemporal information from yeast proteins, later even in bulk [103]. Now armed with the tool of fluorescence microscopy, substantial progress was made for validating the roles for Bud2 and Bud5 [77].

The final confirmation on the mechanisms comes when zooming into the domains of a protein, by expressing truncated versions or intelligently chosen point mutations. For example, the domains responsible for how Bud5 affects whether budding is axial or bipolar are described in [104]. For other proteins, however, it was necessary to consider in greater detail the protein network that the bud genes compose. Therefore, it has taken more time to reverse-engineer the details for Bud3 and Bud4. Authors in [72] combine their findings with earlier literature to demonstrate that septins from the previous division localize Bud4, which can then bind Bud3. Subsequently, Bud4 binds Axl2, and after this Axl1 as well. Domain analysis of Bud3 further uncovered more details [105] to show Bud3 can serve as a GEF for Cdc42 (but early in G1, before Start), which is important for the Bud4–Axl1 link.

This leaves only a few mechanistic unknowns. For example, the exact recruitment of Bud2 as depicted in Figure 2 is putative, as it is unclear how Bud2 is recruited even in the absence of Bud1 or Bud5 [74]. Bud1 can recruit itself by dimerization and Cdc42 at the bud site of the membrane [67]. In the same paper, it is shown that while Bud5 also recruits Bud1, the former may only indirectly be involved in the Bud1 dimerization, considering the full GTP-GDP cycle is critical for appropriate dimerization. This suggests guidance of Bud2 for the self-recruitment, but the exact order for Bud1-binding and hydrolysis of its GTP is open for interpretation, given the alternative model in [74].

As an alternative representation of the current state of research of the bud scar proteins, Figure 3 depicts an ordering in the form of a mind map. For clarity, only Bud1 is considered. As can be seen, the bud scar protein case is relatively well advanced, with ample coverage of all categories.

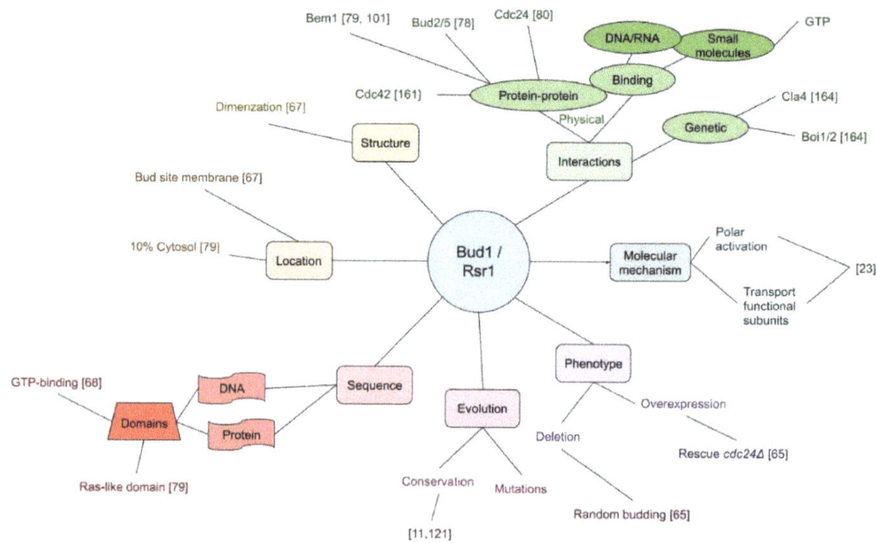

Figure 3. Mind map representation of the bud scar protein Bud1 summarizing its most important properties from literature, such as localization, domains and interactions.

3.2. GAPs

Central to the function of GTPase Cdc42 are proteins that promote GTP hydrolysis, i.e., its GTPase activating proteins. Bem3, was the first to be identified [106], followed by Rga1 and Rga2 [107–109]. A summary of Bem3 studies/information is given in Figure 4. The specific molecular function of GAPs allowed validation in vitro, where Cdc42-GTP was incubated with a GAP to determine whether the amount of GTP indeed decayed faster than without a GAP. In [106] Bem2 was not found to exhibit detectable GAP activity, and not until [110] could Bem2 be convincingly considered a GAP as well.

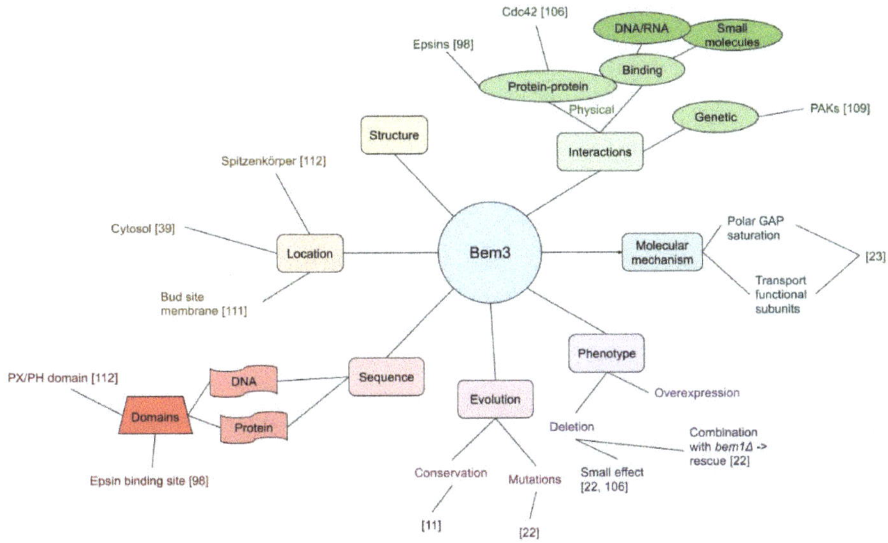

Figure 4. Mind map representation of the GTPase activating protein (GAP) Bem3 summarizing its most important properties from literature, such as localization, domains and interactions.

After their establishment as a GAP, the localization of Bem3, Rga1 and Rga2 was determined using either antibody staining or GFP-tagging [111]. These GAPs colocalized with Cdc42 at the bud site (although Rga1 is slightly more dispersed). Bem3 was also found in Spitzenkörper-like structures, but seemingly after polarity establishment during polarized growth [112]. The GAP localizations affirmed their role in polarity, although they also seemed somewhat redundant, as a triple mutant was still viable [109].

More information on GAPs was gathered through their interactions. Bem2 and Bem3 were found to be associated with the mating pathway [39], Bem3, Rga1 and Rga2 related to the actin pathway [98,113], while [81] suggested a link between Rga1 and the bud scar pathway. However, several open questions remain. For example, why is there more than one GAP? Why and how are these distributed across multiple pathways?

More clues were to follow from the tedious deciphering of the mechanistic action of the GAPs. The difficulty behind this problem was clearly elucidated by the model of [87]. There, authors showed the dominance of the Bem1-mediated positive feedback minimized the phenotypical influence of varying GAP concentration. Consequently, this led to the realization that GAP details only emerged in a Δ*bem1* background.

In [23] a proposed mechanistic model for the GAPs was validated against this background. The idea was that GAPs were temporarily retained on the membrane by Cdc42 during the GTP hydrolysis process. Only in the location in the cell with high membrane concentrations of Cdc42-GTP would this lead to a local depletion of available GAPs, which cannot be compensated by the cytosolic diffusive flux. Elsewhere on the membrane, this was not a problem and Cdc42 was promptly inactivated and recycled, leading to only one spot on the plasma membrane where active Cdc42 accumulated. A generic positive feedback mechanism for Cdc42 (due to, e.g., Cla4) completed the symmetry breaking. This provides a good example of the added value of revealing nonobvious mechanistic details. This GAP model, which is obscured in the presence of Bem1, allowed the authors in [23] to provide a detailed explanation of the evolutionary trajectory observed in [22].

Yet, this may not be the complete role of the GAPs. Their mechanism partly supplements the need for the Bem1-mediated positive feedback, but a further deletion of *CLA4* and *BEM3* reveals that 80% of the positive feedback originates elsewhere [91] (p. 101). Here the interconnectivity with another pathway can surface, which initially seemed elusive and redundant.

As discussed in the actin pathway, there could be a critical link with actin and the GAPs. The aforementioned local depletion of unbound GAPs may be reinforced by actin transport [91] (p. 34). Epsin coatings of endocytic vesicles colocalize with polarized growth and bind GAPs [98,113]. Moreover, active Cdc42 releases the auto-inhibition of kinases Ste20 [54] and Cla4 [114], both of which phosphorylate myosins 3 and 5 to ultimately lead to activation of the Arp2/3 complex [115], critical for endocytosis. In this way, sites of active Cdc42 can promote the endocytosis which may reinforce the stability of the site, providing the feedback needed to establish polarity (model F in [116]). The combination of interactions of recyclable GAPs and active Cdc42 would also fulfil the requirement of actin-mediated recruitment formulated in [117], provided the GAPs diffuse slowly on the membrane [96].

Figure 4 graphically summarizes the available information for one of the GAPs, Bem3. However, there is still room for improvement with experiments whose design can just now be established. As with the bud scar pathway, subtle information may be retrieved through experiments at the domain level, to test, e.g., the GAP trafficking hypothesis. One could remove the link between GAPs and actin through deletion of epsins ENT1 and ENT2 (in the Δ*bem1* Δ*cla4* background) and replacing this by only a weakly expressed ENTH-domain of Ent1 (truncation). Modulating this expression should show how strong this effect is. More information on GAP interactors in this role may also be retrieved by using this mutant as the crippled starting point in an evolution experiment, akin to [22].

Furthermore, the resulting scatter of the GAPs across the other polarity pathways currently leaves room for interpretation and speculation. Given Figure 1, the components that are most shared also seem the most critical. This is most obvious when noting that actin pathway components are not just essential for polarity establishment. For example, Rho1 is needed for cell wall integrity synthesis later on during polarized growth [118–120]. Extrapolating, the location (wedged between pathways) of the GAPs in the Venn diagram may suggest an important (but not essential) function for each of them in establishing polarity.

Hypothetically, the GAPs might serve as an evolutionary control knob to mediate the relative hierarchy between pathways. This could be favorable in situations where different hierarchies are optimal, such as when mating is infrequent (e.g., the diploid state becomes the default), or when the bud scar is not often used (frequent sporulation). Strategic dispersal of multiple GAPs may therefore provide more handles for the cell to optimize the pathways then simply having one GAP in larger copy numbers.

3.3. Nrp1

After discussing two well-studied protein classes, we address an underexposed protein, namely Nrp1. With this we would like to show the difficulties and possibilities that are still open in a case when the most straightforward experiments do not provide obvious, interpretable phenotypes. Although its deletion does not have a detectable phenotype in standard lab conditions, Nrp1 is an important evolutionary [22,121]. Usually, Nrp1 is mentioned merely peripherally in articles as a bycatch in studies with an alternative focus. Therefore, a chronologically ordered literature overview does not make sense here, as very little of the research findings actually builds on previous work. An overview of the Nrp1 knowledge is given in Figure 5, where it is apparent that there are some gaps in our understanding.

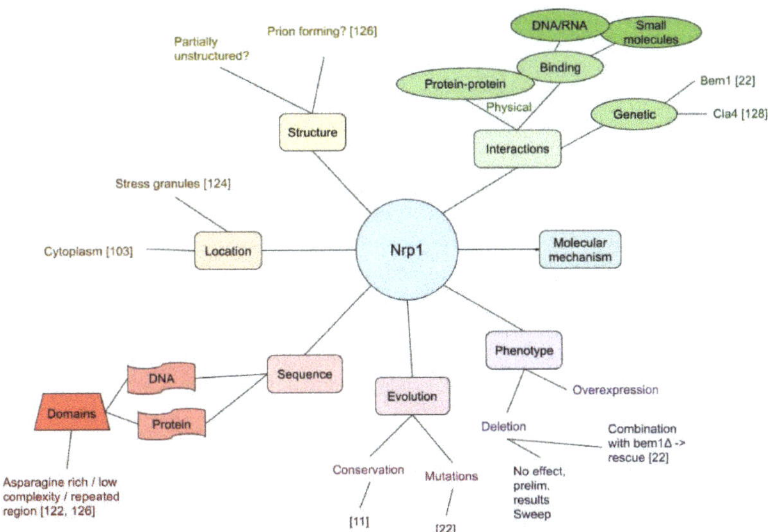

Figure 5. Mind map representation of Nrp1 summarizing its most important properties from literature, such as localization, domains and interactions.

Nrp1 was first described by [122], who have given the protein its name. NRP1 stands for 'Asparagine rich protein', the name refers to the region of the protein sequence that has many asparagines (short name: "N"). Genes are often named for their defining characteristics or functions. Reynaud and coworkers [122] did not find a phenotype or function for Nrp1, thus the seemingly nondescript name.

Nrp1 has been linked to stress response and stress granules. For example, it has been implicated in the response to glucose and oxygen [123–125], although the precise mechanism or function of Nrp1 in this response is not known. A hypothesis is that Nrp1 forms an aggregate or prion (like a stress granule) by its low complexity domains, because the repeated asparagine sequence in Nrp1 is often found to form prions for other proteins [126]. Nrp1 itself also seems to form a prion. In addition, Nrp1 can potentially bind and regulate mRNA. The mRNA regulation occurs through another documented domain, namely an RNA binding motif [126]. This implies that Nrp1 can bind specific mRNA sequences, however no specific mRNAs have been identified until now [127].

The supposed link between Nrp1 and the polarity network was found by authors in [22]. They showed that null mutations in *NRP1* could rescue a *bem1*Δ. This prompted interest in a search for the function of Nrp1. Another connection to the polarity network and a possible explanation for what was found in [22] is the synthetic lethality of *NRP1* with *CLA4* [128].

Diepeveen et al. [11] have found that Nrp1 is highly conserved within the Ascomycota which hints to a function of some importance. One typically expects that essential genes are the most conserved parts as they cannot easily be mutated [3]. However this is not always the case, for example *CDC42* is conserved in most fungal species (and also outside of fungi [129,130]) but it is not present in others [11]. Interestingly, *NRP1* is more conserved than several essential genes [11]. What the reason for this conservation is, is unclear. A conserved sequence does not mean conserved function or interactions. So, although a highly similar Nrp1 protein is present in a species, this does not mean it has the same function in that species.

An interesting look into the kind of experiments and research done for a protein that is similar in sequence to Nrp1 and also probably functionally related to Nrp1 is given by Whi3 [131]. Whi3 like Nrp1 has an RNA recognition motif and a low complexity/repeated region. One difference is that

Whi3 has repeated glutamines and Nrp1 has repeated asparagines, though these amino acids are chemically similar. In the paper of [131] the authors removed these domains and checked functionality and localization of Whi3 in *Ashbya gossypii*. Whi3 in *A.gossypii* and budding yeast share important functionality [33,131]. Whi3 localizes to stress granules, indirectly regulates the G1/S phase transition via Cln3 and affects many mRNAs in yeast [33]. A similar role may be hypothesized for Nrp1. It also localizes to stress granules [124], affects the G1 exit [22], and may bind some mRNAs [127]. Different from Whi3, Nrp1 does not have a clear RNA target (like Cln3) that explains its functioning.

Here we will provide some research paths from different areas of research to test our hypothesis from the previous paragraph. First, from a more chemistry perspective one strategy is to look at the structure of Nrp1. The order of the domains is known, but it is unclear whether the low-complexity domain will fold, as they have been shown to be disordered [132]. This unstructured part of the protein may move about freely. It would be interesting to see if the C-terminal part of the protein does fold specifically again after the unstructured part. One would be able to visualize the structure of the protein by means of NMR [133], however it might be difficult in this case to determine the structure if it is indeed unstructured/moving.

Second, looking at Nrp1 from a cell biological perspective, deletion studies are often used. Unfortunately, this does not work as easily for *NRP1* as the single deletion does not yield any phenotype. However, in a different background a phenotype can be found, the *bem1*Δ [22]. Analyzing what happens in these cells will help understand Nrp1. This can be done on a population level by doing a fitness assay. On a single cell level one can use microscopy. A previous high-throughput study shows Nrp1 present in the cytoplasm [103]. A more detailed study focusing on polarity establishment can give more insight into the functioning of Nrp1, especially when combined with the previous approach (different genetic backgrounds).

By the same token, a more in-depth analysis of the RNA binding ability of Nrp1 is also relevant. Again, using Nrp1 in different environments may give different results and find different specific RNAs that are bound. RNA chip-seq has become easier to execute over the last few years [134] and thus now it might be possible to do the proposed experiments.

The ultimate goal is to find a molecular mechanism for Nrp1, but this goal cannot be reached without knowledge of other aspects of the network. Nrp1 is a good example of a protein that is buried deep in the network, which makes the investigation challenging. However, it is worthwhile to dig deep and find the hidden functionality of proteins like Nrp1 that seem neutral at first glance, but have a significant evolutionary role.

4. Outlook

Protein–protein interaction networks have diverse properties that influence its evolution, such as redundancy, hierarchy and neutrality. We have advocated that studying yeast polarity provides a suitable starting point for general studies on the evolution of these complex networks, as it exhibits many of the aforementioned traits and includes the evolutionary relevant spatiotemporal dynamics [17]. From work in [26], it had become clear that predicting epistasis, which precedes predicting evolution, required deep understanding of the interactions and reactions constituting a protein network like yeast polarity.

However, our case studies demonstrate that obtaining this knowledge in a complex network architecture is far from trivial due to the aforementioned redundancy, hierarchy and neutrality. While some genes generate obvious phenotypes from which the roles of the corresponding gene products are easily deduced, for others the required information is only revealed after multiple rounds of precisely designed experiments. The latter situation can be due to various reasons related to the network architecture, for example the associated protein is found deep down in the hierarchy, forms part of a redundancy in the network or is currently otherwise peripheral to the core function. Regardless of the origin, encountered neutrality is a complication, which is a particularly unresolved feature in the third case study (Nrp1).

By exploring the past and present of the polarity network, we have aimed to determine how to advance for future research, to answer the open questions in the field and underline research opportunities. Although the yeast polarity network has been studied for many years, there still remain many unsolved mysteries. For example, the study by [22] gave much insight into the evolution and possible back-up mechanisms of the polarity network, but gave rise to the question of how Nrp1 is involved.

Much research has been done under perfect lab conditions, which results in specific results. It would be interesting to also design experiments that explore the genotype–phenotype map in different conditions. This would make it possible to find previously hidden components, that do not show up under standard lab conditions. Another possibility is changing the environment together with the expression level of specific genes, which can affect fitness [135], as has been shown for Cdc42 [23].

In vitro work is also an important next step to isolate parts of the network and see if these parts can independently perform a function [136]. As an example in budding yeast, in vitro reconstitution of a She-protein mediated mechanism of asymmetric mRNA transport revealed subtle details that were hard to demonstrate in vivo [137].

From an evolution standpoint it is interesting to see how the network as it is in current strains came to be and what the variation is that can be found in the wild. The variation within a wild population shows the spread that is available in the genotype map. It provides insight into what genotypes are preferable in certain environments. Apart from the genotype showing the history of a population, also, for example, the expression levels of proteins can be inherited [138]. Such epigenetic inheritance can be important for the reaction of an organism to stress and other environmental factors [139].

The yeast polarity examples discussed also delineate a general route forward in dismantling complexly connected protein networks. A graphical overview is depicted in Figure 6. The arrow heads indicate the type of information obtained from proteins inside the network, and higher degrees of information are successively more difficult to obtain, and usually rely on first reaching the previous level. In this way, we work our way deeper into the protein network and slowly but steadily elucidate beyond the obvious phenotypes.

In the top category, much work has been done determining the effects of simple deletions. For budding yeast, a large knock-out database has been present for more than two decades [12]. The ease of experimentally parallelizing the deletion construction even allows for ample double deletion data constituting genetic interactions, which for multiple model systems is bundled in the BioGRID database [140]. Yet, there is a myriad of combinatorial possibilities for gene deletions, and it has become clear from the GAP and Nrp1 examples that these need to be explored intelligently, rather than by brute force. Well-designed starting points for evolution experiments, for example to find that GAPs and Nrp1 genetically interact with Bem1 in [22], or SATAY assays [141] can elucidate interactions of genes of interest with important domains that only surface with strong phenotypes in the right genetic background.

However, genetic interactions can be very indirect. Epistasis is known to act globally even on unrelated networks during adaptation [142], so gathering physical information is a welcome next step. The efficient two-hybrid screens date back more than three decades [143], where two proteins fused to a DNA binding domain and transcriptional activator, respectively, promote transcription of a reporter gene when interacting. To follow interactions across the cell (see also next paragraph), FRET imaging, where two fluorescent fusion proteins cause the emission spectrum to shift when in close proximity, has also been extensively used [144], but this method also relies on the proteins of interest to be tolerant to protein fusions. More subtle modifications for tagging are used for co-immunoprecipitation [145], and are still heavily used in budding yeast [146]. Once physical interactors have been established, these can be further confirmed with their relevant binding sites by point mutation to influence the binding, as in, e.g., [84] for Bem1. This level of precision also means a move towards low-throughput data gathering, but can be useful to conjecture how cells after the deletion of *BEM1* evolved [23].

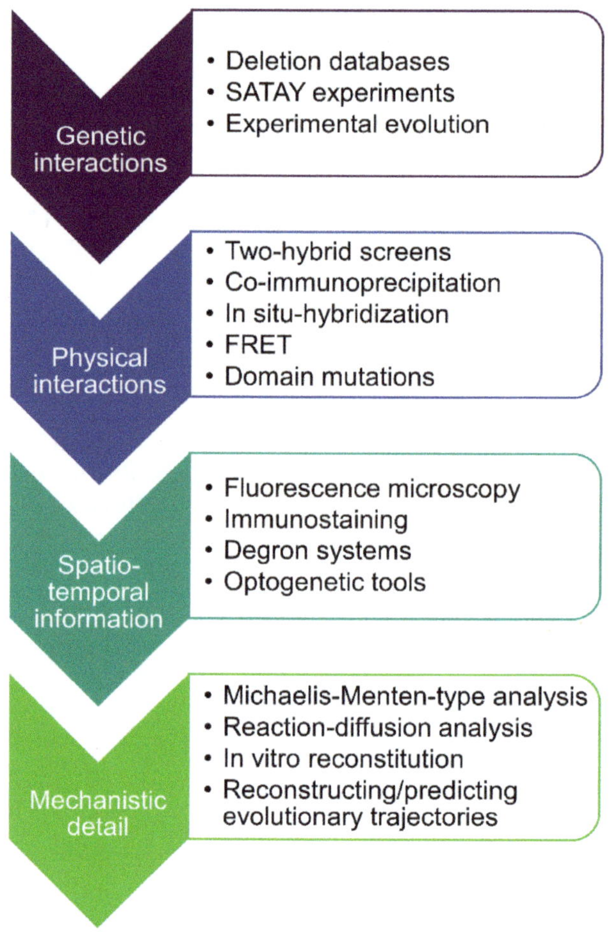

Figure 6. Flow chart for understanding the roles of proteins inside a complex network. Level of detail increases from top to bottom, while the enumerations of accompanying experiments corresponding to a level are ordered in decreasing feasibility of high-throughput data generation.

While the physical interactions constitute a rudimentary form of the protein network, the next step in understanding originates from adding spatiotemporal information. For example, Bud3 is a GEF for Cdc42, but only during early G1 phase [105], before the timing pathway gives the cue for symmetry breaking. If the function of interest is symmetry breaking, this can be excluded from the network overview as in Figure 1. Generally, a high-throughput manner for establishing protein localization in vivo has been with fluorescent protein fusions, particularly with GFP variants [103]. As aforementioned, immunostaining provides a similar option for tagging and hence localization, but requires fixation of the cells, making the temporally transient contributions of components more difficult to trace. If temporal rather than spatial information on the importance of a protein is of the essence, ingenious solutions exist that conditionally disable the protein of interest, after which the results can be swiftly observed. Examples include degron systems [147,148] and optogenetic tools [149]. Finally, as with Nrp1 localization of mRNAs may be the important function, options to trace these are also present with in situ hybridization, as described in, e.g., [150].

Ultimately, when all previous steps have been performed, it becomes possible to make the next step towards complete understanding, which would be mechanistic understanding. As shown in the case of actin, if insufficient information is available, modelling can lead to uncertain and contradictory results, such as is the case with the role of actin in polarity establishment [24,95,96]. If possible, the most unambiguous results would come from bottom-up approaches, such as modelling Michaelis–Menten kinetics for metabolism (e.g., [151] (pp. 165–180)) or solving reaction–diffusion systems, as for polarity done in [23,87] in case proteins cannot be assumed to be uniformly distributed. The more complete understanding of the protein network is then put to the test by the prediction of observed adaptive trajectories in historical or experimental evolution, as done, for example, in the latter case for the path of [22] in [23,26]. Considering the many options and solutions evolution can explore, accurate prediction of evolution is the best guarantee that the full extent of the knowledge of a network has been explored.

In summary, the general path to full network understanding as outlined in Figure 6 brings us from genetic and physical interactions to visualizing precise protein dynamics, modeling and full reconstitution. While initially, even the genetic interaction map was a tedious chore, high-throughput studies and bioinformatics tools continue to facilitate the gathering of information. Considering the speed with which the technological advances occur, the necessary data for network understanding becomes feasible for many more functions and organisms. This marks the relevance of establishing a generalizable and efficient workflow to obtain the right network data and use it for understanding and predicting its evolution.

Author Contributions: Conceptualization, W.K.-G.D., E.S. and L.L.; investigation, W.K.-G.D. and E.S.; writing—original draft preparation, W.K.-G.D. and E.S.; writing—review and editing, W.K.-G.D., E.S. and L.L.; visualization, W.K.-G.D. and E.S.; supervision, L.L. All authors have read and agreed to the published version of the manuscript.

Funding: L.L. and W.K-G.D. gratefully acknowledge support from the Netherlands Organization for Scientific Research (NWO/OCW), as part of the Gravitation Program: Frontiers of Nanoscience. L.L. and E.S. gratefully acknowledge funding from the European Research Council (ERC) under the European Union's Horizon 2020 research and innovation programme (Grant agreement No. [758132]).

Acknowledgments: We would like to thank Enzo Kingma and Leila Iñigo de la Cruz for careful reading of the manuscript.

Conflicts of Interest: The authors declare no conflict of interest. The funders had no role in the design of the study; in the collection, analyses, or interpretation of data; in the writing of the manuscript, or in the decision to publish the results.

Appendix A

Table A1 shows the corresponding literature references for the information in the Venn diagram in Figure 1. In addition, gene essentiality was obtained from [152].

Table A1. Literature supporting information (e.g., bonds and reactions) underlying Figure 1.

Description	Reference
Nrp1 binds Pub1	[153]
Nrp1-Pub1 promotes Whi3-Cln3 bond (putative)	
Pub1 binds Whi3	[154]
Nrp1-Pub1 inhibits Whi3-Cla4 (mRNA) binding (putative)	
Whi3 binds and inhibits Cln3	[33,155]
Whi3 binds Cdc28	[34]
Cln3 binds Cdc28	[34]
Cdc28-Cln3 inhibits activation of Whi5	[36]
Whi5 binds and inhibits activation of Swi4/Swi6/Mbp1	[36]
Cdc28-Cln2 inhibits activation of Whi5 as well	[36,44]
Swi4/Swi6/Mbp1 promotes activity of Cln2	[35]
Whi3 binds Cla4 (mRNA)	[156]
Whi3 can aggregate in stress granules	[33]
Cdc28 binds Cln2	[157]
Cdc28-Cln2 inhibits Ste5	[43]
Cdc28-Cln2 inhibits Far1-Cdc24 binding	[42]
Cdc28-Cln2 inhibits Bem2	[39]
Cdc28-Cln2 inhibits Rga1	[38]
Cdc28-Cln2 inhibits Bem3	[39]
Cdc28-Cln2 inhibits Rga2	[40,41]
Far1 inhibits Cln2	[44,45]
Swi4/Swi6/Mbp1 promotes activity of Bud9	[82]
Ydj1 inhibits Whi3	[32]
Protein synthesis inhibits Ydj1	[32]
Size promotes Ydj1	[32]
Nutrition promotes Cln3	[31]
Pheromone activates Ste2/Ste3	[46,47]
Ste2/Ste3 binds Ste4-Ste18	[48]
Ste4-Ste18 binds Ste20	[158]
Ste4-Ste18 binds Cdc24-Far1	[59]
Ste20 activates Ste11	[51]
Ste5 binds Ste11	[50]
Ste11 activates Ste7	[51]
Ste5 binds Ste7	[50]
Ste7 activates Fus3	[52]
Ste5 binds Fus3	[50]
Ste5 binds Bem1	[57]

Table A1. *Cont.*

Description	Reference
Fus3 inhibits Bem3 activity (putative)	[39]
Fus3 inhibits Bem2 activity (putative)	[39]
Far1 binds Cdc24	[59]
Far1 binds and promotes Cdc24 activity	[159]
Cdc42 binds and promotes Ste20 activity	[54]
Bem1 binds Far1	[58]
Bem1 binds Ste20	[57]
Fus3 binds and promotes Bni1	[61]
Ste5 binds the exocytosis network	[56]
Ste4 binds Ste5	[51]
Ste5 promotes Ste20 activating Ste11	[51]
Ste5 promotes Ste11 activating Ste7	[51]
Ste5 promotes Ste7 activating Fus3	[51,52]
Septins bind Bud4	[73]
Bud3 binds Bud4	[72]
Bud3-Bud4 binds Axl1	[72]
Bud3-Bud4 binds Axl2	[72]
Axl2 binds Bud5	[72]
Septins bind Rga1 (putative)	[81]
Septins bind Bud9	[82]
Axl1 inhibits Rax2 (putative)	[74]
Axl1 inhibits Rax1 (putative)	[74]
Rax1 binds Bud8	[75,76]
Rax1 binds Bud9	[75,76]
Bud8 binds exocytosis network	[82]
Bud2 binds and inhibits Bud1	[78]
Bud5 binds and activates Bud1	[78,160]
Bud1 binds Cdc24	[80]
Bud1 binds Bem1	[79,101]
Bud9 binds exocytosis network	[82]
Bud1 binds Cdc42	[67,161]
Bud3 binds Cdc42	[105]
Bud5 bind Bud8	[76]
Bud5 binds Bud9	[76]
Rdi1 binds Cdc42	[162]
Rga2 binds and deactivates Cdc42	[109]
Rga2 binds Bem1	[40]
Rga1 binds and deactivates Cdc42	[109]
Cdc42 binds Bem1	[163]
Cla4 binds Bem1	[164]

Table A1. *Cont.*

Bem1 binds Cdc24, activates it	[60,84,164]
Cla4 inhibits binding Bem1-Cdc24	[89,163,165]
Cdc24 binds and activates Cdc42	[106,164]
Bem2 binds and deactivates Cdc42	[110]
Cla4 binds and activates Cdc42	[88]
Bem3 binds and deactivates Cdc42	[106]
Cla4 inhibits Cdc42-Rdi1 bond	[88]
Cdc24 binds Bem1-Cla4	[164]
Bni1 binds Act1	[166]
Act1 self-organizes	[167]
Bnr1 binds Act1	[166]
Exocytosis network binds Cdc42	[94]
Act1 binds Arp2/Arp3	[115]
Arp2/Arp3 interacts with intermediate	[168,169]
Intermediates interact with Ent1/Ent2	[168,169]
Ent2 inhibits Bem3 deactivating Cdc42	[113]
Ent1/Ent2 inhibits Rga2 deactivating Cdc42	[98]
Cdc42-Cla4 activates Arp2/3	[115]
Cdc42-Ste20 activates Arp2/3	[115]
Ent1/Ent2 inhibits Rga1 deactivating Cdc42	[98]
Bni1 binds Cdc42	[55]
Act1 binds Bem1	[57]
Bud6 binds Bni1	[93]
Rho1 binds Bni1	[93]
Spa2 binds Bni1	[93]

References

1. Greaves, M.; Maley, C.C. Clonal evolution in cancer. *Nature* **2012**, *481*, 306–313. [CrossRef] [PubMed]
2. Costanzo, M.; VanderSluis, B.; Koch, E.N.; Baryshnikova, A.; Pons, C.; Tan, G.; Wang, W.; Usaj, M.; Hanchard, J.; Lee, S.D. A global genetic interaction network maps a wiring diagram of cellular function. *Science* **2016**, *353*. [CrossRef] [PubMed]
3. Fraser, H.B.; Wall, D.P.; Hirsh, A.E. A simple dependence between protein evolution rate and the number of protein-protein interactions. *BMC Evol. Biol.* **2003**, *3*, 1–6. [CrossRef] [PubMed]
4. Ratmann, O.; Wiuf, C.; Pinney, J.W. From evidence to inference: Probing the evolution of protein interaction networks. *HFSP J.* **2009**, *3*, 290–306. [CrossRef] [PubMed]
5. Zhang, J. Evolution by gene duplication: An update. *Trends Ecol. Evol.* **2003**, *18*, 292–298. [CrossRef]
6. Näsvall, J.; Sun, L.; Roth, J.R.; Andersson, D.I. Real-time evolution of new genes by innovation, amplification, and divergence. *Science* **2012**, *338*, 384–387. [CrossRef]
7. Magadum, S.; Banerjee, U.; Murugan, P.; Gangapur, D.; Ravikesavan, R. Gene duplication as a major force in evolution. *J. Genet.* **2013**, *92*, 155–161. [CrossRef]
8. Eyre-Walker, A.; Keightley, P.D. The distribution of fitness effects of new mutations. *Nat. Rev. Genet.* **2007**, *8*, 610–618. [CrossRef]
9. Thoday, J.M. Non-Darwinian "evolution" and biological progress. *Nature* **1975**, *255*, 675–677. [CrossRef]
10. Khan, A.I.; Dinh, D.M.; Schneider, D.; Lenski, R.E.; Cooper, T.F. Negative epistasis between beneficial mutations in an evolving bacterial population. *Science* **2011**, *332*, 1193–1196. [CrossRef]

11. Diepeveen, E.T.; Gehrmann, T.; Pourquié, V.; Abeel, T.; Laan, L. Patterns of Conservation and Diversification in the Fungal Polarization Network. *Genome Biol. Evol.* **2018**, *10*, 1765–1782. [CrossRef] [PubMed]
12. Winzeler, E.A. Functional Characterization of the S. cerevisiae Genome by Gene Deletion and Parallel Analysis. *Science* **1999**, *285*, 901–906. [CrossRef]
13. Deutschbauer, A.M.; Jaramillo, D.F.; Proctor, M.; Kumm, J.; Hillenmeyer, M.E.; Davis, R.W.; Nislow, C.; Giaever, G. Mechanisms of Haploinsufficiency Revealed by Genome-Wide Profiling in Yeast. *Genetics* **2005**, *169*, 1915–1925. [CrossRef] [PubMed]
14. Martin, S.G.; Arkowitz, R.A. Cell polarization in budding and fission yeasts. *FEMS Microbiol. Rev.* **2014**, *38*, 228–253. [CrossRef] [PubMed]
15. Costanzo, M.; Baryshnikova, A.; Bellay, J.; Kim, Y.; Spear, E.D.; Sevier, C.S.; Ding, H.; Koh, J.L.; Toufighi, K.; Mostafavi, S.; et al. The genetic landscape of a cell. *Science* **2010**, *327*, 425–431. [CrossRef] [PubMed]
16. Gao, J.T.; Guimera, R.; Li, H.; Pinto, I.M.; Sales-Pardo, M.; Wai, S.C.; Rubinstein, B.; Li, R. Modular coherence of protein dynamics in yeast cell polarity system. *Proc. Natl. Acad. Sci. USA* **2011**, *108*, 7647–7652. [CrossRef] [PubMed]
17. Yamada, T.; Bork, P. Evolution of biomolecular networks—Lessons from metabolic and protein interactions. *Nat. Rev. Mol. Cell Biol.* **2009**, *10*, 791–803. [CrossRef] [PubMed]
18. Tuch, B.B.; Galgoczy, D.J.; Hernday, A.D.; Li, H.; Johnson, A.D. The evolution of combinatorial gene regulation in fungi. *PLoS Biol.* **2008**, *6*, e38. [CrossRef]
19. Tanay, A.; Regev, A.; Shamir, R. Conservation and evolvability in regulatory networks: The evolution of ribosomal regulation in yeast. *Proc. Natl. Acad. Sci. USA* **2005**, *102*, 7203–7208. [CrossRef]
20. Dhar, R.; Sagesser, R.; Weikert, C.; Wagner, A. Yeast Adapts to a Changing Stressful Environment by Evolving Cross-Protection and Anticipatory Gene Regulation. *Mol. Biol. Evol.* **2013**, *30*, 573–588. [CrossRef]
21. Slot, J.C.; Rokas, A. Multiple GAL pathway gene clusters evolved independently and by different mechanisms in fungi. *Proc. Natl. Acad. Sci. USA* **2010**, *107*, 10136–10141. [CrossRef] [PubMed]
22. Laan, L.; Koschwanez, J.H.; Murray, A.W. Evolutionary adaptation after crippling cell polarization follows reproducible trajectories. *eLife* **2015**, *4*, e09638. [CrossRef] [PubMed]
23. Brauns, F.; De la Cruz, L.M.I.; Daalman, W.K.-G.; De Bruin, I.; Halatek, J.; Laan, L.; Frey, E. Adaptability and evolution of the cell polarization machinery in budding yeast. *bioRxiv* **2020**. [CrossRef]
24. Freisinger, T.; Klünder, B.; Johnson, J.; Müller, N.; Pichler, G.; Beck, G.; Costanzo, M.; Boone, C.; Cerione, R.A.; Frey, E.; et al. Establishment of a robust single axis of cell polarity by coupling multiple positive feedback loops. *Nat. Commun.* **2013**, *4*, 1807. [CrossRef] [PubMed]
25. Howell, A.S.; Jin, M.; Wu, C.-F.; Zyla, T.R.; Elston, T.C.; Lew, D.J. Negative Feedback Enhances Robustness in the Yeast Polarity Establishment Circuit. *Cell* **2012**, *149*, 322–333. [CrossRef] [PubMed]
26. Daalman, W.K.-G.; Laan, L. Predicting an epistasis-rich genotype-phenotype map with a coarse-grained bottom-up model of budding yeast polarity. *bioRxiv* **2020**. [CrossRef]
27. Breen, M.S.; Kemena, C.; Vlasov, P.K.; Notredame, C.; Kondrashov, F.A. Epistasis as the primary factor in molecular evolution. *Nature* **2012**, *490*, 535–538. [CrossRef]
28. Yu, M.K.; Kramer, M.; Dutkowski, J.; Srivas, R.; Licon, K.; Kreisberg, J.F.; Ng, C.T.; Krogan, N.; Sharan, R.; Ideker, T. Translation of Genotype to Phenotype by a Hierarchy of Cell Subsystems. *Cell Syst.* **2016**, *2*, 77–88. [CrossRef]
29. Park, H.-O.; Bi, E. Central Roles of Small GTPases in the Development of Cell Polarity in Yeast and Beyond. *Microbiol. Mol. Biol. Rev.* **2007**, *71*, 48–96. [CrossRef]
30. Hall, A. The cellular functions of small GTP-binding proteins. *Science* **1990**, *249*, 635–640. [CrossRef]
31. Gallego, C.; Garí, E.; Colomina, N.; Herrero, E.; Aldea, M. The Cln3 cyclin is down-regulated by translational repression and degradation during the G1 arrest caused by nitrogen deprivation in budding yeast. *EMBO J.* **1997**, *16*, 7196–7206. [CrossRef] [PubMed]
32. Vergés, E.; Colomina, N.; Garí, E.; Gallego, C.; Aldea, M. Cyclin Cln3 Is Retained at the ER and Released by the J Chaperone Ydj1 in Late G1 to Trigger Cell Cycle Entry. *Mol. Cell* **2007**, *26*, 649–662. [CrossRef]
33. Cai, Y.; Futcher, B. Effects of the Yeast RNA-Binding Protein Whi3 on the Half-Life and Abundance of CLN3 mRNA and Other Targets. *PLoS ONE* **2013**, *8*, e84630. [CrossRef]
34. Wang, H.; Garí, E.; Vergés, E.; Gallego, C.; Aldea, M. Recruitment of Cdc28 by Whi3 restricts nuclear accumulation of the G1 cyclin–Cdk complex to late G1. *EMBO J.* **2004**, *23*, 180–190. [CrossRef]

35. Koch, C.; Schleiffer, A.; Ammerer, G.; Nasmyth, K. Switching transcription on and off during the yeast cell cycle: Cln/Cdc28 kinases activate bound transcription factor SBF (Swi4/Swi6) at start, whereas Clb/Cdc28 kinases displace it from the promoter in G2. *Genes Dev.* **1996**, *10*, 129–141. [CrossRef] [PubMed]
36. Costanzo, M.; Nishikawa, J.L.; Tang, X.; Millman, J.S.; Schub, O.; Breitkreuz, K.; Dewar, D.; Rupes, I.; Andrews, B.; Tyers, M. CDK activity antagonizes Whi5, an inhibitor of G1/S transcription in yeast. *Cell* **2004**, *117*, 899–913. [CrossRef] [PubMed]
37. Stuart, D.; Wittenberg, C. CLN3, not positive feedback, determines the timing of CLN2 transcription in cycling cells. *Genes Dev.* **1995**, *9*, 2780–2794. [CrossRef] [PubMed]
38. Archambault, V.; Chang, E.J.; Drapkin, B.J.; Cross, F.R.; Chait, B.T.; Rout, M.P. Targeted proteomic study of the cyclin-Cdk module. *Mol. Cell* **2004**, *14*, 699–711. [CrossRef]
39. Knaus, M.; Pelli-Gulli, M.-P.; Van Drogen, F.; Springer, S.; Jaquenoud, M.; Peter, M. Phosphorylation of Bem2p and Bem3p may contribute to local activation of Cdc42p at bud emergence. *EMBO J.* **2007**, *26*, 4501–4513. [CrossRef]
40. McCusker, D.; Denison, C.; Anderson, S.; Egelhofer, T.A.; Yates, J.R.; Gygi, S.P.; Kellogg, D.R. Cdk1 coordinates cell-surface growth with the cell cycle. *Nat. Cell Biol.* **2007**, *9*, 506–515. [CrossRef]
41. Sopko, R.; Huang, D.; Smith, J.C.; Figeys, D.; Andrews, B.J. Activation of the Cdc42p GTPase by cyclin-dependent protein kinases in budding yeast. *EMBO J.* **2007**, *26*, 4487–4500. [CrossRef] [PubMed]
42. Shimada, Y.; Gulli, M.-P.; Peter, M. Nuclear sequestration of the exchange factor Cdc24 by Far1 regulates cell polarity during yeast mating. *Nat. Cell Biol.* **2000**, *2*, 117–124. [CrossRef] [PubMed]
43. Garrenton, L.S.; Braunwarth, A.; Irniger, S.; Hurt, E.; Kunzler, M.; Thorner, J. Nucleus-Specific and Cell Cycle-Regulated Degradation of Mitogen-Activated Protein Kinase Scaffold Protein Ste5 Contributes to the Control of Signaling Competence. *Mol. Cell. Biol.* **2009**, *29*, 582–601. [CrossRef] [PubMed]
44. Doncic, A.; Falleur-Fettig, M.; Skotheim, J.M. Distinct Interactions Select and Maintain a Specific Cell Fate. *Mol. Cell* **2011**, *43*, 528–539. [CrossRef] [PubMed]
45. Doncic, A.; Skotheim, J.M. Feedforward Regulation Ensures Stability and Rapid Reversibility of a Cellular State. *Mol. Cell* **2013**, *50*, 856–868. [CrossRef] [PubMed]
46. Blumer, K.J.; Reneke, J.E.; Thorner, J. The STE2 gene product is the ligand-binding component of the alpha-factor receptor of Saccharomyces cerevisiae. *J. Biol. Chem.* **1988**, *263*, 10836–10842. [PubMed]
47. Hagen, D.C.; McCaffrey, G.; Sprague, G.F. Evidence the yeast STE3 gene encodes a receptor for the peptide pheromone a factor: Gene sequence and implications for the structure of the presumed receptor. *Proc. Natl. Acad. Sci. USA* **1986**, *83*, 1418–1422. [CrossRef]
48. Whiteway, M.; Hougan, L.; Dignard, D.; Thomas, D.Y.; Bell, L.; Saari, G.C.; Grant, F.J.; O'Hara, P.; MacKay, V.L. The STE4 and STE18 genes of yeast encode potential β and γ subunits of the mating factor receptor-coupled G protein. *Cell* **1989**, *56*, 467–477. [CrossRef]
49. Hirschman, J.E.; De Zutter, G.S.; Simonds, W.F.; Jenness, D.D. The Gβγ Complex of the Yeast Pheromone Response Pathway SUBCELLULAR FRACTIONATION AND PROTEIN-PROTEIN INTERACTIONS. *J. Biol. Chem.* **1997**, *272*, 240–248. [CrossRef]
50. Chol, K.-Y.; Satterberg, B.; Lyons, D.M.; Elion, E.A. Ste5 tethers multiple protein kinases in the MAP kinase cascade required for mating in S. cerevisiae. *Cell* **1994**, *78*, 499–512. [CrossRef]
51. Feng, Y.; Song, L.Y.; Kincaid, E.; Mahanty, S.K.; Elion, E.A. Functional binding between Gβ and the LIM domain of Ste5 is required to activate the MEKK Ste11. *Curr. Biol.* **1998**, *8*, 267–282. [CrossRef]
52. Good, M.; Tang, G.; Singleton, J.; Reményi, A.; Lim, W.A. The Ste5 Scaffold Directs Mating Signaling by Catalytically Unlocking the Fus3 MAP Kinase for Activation. *Cell* **2009**, *136*, 1085–1097. [CrossRef] [PubMed]
53. Wang, Y.; Chen, W.; Simpson, D.M.; Elion, E.A. Cdc24 Regulates Nuclear Shuttling and Recruitment of the Ste5 Scaffold to a Heterotrimeric G Protein in *Saccharomyces cerevisiae*. *J. Biol. Chem.* **2005**, *280*, 13084–13096. [CrossRef] [PubMed]
54. Lamson, R.E.; Winters, M.J.; Pryciak, P.M. Cdc42 Regulation of Kinase Activity and Signaling by the Yeast p21-Activated Kinase Ste20. *Mol. Cell. Biol.* **2002**, *22*, 2939–2951. [CrossRef]
55. Evangelista, M.; Blundell, K.; Longtine, M.S.; Chow, C.J.; Adames, N.; Pringle, J.R.; Peter, M.; Boone, C. Bni1p, a yeast formin linking cdc42p and the actin cytoskeleton during polarized morphogenesis. *Science* **1997**, *276*, 118–122. [CrossRef]
56. Qi, M.; Elion, E.A. Formin-induced actin cables are required for polarized recruitment of the Ste5 scaffold and high level activation of MAPK Fus3. *J. Cell Sci.* **2005**, *118*, 2837–2848. [CrossRef]

57. Leeuw, T.; Fourest-Lieuvin, A.; Wu, C.; Chenevert, J.; Clark, K.; Whiteway, M.; Thomas, D.Y.; Leberer, E. Pheromone Response in Yeast: Association of Bem1p with Proteins of the MAP Kinase Cascade and Actin. *Science* **1995**, 1210–1213. [CrossRef]
58. Lyons, D.M.; Mahanty, S.K.; Choi, K.-Y.; Manandhar, M.; Elion, E.A. The SH3-domain protein Bem1 coordinates mitogen-activated protein kinase cascade activation with cell cycle control in Saccharomyces cerevisiae. *Mol. Cell. Biol.* **1996**, *16*, 4095–4106. [CrossRef]
59. Butty, A.-C.; Pryciak, P.M.; Huang, L.S.; Herskowitz, I.; Peter, M. The role of Far1p in linking the heterotrimeric G protein to polarity establishment proteins during yeast mating. *Science* **1998**, *282*, 1511–1516. [CrossRef]
60. Peterson, J. Interactions between the bud emergence proteins Bem1p and Bem2p and Rho- type GTPases in yeast. *J. Cell Biol.* **1994**, *127*, 1395–1406. [CrossRef]
61. Matheos, D.; Metodiev, M.; Muller, E.; Stone, D.; Rose, M.D. Pheromone-induced polarization is dependent on the Fus3p MAPK acting through the formin Bni1p. *J. Cell Biol.* **2004**, *165*, 99–109. [CrossRef]
62. Kim, H.B. Cellular morphogenesis in the Saccharomyces cerevisiae cell cycle: Localization of the CDC3 gene product and the timing of events at the budding site. *J. Cell Biol.* **1991**, *112*, 535–544. [CrossRef]
63. Flescher, E.G. Components required for cytokinesis are important for bud site selection in yeast. *J. Cell Biol.* **1993**, *122*, 373–386. [CrossRef]
64. Freifelder, D. Bud position in Saccharomyces cerevisiae. *J. Bacteriol.* **1960**, *80*, 567. [CrossRef]
65. Chant, J.; Herskowitz, I. Genetic control of bud site selection in yeast by a set of gene products that constitute a morphogenetic pathway. *Cell* **1991**, *65*, 1203–1212. [CrossRef]
66. Chant, J.; Corrado, K.; Pringle, J.R.; Herskowitz, I. Yeast BUD5, encoding a putative GDP-GTP exchange factor, is necessary for bud site selection and interacts with bud formation gene BEM1. *Cell* **1991**, *65*, 1213–1224. [CrossRef]
67. Kang, P.J.; Béven, L.; Hariharan, S.; Park, H.-O. The Rsr1/Bud1 GTPase interacts with itself and the Cdc42 GTPase during bud-site selection and polarity establishment in budding yeast. *Mol. Biol. Cell* **2010**, *21*, 3007–3016. [CrossRef]
68. Bender, A.; Pringle, J.R. Multicopy suppression of the cdc24 budding defect in yeast by CDC42 and three newly identified genes including the ras-related gene RSR1. *Proc. Natl. Acad. Sci. USA* **1989**, *86*, 9976–9980. [CrossRef]
69. Fujita, A.; Oka, C.; Arikawa, Y.; Katagai, T.; Tonouchi, A.; Kuhara, S.; Misumi, Y. A yeast gene necessary for bud-site selection encodes a protein similar to insulin-degrading enzymes. *Nature* **1994**, *372*, 567. [CrossRef]
70. Chant, J.; Mischke, M.; Mitchell, E.; Herskowitz, I.; Pringle, J.R. Role of Bud3p in producing the axial budding pattern of yeast. *J. Cell Biol.* **1995**, *129*, 767–778. [CrossRef]
71. Sanders, S.L.; Herskowitz, I. The BUD4 protein of yeast, required for axial budding, is localized to the mother/BUD neck in a cell cycle-dependent manner. *J. Cell Biol.* **1996**, *134*, 413–427. [CrossRef] [PubMed]
72. Kang, P.J.; Angerman, E.; Jung, C.-H.; Park, H.-O. Bud4 mediates the cell-type-specific assembly of the axial landmark in budding yeast. *J. Cell Sci.* **2012**, *125*, 3840–3849. [CrossRef] [PubMed]
73. Kang, P.J.; Hood-DeGrenier, J.K.; Park, H.-O. Coupling of septins to the axial landmark by Bud4 in budding yeast. *J. Cell Sci.* **2013**, *126*, 1218–1226. [CrossRef] [PubMed]
74. Bi, E.; Park, H.-O. Cell Polarization and Cytokinesis in Budding Yeast. *Genetics* **2012**, *191*, 347–387. [CrossRef] [PubMed]
75. Kang, P.J.; Angerman, E.; Nakashima, K.; Pringle, J.R.; Park, H.-O. Interactions among Rax1p, Rax2p, Bud8p, and Bud9p in marking cortical sites for bipolar bud-site selection in yeast. *Mol. Biol. Cell* **2004**, *15*, 5145–5157. [CrossRef]
76. Krappmann, A.-B.; Taheri, N.; Heinrich, M.; Mösch, H.-U. Distinct domains of yeast cortical tag proteins Bud8p and Bud9p confer polar localization and functionality. *Mol. Biol. Cell* **2007**, *18*, 3323–3339. [CrossRef]
77. Marston, A.L.; Chen, T.; Yang, M.C.; Belhumeur, P.; Chant, J. A localized GTPase exchange factor, Bud5, determines the orientation of division axes in yeast. *Curr. Biol.* **2001**, *11*, 803–807. [CrossRef]
78. Bender, A. Genetic evidence for the roles of the bud-site-selection genes BUD5 and BUD2 in control of the Rsr1p (Bud1p) GTPase in yeast. *Proc. Natl. Acad. Sci. USA* **1993**, *90*, 9926–9929. [CrossRef]
79. Park, H.-O.; Bi, E.; Pringle, J.R.; Herskowitz, I. Two active states of the Ras-related Bud1/Rsr1 protein bind to different effectors to determine yeast cell polarity. *Proc. Natl. Acad. Sci. USA* **1997**, *94*, 4463–4468. [CrossRef]
80. Zheng, Y.; Bender, A.; Cerione, R.A. Interactions among proteins involved in bud-site selection and bud-site assembly in Saccharomyces cerevisiae. *J. Biol. Chem.* **1995**, *270*, 626–630. [CrossRef]

81. Tong, Z.; Gao, X.-D.; Howell, A.S.; Bose, I.; Lew, D.J.; Bi, E. Adjacent positioning of cellular structures enabled by a Cdc42 GTPase-activating protein mediated zone of inhibition. *J. Cell Biol.* **2007**, *179*, 1375–1384. [CrossRef] [PubMed]
82. Schenkman, L.R.; Caruso, C.; Pagé, N.; Pringle, J.R. The role of cell cycle–regulated expression in the localization of spatial landmark proteins in yeast. *J. Cell Biol.* **2002**, *156*, 829–841. [CrossRef] [PubMed]
83. Irazoqui, J.E.; Gladfelter, A.S.; Lew, D.J. Scaffold-mediated symmetry breaking by Cdc42p. *Nat. Cell Biol.* **2003**, *5*, 1062–1070. [CrossRef] [PubMed]
84. Smith, S.E.; Rubinstein, B.; Mendes Pinto, I.; Slaughter, B.D.; Unruh, J.R.; Li, R. Independence of symmetry breaking on Bem1-mediated autocatalytic activation of Cdc42. *J. Cell Biol.* **2013**, *202*, 1091–1106. [CrossRef]
85. Ozbudak, E.M.; Becskei, A.; Van Oudenaarden, A. A System of Counteracting Feedback Loops Regulates Cdc42p Activity during Spontaneous Cell Polarization. *Dev. Cell* **2005**, *9*, 565–571. [CrossRef]
86. Goryachev, A.B.; Pokhilko, A.V. Dynamics of Cdc42 network embodies a Turing-type mechanism of yeast cell polarity. *FEBS Lett.* **2008**, *582*, 1437–1443. [CrossRef]
87. Klünder, B.; Freisinger, T.; Wedlich-Söldner, R.; Frey, E. GDI-Mediated Cell Polarization in Yeast Provides Precise Spatial and Temporal Control of Cdc42 Signaling. *PLoS Comput. Biol.* **2013**, *9*, e1003396. [CrossRef]
88. Tiedje, C.; Sakwa, I.; Just, U.; Höfken, T. The rho gdi rdi1 regulates rho gtpases by distinct mechanisms. *Mol. Biol. Cell* **2008**, *19*, 2885–2896. [CrossRef]
89. Kuo, C.-C.; Savage, N.S.; Chen, H.; Wu, C.-F.; Zyla, T.R.; Lew, D.J. Inhibitory GEF Phosphorylation Provides Negative Feedback in the Yeast Polarity Circuit. *Curr. Biol.* **2014**, *24*, 753–759. [CrossRef]
90. Cvrckova, F.; De Virgilio, C.; Manser, E.; Pringle, J.R.; Nasmyth, K. Ste20-like protein kinases are required for normal localization of cell growth and for cytokinesis in budding yeast. *Genes Dev.* **1995**, *9*, 1817–1830. [CrossRef]
91. Daalman, W.K.-G. *Coupling of Genotype-Phenotype Maps to Noise-Driven Adaptation, Showcased in Yeast Polarity*; TU Delft: Delft, The Netherlands, 2020. [CrossRef]
92. Pruyne, D.; Gao, L.; Bi, E.; Bretscher, A. Stable and dynamic axes of polarity use distinct formin isoforms in budding yeast. *Mol. Biol. Cell* **2004**, *15*, 4971–4989. [CrossRef] [PubMed]
93. Prosser, D.C.; Drivas, T.G.; Maldonado-Báez, L.; Wendland, B. Existence of a novel clathrin-independent endocytic pathway in yeast that depends on Rho1 and formin. *J. Cell Biol.* **2011**, *195*, 657–671. [CrossRef] [PubMed]
94. Wedlich-Soldner, R.; Altschuler, S.; Wu, L.; Li, R. Spontaneous cell polarization through actomyosin-based delivery of the Cdc42 GTPase. *Science* **2003**, *299*, 1231–1235. [CrossRef] [PubMed]
95. Slaughter, B.D.; Unruh, J.R.; Das, A.; Smith, S.E.; Rubinstein, B.; Li, R. Non-uniform membrane diffusion enables steady-state cell polarization via vesicular trafficking. *Nat. Commun.* **2013**, *4*, 1380. [CrossRef] [PubMed]
96. Layton, A.T.; Savage, N.S.; Howell, A.S.; Carroll, S.Y.; Drubin, D.G.; Lew, D.J. Modeling Vesicle Traffic Reveals Unexpected Consequences for Cdc42p-Mediated Polarity Establishment. *Curr. Biol.* **2011**, *21*, 184–194. [CrossRef]
97. Watson, L.J.; Rossi, G.; Brennwald, P. Quantitative Analysis of Membrane Trafficking in Regulation of Cdc42 Polarity. *Traffic* **2014**, *15*, 1330–1343. [CrossRef]
98. Aguilar, R.C.; Longhi, S.A.; Shaw, J.D.; Yeh, L.-Y.; Kim, S.; Schön, A.; Freire, E.; Hsu, A.; McCormick, W.K.; Watson, H.A.; et al. Epsin N-terminal homology domains perform an essential function regulating Cdc42 through binding Cdc42 GTPase-activating proteins. *Proc. Natl. Acad. Sci. USA* **2006**, *103*, 4116–4121. [CrossRef]
99. Gerhart, J.; Kirschner, M. The theory of facilitated variation. *Proc. Natl. Acad. Sci. USA* **2007**, *104*, 8582–8589. [CrossRef]
100. Wedlich-Soldner, R.; Li, R. Yeast and fungal morphogenesis from an evolutionary perspective. *Semin. Cell Dev. Biol.* **2008**, *19*, 224–233. [CrossRef]
101. Miller, K.E.; Lo, W.-C.; Chou, C.-S.; Park, H.-O. Temporal regulation of cell polarity via the interaction of the Ras GTPase Rsr1 and the scaffold protein Bem1. *Mol. Biol. Cell* **2019**, *30*, 2543–2557. [CrossRef]
102. Niedenthal, R.K.; Riles, L.; Johnston, M.; Hegemann, J.H. Green fluorescent protein as a marker for gene expression and subcellular localization in budding yeast. *Yeast* **1996**, *12*, 773–786. [CrossRef]
103. Huh, W.-K.; Falvo, J.V.; Gerke, L.C.; Carroll, A.S.; Howson, R.W.; Weissman, J.S.; O'Shea, E.K. Global analysis of protein localization in budding yeast. *Nature* **2003**, *425*, 686–691. [CrossRef] [PubMed]

104. Kang, P.J.; Lee, B.; Park, H.-O. Specific Residues of the GDP/GTP Exchange Factor Bud5p Are Involved in Establishment of the Cell Type-specific Budding Pattern in Yeast. *J. Biol. Chem.* **2004**, *279*, 27980–27985. [CrossRef] [PubMed]
105. Kang, P.J.; Lee, M.E.; Park, H.-O. Bud3 activates Cdc42 to establish a proper growth site in budding yeast. *J. Cell Biol.* **2014**, *206*, 19–28. [CrossRef]
106. Zheng, Y.; Cerione, R.; Bender, A. Control of the yeast bud-site assembly GTPase Cdc42. Catalysis of guanine nucleotide exchange by Cdc24 and stimulation of GTPase activity by Bem3. *J. Biol. Chem.* **1994**, *269*, 2369–2372.
107. Stevenson, B.J.; Ferguson, B.; De Virgilio, C.; Bi, E.; Pringle, J.R.; Ammerer, G.; Sprague, G.F. Mutation of RGA1, which encodes a putative GTPase-activating protein for the polarity-establishment protein Cdc42p, activates the pheromone-response pathway in the yeast Saccharomyces cerevisiae. *Genes Dev.* **1995**, *9*, 2949–2963. [CrossRef]
108. Gladfelter, A.S.; Bose, I.; Zyla, T.R.; Bardes, E.S.G.; Lew, D.J. Septin ring assembly involves cycles of GTP loading and hydrolysis by Cdc42p. *J. Cell Biol.* **2002**, *156*, 315–326. [CrossRef]
109. Smith, G.R.; Givan, S.A.; Cullen, P.; Sprague, G.F. GTPase-Activating Proteins for Cdc42. *Eukaryot. Cell* **2002**, *1*, 469–480. [CrossRef]
110. Marquitz, A.R.; Harrison, J.C.; Bose, I.; Zyla, T.R.; McMillan, J.N.; Lew, D.J. The Rho-GAP Bem2p plays a GAP-independent role in the morphogenesis checkpoint. *EMBO J.* **2002**, *21*, 4012–4025. [CrossRef]
111. Caviston, J.P.; Longtine, M.; Pringle, J.R.; Bi, E. The role of Cdc42p GTPase-activating proteins in assembly of the septin ring in yeast. *Mol. Biol. Cell* **2003**, *14*, 4051–4066. [CrossRef]
112. Mukherjee, D.; Sen, A.; Boettner, D.R.; Fairn, G.D.; Schlam, D.; Bonilla Valentin, F.J.; McCaffery, J.M.; Hazbun, T.; Staiger, C.J.; Grinstein, S.; et al. Bem3, a Cdc42 GTPase-activating protein, traffics to an intracellular compartment and recruits the secretory Rab GTPase Sec4 to endomembranes. *J. Cell Sci.* **2013**, *126*, 4560–4571. [CrossRef] [PubMed]
113. Mukherjee, D.; Coon, B.G.; Edwards, D.F.; Hanna, C.B.; Longhi, S.A.; McCaffery, J.M.; Wendland, B.; Retegui, L.A.; Bi, E.; Aguilar, R.C. The yeast endocytic protein Epsin 2 functions in a cell-division signaling pathway. *J. Cell Sci.* **2009**, *122*, 2453–2463. [CrossRef]
114. Wild, A.C.; Yu, J.W.; Lemmon, M.A.; Blumer, K.J. The p21-activated Protein Kinase-related Kinase Cla4 Is a Coincidence Detector of Signaling by Cdc42 and Phosphatidylinositol 4-Phosphate. *J. Biol. Chem.* **2004**, *279*, 17101–17110. [CrossRef] [PubMed]
115. Lechler, T.; Shevchenko, A.; Shevchenko, A.; Li, R. Direct Involvement of Yeast Type I Myosins in Cdc42-Dependent Actin Polymerization. *J. Cell Biol.* **2000**, *148*, 363–374. [CrossRef]
116. Goryachev, A.B.; Leda, M. Many roads to symmetry breaking: Molecular mechanisms and theoretical models of yeast cell polarity. *Mol. Biol. Cell* **2017**, *28*, 370–380. [CrossRef]
117. Howell, A.S.; Savage, N.S.; Johnson, S.A.; Bose, I.; Wagner, A.W.; Zyla, T.R.; Nijhout, H.F.; Reed, M.C.; Goryachev, A.B.; Lew, D.J. Singularity in Polarization: Rewiring Yeast Cells to Make Two Buds. *Cell* **2009**, *139*, 731–743. [CrossRef]
118. Kamada, Y.; Qadota, H.; Python, C.P.; Anraku, Y.; Ohya, Y.; Levin, D.E. Activation of yeast protein kinase C by Rho1 GTPase. *J. Biol. Chem.* **1996**, *271*, 9193–9196. [CrossRef]
119. Qadota, H.; Python, C.P.; Inoue, S.B.; Arisawa, M.; Anraku, Y.; Zheng, Y.; Watanabe, T.; Levin, D.E.; Ohya, Y. Identification of yeast Rho1p GTPase as a regulatory subunit of 1, 3-β-glucan synthase. *Science* **1996**, *272*, 279–281. [CrossRef]
120. Saka, A.; Abe, M.; Okano, H.; Minemura, M.; Qadota, H.; Utsugi, T.; Mino, A.; Tanaka, K.; Takai, Y.; Ohya, Y. Complementing Yeast rho1 Mutation Groups with Distinct Functional Defects. *J. Biol. Chem.* **2001**, *276*, 46165–46171. [CrossRef]
121. Diepeveen, E.T.; De la Cruz, L.I.; Laan, L. Evolutionary dynamics in the fungal polarization network, a mechanistic perspective. *Biophys. Rev.* **2017**, *9*, 375–387. [CrossRef]
122. Reynaud, A.; Facca, C.; Sor, F.; Faye, G. Disruption and functional analysis of six ORFs of chromosome IV: YDL103c (QRI1), YDL105w (QRI2), YDL112w (TRM3), YDL113c, YDL116w (NUP84) and YDL167c (NRP1). *Yeast* **2001**, *18*, 273–282. [CrossRef]
123. Outten, C.E.; Falk, R.L.; Culotta, V.C. Cellular factors required for protection from hyperoxia toxicity in Saccharomyces cerevisiae. *Biochem. J.* **2005**, *388*, 93–101. [CrossRef] [PubMed]

124. Buchan, J.R.; Muhlrad, D.; Parker, R. P bodies promote stress granule assembly in *Saccharomyces cerevisiae. J. Cell Biol.* **2008**, *183*, 441–455. [CrossRef] [PubMed]
125. Hoepfner, D.; Helliwell, S.B.; Sadlish, H.; Schuierer, S.; Filipuzzi, I.; Brachat, S.; Bhullar, B.; Plikat, U.; Abraham, Y.; Altorfer, M.; et al. High-resolution chemical dissection of a model eukaryote reveals targets, pathways and gene functions. *Microbiol. Res.* **2014**, *169*, 107–120. [CrossRef]
126. Alberti, S.; Halfmann, R.; King, O.; Kapila, A.; Lindquist, S. A Systematic Survey Identifies Prions and Illuminates Sequence Features of Prionogenic Proteins. *Cell* **2009**, *137*, 146–158. [CrossRef]
127. Hogan, D.J.; Riordan, D.P.; Gerber, A.P.; Herschlag, D.; Brown, P.O. Diverse RNA-Binding Proteins Interact with Functionally Related Sets of RNAs, Suggesting an Extensive Regulatory System. *PLoS Biol.* **2008**, *6*, e255. [CrossRef]
128. Sharifpoor, S.; Van Dyk, D.; Costanzo, M.; Baryshnikova, A.; Friesen, H.; Douglas, A.C.; Youn, J.-Y.; VanderSluis, B.; Myers, C.L.; Papp, B.; et al. Functional wiring of the yeast kinome revealed by global analysis of genetic network motifs. *Genome Res.* **2012**, *22*, 791–801. [CrossRef]
129. Etienne-Manneville, S. Cdc42—The centre of polarity. *J. Cell Sci.* **2004**, *117*, 1291–1300. [CrossRef]
130. Johnson, D.I. Cdc42: An Essential Rho-Type GTPase Controlling Eukaryotic Cell Polarity. *Microbiol. Mol. Biol. Rev.* **1999**, *63*, 54–105. [CrossRef]
131. Lee, C.; Occhipinti, P.; Gladfelter, A.S. PolyQ-dependent RNA–protein assemblies control symmetry breaking. *J. Cell Biol.* **2015**, *208*, 533–544. [CrossRef]
132. Kato, M.; McKnight, S.L. Cross-β Polymerization of Low Complexity Sequence Domains. *Cold Spring Harb. Perspect. Biol.* **2017**, *9*, a023598. [CrossRef]
133. Wuthrich, K. Protein structure determination in solution by nuclear magnetic resonance spectroscopy. *Science* **1989**, *243*, 45–50. [CrossRef]
134. Hrdlickova, R.; Toloue, M.; Tian, B. RNA-Seq methods for transcriptome analysis: RNA-Seq. *Wiley Interdiscip. Rev. RNA* **2017**, *8*, e1364. [CrossRef] [PubMed]
135. Keren, L.; Hausser, J.; Lotan-Pompan, M.; Vainberg Slutskin, I.; Alisar, H.; Kaminski, S.; Weinberger, A.; Alon, U.; Milo, R.; Segal, E. Massively Parallel Interrogation of the Effects of Gene Expression Levels on Fitness. *Cell* **2016**, *166*, 1282–1294.e18. [CrossRef]
136. Vendel, K.J.A.; Tschirpke, S.; Shamsi, F.; Dogterom, M.; Laan, L. Minimal *in vitro* systems shed light on cell polarity. *J. Cell Sci.* **2019**, *132*, jcs217554. [CrossRef] [PubMed]
137. Heym, R.G.; Zimmermann, D.; Edelmann, F.T.; Israel, L.; Ökten, Z.; Kovar, D.R.; Niessing, D. In vitro reconstitution of an mRNA-transport complex reveals mechanisms of assembly and motor activation. *J. Cell Biol.* **2013**, *203*, 971–984. [CrossRef] [PubMed]
138. Cerulus, B.; New, A.M.; Pougach, K.; Verstrepen, K.J. Noise and Epigenetic Inheritance of Single-Cell Division Times Influence Population Fitness. *Curr. Biol.* **2016**, *26*, 1138–1147. [CrossRef]
139. Xue, Y.; Acar, M. Mechanisms for the epigenetic inheritance of stress response in single cells. *Curr. Genet.* **2018**, *64*, 1221–1228. [CrossRef]
140. Stark, C. BioGRID: A general repository for interaction datasets. *Nucleic Acids Res.* **2006**, *34*, D535–D539. [CrossRef]
141. Michel, A.H.; Hatakeyama, R.; Kimmig, P.; Arter, M.; Peter, M.; Matos, J.; De Virgilio, C.; Kornmann, B. Functional mapping of yeast genomes by saturated transposition. *Elife* **2017**, *6*, e23570. [CrossRef]
142. Kryazhimskiy, S.; Rice, D.P.; Jerison, E.R.; Desai, M.M. Global epistasis makes adaptation predictable despite sequence-level stochasticity. *Science* **2014**, *344*, 1519–1522. [CrossRef] [PubMed]
143. Fields, S.; Song, O. A novel genetic system to detect protein–protein interactions. *Nature* **1989**, *340*, 245–246. [CrossRef] [PubMed]
144. Jares-Erijman, E.A.; Jovin, T.M. FRET imaging. *Nat. Biotechnol.* **2003**, *21*, 1387–1395. [CrossRef] [PubMed]
145. Phizicky, E.M.; Fields, S. Protein-protein interactions: Methods for detection and analysis. *Microbiol. Rev.* **1995**, *59*, 94–123. [CrossRef] [PubMed]
146. Foltman, M.; Sanchez-Diaz, A. Studying Protein–Protein Interactions in Budding Yeast Using Co-immunoprecipitation. In *Yeast Cytokinesis: Methods and Protocols*; Sanchez-Diaz, A., Perez, P., Eds.; Springer: New York, NY, USA, 2016; pp. 239–256. ISBN 978-1-4939-3145-3.
147. Nishimura, K.; Fukagawa, T.; Takisawa, H.; Kakimoto, T.; Kanemaki, M. An auxin-based degron system for the rapid depletion of proteins in nonplant cells. *Nat. Methods* **2009**, *6*, 917–922. [CrossRef] [PubMed]

148. Papagiannakis, A.; De Jonge, J.J.; Zhang, Z.; Heinemann, M. Quantitative characterization of the auxin-inducible degron: A guide for dynamic protein depletion in single yeast cells. *Sci. Rep.* **2017**, *7*, 1–13. [CrossRef] [PubMed]
149. Jost, A.P.-T.; Weiner, O.D. Probing Yeast Polarity with Acute, Reversible, Optogenetic Inhibition of Protein Function. *ACS Synth. Biol.* **2015**, *4*, 1077–1085. [CrossRef] [PubMed]
150. Jensen, E. Technical review: In situ hybridization. *Anat. Rec.* **2014**, *297*, 1349–1353. [CrossRef]
151. Pelillo, M.; Poli, I.; Roli, A.; Serra, R.; Slanzi, D.; Villani, M. (Eds.) *Artificial Life and Evolutionary Computation*; Communications in Computer and Information Science; Springer International Publishing: Cham, Switzerland, 2018; Volume 830, ISBN 978-3-319-78657-5.
152. Giaever, G.; Chu, A.M.; Ni, L.; Connelly, C.; Riles, L.; Véronneau, S.; Dow, S.; Lucau-Danila, A.; Anderson, K.; Andre, B. Functional profiling of the Saccharomyces cerevisiae genome. *Nature* **2002**, *418*, 387–391. [CrossRef]
153. Gavin, A.-C.; Bösche, M.; Krause, R.; Grandi, P.; Marzioch, M.; Bauer, A.; Schultz, J.; Rick, J.M.; Michon, A.-M.; Cruciat, C.-M. Functional organization of the yeast proteome by systematic analysis of protein complexes. *Nature* **2002**, *415*, 141. [CrossRef]
154. Tarassov, K.; Messier, V.; Landry, C.R.; Radinovic, S.; Molina, M.M.S.; Shames, I.; Malitskaya, Y.; Vogel, J.; Bussey, H.; Michnick, S.W. An in vivo map of the yeast protein interactome. *Science* **2008**, *320*, 1465–1470. [CrossRef] [PubMed]
155. Garí Marsol, E.; Volpe, T.; Wang, H.; Gallego, C.; Futcher, B.; Aldea, M. Whi3 binds the mRNA of the G (1) cyclin CLN3 to modulate cell fate in budding yeast. *Genes Dev.* **2001**, *15*, 2803–2808.
156. Colomina, N.; Ferrezuelo, F.; Wang, H.; Aldea, M.; Garí, E. Whi3, a Developmental Regulator of Budding Yeast, Binds a Large Set of mRNAs Functionally Related to the Endoplasmic Reticulum. *J. Biol. Chem.* **2008**, *283*, 28670–28679. [CrossRef] [PubMed]
157. Wittenberg, C.; Sugimoto, K.; Reed, S.I. G1-specific cyclins of S. cerevisiae: Cell cycle periodicity, regulation by mating pheromone, and association with the p34CDC28 protein kinase. *Cell* **1990**, *62*, 225–237. [CrossRef]
158. Leeuw, T.; Wu, C.; Schrag, J.D.; Whiteway, M.; Thomas, D.Y.; Leberer, E. Interaction of a G-protein β-subunit with a conserved sequence in Ste20/PAK family protein kinases. *Nature* **1998**, *391*, 191. [CrossRef]
159. Wiget, P.; Shimada, Y.; Butty, A.-C.; Bi, E.; Peter, M. Site-specific regulation of the GEF Cdc24p by the scaffold protein Far1p during yeast mating. *EMBO J.* **2004**, *23*, 1063–1074. [CrossRef]
160. Camus, C.; Boy-Marcotte, E.; Jacquet, M. Two subclasses of guanine exchange factor (GEF) domains revealed by comparison of activities of chimeric genes constructed from CDC25, SDC25 and BUD5 in Saccharomyces cerevisiae. *Mol. Gen. Genet. MGG* **1994**, *245*, 167–176. [CrossRef]
161. Kozminski, K.G.; Beven, L.; Angerman, E.; Tong, A.H.Y.; Boone, C.; Park, H.-O. Interaction between a Ras and a Rho GTPase couples selection of a growth site to the development of cell polarity in yeast. *Mol. Biol. Cell* **2003**, *14*, 4958–4970. [CrossRef]
162. Koch, G.; Tanaka, K.; Masuda, T.; Yamochi, W.; Nonaka, H.; Takai, Y. Association of the Rho family small GTP-binding proteins with Rho GDP dissociation inhibitor (Rho GDI) in Saccharomyces cerevisiae. *Oncogene* **1997**, *15*, 417. [CrossRef]
163. Bose, I.; Irazoqui, J.E.; Moskow, J.J.; Bardes, E.S.G.; Zyla, T.R.; Lew, D.J. Assembly of Scaffold-mediated Complexes Containing Cdc42p, the Exchange Factor Cdc24p, and the Effector Cla4p Required for Cell Cycle-regulated Phosphorylation of Cdc24p. *J. Biol. Chem.* **2001**, *276*, 7176–7186. [CrossRef]
164. Kozubowski, L.; Saito, K.; Johnson, J.M.; Howell, A.S.; Zyla, T.R.; Lew, D.J. Symmetry-Breaking Polarization Driven by a Cdc42p GEF-PAK Complex. *Curr. Biol.* **2008**, *18*, 1719–1726. [CrossRef] [PubMed]
165. Gulli, M.-P.; Jaquenoud, M.; Shimada, Y.; Niederhäuser, G.; Wiget, P.; Peter, M. Phosphorylation of the Cdc42 exchange factor Cdc24 by the PAK-like kinase Cla4 may regulate polarized growth in yeast. *Mol. Cell* **2000**, *6*, 1155–1167. [CrossRef]
166. Sagot, I.; Klee, S.K.; Pellman, D. Yeast formins regulate cell polarity by controlling the assembly of actin cables. *Nat. Cell Biol.* **2002**, *4*, 42. [CrossRef] [PubMed]
167. Adams, A.E. Relationship of actin and tubulin distribution to bud growth in wild- type and morphogenetic-mutant Saccharomyces cerevisiae. *J. Cell Biol.* **1984**, *98*, 934–945. [CrossRef]

168. Wendland, B. Epsins: Adaptors in endocytosis? *Nat. Rev. Mol. Cell Biol.* **2002**, *3*, 971. [CrossRef]
169. Goode, B.L.; Eskin, J.A.; Wendland, B. Actin and Endocytosis in Budding Yeast. *Genetics* **2015**, *199*, 315–358. [CrossRef]

Publisher's Note: MDPI stays neutral with regard to jurisdictional claims in published maps and institutional affiliations.

MDPI
St. Alban-Anlage 66
4052 Basel
Switzerland
Tel. +41 61 683 77 34
Fax +41 61 302 89 18
www.mdpi.com

Cells Editorial Office
E-mail: cells@mdpi.com
www.mdpi.com/journal/cells

www.ingramcontent.com/pod-product-compliance
Lightning Source LLC
LaVergne TN
LVHW071535170726
843515LV00008B/2148

* 9 7 8 3 0 3 6 5 0 3 3 8 7 *